Modern General Relativity
Black Holes, Gravitational Waves, and Cosmology

Einstein's general theory of relativity is widely considered to be one of the most elegant and successful scientific theories ever developed, and it is increasingly being taught in a simplified form at advanced undergraduate level within both physics and mathematics departments. Due to the increasing interest in gravitational physics, in both the academic and the public sphere, driven largely by widely-publicised developments such as the recent observations of gravitational waves, general relativity is also one of the most popular scientific topics pursued through self-study. *Modern General Relativity* introduces the reader to the general theory of relativity using an example-based approach, before describing some of its most important applications in cosmology and astrophysics, such as gamma-ray bursts, neutron stars, black holes, and gravitational waves. With hundreds of worked examples, explanatory boxes, and end-of-chapter problems, this textbook provides a solid foundation for understanding one of the towering achievements of twentieth-century physics.

Mike Guidry is Professor of Physics and Astronomy at the University of Tennessee, Knoxville. His current research is focused on the development of new algorithms to solve large sets of differential equations, and applications of Lie algebras to strongly-correlated electronic systems. He has written five textbooks and authored more than 120 journal publications on a broad variety of topics. He previously held the role of Lead Technology Developer for several major college textbooks in introductory physics, astronomy, biology, genetics, and microbiology. He has won multiple teaching awards and is responsible for a variety of important science outreach initiatives.

Modern General Relativity

Black Holes, Gravitational Waves, and Cosmology

MIKE GUIDRY
University of Tennessee, Knoxville

CAMBRIDGE
UNIVERSITY PRESS

University Printing House, Cambridge CB2 8BS, United Kingdom

One Liberty Plaza, 20th Floor, New York, NY 10006, USA

477 Williamstown Road, Port Melbourne, VIC 3207, Australia

314–321, 3rd Floor, Plot 3, Splendor Forum, Jasola District Centre, New Delhi – 110025, India

79 Anson Road, #06–04/06, Singapore 079906

Cambridge University Press is part of the University of Cambridge.

It furthers the University's mission by disseminating knowledge in the pursuit of education, learning, and research at the highest international levels of excellence.

www.cambridge.org
Information on this title: www.cambridge.org/9781107197893
DOI: 10.1017/9781108181938

© Mike Guidry 2019

This publication is in copyright. Subject to statutory exception and to the provisions of relevant collective licensing agreements, no reproduction of any part may take place without the written permission of Cambridge University Press.

First published 2019

Printed in the United Kingdom by TJ International Ltd. Padstow Cornwall

A catalog record for this publication is available from the British Library.

Library of Congress Cataloging-in-Publication Data
Names: Guidry, M. W., author.
Title: Modern general relativity : black holes, gravitational waves, and cosmology / Mike Guidry (University of Tennessee, Knoxville).
Description: Cambridge ; New York, NY : Cambridge University Press, 2019. | Includes bibliographical references and index.
Identifiers: LCCN 2018034213 | ISBN 9781107197893 (hardback : alk. paper)
Subjects: LCSH: General relativity (Physics)–Textbooks. | Astronomy–Textbooks. | Physics–Textbooks.
Classification: LCC QC173.6 .G84 2019 | DDC 523.01–dc23
LC record available at https://lccn.loc.gov/2018034213

ISBN 978-1-107-19789-3 Hardback

Cambridge University Press has no responsibility for the persistence or accuracy of URLs for external or third-party internet websites referred to in this publication and does not guarantee that any content on such websites is, or will remain, accurate or appropriate.

For
Jo Ann

Brief Contents

1	Introduction	*page* 3
2	Coordinate Systems and Transformations	14
3	Tensors and Covariance	32
4	Lorentz Covariance and Special Relativity	69
5	Lorentz-Invariant Dynamics	92
6	The Principle of Equivalence	106
7	Curved Spacetime and General Covariance	125
8	The General Theory of Relativity	145
9	The Schwarzschild Spacetime	159
10	Neutron Stars and Pulsars	193
11	Spherical Black Holes	213
12	Quantum Black Holes	229
13	Rotating Black Holes	243
14	Observational Evidence for Black Holes	257
15	Black Holes as Central Engines	280
16	The Hubble Expansion	327
17	Energy and Matter in the Universe	341
18	Friedmann Cosmologies	365
19	Evolution of the Universe	386
20	The Big Bang	412
21	Extending Classical Big Bang Theory	448
22	Gravitational Waves	465
23	Weak Sources of Gravitational Waves	489
24	Strong Sources of Gravitational Waves	500
25	Tests of General Relativity	539
26	Beyond Standard Models	547

Contents

Preface		page xxiii
Part I General Relativity		1
1	**Introduction**	3
	1.1 Gravity and the Universe on Large Scales	3
	1.2 Classical Newtonian Gravity	4
	1.3 Transformations between Inertial Systems	5
	1.4 Maxwell, the Aether, and Galileo	6
	1.5 The Special Theory of Relativity	7
	1.6 Minkowski Space	8
	1.7 A New Theory of Gravity	9
	1.8 The Equivalence Principle	10
	1.9 General Relativity	11
	Background and Further Reading	12
	Problems	12
2	**Coordinate Systems and Transformations**	14
	2.1 Coordinate Systems in Euclidean Space	14
	2.1.1 Parameterizing in Different Coordinate Systems	14
	2.1.2 Basis Vectors	15
	2.1.3 Expansion of Vectors and Dual Vectors	20
	2.1.4 Vector Scalar Product and the Metric Tensor	20
	2.1.5 Relationship of Vectors and Dual Vectors	21
	2.1.6 Properties of the Metric Tensor	23
	2.1.7 Line Elements	24
	2.1.8 Euclidean Line Element	25
	2.2 Integration	26
	2.3 Differentiation	26
	2.4 Non-euclidean Geometry	27
	2.5 Transformations	28
	2.5.1 Rotational Transformations	29
	2.5.2 Galilean Transformations	29
	Background and Further Reading	30
	Problems	30

3 Tensors and Covariance — 32
- 3.1 Invariance and Covariance — 32
- 3.2 Spacetime Coordinates — 32
- 3.3 Vectors in Non-euclidean Space — 35
- 3.4 Coordinates in Spacetime — 36
  - 3.4.1 Coordinate and Non-coordinate Bases — 37
  - 3.4.2 Utility of Coordinate and Non-coordinate Bases — 40
- 3.5 Tensors and Coordinate Transformations — 40
- 3.6 Tensors as Linear Maps — 41
  - 3.6.1 Linear Maps to Real Numbers — 41
  - 3.6.2 Vectors and Dual Vectors — 42
  - 3.6.3 Tensors of Higher Rank — 45
  - 3.6.4 Identification of Vectors and Dual Vectors — 46
  - 3.6.5 Index-free versus Component Transformations — 47
- 3.7 Tensors Specified by Transformation Laws — 48
  - 3.7.1 Scalar Transformation Law — 48
  - 3.7.2 Dual Vector Transformation Law — 49
  - 3.7.3 Vector Transformation Law — 49
  - 3.7.4 Duality of Vectors and Dual Vectors — 51
- 3.8 Scalar Product of Vectors — 51
- 3.9 Tensors of Higher Rank — 52
- 3.10 The Metric Tensor — 53
- 3.11 Symmetric and Antisymmetric Tensors — 54
- 3.12 Summary of Algebraic Tensor Operations — 55
- 3.13 Tensor Calculus on Curved Manifolds — 56
  - 3.13.1 Invariant Integration — 56
  - 3.13.2 Partial Derivatives — 57
  - 3.13.3 Covariant Derivatives — 58
  - 3.13.4 Absolute Derivatives — 61
  - 3.13.5 Lie Derivatives — 62
- 3.14 Invariant Equations — 64
- Background and Further Reading — 65
- Problems — 65

4 Lorentz Covariance and Special Relativity — 69
- 4.1 Minkowski Space — 69
  - 4.1.1 The Indefinite Metric of Spacetime — 69
  - 4.1.2 Scalar Products and the Metric Tensor — 70
  - 4.1.3 The Line Element — 70
  - 4.1.4 Invariance of the Spacetime Interval — 71
- 4.2 Tensors in Minkowski space — 72
- 4.3 Lorentz Transformations — 73
  - 4.3.1 Rotations in Euclidean Space — 73
  - 4.3.2 Generalized 4D Minkowski Rotations — 74

	4.3.3	Lorentz Spatial Rotations	74
	4.3.4	Lorentz Boost Transformations	75
4.4	Lightcone Diagrams		77
4.5	The Causal Structure of Spacetime		79
4.6	Lorentz Transformations in Spacetime Diagrams		80
	4.6.1	Lorentz Boosts and the Lightcone	82
	4.6.2	Spacelike and Timelike Intervals	84
4.7	Lorentz Covariance of the Maxwell Equations		84
	4.7.1	Maxwell Equations in Noncovariant Form	85
	4.7.2	Scalar and Vector Potentials	85
	4.7.3	Gauge Transformations	86
	4.7.4	Maxwell Equations in Manifestly Covariant Form	87
Background and Further Reading			88
Problems			88

5 Lorentz-Invariant Dynamics 92

5.1	A Natural Set of Units	92
5.2	Velocity and Momentum for Massive Particles	94
5.3	Geodesics and a Variational Principle	95
5.4	Light and other Massless Particles	98
	5.4.1 Affine Parameters	98
	5.4.2 Energy and Momentum	99
5.5	Observers	99
5.6	Isometries and Killing Vectors	100
	5.6.1 Symmetries of the Metric	101
	5.6.2 Quantities Conserved along Geodesics	103
Background and Further Reading		103
Problems		103

6 The Principle of Equivalence 106

6.1	Einstein and Equivalence	106
6.2	Inertial and Gravitational Mass	107
6.3	The Strong Equivalence Principle	108
	6.3.1 Elevators, Gravity, and Acceleration	108
	6.3.2 Alternative Statements of the Equivalence Principle	109
	6.3.3 Equivalence and the Path to General Relativity	110
6.4	Deflection of Light in a Gravitational Field	110
	6.4.1 A Thought Experiment	111
	6.4.2 Curvature Radius and the Strength of Gravity	111
6.5	The Gravitational Redshift	112
	6.5.1 A Second Thought Experiment	112
	6.5.2 The Total Redshift in a Gravitational Field	113
	6.5.3 Gravitational Time Dilation	115

	6.6	Equivalence and Riemannian Manifolds	116
	6.7	Local Inertial Frames and Inertial Observers	118
		6.7.1 Locality and Tidal Forces	119
		6.7.2 Inertial Observers	119
		6.7.3 Definition of Local Inertial Frames	119
	6.8	Lightcones in Curved Spacetime	120
	6.9	The Road to General Relativity	121
	Background and Further Reading		122
	Problems		122
7	**Curved Spacetime and General Covariance**		**125**
	7.1	General Covariance	125
	7.2	Curved Spacetime	126
		7.2.1 Coordinate Systems	126
		7.2.2 Gaussian Curvature	126
		7.2.3 Distance Intervals	129
	7.3	A Covariant Description of Matter	129
		7.3.1 Stress–Energy for Perfect Fluids	131
		7.3.2 Local Conservation of Energy	132
	7.4	Covariant Derivatives and Parallel Transport	132
		7.4.1 Parallel Transport of Vectors	134
		7.4.2 The Affine Connection and Covariant derivatives	135
		7.4.3 Absolute Derivatives and Parallel Transport	136
		7.4.4 Geometry and Covariant Derivatives	137
	7.5	Gravity and Curved Spacetime	137
		7.5.1 Free Particles	137
		7.5.2 The Geodesic Equation	138
	7.6	The Local Inertial Coordinate System	139
	7.7	The Affine Connection and the Metric Tensor	139
	7.8	Uniqueness of the Affine Connection	140
	Background and Further Reading		141
	Problems		141
8	**The General Theory of Relativity**		**145**
	8.1	Weak-Field Limit	145
	8.2	Recipe for Motion in a Gravitational Field	147
	8.3	Towards a Covariant Theory of Gravitation	147
	8.4	The Riemann Curvature Tensor	148
	8.5	The Einstein Equations	150
	8.6	Limiting Behavior of the Einstein Tensor	154
	8.7	Sign Conventions	154
	8.8	Solving the Einstein Equations	154
		8.8.1 Solutions in the Limit of Weak Fields	155

		8.8.2	Solutions with a High Degree of Symmetry	155
		8.8.3	Solutions by Numerical Relativity	156
	Background and Further Reading			156
	Problems			157

9 The Schwarzschild Spacetime — 159

- 9.1 The Form of the Metric — 159
  - 9.1.1 The Schwarzschild Solution — 159
  - 9.1.2 The Schwarzschild Radius — 160
  - 9.1.3 Measuring Distance and Time — 162
  - 9.1.4 Embedding Diagrams — 164
- 9.2 The Gravitational Redshift — 165
  - 9.2.1 Exploiting a Symmetry of the Metric — 166
  - 9.2.2 Recovering the Weak-Field Limit — 167
- 9.3 Particle Orbits in the Schwarzschild Metric — 167
  - 9.3.1 Conserved Quantities — 167
  - 9.3.2 Equation of Motion — 168
  - 9.3.3 Classification of Orbits — 169
  - 9.3.4 Stable Circular Orbits — 171
- 9.4 Precession of Orbits — 172
  - 9.4.1 The Change in Perihelion Angle — 173
  - 9.4.2 Testing the Prediction — 174
- 9.5 Escape Velocity — 174
- 9.6 Radial Fall of a Test Particle — 175
- 9.7 Orbits for Light Rays — 177
- 9.8 Deflection of Light in the Gravitational Field — 178
- 9.9 Shapiro Time Delay of Light — 179
- 9.10 Gyroscopes in Curved Spacetime — 180
- 9.11 Geodetic Precession — 180
- 9.12 Gyroscopes in Rotating Spacetimes — 183
  - 9.12.1 Slow Rotation in the Schwarzschild Metric — 183
  - 9.12.2 Dragging of Inertial frames — 186

Background and Further Reading — 188
Problems — 188

10 Neutron Stars and Pulsars — 193

- 10.1 A Qualitative Picture of Neutron Stars — 193
- 10.2 Solutions inside Spherical Mass Distributions — 194
  - 10.2.1 Simplifying Assumptions — 194
  - 10.2.2 Solving the Einstein Equations — 195
  - 10.2.3 The Oppenheimer–Volkov Equations — 196
  - 10.2.4 Interpretation of Oppenheimer–Volkov Equations — 197
- 10.3 Interpretation of the Mass Parameter — 198

	10.3.1	Total Mass–Energy for a Relativistic Star	199
	10.3.2	Gravitational Mass and Baryonic Mass	199
10.4	Pulsars and Tests of General Relativity		200
	10.4.1	The Binary Pulsar	200
	10.4.2	Precision Tests of General Relativity	202
	10.4.3	Origin and Fate of the Binary Pulsar	203
	10.4.4	The Double Pulsar	204
	10.4.5	The Pulsar–White Dwarf Binary PSR J0348+0432	205
	10.4.6	The Pulsar–WD–WD Triplet PSR J0337+1715	207
Background and Further Reading			208
Problems			208

Part II Black Holes   211

11 Spherical Black Holes   213

11.1	Schwarzschild Black Holes		213
	11.1.1	Event Horizons	213
	11.1.2	Approaching the Horizon: Outside View	215
	11.1.3	Approaching the Horizon: Spacecraft View	215
11.2	Lightcone Description of a Trip to a Black Hole		216
	11.2.1	Worldline Exterior to the Event Horizon	216
	11.2.2	Worldline Interior to the Event Horizon	217
	11.2.3	You Can't Get There From Here	218
11.3	Solution in Eddington–Finkelstein Coordinates		218
	11.3.1	Eddington–Finkelstein Coordinates	219
	11.3.2	Behavior of Radial Light Rays	219
	11.3.3	The Event Horizon	220
11.4	Solution in Kruskal–Szekeres Coordinates		221
	11.4.1	Kruskal–Szekeres Coordinates	221
	11.4.2	Kruskal Diagrams	221
	11.4.3	The Event Horizon	223
11.5	Black Hole Theorems and Conjectures		224
Background and Further Reading			226
Problems			226

12 Quantum Black Holes   229

12.1	Geodesics and Uncertainty		229
12.2	Hawking Radiation		230
	12.2.1	4-Momentum Conservation	230
	12.2.2	Black Hole Evaporation	231
	12.2.3	Relative Importance of Quantum Fluctuations	231
12.3	Black Hole Temperatures		232
12.4	Miniature Black Holes		234

12.5		Black Hole Thermodynamics	236
	12.5.1	Entropy of a Black Hole	236
	12.5.2	The Generalized Second Law	236
	12.5.3	The Four Laws of Black Hole Dynamics	237
12.6		The Planck Scale and Quantum Gravity	238
12.7		Black Holes and Information	239
	12.7.1	The Holographic Principle	240
	12.7.2	The Holographic Universe	240
Background and Further Reading			240
Problems				241

13 Rotating Black Holes — 243

13.1		The Kerr Solution	243
	13.1.1	The Kerr Metric	243
	13.1.2	Extremal Kerr Black Holes	245
	13.1.3	Cosmic Censorship	246
	13.1.4	The Kerr Horizon	247
13.2		Particle and Photon Motion	248
	13.2.1	Orbits in the Kerr Metric	248
	13.2.2	Frame Dragging	249
	13.2.3	The Ergosphere	251
	13.2.4	Motion of Photons in the Ergosphere	252
13.3		Extracting Rotational Energy from Black Holes	253
	13.3.1	Penrose Processes	253
	13.3.2	Practical Energy Extraction	255
Background and Further Reading			255
Problems				255

14 Observational Evidence for Black Holes — 257

14.1		Gravitational Collapse and Observations	257
14.2		Singularity Theorems and Black Holes	257
	14.2.1	Global Methods in General Relativity	258
	14.2.2	Singularities and Trapped Surfaces	258
	14.2.3	Generalized Singularity Theorems	261
14.3		Observing Black Holes	261
14.4		Stellar-Mass Black Holes	262
	14.4.1	Masses for Compact Objects in X-Ray Binaries	262
	14.4.2	Masses from Mass Functions	263
	14.4.3	An Example: A0620–00	264
	14.4.4	Some Black Hole Candidates	266
14.5		Supermassive Black Holes	266
	14.5.1	The Black Hole at Sgr A*	268
	14.5.2	The Water Masers of NGC 4258	270

		14.5.3 The Virial Theorem and Gravitating Mass	270
	14.6	Intermediate-Mass Black Holes	272
	14.7	Black Holes in the Early Universe	274
	14.8	Show Me an Event Horizon!	276
	14.9	A Circumstantial but Strong Case	278
	Background and Further Reading		278
	Problems		278
15	**Black Holes as Central Engines**		**280**
	15.1	Black Hole Energy Sources	280
	15.2	Accretion and Energy Release for Black Holes	281
		15.2.1 Maximum Energy Release for Spherical Accretion	281
		15.2.2 Limits on Accretion Rates	282
		15.2.3 Accretion Efficiencies	283
		15.2.4 Accretion onto Rotating Black Holes	283
	15.3	Jets and Magnetic Fields	285
	15.4	Quasars	285
		15.4.1 "Radio Stars" and a Spectrum in Disguise	288
		15.4.2 Quasar Characteristics	288
		15.4.3 Quasar Energy Sources	290
	15.5	Active Galactic Nuclei	292
		15.5.1 Radio Galaxies	292
		15.5.2 Seyfert Galaxies	293
		15.5.3 BL Lac Objects	295
	15.6	A Unified Model of AGN and Quasars	295
		15.6.1 The AGN Black Hole Central Engine Model	297
		15.6.2 Anisotropic Ionization Cones	298
		15.6.3 A Unified Model	299
		15.6.4 Example: Feeding a Nearby Monster	301
		15.6.5 High-Energy Photons from AGN	302
	15.7	Gamma-Ray Bursts	304
		15.7.1 The Gamma-Ray Sky	304
		15.7.2 Two Classes of Gamma-Ray Bursts	306
		15.7.3 Localization of Gamma-Ray Bursts	307
		15.7.4 Necessity of Ultrarelativistic Jets	308
		15.7.5 Association of GRBs with Galaxies	312
		15.7.6 Long-Period GRBs and Supernovae	312
		15.7.7 Characteristics of Gamma-Ray Bursts	314
		15.7.8 Mechanisms for the Central Engine	315
		15.7.9 Gamma-Ray Bursts and Gravitational Waves	318
	Background and Further Reading		321
	Problems		322

Part III Cosmology   325

16 The Hubble Expansion   327
- 16.1 The Standard Picture   327
  - 16.1.1 Mass Distribution on Large Scales   327
  - 16.1.2 The Universe is Expanding   327
  - 16.1.3 The Expansion Is Governed by General Relativity   328
  - 16.1.4 There is a Big Bang in Our Past   329
  - 16.1.5 Particle Content Influences the Evolution   329
  - 16.1.6 There is a Cosmic Microwave Background   330
- 16.2 The Hubble Law   331
  - 16.2.1 The Hubble Parameter   331
  - 16.2.2 Redshifts   331
  - 16.2.3 Expansion Interpretation of Redshifts   332
  - 16.2.4 The Hubble Time   335
  - 16.2.5 A 2-Dimensional Hubble Expansion Model   336
  - 16.2.6 Measuring the Hubble Constant   337
- 16.3 Limitations of the Standard Picture   337
- Background and Further Reading   338
- Problems   339

17 Energy and Matter in the Universe   341
- 17.1 Expansion and Newtonian Gravity   341
- 17.2 The Critical Density   342
- 17.3 The Cosmic Scale Factor   343
- 17.4 Possible Expansion Histories   344
- 17.5 Lookback Times   346
- 17.6 The Inadequacy of Dust Models   348
- 17.7 Evidence for Dark Matter   348
  - 17.7.1 Rotation Curves for Spiral Galaxies   348
  - 17.7.2 The Mass of Galaxy Clusters   349
  - 17.7.3 Hot Gas in Clusters of Galaxies   350
  - 17.7.4 Gravitational Lensing   350
  - 17.7.5 Dark Matter in Ultra-diffuse Galaxies   354
- 17.8 The Amount of Baryonic Matter   355
- 17.9 Baryonic Candidates for Dark Matter   356
- 17.10 Candidates for Nonbaryonic Dark Matter   356
  - 17.10.1 Cold Dark Matter   357
  - 17.10.2 Hot Dark Matter   357
- 17.11 Dark Energy   358
- 17.12 Radiation   358
- 17.13 The Scale Factor and Density Parameters   358
- 17.14 The Deceleration Parameter   359

	17.14.1 Deceleration and Density Parameters	361
	17.14.2 Deceleration and Cosmology	361
17.15	Problems with Newtonian Cosmology	361
Background and Further Reading		362
Problems		363

18 Friedmann Cosmologies   365

18.1	The Cosmological Principle	365
18.2	Homogeneous and Isotropic 2D Spaces	366
18.3	Homogeneous and Isotropic 3D Spaces	368
	18.3.1 Constant Positive Curvature	368
	18.3.2 Constant Negative Curvature	369
	18.3.3 Zero Curvature	369
18.4	The Robertson–Walker Metric	369
18.5	Comoving Coordinates	371
18.6	Proper Distances	373
18.7	The Hubble Law and the RW Metric	374
18.8	Particle and Event Horizons	375
	18.8.1 Particle Horizons in the RW Metric	375
	18.8.2 Event Horizons in the RW Metric	376
18.9	Einstein Equations for the RW Metric	378
	18.9.1 The Metric and Stress–Energy Tensor	378
	18.9.2 The Connection Coefficients	379
	18.9.3 The Ricci Tensor and Ricci Scalar	380
	18.9.4 The Friedmann Equations	381
	18.9.5 Static Solutions and the Cosmological Constant	381
18.10	Resolution of Newtonian Difficulties	383
Background and Further Reading		383
Problems		384

19 Evolution of the Universe   386

19.1	Friedmann Cosmologies	386
	19.1.1 Reformulation of the Friedmann Equations	386
	19.1.2 Equations of State	387
19.2	Friedmann Equations in Concise Form	389
	19.2.1 Evolution and Scaling of Density Components	389
	19.2.2 A Standard Model	390
19.3	Flat, Single-Component Universes	391
	19.3.1 Special Solution: Vacuum Energy Domination	392
	19.3.2 General Solutions	394
	19.3.3 Flat Universes with Radiation or Matter	396
19.4	Full Solution of the Friedmann Equations	397
	19.4.1 Evolution Equations in Dimensionless Form	397

19.4.2	Algorithm for Numerical Solution	398
19.4.3	Examples: Single Component with Curvature	399
19.4.4	Examples: Multiple Components	401
19.4.5	Parameters for a Realistic Model	402
19.4.6	Concordance of Cosmological Parameters	406
19.4.7	Calculations with Benchmark Parameters	409

Background and Further Reading 410
Problems 410

20 The Big Bang 412

20.1 Radiation- and Matter-Dominated Universes 412
    20.1.1 Evolution of the Scale Factor 412
    20.1.2 Matter and Radiation Density 413
20.2 Evolution of the Early Universe 414
    20.2.1 Thermodynamics of the Big Bang 414
    20.2.2 Equilibrium in an Expanding Universe 416
    20.2.3 A Timeline for the Big Bang 419
20.3 Nucleosynthesis and Cosmology 424
    20.3.1 The Neutron to Proton Ratio 424
    20.3.2 Elements Synthesized in the Big Bang 424
    20.3.3 Constraints on Baryon Density 425
20.4 The Cosmic Microwave Background 426
    20.4.1 The Microwave Background Spectrum 427
    20.4.2 Anisotropies in the Microwave Background 429
    20.4.3 The Origin of CMB Fluctuations 433
    20.4.4 Acoustic Signature in the CMB 437
    20.4.5 Acoustic Signature in Galaxy Distributions 438
    20.4.6 Precision Cosmology 439
    20.4.7 Seeds for Structure Formation 441
20.5 Accelerated Structure Formation 442
20.6 Dark Matter, Dark Energy, and Structure 442

Background and Further Reading 444
Problems 445

21 Extending Classical Big Bang Theory 448

21.1 Successes of the Big Bang Theory 448
21.2 Problems with the Big Bang 449
    21.2.1 The Horizon Problem 449
    21.2.2 The Flatness Problem 451
    21.2.3 The Magnetic Monopole Problem 451
    21.2.4 The Structure and Smoothness Dichotomy 452
    21.2.5 The Vacuum Energy Problem 452
    21.2.6 The Matter–Antimatter Problem 453

		21.2.7 Modifying the Classical Big Bang	453
	21.3	Cosmic Inflation	453
		21.3.1 The Basic Idea and Generic Consequences	454
		21.3.2 Taking the Inflationary Cure	455
		21.3.3 Inflation Doesn't Replace the Big Bang	457
	21.4	The Origin of the Baryons	457
		21.4.1 Conditions for a Baryon Asymmetry	457
		21.4.2 Grand Unified Theories	459
		21.4.3 Leptogenesis	459
	Background and Further Reading		459
	Problems		460

Part IV Gravitational Wave Astronomy   463

22 Gravitational Waves   465

22.1	Significance of Gravitational Waves		465
	22.1.1	Unprecedented Tests of General Relativity	466
	22.1.2	A Probe of Dark Events	467
	22.1.3	The Deepest Probe	467
	22.1.4	Technology and the Quest for Gravitational Waves	468
22.2	Linearized Gravity		468
	22.2.1	Linearized Curvature Tensor	470
	22.2.2	Wave Equation	470
	22.2.3	Coordinates and Gauge Transformations	471
	22.2.4	Choice of Gauge	471
22.3	Weak Gravitational Waves		473
	22.3.1	Polarization Tensor in TT Gauge	473
	22.3.2	Helicity Components	474
	22.3.3	General Solution in TT Gauge	474
22.4	Gravitational versus Electromagnetic Waves		476
	22.4.1	Interaction with Matter	476
	22.4.2	Wavelength Relative to Source Size	476
	22.4.3	Phase Coherence	477
	22.4.4	Field of View	477
22.5	The Response of Test Particles		477
	22.5.1	Response of Two Test Masses	477
	22.5.2	The Effect of Polarization	480
22.6	Gravitational Wave Detectors		480
	22.6.1	Operating and Proposed Detectors	482
	22.6.2	Strain and Frequency Windows	483
	22.6.3	Detecting Very Long Wavelengths	484
	22.6.4	Reach of Advanced LIGO and Advanced VIRGO	486
Background and Further Reading			487
Problems			487

23 Weak Sources of Gravitational Waves — 489
- 23.1 Production of Weak Gravitational Waves — 489
  - 23.1.1 Energy Densities — 489
  - 23.1.2 Multipolarities — 490
  - 23.1.3 Linearized Einstein Equation with Sources — 490
  - 23.1.4 Gravitational Wave Amplitudes — 491
  - 23.1.5 Amplitudes and Event Rates — 492
  - 23.1.6 Power in Gravitational Waves — 493
- 23.2 Gravitational Radiation from Binary Systems — 494
  - 23.2.1 Gravitational Wave Luminosity — 494
  - 23.2.2 Gravitational Radiation and Binary Orbits — 496
  - 23.2.3 Gravitational Waves from the Binary Pulsar — 497
- Background and Further Reading — 498
- Problems — 499

24 Strong Sources of Gravitational Waves — 500
- 24.1 A Survey of Candidate Sources — 500
  - 24.1.1 Merger of a Neutron Star Binary — 500
  - 24.1.2 Stellar Black Hole Mergers — 502
  - 24.1.3 Merger of a Black Hole and a Neutron Star — 503
  - 24.1.4 Core Collapse in Massive Stars — 503
  - 24.1.5 Merging Supermassive Black Holes — 504
  - 24.1.6 Sample Gravitational Waveforms — 504
- 24.2 The Gravitational Wave Event GW150914 — 506
  - 24.2.1 Observed Waveforms — 506
  - 24.2.2 Source Localization — 508
  - 24.2.3 Comparisons with Candidate Events — 508
  - 24.2.4 Binary Black Hole Mergers — 510
- 24.3 Additional Gravitational Wave Events — 514
  - 24.3.1 GW151226 and LVT151012 — 514
  - 24.3.2 Matched Filtering — 515
  - 24.3.3 Binary Masses and Inspiral Cycles — 518
  - 24.3.4 Increasing Sensitivity — 519
  - 24.3.5 LIGO–Virgo Triple Coincidences — 519
- 24.4 Testing General Relativity in Strong Gravity — 520
- 24.5 A New Window on the Universe — 521
- 24.6 Multimessenger Astronomy — 522
- 24.7 Gravitational Waves from Neutron Star Mergers — 522
  - 24.7.1 New Discoveries Associated with GW170817 — 524
  - 24.7.2 The Kilonova — 528
- 24.8 Gravitational Waves and Stellar Evolution — 530
  - 24.8.1 A Possible Evolutionary Scenario for GW150914 — 530
  - 24.8.2 Measured Stellar Black Hole Masses — 532

		24.8.3 Are Stellar and Supermassive Black Holes Related?	534

Background and Further Reading — 534
Problems — 535

Part V General Relativity and Beyond — 537

25 Tests of General Relativity — 539
25.1 The Classical Tests — 539
25.2 The Modern Tests — 540
  25.2.1 The PPN Formalism — 540
  25.2.2 Results of Modern Tests — 543
25.3 Strong-Field Tests — 544
25.4 Cosmological Tests — 545
Background and Further Reading — 546
Problems — 546

26 Beyond Standard Models — 547
26.1 Supersymmetry — 547
  26.1.1 Fermions and Bosons — 548
  26.1.2 Normal Symmetries — 548
  26.1.3 Symmetries Relating Fermions and Bosons — 549
26.2 Vacuum Energy from Quantum Fluctuations — 549
  26.2.1 Vacuum Energy for Bosonic Fields — 550
  26.2.2 Vacuum Energy for Fermionic Fields — 552
  26.2.3 Supersymmetry and Dark Energy — 552
26.3 Quantum Gravity — 553
  26.3.1 Superstrings and Branes — 554
  26.3.2 How Many Dimensions? — 556
  26.3.3 Spacetime Foam, Wormholes, and Such — 556
  26.3.4 The Ultimate Free Lunch — 557
  26.3.5 Does the Planck Scale Matter? — 558
Background and Further Reading — 560
Problems — 560

Appendix A Constants — 562
Appendix B Natural Units — 565
Appendix C Einstein Tensor for a General Spherical Metric — 569
Appendix D Using arXiv and ADS — 571
References — 573
Index — 583

Preface

This book contains material used in an astronomy course on general relativity, black holes, gravitational waves, and cosmology that I teach regularly at the University of Tennessee for advanced undergraduates and beginning graduate students. The goal of the course and of the book is to provide an introduction that is topically current and accessible to a reader with some physics but minimal astronomy, astrophysics, and advanced mathematics background.

The reader is expected to have physics experience commensurate with that of a third or fourth year undergraduate physics major in a US university, and to be familiar with the material typically covered in an introductory descriptive course in astronomy. Readers are assumed to be conversant with special relativity and quantum mechanics only at the level typically covered in first or second year introductions to modern physics. Mathematically I assume the reader to be familiar with basic algebra, geometry, calculus, and differential equations, and the rudiments of matrices. I introduce sufficient differential geometry and tensor calculus to understand the topics to be covered as an integral part of the presentation. Given the target audience I opt for practicality over beauty, emphasizing the physicist's utilitarian "engineering" approach to these topics. However, I try to provide at various places a glimpse of the more elegant but abstract formulation that mathematicians would favor. Specifically, I have tried to present tensors both from the mathematical point of view as maps from vectors to the real numbers, and from the practical physics perspective in terms of transformation properties.

Our approach will be to build from the familiar to the unfamiliar, introducing central concepts first in euclidean (flat) space in two and three dimensions, where familiar mathematics will be cast in a form more useful for what follows. Then these ideas will be extended to still-flat but now 4-dimensional "pseudo-euclidean" spacetime, deriving in the process the theory of special relativity. Finally these concepts will be extended to 4-dimensional curved spacetime and, with the aid of the equivalence principle, used to formulate the theory of general relativity in terms of the Einstein field equations. With these tools in hand, some of the most interesting problems in astrophysics and cosmology will be addressed: spherical and rotating black holes, quantum black holes, the modern view of cosmology, dark matter, dark energy, and gravitational waves.

To aid in comprehension, a number of worked examples and supplementary information boxes are scattered throughout each chapter. These serve two general functions: to illustrate how to do some essential tasks, like conversion between geometrized ($G = c = 1$) units and normal units, or to set in context and provide broader perspective, like a concise overview of group theory as a mathematical and conceptual framework for dealing with symmetries.

A total of 273 problems of varying complexity and difficulty may be found at the ends of the chapters, each chosen to familiarize the reader with basic concepts, illustrate important points, fill in details, or prove assertions made in the text. Where appropriate, when teaching this material I encourage the use of programming tools such as MatLab, Mathematica, or Maple, or more formal programming languages like C/C++ or Java in solving problems. While very helpful, none of these tools is essential for working the problems. The solutions for all 273 problems are available at www.cambridge.org/guidryGR as a PDF file in typeset book format for instructors, and a subset of 107 problem solutions is available to students in the same format. Those problems with solutions available to students are marked by the symbol *** at the end of the problem.

A book dealing with astrophysics at an intermediate level must make a no-win decision concerning units. It is desirable to standardize units and in introductory astronomy it makes some sense to use the SI (MKS) system of units. However, professionals in the field routinely employ the CGS (centimeter–gram–second) system, or natural units that are defined such that fundamental constants like the speed of light, gravitational constant, Planck's constant, or Boltzmann constant take the value of one. Since one of the purposes of the present material is to address the significance of general relativity for cutting-edge research in astronomy and astrophysics, and to encourage students to use and explore the corresponding literature, I have adopted a policy of generally using the CGS system, or natural units.

Many papers referenced in this book are published in journals with limited free public access. To help ensure broad availability to these references for readers I have included for journal articles information allowing free access through the preprint server *arXiv* or the *ADS Astronomy Abstract Service*. More details may be found at the beginning of the References in Ref. [1], and instructions for using *arXiv* and *ADS* to retrieve articles are given in Appendix D.

Let me make some comments on my approach to teaching from this material. There is too much information to cover in full depth in a one-semester course, so some choices have to be made if only a single semester is available. I will suggest several possibilities that may be useful in guiding that choice. To be definite I will assume a traditional lecture format but my remarks should be adaptable to other teaching modes as well.

Survey track: This is my usual approach, where I try to cover most of the material, but with some only outlined, or assigned as reading and self-study. I would typically have the students read Chapters 1 and 2 as introduction, cover Chapters 3–9 in some depth since they provide the essential foundation, assign Chapter 10 as self-study, cover Chapters 10–13 in lecture, and lecture on only selected examples in Chapters 14–15. I would then assign Chapter 16 as self-study since it is introductory, cover Chapters 17–24 in as much depth as time permits, and assign Chapters 25–26 as reading.

Black hole track: Instructors wishing to emphasize black holes can assign Chapter 1 as reading, cover Chapter 2 or assign it as reading, depending on the background of the class, cover Chapters 3–13 in some depth, cover selected topics in Chapters 14–15 as evidence for the existence of black holes and their efficacy as power sources, and conclude with a brief introduction to gravitational waves with emphasis on black holes: Chapter 24 (with Chapters 22–23 as introduction if time permits).

Gravitational wave track: A course emphasizing gravitational waves can be constructed by assigning Chapters 1 and 2 as reading, covering in depth Chapters 3–13 as a background in general relativity, black holes, and neutron stars, and then concluding with gravitational waves from merging black holes and neutron stars in Chapters 22–24, and the discussion of tests of general relativity – particularly in the strong-field limit – in Chapter 25.

Cosmology track: Instructors wishing to emphasize cosmology can assign Chapter 1 as reading, cover Chapter 2 or assign it as reading, depending on the background of the class, cover Chapters 3–9 in some depth to provide essential foundation, cover Chapters 16–21 in depth, and finish with, or assign as reading, Chapter 26. Those wishing to at least mention gravitational waves can add Chapter 24, since it is relatively self-contained and highly observationally oriented.

Tuning a more-mathematical flavor: All of the problems at the ends of chapters have solutions available (often with some explanatory detail), and some of those problems cover material in more technical depth than the text itself. Therefore, it is possible to tune any of the above tracks in a more mathematical direction by incorporating into the classroom more-technical material from the exercise solutions.

For those wishing to teach from this book, several additional resources are available from the publisher at www.cambridge.org/guidryGR for instructors and for students:

1. *Instructor Solutions Manual for Modern General Relativity,* which is a PDF file typeset in the format of the book that presents the solutions for all 273 problems at the ends of chapters. This manual is available only to instructors.
2. *Student Solutions Manual for Modern General Relativity,* which is a PDF file typeset in the format of the book that contains the solutions for a subset of 107 of the 273 problems at the ends of chapters. This manual is available to students and instructors. As noted above, the problems contained in this solutions manual for students are marked by *** at the end of the problem in the text.
3. *Modern General Relativity Lecture Notes,* which presents a synopsis in PDF format appropriate for projection and presentation of the essential material in each chapter. Individual slides are organized in a presentation format suitable for teaching, with text formatted in larger fonts and in color. These are the slides that I use myself when teaching this material.

We conclude this list by noting that the inclusion of DOI or *arXiv* numbers for all journal references – which allows easy browser access through *arXiv* and *ADS* for most articles (see Appendix D) – may be viewed as an additional resource permitting creative literature-based projects to be assigned with minimal bother, if an instructor is so inclined.

Finally, I would like to extend my thanks to the many students and colleagues whose questions and comments sharpened this presentation, to Nicholas Gibbons, Ilaria Tassistro, Jon Billam, and Dominic Stock at Cambridge University Press for all their help in shepherding this book to publication, and especially to my wife Jo Ann for her patience and support over many years.

PART I

GENERAL RELATIVITY

1 Introduction

This is a book about gravity. Of the four fundamental interactions (strong, weak, electromagnetic, and gravitational), gravity is by far the weakest, characterized by a force that is intrinsically $\sim 10^{36}$ times more feeble than the electromagnetic force. Yet gravity determines *completely* the large-scale structure of the Universe. How can this be? In this introductory chapter we will look qualitatively at how gravity sets itself apart from all other fundamental interactions, how it can be best described in mathematical terms, and how Einstein's theory of general relativity revised its fundamental meaning and interpretation.

1.1 Gravity and the Universe on Large Scales

Gravity is intrinsically weak but it has some properties that distinguish it from all the other fundamental interactions.

1. *Gravity is long-ranged.* It is one of only two fundamental interactions that are long-ranged, the other being electromagnetism, with the gravitational force and the electromagnetic force each varying as the inverse square of the distance to the source of the corresponding field. In contrast, the strong and weak interactions act only over distances comparable to the size of a nucleus, a very short range indeed! It follows that the strong and weak forces are fundamental in determining the microscopic properties of matter but they have no direct bearing on the large-scale structure of the Universe. The race to determine that structure is now down to electromagnetism and gravity with, in the language of *The Tortoise and the Hare* from *Aesop's Fables,* the sleek, fast rabbit of electromagnetism pitted against the plodding, methodical tortoise of gravity (with the rabbit sporting a top speed 10^{36} times that of the tortoise). Surely only a fool would bet against the rabbit. But wait; I haven't told you everything yet!

2. *Gravity is unscreened.* Electrical charges can be positive or negative. Thus although in principle electromagnetic forces are long-ranged, in practice they tend to be short-ranged because positive and negative charges partially offset each other at shorter range and completely cancel each other at longer range; this is *screening,* and it implies that matter on larger scales (moons, planets, stars, galaxies, ...) may under normal conditions be assumed *completely electrically neutral.* In contrast, a comparison of the equations for Newtonian gravity and for electrostatics indicates that *mass* is the gravitational "charge," but mass

has only one sign so the gravitational interaction is *unscreened* and *always attractive*.[1] An electron on the Moon feels no electrical force from a proton on the Earth because the force is completely screened by intervening matter. In contrast, the electron on the Moon feels the full gravitational force exerted by that same proton on the Earth because there is no screening of the gravitational force, even if there is intervening matter. Advantage tortoise!

3. *Gravity is universal,* acting between all masses (and energy, since $E = mc^2$) with the same attractive sign. This is fundamentally different from electromagnetism, where the Coulomb interaction between two objects depends on their charges, which can be positive, negative, or zero (even for unscreened matter). Advantage tortoise!

Because of points 1–3, the plodding tortoise carrying the banner of gravity easily wins the race to determine the large-scale structure of the Universe over the swift electromagnetic hare. The reason is the same reason that the tortoise wins in the original *Tortoise and Hare* fable: the relentless pursuit of a singular goal. Gravity can do only one thing, but it does it tirelessly and methodically.

On the other hand, the extreme weakness of gravity means that it can be neglected completely for the microscopic structure of matter: that of molecules, atoms, or nuclei. The lone caveat to this statement is that on incredibly short distance scales (many, many orders of magnitude below present measurement capabilities) gravity can become strong enough that it cannot be ignored in considering the quantum structure of matter. This *Planck scale* is the regime of *quantum gravity,* for which we do not yet have an adequate theory and are reduced to speculation and analogy.

1.2 Classical Newtonian Gravity

Having established the dominance of gravity in determining how the Universe operates on all but the shortest distance scales, it is of importance to ask how gravity can best be described in mathematical terms. A quite serviceable option has been available for three centuries. *Newtonian gravity* works remarkably well for just about everything. It describes the motion of rocks thrown at the Earth's surface and the orbits of the planets and moons and asteroids of the Solar System with almost arbitrary precision, and NASA engineers with confidence send astronauts to the Moon and back, and spacecraft to a precise rendezvous with bodies in the far reaches of the Solar System, based on its prowess. These are remarkable technical achievements, so why would anyone want anything better?

The basic answer is that the motivation and successful quest for a better theory of gravity grew from the remarkable physical intuition and work of one person, Albert Einstein, in the early years of the twentieth century. That better theory of gravity is called *general relativity.* The development of general relativity was different from the development of almost any other new scientific theory in two regards: (1) As just suggested, it was very much the work

---

[1] The discussion in this Introduction assumes the gravity of everyday experience. Later it will be shown that on cosmological scales it is possible for gravity to become effectively repulsive. But that is a story for later that has no bearing on daily life in our little corner of the Universe.

of a single person, unlike most scientific breakthroughs, which involve the direct work of more than one person "standing on the shoulders of giants" (in the words of Newton) who had paved the way before them. (2) Unlike for many paradigm shifts in science, there was no crying need for general relativity brought about by new experiments or observations (quite different from, say, quantum mechanics, which arose also in the early part of the twentieth century in response to a crucial need to understand what measurements in the new field of atomic physics implied about the structure of atoms).

For the theory of gravity it may fairly be argued that at the beginning of the twentieth century there was but a single fly in the ointment of Newtonian gravity, and it was an extremely tiny and arcane fly: the measured orbit of the planet Mercury showed a discrepancy with the predictions of Newtonian gravity in a certain measured angle called the *perihelion shift* that corresponded to a difference of 43 arcseconds per century.[2] You read correctly, *per century!* While this was a puzzling anomaly, one can imagine that very few scientists of the time lost sleep over this discrepancy in the perihelion advance of Mercury, and even fewer would have guessed that the resolution of this tiny anomaly would entail a seismic shift in our understanding of gravity and the nature of space and time.

The precession of the perihelion of Mercury was the first problem to which Einstein applied his new theory of general relativity, and Einstein himself said that he was so overcome with joy when he found that his new theory predicted exactly 43 arcseconds per century of precession over that of Newtonian theory that for several days he could hardly function and experienced heart palpitations [178]. However, the resolution of this problem in Mercury's orbit was *not* Einstein's motivation for developing general relativity. Instead, it seems that Einstein was motivated by more abstract reasoning to develop a new theory of gravity, only later applying the new theory to practical problems like Mercury's orbit. To understand this reasoning it is necessary to first consider the *special theory of relativity,* and to do that we must address the effect of transformations between coordinate systems on the laws of physics.

1.3 Transformations between Inertial Systems

In 1905 Einstein published the *special theory of relativity*, which revolutionized our understanding of space and time with its concepts of space contraction and time dilation, and that the simultaneity of two events was not an absolute thing but rather depended on the relative velocity of the observer. The motivation for the special theory was the central conviction of Einstein that the laws of physics cannot depend on the observer (the *principle of relativity*), meaning that the laws of physics must not depend on the coordinate system in which they are formulated, and that the transformations between inertial frames (coordinate systems not accelerated with respect to each other in which Newton's first law is valid) should be the same for particles and for light.

---

[2] An arcsecond is 1/3600 of an angular degree.

Newtonian mechanics already contained a principle of relativity for space (but not for time), in that the Newtonian laws of mechanics were unchanged by a transformation between inertial frames called a *Galilean transformation*. For motion along the x axis a Galilean transformation takes the form

$$x' = x - vt \qquad y' = y \qquad z' = z \qquad t' = t, \tag{1.1}$$

where primed coordinates and unprimed coordinates represent the two different inertial frames, the velocity in the x direction is v, and a single universal time $t = t'$ has been assumed for all observers. This is just the "common sense" notion that if you are moving in the x direction at constant speed on a railroad flatcar and throw a ball forward with some velocity as measured from the flatcar, the velocity of the ball in the x direction measured by an observer on the ground is the sum of velocities for the train and the ball relative to the train, the transverse y and z directions are unaffected, and the time t' measured on the train and the time t measured on the ground are the same. Obvious, right? And it *is obvious* for balls and trains moving at relatively low velocities, as has been confirmed by many experiments. But Einstein (and others) realized that there is a problem in that these common sense notions of relative motion between inertial frames were inconsistent with the theory of light, which was well understood in 1905 to be an electromagnetic wave described by the Maxwell equations (first published by James Clerk Maxwell in 1861, but put in their more modern form in the 1880s by Oliver Heaviside).

1.4 Maxwell, the Aether, and Galileo

One of the revolutionary features of the Maxwell theory was that wave disturbances could propagate in the electromagnetic field, and these waves traveled with a speed that was a constant of the theory and thus independent of inertial frames. When the constant was evaluated it was found to be equal numerically to the measured speed of light, which caused the Maxwell electromagnetic waves to be identified with light. The beauty of Maxwell's equations greatly impressed Einstein, but they presented a problem of interpretation for classical physics.

By the Galilean transformations, which worked well for mechanical problems, the speed of light should certainly depend upon the frame from which it was measured; but by Maxwell's equations it should not because it is a *constant* of the theory. Since the Maxwell equations and the Galilean transformations were valid in their respective domains, it was desirable to keep both. The standard interpretation that emerged to permit this was that there must be a medium through which the electromagnetic waves moved (Obvious, right? A wave can't just travel through nothing, can it?) This medium was called the *luminiferous aether* or *aether* for short, and it was assumed to be an invisible, rigid (because light waves are transverse and transverse waves don't propagate through fluids) substance permeating all of space but relevant only for light propagation. Then it was proposed that the constant speed of light was an artifact of the special aether rest frame in which light propagated.

The aether of course is fictitious and it is now well understood that light waves are propagating disturbances in electric and magnetic fields that do not require a physical medium, but in the latter part of the nineteenth century the aether was widely believed to be real and various attempts were made to detect the motion of the Earth relative to the aether.[3] The definitive experiments were those of Michelson and Morley, who showed in 1887 using light interferometry that there was no evidence for a different drift of the Earth with respect to the hypothetical aether when it was moving in different directions on its orbit around the Sun, thus casting serious doubt on the existence of the aether.[4]

1.5 The Special Theory of Relativity

With the aether discounted, the Maxwell equations and their constant speed of light were clearly incompatible with the Galilean invariance exhibited by material particles. Others had proposed hints of a solution (most notably Hendrik Lorentz and Henri Poincaré) but it was Einstein who bridged this impasse with the bold hypothesis that the Maxwell theory was correct and that it was the *Galilean transformations* that needed modification to bring them into accord with the Maxwell equations. Thus he proposed that the speed of light was constant for all observers, with no qualifications. This, along with the assumption of relativity (physical law does not depend on the coordinate system) yielded in 1905 the special theory of relativity. The special theory assumes the existence of global inertial frames, so it could not be applied to gravity, which is incompatible with the existence of global inertial frames because it is associated with curved spacetime.

In the special theory (of relativity), the requirement that the speed of light c be an invariant in all inertial frames necessitated replacement of the Galilean transformations with the *Lorentz transformations*,

$$x' = \gamma(x - vt) \qquad y' = y \qquad z' = z \qquad t' = \gamma\left(t - \frac{vx}{c^2}\right) \qquad \gamma \equiv \frac{1}{\sqrt{1 - v^2/c^2}}, \tag{1.2}$$

---

[3] This was called the *aether drift*. Einstein himself apparently was interested in constructing an experiment to measure the aether drift as a student. This never came to fruition because of lack of funds and equipment, and the opposition of his teachers, and in retrospect his methods likely would not have worked, even if the aether existed. It is not clear when Einstein abandoned the idea of the aether, but it was certainly before the publication of the special theory of relativity in 1905 [178].

[4] Nevertheless, efforts persisted for decades to salvage the aether hypothesis. For example, one idea was that a piece of the aether had been somehow trapped in the basement laboratory in which the Michelson–Morley experiment was carried out, thus explaining the null result. The reader will not be surprised to learn that experiments carried out at other locations also found no evidence for the aether [178]. It is not clear to what degree Einstein was influenced by the results of the Michelson–Morley experiment. Einstein claimed at various times that it had little effect on his reasoning. Scientific journals were not nearly as widely available then as they are today, and there is evidence that in his earlier years Einstein sometimes did not have access to important scientific papers. But it seems likely that Einstein knew of the Michelson–Morley results. Perhaps the most consistent interpretation is that Einstein knew of the null aether results but felt that this was not as important as his own reasoning in coming to the special theory of relativity. For example, Einstein placed special emphasis on the role of thought experiments that he began carrying out as a student concerning whether one could travel fast enough to catch up with a light wave, and what the observational consequences would be.

where γ is called the *Lorentz γ-factor*. Notice that in the limit $v/c \to 0$ the factor $\gamma \to 1$ and the Lorentz transformations (1.2) become equivalent to the Galilean transformations (1.1), so the Galilean transformations are quite correct in the low-velocity world of our normal experience. Notice also that, unlike in the Galilean transformation where there is a universal time shared by observers, time transforms non-trivially under Lorentz transformations with the consequence that the Lorentz transformations mix the space and time coordinates.

The mathematician Hermann Minkowski (who was once Einstein's teacher at what is now ETH Zurich) then noted that it is most natural to abandon separate notions of space and time and instead view special relativity in terms of a *4-dimensional spacetime* parameterized by *spacetime coordinates*

$$(x^0, x^1, x^2, x^3) \equiv (ct, x, y, z), \tag{1.3}$$

where the superscripts are indices (not exponents).[5] In a 1908 presentation entitled *Raum und Zeit* (*Space and Time*) that was delivered to the *80th Assembly of German Natural Scientists and Physicians* in Cologne, Minkowski introduced the idea of 4-dimensional spacetime using a phrasing that has now become legendary:[6]

> The views of space and time which I wish to lay before you have sprung from the soil of experimental physics, and therein lies their strength. They are radical. Henceforth space by itself, and time by itself, are doomed to fade away into mere shadows, and only a kind of union of the two will preserve an independent reality. (Hermann Minkowski (1908))

Now time (more precisely time scaled by the speed of light, so that it has the same units as the other three coordinates) becomes just another coordinate in the 4-dimensional spacetime. As Minkowski noted, special relativity is simple when viewed in 4-dimensional spacetime, but becomes more complicated when projected onto 3-dimensional space.

Minkowski also introduced the tensor formalism for special relativity (Einstein's 1905 paper did not use tensors), and introduced terminology such as *worldline* that is now standard. Einstein at first viewed Minkowski's formulation of special relativity using tensors as just a mathematical trick, but soon realized the power of these methods and adopted many of them in his later formulation of the general theory of relativity. Minkowski undoubtedly would have made further contributions to the development of relativity but he died unexpectedly of peritonitis only months after his famous Space and Time lecture.

1.6 Minkowski Space

The surface traced out by allowing the coordinates (x^0, x^1, x^2, x^3) to range over all their possible values defines the manifold of 4-dimensional spacetime. The resulting space is

---

[5] The reason that the indices are in an upper position will be made clear in due time; for now just view them as labels with a possibly eccentric placement, and don't confuse them with exponents.

[6] An English translation of the full presentation may be found at https://en.wikisource.org/wiki/Translation:Space_and_Time.

commonly called *Minkowski spacetime,* which is often shortened to *Minkowski space* or just *spacetime.* In Minkowski space the square of the infinitesimal distance ds^2 between two points (ct, x, y, z) and $(ct + cdt, x + dx, y + dy, z + dz)$ is given by

$$ds^2 = \sum_{\mu\nu} \eta_{\mu\nu} dx^\mu dx^\nu = -c^2 dt^2 + dx^2 + dy^2 + dz^2,$$

$$= -(dx^0)^2 + (dx^1)^2 + (dx^2)^2 + (dx^3)^2, \qquad (1.4)$$

which is called the *line element* of the Minkowski space.[7] The quantity $\eta_{\mu\nu}$ may be expressed as the diagonal matrix

$$\eta_{\mu\nu} = \begin{pmatrix} -1 & 0 & 0 & 0 \\ 0 & 1 & 0 & 0 \\ 0 & 0 & 1 & 0 \\ 0 & 0 & 0 & 1 \end{pmatrix}, \qquad (1.5)$$

and is termed the *metric tensor* of the Minkowski space. The line element (1.4) or the metric tensor $\eta_{\nu\mu}$ determine the geometry of Minkowski space because they specify distances, distances can be used to define angles, and that is geometry. The pattern of signs on the right side of Eq. (1.4) defines the *signature of the metric.* For Minkowski space the signature is $(-+++)$.[8]

The geometry of 4-dimensional Minkowski space differs from that of 4-dimensional euclidean space, so 4-dimensional Minkowski spacetime is *not* "just like ordinary space but with more dimensions." The difference is encoded in the signature of the metric, which for 4-dimensional euclidean space is $(+++ +)$, compared with the signature $(-+++)$ for the Minkowski metric. (That is, the metric tensor of 4-dimensional euclidean space is just the 4×4 unit matrix.) That change in sign for the first entry makes all the difference. Most of the unusual features of special relativity (space contraction, time dilation, relativity of simultaneity, the twin "paradox," ...) follow from this difference in geometry between 4-dimensional Minkowski spacetime and 4-dimensional euclidean space.

1.7 A New Theory of Gravity

Now that the transformation laws between inertial systems for both light and mechanical particles had been unified in special relativity, Einstein turned his attention to how this

---

[7] In this notation ds^2 means the square of ds [that is, $(ds)^2$], and dx^2 means $(dx)^2$, but in (x^0, x^1, x^2, x^3) the superscripts are indices and not powers. This is standard notation and you quickly will learn to distinguish whether a superscript is meant as an index or an exponent from the context. If there is potential ambiguity, use parentheses to make the intention clear, as in the second line of Eq. (1.4).

[8] The pattern of $+$ and -1 signs is referred to here as the signature. Some authors define the signature to be an integer that is the difference of the number of $+$ and $-$ signs. The two definitions convey similar information. A metric such as that of Minkowski space in which the signs in the sign pattern are not all the same is termed an *indefinite metric.* Some authors use instead the signature $(+ - - -)$ for the sign pattern in Eq. (1.4). This leads to the same physical results as our choice as long as all signs are carried through consistently with either choice. The important point is that for Minkowski spacetime the last three terms have the same sign and the sign of the first term is different from that of the other three (provided that the usual modern convention of displaying the timelike coordinate in the first position and the spacelike coordinates in the last three positions is employed).

"special" relativity, which applied only to the restricted case of flat spacetime in which global inertial frames were valid, could be generalized to apply the same principles to the gravitational problem. This was a much more difficult task, so much so that it took Einstein almost a decade to solve it. (During this decade Einstein also made fundamental contributions to quantum mechanics and statistical mechanics, but that is not relevant for the present discussion.)

What Einstein sought was a new gravitational theory that would remove the inconsistencies of Newtonian gravity with respect to the principles of special relativity, but still recover the documented success of Newtonian gravity in a suitable limit. Newtonian gravity and special relativity are at odds in several respects. The most important are that Newtonian theory ascribes a physical reality to space and time coordinates, it treats space and time in an asymmetric way, and it implies that the gravitational force exerted by one mass on another is felt instantaneously by the second mass, no matter how large the distance between them. The first is inconsistent with the Einstein principle of relativity, the second is inconsistent with the Lorentz transformations, and the third is inconsistent with the constant (finite) speed of light in special relativity, which implies that the gravitational interaction cannot act instantaneously across space (no "action at a distance"). Einstein tried various approaches to the generalization of special relativity to incorporate gravity without much success until in late 1907 he hit upon the idea that would eventually lead to general relativity, though it was not until 1915 that he was able to elaborate the idea mathematically into a complete theory.

1.8 The Equivalence Principle

The starting point for this new gravitational theory is that the universality of gravity alluded to above is even more curious than just the generic statement that gravity differs from all other forces in that it acts attractively on all matter. It has been known since the days of Galileo – before Newton was even born – that objects of different mass and/or different composition *fall at the same rate in Earth's gravitational field* (neglecting the effect of friction with the air, of course). This may be stated somewhat more esoterically in terms of the *weak equivalence principle:* the *gravitational mass* of an object (the mass determined from Newton's law of gravity by observing its interaction with a gravitational field) and its *inertial mass* (the mass determined from Newton's second law of motion by pushing the object) are to high experimental precision *equivalent*.[9]

This is at first glance surprising, since there is no a priori reason to expect the two definitions of mass to coincide. In Newtonian theory the equivalence of gravitational and inertial mass is an interesting but unexplained coincidence that mostly gets ignored. For Einstein it became the key to understanding the true nature of gravity and thence to the formulation of general relativity. Einstein realized (what is in retrospect) the obvious

---

[9] More precisely they are proportional, but with a suitable choice of units the constant of proportionality can be chosen to be one.

implication of the gravitational acceleration being independent of any specific property of the mass being accelerated:

> If gravity acts universally on all mass, irrespective of its specific characteristics, then the gravitational force cannot be a property of the masses themselves and therefore must be a universal property of the spacetime in which gravity acts.

Specifically, Einstein realized that if he were in free fall in a gravitational field he would not be able to feel his own weight, so in a small freely falling reference frame the effects of gravity may be transformed away.[10] This led to the realization that (in a small region of spacetime) there was no operational way to distinguish an arbitrary acceleration from the effects of gravity, and this set of ideas came to be called the *(strong) equivalence principle.* By various thought experiments using the equivalence principle it became apparent to Einstein that gravity was associated with the *geometry of spacetime,* specifically through its *curvature:* In the absence of gravity spacetime is flat (Minkowski space); in the presence of gravity, spacetime becomes curved. Using these ideas Einstein was able to find some essential features of general relativity such as the gravitational deflection and redshift of light but the field equations describing the full effects of general relativity required substantial additional mathematical development and were revealed by Einstein for the first time only in late 1915, in a presentation to the Prussian Academy of Sciences.[11]

1.9 General Relativity

The theory of general relativity published by Einstein beginning in 1915 represents a radical new view of space, time, and gravity relative to our "common sense" intuition. It supersedes Newtonian mechanics and Newtonian gravity, but reduces to those theories in the limit of velocities that are small with respect to the speed of light and gravitational fields that are weak (with respect to criteria that will be specified later). It reduces to the theory of special relativity in the limit that gravity vanishes or, in a sense that will be specified more precisely later, for sufficiently local regions of spacetime even in the presence of strong gravitational fields.

General relativity revises fundamentally the very meaning of space, time, and gravity because the effects of gravity no longer appear as a force but as the motion of *free*

---

[10] This is presumably far more obvious to children of the space age accustomed to seeing weightless astronauts in orbital free fall on television than it would have been to Einstein in the early 1900s.

[11] It is sometimes claimed that the mathematician David Hilbert published the field equations for general relativity shortly before Einstein's presentation to the Prussian Academy, and thus should be given full or joint credit for the theory of general relativity. Historical research indicates that Hilbert and Einstein were in a race to complete the field equations, but that the version published by Hilbert before Einstein's presentation was later modified to be consistent with Einstein's correct version of the field equations. At any rate, Hilbert's contribution to the mathematics of general relativity was substantial but he had been motivated to work on the theory by lectures that Einstein gave in Göttingen, and Hilbert gave full credit to Einstein as the author of general relativity. Although Hilbert was a far better mathematician than Einstein, he understood that it was Einstein's deep physical intuition that was most essential to the basic formulation of general relativity.

particles that move in the straightest paths possible (again, a concept that will be quantified in due course), but in a curved spacetime. That is, general relativity will identify the effects of gravity as arising from curvature of space itself. In the classic statement of John Wheeler: *mass tells space how to curve; curved space tells matter how to move.* Note that implied in the circularity of this statement is another basic feature of general relativity: it is a highly nonlinear theory. Speaking loosely, only when the curvature of space is known can the distribution and motion of matter be determined, but the curvature of space can be determined only when the distribution and motion of the matter are known.

As a result of the nonlinear nature of general relativity and its formulation on a 4-dimensional spacetime manifold with indefinite metric, it is notoriously difficult to find exact solutions for the theory in physically meaningful situations and only a few are known. In the general case the nonlinear equations must be solved numerically (a technically demanding task), which is the purview of *numerical relativity*. However, it will be found that the simplest known solutions of general relativity with potential physical application may be formulated in remarkably transparent and elegant mathematical terms, and these formulations may then be used to understand some of the most intriguing aspects of our Universe: black holes, gravitational waves, dark matter, dark energy, and the *new cosmology.*

This Introduction has been a whirlwind tour but you now should understand the essence of special relativity and general relativity. The only task that remains is to fill in a few of the details!

Background and Further Reading

Most introductions to general relativity listed in the references give an overview of how general relativity developed with varying levels of detail. A particularly useful historical reference is the scientific biography of Einstein written by Abraham Pais [178], himself an accomplished theoretical physicist who knew Einstein. The book by Thorne [234] is a comprehensive but non-mathematical account of the development of modern general relativity by one of its leading practitioners.

Problems

1.1 Einstein credited a thought experiment about whether he could catch up with a light wave, and what he would observe if he did, as being important in arriving at the special theory of relativity. What do the Lorentz transformations of special relativity in Eq. (1.2) have to say about "catching up" with a light wave?

1.2 In the first upgrade of the Large Hadron Collider each of the colliding proton beams was designed to reach an energy of 7 TeV (7×10^{12} eV). What is the corresponding Lorentz γ-factor and what is v/c for a proton beam of this energy, given that in special relativity the total energy E is given by $E = \gamma mc^2$, where m is the rest mass?

1.3 From the discussion in this chapter it may be surmised that curvature (of spacetime) will be crucial to understanding general relativity. As conceptual preparation for that discussion, are the following surfaces flat or curved: (a) a piece of paper lying on a table, (b) a sphere, (c) a cylinder? ***

2 Coordinate Systems and Transformations

A physical system is said to have a symmetry under some operation if the system after the operation is indistinguishable from the system before the operation. For example, a perfectly uniform sphere has a symmetry under rotation about any axis because after the rotation the sphere looks the same as before the rotation. The theory of relativity may be viewed as a symmetry under coordinate transformations. As we shall see, relativity is ultimately a statement that physics is independent of coordinate system choice: two observers, referencing their measurements of the same physical phenomena to two different coordinate systems should deduce the *same laws of physics* from their observations. The distinction between special and general relativity is just that special relativity requires a symmetry under only a subset of possible coordinate transformations (those between inertial frames), while general relativity requires that the laws of physics be invariant under the most general physically and mathematically reasonable coordinate transformations. We shall elaborate substantially on the preceding statements in the material that follows. However, it should be obvious already that to understand general relativity it is important to begin by examining coordinate systems, the transformations that are possible between coordinate systems, and the properties of those transformations.

2.1 Coordinate Systems in Euclidean Space

Our goal is to describe coordinates and transformations between coordinates in a general curved space having three spacelike coordinates and one timelike coordinate, since observations imply that this is the nature of the Universe. However, to introduce these concepts it is useful to begin with the simpler and more familiar case of vector fields defined in euclidean space [89].

2.1.1 Parameterizing in Different Coordinate Systems

Assume a 3-dimensional euclidean (that is, flat) space having a cartesian coordinate system (x, y, z), and an associated set of mutually orthogonal unit vectors $(\boldsymbol{i}, \boldsymbol{j}, \boldsymbol{k})$ pointing in the x, y, and z directions, respectively. Assume also that there is an alternative coordinate system (u, v, w), not necessarily cartesian, with the (x, y, z) coordinates related to the (u, v, w) coordinates by functional relationships

$$x = x(u, v, w) \qquad y = y(u, v, w) \qquad z = z(u, v, w), \tag{2.1}$$

where the transformation is assumed to be invertible so that it is possible also to solve for (u, v, w) in terms of (x, y, z) if we choose.

Example 2.1 Take the (u, v, w) system to be the spherical coordinates (r, θ, φ), in which case Eqs. (2.1) takes the familiar form

$$x = r \sin\theta \cos\varphi \qquad y = r \sin\theta \sin\varphi \qquad z = r \cos\theta, \qquad (2.2)$$

with the ranges of values $r \geq 0$ and $0 \leq \theta \leq \pi$ and $0 \leq \varphi \leq 2\pi$.

It will be useful to combine Eqs. (2.1) into a vector equation that gives a position vector $\boldsymbol{r}$ for a point in the space in terms of the (u, v, w) coordinates:

$$\boldsymbol{r} = x(u, v, w)\boldsymbol{i} + y(u, v, w)\boldsymbol{j} + z(u, v, w)\boldsymbol{k}. \qquad (2.3)$$

For example, in terms of the spherical coordinates (r, θ, φ),

$$\boldsymbol{r} = (r \sin\theta \cos\varphi)\boldsymbol{i} + (r \sin\theta \sin\varphi)\boldsymbol{j} + (r \cos\theta)\boldsymbol{k}. \qquad (2.4)$$

The second coordinate system in these examples generally is not cartesian but the space still is assumed to be intrinsically euclidean (not curved). The transformation from the (x, y, z) coordinates to the (r, θ, φ) coordinates just gives two different schemes to label points in the same flat space. This distinction is important because shortly we will generalize this discussion to consider coordinate transformations in spaces that are intrinsically curved (non-euclidean geometry).

2.1.2 Basis Vectors

Vectors are *geometrical objects*: they have an abstract mathematical existence independent of their representation in any particular coordinate system. However, it is often of practical utility to express vectors in terms of components within a specific coordinate system by defining a set of basis vectors that permit arbitrary vectors to be expanded in terms of that basis. Equations (2.3) and (2.4) are familiar examples, where an arbitrary vector has been expanded in terms of the three orthogonal cartesian unit vectors $(\boldsymbol{i}, \boldsymbol{j}, \boldsymbol{k})$ pointing in the x, y, and z directions, respectively. If the space is flat, a single universal basis can be chosen that applies to all points in the space [for example, the cartesian unit vectors $(\boldsymbol{i}, \boldsymbol{j}, \boldsymbol{k})$]. If the space is curved, as will be the case in general relativity, it is no longer possible to define a single coordinate system valid for the whole space and it is often useful to define basis vectors at individual points of the space, as we shall now describe.

Parameterized curves and surfaces: At any point $P(u_0, v_0, w_0)$ defined at specific coordinates (u_0, v_0, w_0) of a space, three surfaces pass. They are defined by setting $u = u_0$, $v = v_0$, or $w = w_0$, respectively. The intersections of these three surfaces define three curves passing through $P(u_0, v_0, w_0)$. From Eq. (2.3) we may obtain general parametric equations for coordinate surfaces by setting one of the variables (u, v, w) equal to a constant, and for curves by setting two of the variables to constants. For example, setting

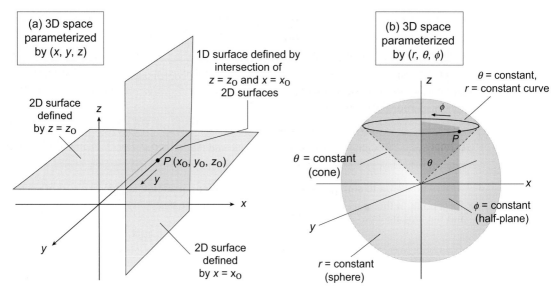

Fig. 2.1 Examples of surfaces and curves arising from constraints. (a) In 3D euclidean space parameterized by cartesian coordinates (x, y, z), the constraints $x = x_0$ and $z = z_0$ define 2D planes and the intersection of these planes defines a 1D surface parameterized by the variable y. (b) In 3D space described in spherical coordinates (r, θ, φ), the constraint $r =$ constant defines a 2D sphere, the constraint $\theta =$ constant defines a cone, and the constraint $\varphi =$ constant defines a half-plane. The intersection of any two of these surfaces defines a curve parameterized by the variable not being held constant.

v and w to constant values, $v = v_0$ and $w = w_0$, yields a parametric equation for a curve given by the intersection of $v = v_0$ and $w = w_0$,

$$\boldsymbol{r}(u) = x(u, v_0, w_0)\boldsymbol{i} + y(u, v_0, w_0)\boldsymbol{j} + z(u, v_0, w_0)\boldsymbol{k}, \tag{2.5}$$

where u acts as a coordinate along the curve defined by the constraints $v = v_0$ and $w = w_0$. Figure 2.1 illustrates these ideas for a space parameterized by cartesian and spherical coordinate systems. For example, in Fig. 2.1(b) the surface corresponding to $r =$ constant is a sphere parameterized by the variables θ and φ, the constraint $\theta =$ constant corresponds to a cone parameterized by the variables r and φ, and the constraint $\varphi =$ constant defines a half-plane parameterized by r and θ. Setting all three variables to constants defines a point P within the space. Through any such point three curves pass that are determined by the pairwise intersections of the three surfaces just defined. (1) Setting both r and θ to constants specifies a curve that is the intersection of the sphere and the cone, which is parameterized by the variable φ. (2) Setting r and φ to constants specifies an arc along the spherical surface parameterized by θ. (3) Setting θ and φ to constants specifies a ray along the conic surface parameterized by r.

The tangent basis: Partial differentiation of (2.3) with respect to u, v, and w, respectively, gives tangents to the three coordinate curves passing though a point P. These may be used to define a set of basis vectors $\boldsymbol{e}_i$ through

$$e_u \equiv \frac{\partial r}{\partial u} \qquad e_v \equiv \frac{\partial r}{\partial v} \qquad e_w \equiv \frac{\partial r}{\partial w}, \qquad (2.6)$$

with the understanding that all partial derivatives are to be evaluated at the point $P = (u_0, v_0, w_0)$. We shall term the basis generated by the tangents to the coordinate curves the *tangent basis*. Example 2.2 illustrates construction of the tangent basis for a spherical coordinate system.

Example 2.2 Consider the spherical coordinate system defined in Eq. (2.2) and illustrated in Fig. 2.1(b). The position vector r is

$$r = (r \sin\theta \cos\varphi)i + (r \sin\theta \sin\varphi)j + (r \cos\theta)k$$

and the tangent basis is obtained from Eq. (2.6):

$$e_1 \equiv e_r = \frac{\partial r}{\partial r} = (\sin\theta \cos\varphi)i + (\sin\theta \sin\varphi)j + (\cos\theta)k,$$

$$e_2 \equiv e_\theta = \frac{\partial r}{\partial \theta} = (r \cos\theta \cos\varphi)i + (r \cos\theta \sin\varphi)j - (r \sin\theta)k,$$

$$e_3 \equiv e_\varphi = \frac{\partial r}{\partial \varphi} = -(r \sin\theta \sin\varphi)i + (r \sin\theta \cos\varphi)j.$$

These basis vectors are mutually orthogonal because $e_1 \cdot e_2 = e_2 \cdot e_3 = e_3 \cdot e_1 = 0$. For example,

$$e_1 \cdot e_2 = r \sin\theta \cos\theta \cos^2\varphi + r \sin\theta \cos\theta \sin^2\varphi - r \cos\theta \sin\theta$$
$$= r \sin\theta \cos\theta (\cos^2\varphi + \sin^2\varphi) - r \cos\theta \sin\theta = 0.$$

From the scalar products of the basis vectors with themselves, their lengths are

$$|e_1| = 1 \qquad |e_2| = r \qquad |e_3| = r \sin\theta,$$

and for convenience these can be used to define a normalized basis,

$$\hat{e}_1 \equiv \frac{e_1}{|e_1|} = (\sin\theta \cos\varphi)i + (\sin\theta \sin\varphi)j + (\cos\theta)k,$$

$$\hat{e}_2 \equiv \frac{e_2}{|e_2|} = (\cos\theta \cos\varphi)i + (\cos\theta \sin\varphi)j - (\sin\theta)k,$$

$$\hat{e}_3 \equiv \frac{e_3}{|e_3|} = -(\sin\varphi)i + (\cos\varphi)j.$$

These basis vectors are now mutually orthogonal and of unit length. Their geometry is illustrated in Fig. 2.2.

For many applications in elementary physics it is standard to use an orthogonal coordinate system so that the basis vectors are mutually orthogonal, and to normalize them to unit length, as in the preceding example. In the more general applications of interest here, the tangent basis defined by the partial derivatives as in Eq. (2.6) need not be orthogonal or normalized to unit length (though for the example shown in Fig. 2.2 and discussed in Example 2.2 the tangent basis happens to be orthogonal).

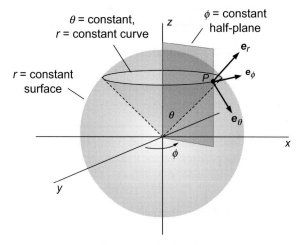

Fig. 2.2 Unit vectors in the tangent basis at point P for Example 2.2. The three basis vectors are tangents to the curves passing through P that are defined by setting any two of the three variables to a constant.

The dual basis: It is also perfectly legitimate to construct a basis at a point P by using the normals to coordinate surfaces rather than the tangents to coordinate curves to define the basis vectors. Since (2.1) was assumed to be invertible, we may solve for

$$u = u(x, y, z) \qquad v = v(x, y, z) \qquad w = w(x, y, z),$$

and a set of basis vectors (e^u, e^v, e^w) may be defined through the gradients

$$e^u \equiv \nabla u = \frac{\partial u}{\partial x}i + \frac{\partial u}{\partial y}j + \frac{\partial u}{\partial z}k,$$
$$e^v \equiv \nabla v = \frac{\partial v}{\partial x}i + \frac{\partial v}{\partial y}j + \frac{\partial v}{\partial z}k, \qquad (2.7)$$
$$e^w \equiv \nabla w = \frac{\partial w}{\partial x}i + \frac{\partial w}{\partial y}j + \frac{\partial w}{\partial z}k,$$

which are normal to the three coordinate surfaces through P defined by $u = u_0$, $v = v_0$, and $w = w_0$, respectively. We say that this basis (e^u, e^v, e^w) defined in terms of normals to surfaces is the *dual* of the basis (2.6), which is defined in terms of tangents to curves. Notice that the basis (2.7) and the basis (2.6) have been distinguished by the use of *superscript indices* on the basis vectors (2.7) and *subscript indices* on the basis vectors (2.6).

Orthogonal and non-orthogonal coordinate systems: The tangent basis and dual basis are of equal validity. For orthogonal coordinate systems as in Example 2.2 the set of normals to the planes corresponds to the set of tangents to the curves in orientation, differing possibly only in the lengths of basis components. Thus, if the basis vectors are normalized the tangent basis and the dual basis for orthogonal coordinates are equivalent and the preceding distinctions have little practical significance. However, for non-orthogonal coordinate systems the two bases generally are not equivalent and the distinction between upper and lower indices is relevant. Example 2.3 illustrates.

Example 2.3 Define a coordinate system (u, v, w) in terms of cartesian coordinates (x, y, z) through [89]

$$x = u + v \qquad y = u - v \qquad z = 2uv + w.$$

The position vector for a point r is then

$$r = xi + yj + zk = (u+v)i + (u-v)j + (2uv+w)k$$

and from Eq. (2.6) the tangent basis is

$$e_1 \equiv e_u = \frac{\partial r}{\partial u} = i + j + 2vk \qquad e_2 \equiv e_v = \frac{\partial r}{\partial v} = i - j + 2uk \qquad e_3 \equiv e_w = \frac{\partial r}{\partial w} = k.$$

Solving the original equations for (u, v, w),

$$u = \tfrac{1}{2}(x+y) \qquad v = \tfrac{1}{2}(x-y) \qquad w = z - \tfrac{1}{2}(x^2 - y^2),$$

and Eq. (2.7) gives for the dual basis

$$e^1 \equiv e^u = \frac{\partial u}{\partial x}i + \frac{\partial u}{\partial y}j + \frac{\partial u}{\partial z}k = \tfrac{1}{2}(i+j) \qquad e^2 \equiv e^v = \frac{\partial v}{\partial x}i + \frac{\partial v}{\partial y}j + \frac{\partial v}{\partial z}k = \tfrac{1}{2}(i-j)$$

$$e^3 \equiv e^w = \frac{\partial w}{\partial x}i + \frac{\partial w}{\partial y}j + \frac{\partial w}{\partial z}k = -(u+v)i + (u-v)j + k.$$

For the tangent basis the preceding expressions give

$$e_1 \cdot e_2 = 4uv \qquad e_2 \cdot e_3 = 2u \qquad e_3 \cdot e_1 = 2v,$$

where the orthonormality of the basis (i, j, k) has been used. Thus the tangent basis is non-orthogonal. Taking scalar products of tangent basis vectors with themselves gives

$$e_1 \cdot e_1 = 2 + 4v^2 \qquad e_2 \cdot e_2 = 2 + 4u^2 \qquad e_3 \cdot e_3 = 1,$$

so the tangent basis in this example is neither orthogonal nor normalized to unit length. It is also clear from the above expressions that generally e_i is not parallel to e^i, so in this non-orthogonal case the normal basis and the dual basis clearly are distinct.

The preceding example illustrates that Eqs. (2.6) and (2.7) define *different but equally valid bases* that are physically distinguishable in the general case of non-cartesian coordinate systems, and hence that the placement of indices in upper or lower positions matters. Since in formulating general relativity it will be necessary to deal with curvilinear coordinate systems, it should be assumed henceforth that the vertical placement of indices in equations (upper or lower positions) is significant.

2.1.3 Expansion of Vectors and Dual Vectors

An arbitrary vector $\boldsymbol{V}$ may be expanded in terms of the tangent basis $\{\boldsymbol{e}_i\}$ and an arbitrary dual vector $\boldsymbol{\omega}$ may be expanded in terms of the dual basis $\{\boldsymbol{e}^i\}$:

$$\boldsymbol{V} = V^1 \boldsymbol{e}_1 + V^2 \boldsymbol{e}_2 + V^3 \boldsymbol{e}_3 = \sum_i V^i \boldsymbol{e}_i \equiv V^i \boldsymbol{e}_i, \qquad (2.8)$$

$$\boldsymbol{\omega} = \omega_1 \boldsymbol{e}^1 + \omega_2 \boldsymbol{e}^2 + \omega_3 \boldsymbol{e}^3 = \sum_i \omega_i \boldsymbol{e}^i \equiv \omega_i \boldsymbol{e}^i, \qquad (2.9)$$

where in the last step of each equation the *Einstein summation convention* has been introduced: an index appearing twice on one side of an equation, once in a lower position and once in an upper position, implies a summation on that repeated index. The index that is summed over is termed a *dummy index;* notice that summation on a dummy index on one side of an equation implies that it does not appear on the other side of the equation. Generally also, if the same index appears more than twice on the same side of an equation, or it appears more than once in an upper position or more than once in a lower position, we have likely made a mistake. Since the dummy (repeated) index is summed over, it should be apparent also that it does not matter what the repeated index is, as long as it is not equivalent to another index in the equation. From this point onward the Einstein summation convention usually will be assumed because it leads to more compact, easier to read equations.

The upper-index coefficients V^i appearing in Eq. (2.8) are termed the *components of the vector* in the basis $\boldsymbol{e}_i = \{\boldsymbol{e}_1, \boldsymbol{e}_2, \boldsymbol{e}_3\}$, while the lower-index coefficients ω_i appearing in Eq. (2.9) are termed the *components of the dual vector* in the basis $\boldsymbol{e}^i = \{\boldsymbol{e}^1, \boldsymbol{e}^2, \boldsymbol{e}^3\}$. The preceding discussion indicates that the vector and dual vector components generally are distinct because they are components in two different bases. However, as we will now discuss the vector and dual vector spaces are related in a fundamental way that permits vector components V^i and dual vector components ω_i to be treated operationally as if they were different components of the same vector. The first step in establishing this relationship is to introduce the ideas of a *scalar product* and a *metric*.

2.1.4 Vector Scalar Product and the Metric Tensor

Utilizing Eq. (2.8), we can write the scalar product of two vectors $\boldsymbol{A}$ and $\boldsymbol{B}$ as

$$\boldsymbol{A} \cdot \boldsymbol{B} = (A^i \boldsymbol{e}_i) \cdot (B^j \boldsymbol{e}_j) = \boldsymbol{e}_i \cdot \boldsymbol{e}_j A^i B^j = g_{ij} A^i B^j, \qquad (2.10)$$

where the *metric tensor* components g_{ij} in this basis are defined by

$$g_{ij} \equiv \boldsymbol{e}_i \cdot \boldsymbol{e}_j. \qquad (2.11)$$

Two vectors alone cannot form a scalar product (which is a real number), but the scalar product of two vectors can be computed with the aid of the metric tensor, as Eq. (2.10) illustrates. Equivalently, the scalar product of dual vectors $\boldsymbol{\alpha}$ and $\boldsymbol{\beta}$ may be expressed as

$$\boldsymbol{\alpha} \cdot \boldsymbol{\beta} = \alpha_i \boldsymbol{e}^i \cdot \beta_j \boldsymbol{e}^j = g^{ij} \alpha_i \beta_j, \qquad (2.12)$$

where the metric tensor component g^{ij} with two upper indices is defined by

$$g^{ij} \equiv \boldsymbol{e}^i \cdot \boldsymbol{e}^j, \qquad (2.13)$$

and the scalar product of dual vectors and vectors as

$$\boldsymbol{\alpha} \cdot \boldsymbol{B} = \alpha_i \boldsymbol{e}^i \cdot B^j \boldsymbol{e}_j = g^i_j \alpha_i B^j, \qquad (2.14)$$

where the metric tensor component g^i_j with mixed upper and lower indices is defined by

$$g^i_j \equiv \boldsymbol{e}^i \cdot \boldsymbol{e}_j. \qquad (2.15)$$

General properties of the metric tensor will be discussed further below but first let us use it to establish a relationship called *duality* between the vector and dual vector spaces.

2.1.5 Relationship of Vectors and Dual Vectors

There is little practical difference between vectors and dual vectors in euclidean space with cartesian coordinates. However, in a curved space the situation is more complex. The essential issue is how to define a vector or dual vector in a curved space, and what that implies. Although the examples in this chapter are primarily from non-curved spaces where it is possible to finesse the issue, it is important not to build into the discussion at this stage methods and terminologies that will not serve us well in curved spacetime. The essential mathematics will be discussed in more depth later, primarily in Section 3.6, but the salient points are that

1. Vectors are not defined directly in the space (more precisely, in the manifold; see Box 3.1), but instead are defined in a euclidean vector space (see Box 3.3) attached to the (possibly curved) manifold at each spacetime point that is called the *tangent space.*
2. Dual vectors are not defined directly in the manifold, but instead are defined in a euclidean vector space attached to the (possibly curved) manifold at each spacetime point that is called the *cotangent space.*
3. The tangent space of vectors and the cotangent space of dual vectors at a point P of the manifold are different but *dual to each other* in a manner that will be made precise below, and this duality allows objects in the two different spaces to be treated as effectively the same kinds of objects.

As will be discussed further in Chapter 3, vectors and dual vectors are special cases of more general objects called *tensors,* and this permits an abstract definition in terms of mappings from vectors and dual vectors to the real numbers.[1] To be specific,

1. Dual vectors $\boldsymbol{\omega}$ are linear maps of vectors $\boldsymbol{V}$ to the real numbers: $\boldsymbol{\omega}(\boldsymbol{V}) = \omega_i V^i \in \mathbb{R}$.
2. Vectors $\boldsymbol{V}$ are linear maps of dual vectors $\boldsymbol{\omega}$ to the real numbers: $\boldsymbol{V}(\boldsymbol{\omega}) = V^i \omega_i \in \mathbb{R}$.

---

[1] A *mapping* is a generalization of a *function.* For example, $f(x) = x^2$ is a map that associates the real number x with the real number x^2. In this example the mapping is from a space to the same space (from real numbers to real numbers). More generally the mapping can be between different spaces, such as from vectors to real numbers. For purposes of the present discussion, no harm will be done if you prefer to think "function" when a map is mentioned.

In these definitions expressions like $\omega(V) = \omega_i V^i \in \mathbb{R}$ can be read as "the dual vectors ω act linearly on the vectors V to produce $\omega_i V^i \equiv \sum_i \omega_i V^i$, which are elements of the real numbers," or "dual vectors ω are functions (maps) that take vectors V as arguments and yield $\omega_i V^i$, which are real numbers", while linearity of the mapping means, for example,

$$\omega(\alpha A + \beta B) = \alpha\omega(A) + \beta\omega(B),$$

where ω is a dual vector, α and β are arbitrary real numbers, and A and B are arbitrary vectors. It is easy to show (see Box 3.3) that, just as the space of vectors satisfies the conditions required of a vector space, the space of dual vectors as defined by the map above satisfies the same conditions and also is a vector space. The vector space of vectors and corresponding vector space of dual vectors are said to be *dual to each other* because they are related by

$$\omega(V) = V(\omega) = V^i \omega_i \in \mathbb{R}. \tag{2.16}$$

Notice further that the expression $A \cdot B = g_{ij} A^i B^j$ from Eq. (2.10) defines a linear map from the vectors to the real numbers, since it takes two vectors A and B as arguments and returns the scalar product, which is a real number. Thus one may write

$$A(B) = A \cdot B \equiv A_i B^i = g_{ij} A^i B^j. \tag{2.17}$$

But since in $A_i B^i = g_{ij} A^i B^j$ the vector B is arbitrary, in general

$$A_i = g_{ij} A^j, \tag{2.18}$$

which specifies a correspondence between a vector with components A^i in the tangent space of vectors and a dual vector with components A_i in the cotangent space of dual vectors. Likewise, Eq. (2.18) can be inverted using that the inverse of g_{ij} is g^{ij} (see Section 2.1.6) to give

$$A^i = g^{ij} A_j. \tag{2.19}$$

Hence, using the metric tensor to raise and lower indices by summing over a repeated index (an operation called *contraction*) as in Eqs. (2.18) and (2.19), we see that the vector and dual vector components are related through contraction with the metric tensor. This is the precise sense in which the tangent and cotangent spaces are dual: they are different, but closely related through the metric tensor. The duality of the vector and dual vector spaces may be incorporated concisely by requiring that for the basis vectors $\{e_i\}$ and basis dual vectors $\{e^i\}$ in Eqs. (2.8) and (2.9)

$$e^i(e_j) = e^i \cdot e_j = \delta^i_j, \tag{2.20}$$

where the Kronecker delta is defined by

$$\delta^i_j = \begin{cases} 1 & i = j \\ 0 & i \neq j \end{cases}. \tag{2.21}$$

This implies that the basis vectors can be used to project out the components of a vector V by taking the scalar product with the vector,

$$V^i = e^i \cdot V \qquad V_i = e_i \cdot V, \tag{2.22}$$

which you are asked to prove in Problem 2.8.

A lot of important mathematics has transpired in the last few equations, so let's pause for a moment and take stock. For a space with a metric tensor defined, Eqs. (2.16)–(2.22) imply that vectors and dual vectors are in a one-to-one relationship that permits them to be manipulated effectively as if a dual vector component were just a vector component with a lower index, and component indices can be raised or lowered as desired by contraction with the metric tensor. Since all spaces of interest here will have metrics, this reduces the practical implications of the fundamental distinction between vectors and dual vectors to a simple matter of keeping proper track of upper and lower positions for indices.

2.1.6 Properties of the Metric Tensor

The metric tensor will play a starring role in formulating the general theory of relativity. Accordingly, let us summarize some of its properties in the simple euclidean spaces investigated to this point, since most will carry over (suitably generalized) to 4-dimensional curved spacetime. Because it may be defined through scalar products of basis vectors, the metric tensor must be symmetric in its indices [see also Problem 3.3(b)]:

$$g^{ij} = g^{ji} \qquad g_{ij} = g_{ji}. \tag{2.23}$$

From Eqs. (2.18) and (2.19)

$$g_{ij} A^j = A_i \qquad g^{ij} A_j = A^i. \tag{2.24}$$

That is, *contraction with the metric tensor may be used to raise or lower an index*. The scalar product of vectors may be written in any of the following equivalent ways,

$$\mathbf{A} \cdot \mathbf{B} = g_{ij} A^i B^j = g^{ij} A_i B_j = g^i_j A_i B^j = A^i B_i = A_i B^i. \tag{2.25}$$

From the two expressions in (2.24), $A^i = g^{ij} A_j = g^{ij} g_{jk} A^k$, and since this is valid for arbitrary components A^i it follows that the metric tensor obeys

$$g^{ij} g_{jk} = g_{kj} g^{ji} = \delta^i_k. \tag{2.26}$$

Viewing g^{ij} as the elements of a matrix $\tilde{G}$ and g_{ij} as the elements of a matrix G, Eqs. (2.23) are equivalent to the matrix equations $G = G^{\mathrm{T}}$ and $\tilde{G} = \tilde{G}^{\mathrm{T}}$, where T denotes the transpose of the matrix. The Kronecker delta is just the unit matrix I, implying that Eq. (2.26) may be written as the matrix equations

$$\tilde{G} G = G \tilde{G} = I. \tag{2.27}$$

Therefore, the matrix corresponding to the metric tensor with two lower indices is the *inverse* of the matrix corresponding to the metric tensor with two upper indices, and one may be obtained from the other by matrix inversion. Some fundamental properties of the metric tensor for 3-dimensional euclidean space are summarized in Box 2.1.

> **Box 2.1**   **The Metric Tensor for 3-Dimensional Euclidean Space**
>
> Components: $g_{ij} \equiv e_i \cdot e_j$ $g^{ij} \equiv e^i \cdot e^j$ $g^i_j \equiv e^i \cdot e_j = \delta^i_j$
>
> Scalar product: $A \cdot B = g_{ij}A^iB^j = g^{ij}A_iB_j = g^i_j A_i B^j = A^iB_i = A_iB^i$
>
> Symmetry: $g^{ij} = g^{ji}$ $g_{ij} = g_{ji}$
>
> Contractions: $g_{ij}A^j = A_i$ $g^{ij}A_j = A^i$
>
> Orthogonality: $g^{ij}g_{jk} = g_{kj}g^{ji} = \delta^i_k$
>
> Matrix properties: $\tilde{G}G = G\tilde{G} = I$ $\tilde{G} \equiv [g^{ij}]$ $G \equiv [g_{ij}]$

2.1.7 Line Elements

Assume coordinates $u^1(t)$, $u^2(t)$, and $u^3(t)$ parameterized by the variable t. As the parameter t varies the points characterized by the specific values of the coordinates

$$u^1 = u^1(t) \qquad u^2 = u^2(t) \qquad u^3 = u^3(t)$$

will trace out a curve in the 3-dimensional space. From Eq. (2.3), the position vector for these points as a function of t is

$$r(t) = x\big(u^1(t), u^2(t), u^3(t)\big)i + y\big(u^1(t), u^2(t), u^3(t)\big)j + z\big(u^1(t), u^2(t), u^3(t)\big)k,$$

and by the chain rule

$$\frac{dr}{dt} = \frac{\partial r}{\partial u^1}\frac{du^1}{dt} + \frac{\partial r}{\partial u^2}\frac{du^2}{dt} + \frac{\partial r}{\partial u^3}\frac{du^3}{dt} = \dot{u}^1 e_1 + \dot{u}^2 e_2 + \dot{u}^3 e_3, \tag{2.28}$$

where (2.6) has been used and $du^i/dt \equiv \dot{u}^i$. In summation convention this equation is $\dot{r} = \dot{u}^i e_i$, which may be expressed in differential form as $dr = du^i e_i$, so the squared infinitesimal distance along the curve is

$$\begin{aligned} ds^2 = dr \cdot dr &= du^i e_i \cdot du^j e_j \\ &= e_i \cdot e_j \, du^i du^j \\ &= g_{ij} \, du^i du^j, \end{aligned} \tag{2.29}$$

where Eq. (2.11) has been used. Notice that in writing the line element a standard notational convention $d\alpha^2 \equiv (d\alpha)^2$ has been used; that is, $d\alpha^2$ means the square of $d\alpha$, not the differential of α^2. Thus $ds^2 = g_{ij} du^i du^j$ is the infinitesimal line element for the space described by the metric g_{ij}, and the length s of a finite segment between points a and b is obtained from the integral

$$s = \int_a^b \left(g_{ij}\frac{du^i}{dt}\frac{du^j}{dt}\right)^{1/2} dt, \tag{2.30}$$

where t parameterizes the position along the segment, as illustrated in Fig. 2.3.

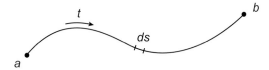

Fig. 2.3 Distance between points *a* and *b* integrated along a curve parameterized by *t*.

2.1.8 Euclidean Line Element

The line element for 2-dimensional euclidean space in cartesian coordinates (x, y) is given by

$$ds^2 = dx^2 + dy^2, \tag{2.31}$$

which is just the Pythagorean theorem for right triangles having infinitesimal sides. The corresponding line element expressed in plane polar coordinates (r, φ) is then the familiar

$$ds^2 = dr^2 + r^2 d\varphi^2, \tag{2.32}$$

which is worked out in Example 2.4.

Example 2.4 For plane polar coordinates (r, φ)

$$x = r \cos \varphi \qquad y = r \sin \varphi,$$

so the position vector (2.3) takes the form

$$\boldsymbol{r} = (r \cos \varphi)\boldsymbol{i} + (r \sin \varphi)\boldsymbol{j}.$$

Then from Eq. (2.6) the basis vectors in the tangent basis are

$$\boldsymbol{e}_1 = \frac{\partial \boldsymbol{r}}{\partial r} = (\cos \varphi)\boldsymbol{i} + (\sin \varphi)\boldsymbol{j} \qquad \boldsymbol{e}_2 = \frac{\partial \boldsymbol{r}}{\partial \varphi} = -r(\sin \varphi)\boldsymbol{i} + r(\cos \varphi)\boldsymbol{j}.$$

The elements of the metric tensor then follow from Eq. (2.11):

$$g_{11} = \cos^2 \varphi + \sin^2 \varphi = 1 \qquad g_{22} = r^2(\cos^2 \varphi + \sin^2 \varphi) = r^2$$

and $g_{12} = g_{21} = 0$, or written as a matrix,

$$g_{ij} = \begin{pmatrix} 1 & 0 \\ 0 & r^2 \end{pmatrix}.$$

Then from Eq. (2.29) the line element is

$$ds^2 = g_{11}(du^1)^2 + g_{22}(du^2)^2 = dr^2 + r^2 d\varphi^2,$$

where $u^1 = r$ and $u^2 = \varphi$. Equivalently,

$$ds^2 = (dr \; d\varphi) \begin{pmatrix} 1 & 0 \\ 0 & r^2 \end{pmatrix} \begin{pmatrix} dr \\ d\varphi \end{pmatrix} = dr^2 + r^2 d\varphi^2,$$

in matrix form.

The form of the line element is different for cartesian and polar coordinates, but for any two nearby points the distance between them is given by ds, independent of the coordinate system. Thus, the line element ds is unchanged by coordinate transformations. Since the distance between any two points that are not nearby can be obtained by integrating ds along the curve, this means that *the distance between any two points is invariant under coordinate transformations for metric spaces.*[2]

The line element, which is specified in terms of the metric tensor, characterizes the geometry of the space because integrals of the line element define distances and angles can be defined in terms of ratios of distances. For instance, in Problem 2.1 you are asked to use the line element to verify the relationship between the circumference of a circle and its radius in euclidean space. Indeed, all the axioms of euclidean geometry could be verified starting from the line elements (2.31) or (2.32).

2.2 Integration

Integration enters into physical theories in various ways, for example in the formulation of conservation laws. It is important to understand how the volume element for integrals behaves under change of coordinate systems. In euclidean space with orthonormal coordinates this is trivial. It becomes nontrivial in curved spaces, or even in flat spaces parameterized in non-cartesian coordinates. Let us illustrate in flat 2D space with coordinates (x^1, x^2) and basis vectors $(\boldsymbol{e}_1, \boldsymbol{e}_2)$, assuming an angle θ between the basis vectors. As you are asked to demonstrate in Problem 2.9, the 2D volume (area) element is in this case

$$dA = \sqrt{\det g}\, dx^1 dx^2, \qquad (2.33)$$

where $\det g$ is the determinant of the metric tensor g_{ij} represented as a 2×2 matrix. For orthonormal coordinates g_{ij} is the unit matrix and $(\det g)^{1/2} = 1$, but in the general case the $(\det g)^{1/2}$ factor is not unity and its presence is essential to making integration invariant under change of coordinates. As we will show in Section 3.13.1, this 2D example generalizes easily to define invariant integration in 4D spacetime.

2.3 Differentiation

Derivatives of vectors in spaces defined by position-dependent metrics will be crucial in formulating general relativity. We introduce the issue using the simpler case of taking the derivative of a vector in a euclidean space, but one parameterized with a vector basis that may depend on the coordinates. A vector $\boldsymbol{V}$ may be expanded in a convenient basis $\boldsymbol{e}_i$,

---

[2] Mathematically, spaces (more precisely, manifolds; see Box 3.1) are equipped with a hierarchy of characteristics and a prescription for measuring distances (a metric) need not be one of them. If such a prescription is defined, the space is termed a *metric space*. This distinction might seem a bit pedantic to a physicist since almost all spaces employed in physics are metric spaces, but it is important to mathematicians.

$$V = V^i e_i. \tag{2.34}$$

By the usual product rule, the partial derivative is given by a sum of two terms,

$$\frac{\partial V}{\partial x^j} = \frac{\partial V^i}{\partial x^j} e_i + V^i \frac{\partial e_i}{\partial x^j}, \tag{2.35}$$

where the first term represents the change in the component V^i and the second term represents the change in the basis vectors e_i. For a basis independent of coordinates the second term is zero and the usual formula is recovered. However, if the basis depends on the coordinates the second term will generally not be zero. In the second term the factor $\partial e_i/\partial x^j$ resulting from the action of the derivative operator on the basis vectors is itself a vector and therefore can be expanded in the vector basis,

$$\frac{\partial e_i}{\partial x^j} = \Gamma^k_{ij} e_k. \tag{2.36}$$

The Γ^k_{ij} appearing in Eq. (2.36) are called *connection coefficients* or *Christoffel symbols*; their generalization to 4-dimensional spacetime will be discussed in Ch. 7. There it will be shown that the connection coefficients can be evaluated from the metric tensor and are central to the definition of derivatives and to specifying a prescription for parallel transport of vectors in curved spacetime.

2.4 Non-euclidean Geometry

Now let us take a brief initial peek at *non-euclidean geometry*, by which we will mean a geometry for a space that does not satisfy Euclid's postulates. For example, in a *euclidean* or *flat* space lines that are parallel locally never meet, but in spaces described by non-euclidean geometries locally parallel lines may cross globally. Such spaces are *intrinsically curved*, in a sense that will be made precise later. A simple example of non-euclidean geometry is afforded by a sphere,[3] for which locally parallel lines do not remain parallel globally: lines of longitude on the surface of the Earth appear to be parallel at the equator but cross at the poles.

Imposing a standard polar coordinate system (θ, φ) on the sphere S^2, the results of Problem 2.4 indicate that the line element is given by

$$ds^2 = R^2(d\theta^2 + \sin^2\theta \, d\varphi^2), \tag{2.37}$$

where R is the radius of the sphere. The geometry of S^2 may be investigated by calculating the ratio of the circumference of a circle to its radius. A circle may be defined in the 2D space by marking a locus of points lying a constant distance S from a reference point, which may be chosen to be the north pole, as illustrated in Fig. 2.4. The angle subtended

---

[3] The notation S^n denotes a sphere of n dimensions. Thus the ordinary sphere S^2 is 2-dimensional (the interior when embedded in 3D space is not part of the sphere) and a sphere S^1 is a circle, which is 1-dimensional. (S^0 corresponds to two points.) Spheres with $n > 2$ are not easily visualized but are well-defined mathematically.

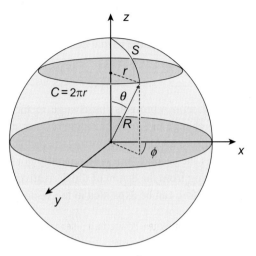

Fig. 2.4 Measuring the circumference of a circle in a curved 2D space S^2.

by S is S/R and $r = R\sin(S/R)$. Then from the geometry in Fig. 2.4 the circumference of the circle is

$$C = 2\pi r = 2\pi R \sin \frac{S}{R} = 2\pi S \left(1 - \frac{S^2}{6R^2} + \cdots \right). \tag{2.38}$$

Alternatively, the same result may be obtained by integrating the line element (2.37)

$$C = \oint ds = \int_0^{2\pi} R \sin \frac{S}{R} d\varphi = 2\pi R \sin \frac{S}{R}. \tag{2.39}$$

If S is much less than R, higher-order terms in the expansion of Eq. (2.38) may be ignored and the standard euclidean result $C = 2\pi S$ is obtained. But more generally the difference between the circumference of a circle drawn on the sphere and $2\pi S$ is a measure of how much the sphere deviates from euclidean geometry. At the present introductory level the geometry of Fig. 2.4 has been visualized by embedding the 2D sphere in 3D euclidean space, but in Section 7.2.2 we will show that a quantitative measure of curvature for a non-euclidean space can be determined entirely by measurements within the space, with no need of embedding in higher dimensions.

2.5 Transformations

Often it proves useful (or even essential) to express physical quantities in more than one coordinate system. Therefore, we need to understand how to transform between coordinate systems. This issue is particularly important in general relativity, where it is essential to ensure that the laws of physics are not altered by the most general transformation between coordinate systems. As illustration of the principles involved, consider two familiar examples: spatial rotations and Galilean boosts.

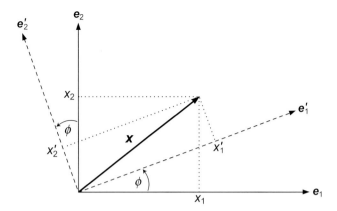

Fig. 2.5 Rotation of the coordinate system for a vector x. The vector is invariant under the rotation but its components in the original and rotated coordinate systems are different.

2.5.1 Rotational Transformations

Consider the description of a vector under rotation of a coordinate system about the z axis by an angle φ, as in Fig. 2.5. In terms of the original basis vectors $\{e_i\}$ the vector x has the components x_1 and x_2. After rotation of the coordinate system by the angle φ to give the new basis vectors $\{e'_i\}$, the vector x has the components x'_1 and x'_2 in the new coordinate system. The vector x can be expanded in terms of the components for either basis:

$$x = x^i e_i = x'^i e'_i, \tag{2.40}$$

and the geometry of Fig. 2.5 may be used to find that the components in the two bases are related by the transformation (assuming a clockwise rotation; for counterclockwise the signs of the $\sin \varphi$ entries would be reversed)

$$\begin{pmatrix} x'^1 \\ x'^2 \\ x'^3 \end{pmatrix} = \begin{pmatrix} \cos\varphi & \sin\varphi & 0 \\ -\sin\varphi & \cos\varphi & 0 \\ 0 & 0 & 1 \end{pmatrix} \begin{pmatrix} x^1 \\ x^2 \\ x^3 \end{pmatrix}, \tag{2.41}$$

which may be written compactly as

$$x'^i = R^i_{\ j} x^j, \tag{2.42}$$

where the $R^i_{\ j}$ are the elements of the matrix in Eq. (2.41). This transformation law holds for any vector. In fact, a vector in the x–y plane may be *defined* to be a quantity that obeys this transformation law.

2.5.2 Galilean Transformations

Another simple example of a transformation is that between inertial frames in classical mechanics. Transformations between inertial frames with the same orientation but different relative velocities are called *boosts*. In Newtonian physics time is considered an absolute quantity that is the same for all observers and boosts for motion along the x axis take the Galilean form

$$x' = x'(x,t) = x - vt \qquad t' = t'(x,t) = t. \tag{2.43}$$

The Newtonian version of relativity asserts that the laws of physics are invariant under such Galilean transformations. Although the laws of mechanics at low velocity are approximately invariant under (2.43), the laws of electromagnetism (Maxwell's equations) and the laws of mechanics for velocities near that of light are not. Indeed, as we discussed in Ch. 1, the failure of Galilean invariance for the Maxwell equations was a large initial motivation for Einstein's demonstration that the existing laws of mechanics required modification, leading to the special theory of relativity.

As will be discussed further in Chapter 4, in the absence of gravity the laws of both special relativistic mechanics and of electromagnetism are invariant under Lorentz transformations but not under Galilean transformations. In the presence of a gravitational field, neither Galilean nor Lorentz invariance holds globally and it is necessary to seek a more comprehensive invariance to describe systems that are subject to gravitational forces. This quest eventually leads to the theory of general relativity.

Background and Further Reading

Discussion of the material in this chapter at a comparable level may be found in Foster and Nightingale [89] (from which this chapter has adapted various examples), Cheng [65], Lambourne [142], and d'Inverno [76].

Problems

2.1 Using only the 2-dimensional euclidean line element of Eq. (2.31) expressed in cartesian coordinates, prove that the circumference of a circle C and its radius R are related by $C = 2\pi R$. Show that the same result follows if the line element is expressed in plane polar coordinates [Eq. (2.32)]. ***

2.2 Prove that Newton's second law and Newton's law of universal gravitation are invariant under Galilean boosts.

2.3 Use tangent and dual basis vectors to construct metric tensor components g^{ij} and g_{ij}, and the line element, for the coordinate system (u, v, w) of Example 2.3. Verify that the matrices g^{ij} and g_{ij} found for this problem are inverses of each other.

2.4 For a sphere described in the spherical coordinate system of Eq. (2.2), determine the basis vectors in the tangent representation and use this to construct the metric tensor g_{ij} and the line element ds^2. ***

2.5 Consider the following torus:

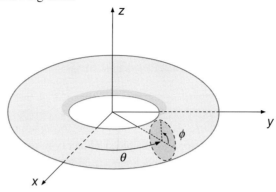

Taking θ and φ as parameters, points on the torus are defined by

$$x = (a + b\cos\varphi)\cos\theta \qquad y = (a + b\cos\varphi)\sin\theta \qquad z = b\sin\varphi,$$

where a and b are constants. Construct the tangent basis vectors for θ and φ, and the corresponding metric tensor.

2.6 For the plane polar coordinate system of Example 2.4, find the basis vectors dual to the tangent basis $e_r \equiv e_1$ and $e_\varphi \equiv e_2$.

2.7 Show that the euclidean line element $ds^2 = dx^2 + dy^2 + dz^2$ is invariant under constant displacements and under rotations.

2.8 Prove that the vector components are given by the projections in Eq. (2.22). ***

2.9 Consider a 2D coordinate system for a planar surface with coordinates (x^1, x^2) and basis vectors (e_1, e_2). Assume that the basis vectors are not necessarily orthogonal, with an angle θ between them. Show that the area of the infinitesimal parallelogram $dA = |e_1||e_2|\sin\theta\, dx^1 dx^2$ is given by Eq. (2.33), which is a 2D version of the 4D invariant volume element that will be given in Eq. (3.48). ***

3 Tensors and Covariance

The *principle of relativity,* which is the cornerstone of both special and general relativity, implies that coordinates in a physical theory must be viewed as *arbitrary labels,* so that the laws of physics are independent of the coordinate system in which they are formulated. The simplest way to ensure this coordinate independence is to implement the laws of physics in terms of equations that are *covariant* with respect to the most general coordinate transformations. The term *covariance* describes a formalism in which a set of equations maintains the same mathematical form under a specified set of transformations. Covariance is stated most concisely in terms of *tensors,* which we may think of as generalizing the idea of vectors. This chapter gives an introduction to tensors, tensor notation, and tensor properties. In the following two chapters we will illustrate with an application of the tensor formalism to Lorentz covariance, which will lead to the theory of special relativity. In subsequent chapters curved spacetime and the extension of Lorentz covariance to a more general covariance will be introduced, which will lead to the theory of general relativity.

3.1 Invariance and Covariance

Notice that covariance is defined relative to a particular set of transformations. You should notice also the subtle difference between *invariance* and *covariance* with respect to a set of transformations. Invariance means that the physical observables of the system are not changed by the transformations (think of the rotation of a featureless sphere). Covariance means that the system is formulated mathematically so that the form of the equations does not change under the transformations (think of an equation formulated in terms of vectors in 3D euclidean space). One sometimes also encounters the terminology "manifest covariance," meaning that the invariance is *manifest* in the formulation of the theory (can be "seen at a glance"). A system that is not formulated in a manifestly covariant way might still be invariant, but it may not be obvious (not manifest) without detailed examination. These distinctions can be confusing, but this is by now established terminology.

3.2 Spacetime Coordinates

Because relativity implies that space and time enter descriptions of nature on comparable footings, it will be useful to unify them into a 4-dimensional continuum termed *spacetime* (see Section 1.5). Spacetime is an example of a *differentiable manifold,* which is discussed

Box 3.1 Manifolds

Loosely, a manifold is a space. More formally, an *n-dimensional manifold* is a set that can be parameterized continuously by *n* independent real coordinates for each point (member of the set). Physics is usually concerned with *differentiable manifolds,* which are continuous and differentiable (to a suitable order). Again, more formally: a manifold is continuous if for every point there are neighboring points having infinitesimally different coordinates, and differentiable if a scalar field can be defined on the manifold that is everywhere differentiable. The preceding definitions still lack a crucial ingredient for physical applications: *geometry.* This will be remedied in Section 6.6, where it will be shown that general relativity assumes spacetime to be a *Riemannian manifold*: a continuous, differentiable manifold with geometry described by a metric tensor of quadratic form that may depend on spacetime coordinates.

Coordinates, Charts, and Atlases

A *coordinate system* or *chart* associates *n* real parameter values (labels) uniquely with each point of an *n*-dimensional manifold *M* through a one-to-one mapping from $\mathbb{R}^n$ (cartesian product of *n* copies of the real numbers $\mathbb{R}$; see the footnote in Box 3.2) to *M*. For example, the set of continuous rotations about a single axis defines a one-dimensional manifold parameterized by an angle $\varphi \in \mathbb{R}$. The one-to-one association of points in the *n*-dimensional manifold with the values of their parameter labels is analogous to mapping points of the manifold to points of an *n*-dimensional euclidean space. Thus, *locally* a manifold looks like euclidean space in its most general properties, such as dimensionality and differentiability.

A single coordinate system is usually insufficient to give a unique correspondence between points and coordinate labels for all but the simplest manifolds. For example, the latitude–longitude system for the Earth (viewed as a 2-sphere, S^2) is degenerate at the poles where all values of longitude correspond to a single point. In such cases the manifold must be parameterized by overlapping *coordinate patches* (*charts*), with *transition functions* between the different sets of coordinates for points in each overlap region. An *atlas* is a collection of charts sufficient to parameterize an entire manifold. For the latitude–longitude example it may be shown that the atlas must contain at least two overlapping charts to parameterize the full manifold uniquely.

Curves and Surfaces

Subsets of points within a manifold can be used to define *curves* and *surfaces*, which represent *submanifolds* of the full manifold. Often it is convenient to represent these parametrically (see Chapter 2), with an *m*-dimensional submanifold parameterized by *m* parameters. A curve is a one-dimensional submanifold parameterized by a single parameter, while a *hypersurface* is a surface of one less dimension than the full manifold, parameterized by $n - 1$ real numbers.

---

further in Box 3.1. In spacetime points will be defined by coordinates having four components, the first labeling the time multiplied by the speed of light c, the other three labeling the spatial coordinates:

$$x \equiv x^\mu = (x^0, x^1, x^2, x^3) = (ct, \boldsymbol{x}), \tag{3.1}$$

where x denotes a vector with three components (x^1, x^2, x^3) labeling the spatial position. The first component x^0 is termed *timelike* and the last three components (x^1, x^2, x^3) are termed *spacelike*. As for the discussion in Chapter 2, the placement of indices in upper or lower positions is meaningful. Bold symbols will be used to denote (ordinary) vectors defined in the three spatial degrees of freedom, with 4-component vectors in spacetime denoted in non-bold symbols. For spacetime the modern convention is to number the indices beginning with zero rather than one. The coordinate systems of interest will be assumed to be quite general, subject only to the requirement that they assign a coordinate uniquely to every point of spacetime, and that they be differentiable to sufficient order for the task at hand at every spacetime point.

A point within an n-dimensional manifold can be identified using a coordinate system of n parameters but the choice of coordinate system is arbitrary. Hence the points may be relabeled by a *passive coordinate transformation* that switches the coordinate labels of the points, $x^\mu \to x'^\mu$, with x^μ and x'^μ labeling the *same point* but in two different coordinate systems. Therefore, our generic concern is with a transformation between one set of spacetime coordinates denoted by (x^0, x^1, x^2, x^3), and a new set

$$x'^\mu = x'^\mu(x) \qquad (\mu = 0, 1, 2, 3), \tag{3.2}$$

where $x = x^\mu$ denotes the original (untransformed) coordinates. This notation is an economical form of

$$x'^\mu = f^\mu(x^0, x^1, x^2, x^3) \qquad (\mu = 0, 1, \ldots), \tag{3.3}$$

where the single-valued, continuously differentiable function f^μ assigns new (primed) coordinates (x'^0, x'^1, x'^2, x'^3) to a point of the manifold with old coordinates (x^0, x^1, x^2, x^3). This transformation may be abbreviated to $x'^\mu = f^\mu(x)$ and, even more tersely, to Eq. (3.2). Equation (3.2) is a generalization of the transformations between coordinate systems in two dimensions discussed in Section 2.5 to transformations in 4-dimensional spacetime. The coordinates in Eq. (3.2) are just labels so the laws of physics cannot depend on them. This implies that the system x'^μ is not privileged and therefore Eq. (3.2) should be invertible.

The Einstein summation convention introduced in Section 2.1.3 will be assumed for all subsequent equations: for any term on one side of an equation any index that is repeated, once as a superscript and once as a subscript, implies a summation over that index. Thus, such an index is a *dummy index* that is removed by the summation and should not appear on the other side of the equation. In this summation convention a superscript (subscript) in a denominator counts as a subscript (superscript) in a numerator. For example, in x^μ the index μ acts as a superscript in the summation convention but in $1/x^\mu$ the index μ acts as a subscript in the summation convention. Because it is important in some contexts to distinguish notationally whether an index represents a timelike or spacelike component, we adopt also the convention that Greek indices $(\alpha, \beta, \ldots)$ will be used to denote the full set of spacetime indices running over 0, 1, 2, 3, while roman indices $(i, j, \ldots)$ will be used to denote the indices 1, 2, 3 running only over the spatial coordinates. Thus x^μ means any of the components x^0, x^1, x^2, x^3, but x^i means any of the components x^1, x^2, x^3.

3.3 Vectors in Non-euclidean Space

Spacetime is characterized by a manifold that is not euclidean. In euclidean space we are used to representing vectors as directed line segments of finite length, with an arrow indicating the direction and the length of the line segment indicating the magnitude of the vector. This picture will not do in curved spacetime, which is locally but not globally euclidean, so extended straight lines have no clear meaning. Thus in non-euclidean manifolds one of the first questions that we need to address is how to define a vector at some spacetime point. The answer is that vectors are not defined in the curved manifold itself but rather in a *tangent space* that may be visualized for a 2D manifold as a plane tangent to the point on the curved surface, as discussed further in Box 3.2 and illustrated in Fig. 3.1 for a 2D sphere S^2. The idea conveyed by Fig. 3.1 in which planes tangent to a 2D surface are shown embedded in a 3D space is useful conceptually. However, defining the tangent space at each point is an *intrinsic process* with respect to a manifold and does not require embedding it in a higher-dimensional manifold, as will be shown below in Section 3.4.1.

Example 3.1 Box 3.2 describes how the tangent spaces of a manifold may be combined into a *tangent bundle*, which is the prototype of a mathematical construction called a *fiber bundle*. Fiber bundles also are useful in describing quantum field theories for which elementary particles have an internal (not spacetime) degree of freedom that depends on the spacetime coordinates (such as *local gauge invariance*). Then it can be useful to construct fiber bundles in which the base space is Minkowski space and the fibers represent the internal degrees of freedom at each spacetime point. Nonabelian gauge field theories have been given an elegant description in such terms.

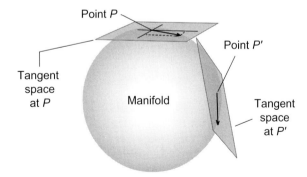

Fig. 3.1 Tangent spaces and vectors in curved spaces, illustrated for the manifold S^2. Vectors (indicated by arrows) are defined in the tangent spaces at each point, not in the curved manifold. Embedding the 2D manifold in 3D euclidean space is for visualization purposes only; the tangent space has a specification that is intrinsic to the 2D manifold.

> **Box 3.2** **Tangent and Cotangent Bundles**
>
> An n-dimensional Riemannian manifold (see Section 6.6) has associated with each point P an n-dimensional euclidean vector space T_P with a basis defined by the directional derivatives evaluated at P for coordinate curves passing through P (see Section 3.4.1). This is termed the *tangent space* and vectors at P are defined within that space. Likewise, at each point P the manifold has an intrinsically defined n-dimensional euclidean vector space termed the *cotangent space* T_P^*, in which dual vectors at P are defined. The tangent space T_P and the cotangent space T_P^* are *dual to each other,* in the sense defined in Sections 2.1.5 and 3.6.4.
>
> **Tangent and Cotangent Spaces are Intrinsic to a Manifold**
> Tangent spaces are illustrated in Fig. 3.1 for an $n = 2$ manifold by embedding the manifold in 3D euclidean space and displaying the tangent space as a plane tangent to the surface at a given point. While this figure is useful for intuition, the definitions given above make clear that vectors and dual vectors are local to a point in the manifold, and that they are not defined in the manifold itself but that the tangent and cotangent spaces in which they are defined may be constructed from the properties of the manifold alone. Thus the tangent space and cotangent space at each point of a manifold have an *intrinsic meaning* that is independent of whether the manifold is embedded in some higher-dimensional space.
>
> **Fiber Bundles**
> The *tangent bundle TM* of the manifold M is a manifold consisting of all the tangent vectors defined for M, which is given by the disjoint union of all of its tangent spaces T_P. Likewise, a *cotangent bundle* T^*M of the manifold M is defined by the disjoint union of all the cotangent spaces T_P^* of M. A simple example of of a tangent bundle is illustrated in Fig. 3.2 for the 1-dimensional sphere (circle) S^1.
>
> Tangent bundles are examples of a *fiber bundle,* which is a manifold E that is *locally* the *cartesian product* $E = F \times B$ of two spaces,[a] the *base space* B and the *fiber space* F (with the fiber at P corresponding to the tangent space T_P at P), but that *globally* may have a different topological structure. For a manifold of dimension n the tangent and cotangent fiber bundles are of dimension $2n$. If a bundle is both locally and globally a product, it is said to be *trivial*. The tangent bundle on S^1 illustrated in Fig. 3.2(b) is a trivial bundle that is both locally and globally $\mathbb{R}^1 \times S^1$. However, a circular strip with a twist (*Möbius strip*) leads to a bundle that is $\mathbb{R}^1 \times S^1$ locally, but that has a nontrivial global topology; see Figs. 3.2 (c)–(d).
>
> [a] A cartesian product $X \times Y$ of two sets X and Y is the set of all possible ordered pairs (x, y) with x an element of X and y an element of Y. The prototype is $\mathbb{R}^2 = \mathbb{R}^1 \times \mathbb{R}^1$, where $\mathbb{R}^1$ is 1D euclidean space (a line) and $\mathbb{R}^2$ is 2D euclidean space (a plane). If X is a manifold of dimension m and Y is a manifold of dimension n, then the cartesian product $X \times Y$ is a manifold of dimension $m \times n$.

3.4 Coordinates in Spacetime

A universal coordinate system can be chosen in flat space, with basis vectors that are mutually orthogonal and constant. Furthermore, these constant basis vectors can be normalized once and for all to unit length. Much of ordinary physics is conveniently described using such *orthonormal bases*. The situation is more complicated in curved

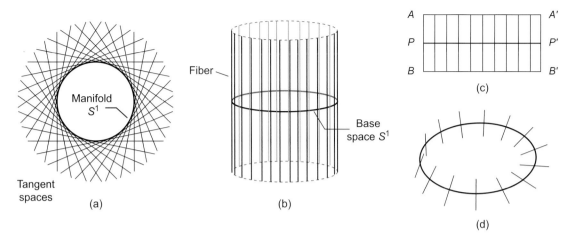

Fig. 3.2 Tangent bundle TS^1 for the 1-dimensional manifold S^1. (a) The manifold S^1 (the circle) and some of its tangent spaces (lines tangent to the circle). (b) The corresponding tangent bundle (locally and globally $R^1 \times S^1$). (c) Figure (b) cut vertically and rolled out flat. Figure (b) corresponds to identifying $A \leftrightarrow A'$ and $B \leftrightarrow B'$. (d) Nontrivial topology (Möbius band) generated by identifying $A \leftrightarrow B'$ and $B \leftrightarrow A'$ in (c) (locally $R^1 \times S^1$, but not globally). The lines corresponding to the tangent spaces (fibers) should be imagined extending to infinity to accommodate vectors of arbitrary length. The overlaps of these lines in (a) are meaningless since each tangent space is defined independently at a different point of the manifold. Therefore, in (b) the lines (fibers) corresponding to tangent spaces have been arrayed perpendicular to the base manifold so that they do not overlap [(b) can be constructed from (a) by rotating all the tangent lines such that they are perpendicular to the plane of the circle, keeping the intersection with the circle fixed]. A location y on a fiber may be interpreted as existing at the point $P = x$ where the fiber intersects the base space, with y (vector length) given by the distance to the intersection of the fiber with the base space. Thus (x, y) identifies a point in the fiber bundle manifold uniquely. Cases (b) and (d) are equivalent locally but distinct topologically: the orientation of the fiber winds through π for once around the base space in (d), so that (d) cannot be deformed continuously into (b).

manifolds. Because of the position-dependent metric of curved spacetime it is most convenient in general relativity to choose basis vectors that depend on position and that need not be orthogonal. Since such basis vectors are position-dependent, it usually is not useful to normalize them.

3.4.1 Coordinate and Non-coordinate Bases

As discussed in Section 3.3, the standard conception of a vector as a directed line segment has ill-defined meaning in a curved manifold. The key to specifying vectors in curved space is to separate the "directed" part from the "line segment" part of the usual definition, because the direction for vectors of infinitesimal length can be defined consistently in curved or flat spaces using *directional derivatives.* Consider a curve in a differentiable manifold along which one of the coordinates x^μ varies while all others $x^\nu (\nu \neq \mu)$ are held constant. This curve will be termed *the coordinate curve* x^μ. Through any point P in a spacetime manifold four such curves will pass, corresponding to the coordinate curves x^μ with $\mu = (0, 1, 2, 3)$. A convenient set of position-dependent basis vectors $e_\mu (\mu = 0, 1, 2, 3)$ can be defined at an arbitrary point P in the manifold by

$$e_\mu = \lim_{\delta x^\mu \to 0} \frac{\delta s}{\delta x^\mu}, \tag{3.4}$$

where δs is the infinitesimal displacement along the coordinate curve x^μ between the point P with coordinate x^μ and a nearby point Q with coordinate $x^\mu + \delta x^\mu$. For a parameterized curve $x^\mu(\lambda)$ having a tangent vector t with components $t^\mu = dx^\mu/d\lambda$,

$$t = t^\mu e_\mu = \frac{dx^\mu}{d\lambda} e_\mu,$$

the directional derivative of an arbitrary scalar function $f(x^\mu)$ that is defined in the neighborhood of the curve is

$$\frac{df}{d\lambda} \equiv \lim_{\epsilon \to 0} \left[\frac{f(x^\mu(\lambda + \epsilon)) - f(x^\mu(\lambda))}{\epsilon} \right] = \frac{dx^\mu}{d\lambda} \frac{\partial f}{\partial x^\mu} = t^\mu \frac{\partial f}{\partial x^\mu},$$

and since $f(x)$ is arbitrary this implies the operator relation

$$\frac{d}{d\lambda} = \frac{dx^\mu}{d\lambda} \frac{\partial}{\partial x^\mu} = t^\mu \frac{\partial}{\partial x^\mu}. \tag{3.5}$$

Hence the components t^μ are associated with a unique directional derivative and the partial derivative operators $\partial/\partial x^\mu$ may be identified with the basis vectors e_μ,

$$e_\mu = \frac{\partial}{\partial x^\mu} \equiv \partial_\mu, \tag{3.6}$$

which permits an arbitrary vector to be expanded as

$$V = V^\mu e_\mu = V^\mu \frac{\partial}{\partial x^\mu} = V^\mu \partial_\mu. \tag{3.7}$$

Position-dependent basis vectors specified in this way (that generally are neither orthogonal nor normalized) define a *coordinate basis* or *holonomic basis;* a basis using orthonormal coordinates is then termed a *non-coordinate basis* or an *anholonomic basis*. A coordinate basis is illustrated schematically in Fig. 3.3 for a generic curved 2D manifold.

The definition of a vector in terms of directional derivatives is valid in any curved or flat differentiable manifold. It replaces the standard idea of a vector as the analog of a displacement vector between two points, which is useful in flat space but does not generalize to curved manifolds. From Eq. (3.4) the separation between nearby points is $ds = e_\mu(x) dx^\mu$, from which

$$ds^2 = ds \cdot ds = (e_\mu \cdot e_\nu) dx^\mu dx^\nu = g_{\mu\nu} dx^\mu dx^\nu,$$

with the metric tensor $g_{\mu\nu}$ defined by

$$e_\mu(x) \cdot e_\nu(x) \equiv g_{\mu\nu}(x), \tag{3.8}$$

which implies that in a coordinate basis the scalar product of vectors A and B is given by

$$A \cdot B = (A^\mu e_\mu) \cdot (B^\nu e_\nu) = g_{\mu\nu} A^\mu B^\nu. \tag{3.9}$$

Equation (3.8) may be taken as a definition of a vector coordinate basis $\{e_\mu\}$.

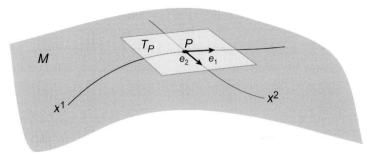

Fig. 3.3 Tangent space T_P at a point P for a curved 2D manifold M. The vectors tangent to the coordinate curves at each point define a coordinate or holonomic basis. This figure is a generalization of Fig. 3.1 to an arbitrary curved 2D manifold with a position-dependent, non-orthogonal (coordinate) basis. This embedding of M in 3D euclidean space is for visualization purposes only; the basis vectors e_1 and e_2 of the tangent space are specified by directional derivatives of the coordinate curves evaluated entirely in M at the point P, as described in Eqs. (3.4)–(3.7).

The preceding discussion has been specifically for vectors and involves defining a basis for the tangent space T_P at each point P using the tangents $\partial/\partial x^\mu$ to coordinate curves passing through P. By analogy with the discussion in Section 2.1.2, a similar intrinsic procedure can be invoked to construct a basis for dual vectors in the cotangent space T_P^* at a point P. This leads to equations analogous to (3.8)–(3.9), but with the indices of the basis vectors in the upper position. A set of dual basis vectors e^μ may be used to expand dual vectors ω as[1]

$$\omega = \omega_\mu e^\mu, \qquad (3.10)$$

allowing the metric tensor with upper indices to be defined through

$$e^\mu(x) \cdot e^\nu(x) \equiv g^{\mu\nu}(x), \qquad (3.11)$$

with the scalar product of arbitrary dual vectors α and β given by

$$\alpha \cdot \beta = g^{\mu\nu} \alpha_\mu \beta_\nu. \qquad (3.12)$$

Equation (3.11) may be taken as a definition of a dual-vector coordinate basis $\{e^\mu\}$.

Just as Eqs. (3.8) or (3.11) are characteristic of a coordinate basis, an orthonormalized non-coordinate basis is specified by the requirement

$$e_{\hat{\mu}}(x) \cdot e_{\hat{\nu}}(x) = \eta_{\hat{\mu}\hat{\nu}}, \qquad (3.13)$$

where $\eta = \text{diag}\{-1, 1, 1, 1\}$. In this expression a common notational convention has been employed of using hats on indices to indicate explicitly that this is an orthonormal and not coordinate basis. As elaborated further in Problems 3.20 and 3.22, the basis vectors of a coordinate basis have a vanishing *Lie bracket*, $[e_\mu, e_\nu] = 0$, while for a non-coordinate basis $[e_{\hat{\mu}}, e_{\hat{\nu}}] \neq 0$, where the Lie bracket of two vector fields A and B is defined by the commutator $[A, B] \equiv AB - BA$. This provides a formal way to identify a coordinate basis.

---

[1] As in Chapter 2, the same symbol e will be used for vector and dual vector basis vectors, with the index in the lower position for a vector basis and in the upper position for a dual vector basis. The justification for this notation is given in Sections 2.1.5 and 3.6.4.

3.4.2 Utility of Coordinate and Non-coordinate Bases

Our discussion will seldom require display of explicit basis vectors but usually we will assume implicitly the use of a coordinate basis such that Eqs. (3.4)–(3.12) are valid. However, the equivalence principle to be discussed in Chapter 6 asserts that even curved spacetime is locally flat; hence locally a constant orthonormal coordinate system can be useful. For example, observers occupy a local laboratory in spacetime for which a constant, orthonormal coordinate system is a natural choice (see Section 5.5). Thus, it is convenient to employ a coordinate basis for the overall formulation of general relativity but to invoke a non-coordinate basis in some contexts, particularly for physical interpretation.

3.5 Tensors and Coordinate Transformations

In formulating general relativity we are interested in how quantities that enter the physical description of the Universe change when the spacetime coordinates are transformed as in Eq. (3.2). This requires understanding the transformations of fields, their derivatives, and their integrals, since the first two are necessary to formulate equations of motion and the latter enters into various conservation laws. To facilitate this task, it is useful to introduce a set of mathematical objects called *tensors*. These have a fundamental definition without reference to specific coordinate systems that was introduced in Section 2.1.5 and will be discussed further in Section 3.6. However, for physical applications it often proves convenient to view tensors as objects expressed in a basis with components that carry some number of upper and lower indices, and that change in a precise way under coordinate transformations. This more practical interpretation of tensors will be developed in Section 3.7. Let us note also that our interest here almost always will be in *tensor fields*, which correspond to tensors of a given type defined at every point of the manifold. Since this is a rather trivial generalization of a tensor defined at a point, the discussion will for brevity often use shorthand like "vector" or "tensor" to mean "vector field" or "tensor field," respectively. This is unlikely to engender confusion, since the meaning should be clear from the context.

The *rank* of a tensor will be given a more fundamental definition below but practically it is equal to the total number of indices required to specify its components when expressed in some basis. Thus scalars are tensors of rank zero and vectors or dual vectors are tensors of rank one. This may be generalized to tensors carrying more than one index. As for vectors and dual vectors, the indices may either be upper (*contravariant*) or lower (*covariant*); see Box 3.4 for the origin of this terminology. Tensors carrying only lower indices are termed *covariant tensors,* tensors carrying only upper indices are termed *contravariant tensors,* and tensors carrying both lower and upper indices are termed *mixed tensors.* It is convenient to indicate the *type* of a tensor by the ordered pair (p, q), where p is the number of contravariant (upper) indices, q is the number of covariant (lower) indices, and the rank of the tensor is $p + q$. Thus a dual vector is a tensor of type $(0, 1)$ with a rank of one, while the Kronecker delta δ^ν_μ is a mixed tensor of type (1, 1) and rank 2. Not all quantities with indices are tensors; it is their *mathematical properties* (their transformation laws, or that they define linear maps from vectors and dual vectors to the real numbers) that mark objects as tensors, not merely that they carry indices.

3.6 Tensors as Linear Maps

The characterization of tensors in terms of their transformation properties that will be discussed in §3.7 is particularly useful from a physical perspective, and is the most pragmatic approach to the mathematics required to achieve a physicist-level understanding and to solve realistic problems in general relativity. However, mathematicians prefer to define tensors in a more elegant and abstract manner that makes manifest the essential property of tensors that they are independent of representation in a particular coordinate system, and that often is more precise in defining some essential underlying mathematical concepts. This approach, which frequently goes by the name *index-free formalism*, is described in this section.[2]

3.6.1 Linear Maps to Real Numbers

From a fundamental perspective a tensor of type (n, m) has input slots for n vectors and m dual vectors, and acts linearly on these inputs to produce a real number. For example, if ω is a (0, 1) tensor (dual vector) and A and B are (1, 0) tensors (vectors), linearity implies

$$\omega(aA + bB) = a\omega(A) + b\omega(B) \in \mathbb{R}, \tag{3.14}$$

where a and b are arbitrary scalars and $\mathbb{R}$ denotes the set of real numbers. Since this definition makes no reference to components of the vectors or dual vectors, the tensor map must give the same real number, irrespective of any choice of coordinate system. Thus, a tensor may be viewed as a *function of the vectors and dual vectors themselves,* rather than as a function of their components, or as an *operator* that accepts some number of vectors and dual vectors as input and outputs a real number.

Example 3.2 As a warmup exercise, consider a real-valued function of the coordinates $f(x)$. Since this function takes no vectors or dual vectors as input and yields a real number (the value of the function at x) as output, it is a tensor of rank zero (a scalar).

---

[2] The language of this section assumes familiarity with the definitions of vectors in the tangent space T_P and dual vectors in the cotangent space T_P^* of curved manifolds, which was discussed in Section 3.3, and with the associated concepts of tangent bundles and cotangent bundles, as described in Box 3.2. The reader also is assumed to be familiar with Section 2.1.5, which introduces these concepts in the simpler euclidean context.

> **Box 3.3** **Vector Spaces**
>
> A *vector space* has a precise axiomatic definition in mathematics, but for our purposes it will be sufficient to view it more loosely as a set of objects (the vectors) that can be multiplied by real numbers and added together in a linear way while exhibiting *closure*: any such operations on elements of the set give back a linear combination of elements. For arbitrary vectors A and B, and arbitrary scalars a and b, one expects then that expressions like
>
> $$(a+b)(A+B) = aA + aB + bA + bB$$
>
> should be satisfied. A few other things are necessary: a zero vector that serves as an identity under vector addition, an inverse for every vector, and the usual assortment of associativity, distributivity, and commutativity rules, for example; but it is clear that it isn't very hard to be a vector space. Vector spaces of use in physics often have additional structure like a norm and inner product, but that is over and above the minimal requirements for being a bona-fide vector space.
>
> A *basis* for a vector space is a set of vectors that *span the space* (any vector is a linear combination of basis vectors) and that are *linearly independent* (no basis vector is a linear combination of other basis vectors). The number of basis vectors is the *dimension* of the space. For spacetime the vector spaces of interest will be defined at each point of the manifold and will be of dimension four.
>
> Be aware that the term "vector" is being used here in three senses: (1) as an arbitrary element of a vector space, according to the definition given above; (2) as an element of a vector space that *also* is a vector in the precise sense defined in Section 3.6.2 (a tensor with one upper index when evaluated in a basis); and (3) sometimes as a generic term for an element of a vector space that *is either* a vector or a dual vector. This does not normally lead to confusion because the meaning is usually clear from the context.

Let's now give a few less-trivial examples of how this approach to tensors works, beginning with vectors and dual vectors.

3.6.2 Vectors and Dual Vectors

Vector spaces are discussed in Box 3.3. Suppose a vector field to be defined on a manifold such that each point P has associated with it a vector V that may be expanded in a vector basis e_μ,

$$V = V^\mu e_\mu, \qquad (3.15)$$

and that there is a corresponding dual vector field ω defined at each point P that may be expanded in a dual-vector basis e^μ,

$$\omega = \omega_\mu e^\mu, \qquad (3.16)$$

where the basis vectors e_μ are defined in the tangent space T_P and the basis dual vectors e^μ are defined in the cotangent space T_P^* at each point of the manifold, as described in Section 3.3. Hence the e_μ are basis vectors in the tangent bundle and the e^μ are basis vectors in the cotangent bundle of the manifold, as elaborated in Box 3.2.

As discussed in Section 2.1.5, the vector spaces for V and ω are said to be *dual* in the following sense. The space of vectors (tangent bundle) consists of all linear maps of dual vectors to the real numbers; conversely, the space of dual vectors (cotangent bundle) consists of all linear maps of vectors to the real numbers. This *duality* of vector and dual vector spaces can be implemented systematically by requiring the basis vectors to satisfy

$$e^\mu(e_\nu) = e^\mu \cdot e_\nu = \delta^\mu_\nu, \tag{3.17}$$

where the Kronecker delta is given by

$$\delta^\mu_\nu = \begin{cases} 1 & \mu = \nu \\ 0 & \mu \neq \nu \end{cases}.$$

Note that $A(B)$, which indicates the action of A on B, can be expressed in the alternative form $\langle A, B \rangle$, so Eq. (3.17) is also commonly written as $\langle e^\mu, e_\nu \rangle = \delta^\mu_\nu$. From Eqs. (3.14)–(3.17) it follows that a dual vector ω acts on a vector V in the manner

$$\begin{aligned} \omega(V) = \langle \omega, V \rangle &= \omega_\mu e^\mu (V^\nu e_\nu) \\ &= \omega_\mu V^\nu e^\mu(e_\nu) \\ &= \omega_\mu V^\nu \delta^\mu_\nu \\ &= \omega_\mu V^\mu \in \mathbb{R}, \end{aligned} \tag{3.18}$$

where $\mathbb{R}$ denotes the real numbers.[3] This illustrates clearly that a dual vector is an operator that accepts a vector as an argument and produces a real number (the scalar product $\omega_\mu V^\mu$, which is unique and independent of basis) as output. Conversely, a vector is an operator that accepts a dual vector as an argument and produces a real number equal to the scalar product, $V(\omega) = \omega_\mu V^\mu$. It is also clear that these definitions involve no uncontracted indices, so the results are independent of any basis choice. This suggests that vectors and dual vectors may be defined fundamentally in terms of linear maps to the real numbers:

1. A *dual vector* is an operator that acts linearly on a vector to return a real number.
2. A *vector* is as an operator that acts linearly on a dual vector to return a real number.

For those having a knowledge of linear algebra or quantum mechanics this may sound vaguely familiar, as suggested by the following.

Example 3.3 In the language of linear algebra, vectors may be represented as column vectors and dual vectors as row vectors, and their matrix product is a number. For example,

$$A \equiv (a\ b) \qquad B \equiv \begin{pmatrix} c \\ d \end{pmatrix} \qquad AB = (a\ b)\begin{pmatrix} c \\ d \end{pmatrix} = ac + bd \in \mathbb{R}$$

---

[3] As has been noted previously, basis vectors are defined in the tangent bundle and basis dual vectors are defined in the cotangent bundle of the manifold. Thus for applications in spacetime our concern is really with the action of vector fields on dual vector fields and vice versa, in which case what is returned is not a real number but rather a scalar field of real numbers defined over the manifold. This is just another example of many where for simplicity our presentation has been careless about speaking of some thing when what is really meant is a field of those things. We trust that the reader is sophisticated enough at this point to realize when "thing" really means "field of things."

may be regarded as the dual vector A acting linearly on the vector B to produce the real number $ac + bd$: $A(B) \in \mathbb{R}$. For readers familiar with Dirac notation for matrix elements in quantum mechanics, a *ket* $|a\rangle$ may be viewed as representing a vector and a *bra* $\langle a|$ as representing a dual vector (in the linear vector space of quantum theory called *Hilbert space*), and mathematically the vector space of bras is the dual of the vector space of kets (see Ref. [220]). Thus the overlap $\langle f|i\rangle$ is a number (a *c-number* in quantum lingo).

These definitions of vectors and dual vectors are independent of any choice of basis but practically it often is convenient to work in a basis. The components of a vector or dual vector in a specific basis are obtained by evaluating them with respect to the corresponding basis vectors and dual basis vectors; for example,

$$V^\mu = V(e^\mu) = e^\mu \cdot V \qquad \omega_\mu = \omega(e_\mu) = e_\mu \cdot \omega, \qquad (3.19)$$

which follows from Eq. (3.17). Equations such as (3.19) may be interpreted (for example) as a vector accepting a basis vector e^μ as input and acting linearly on it to return a real number that is the component of the vector evaluated in that basis.

Example 3.4 The validity of Equation (3.19) may be checked easily:

$$e^\mu \cdot V = e^\mu \cdot (V^\alpha e_\alpha) = V^\alpha e^\mu(e_\alpha) = V^\alpha \delta^\mu_\alpha = V^\mu,$$
$$e_\mu \cdot \omega = e_\mu \cdot (\omega_\alpha e^\alpha) = \omega_\alpha e_\mu(e^\alpha) = \omega_\alpha \delta^\alpha_\mu = \omega_\mu,$$

where the expansions (3.15)–(3.16), the linearity requirement (3.14), and the orthogonality condition (3.17) were employed.

Vector and dual vector components as in Eq. (3.19), and more generally tensor components, are then found to obey the same transformation laws and tensor calculus that will be presented in Sections 3.7–3.10 as alternative defining characteristics of tensors. Example 3.5 illustrates for dual vectors and vectors.

Example 3.5 Consider a coordinate transformation $x^\mu \to x'^\mu$ on a dual vector $\omega = \omega_\mu e^\mu$ and on a vector $V = V^\mu e_\mu$. The basis dual vectors e^μ and basis vectors e_μ transform as (Problem 3.24)

$$e^\mu \to e'^\mu = \frac{\partial x'^\mu}{\partial x^\nu} e^\nu \qquad e_\mu \to e'_\mu = \frac{\partial x^\nu}{\partial x'^\mu} e_\nu.$$

How do the components ω_μ transform? This may be determined by noting that the dual vector ω is a geometrical object having an existence independent of representation in a specific coordinate system, so it must be invariant under coordinate transformations. This will be ensured only if the components of ω transform as

$$\omega_\nu \to \omega'_\nu = \frac{\partial x^\alpha}{\partial x'^\nu} \omega_\alpha,$$

since then the dual vector ω is invariant under $x^\mu \to x'^\mu$:

$$\begin{aligned}
\omega' &= \omega'_\mu e'^\mu \\
&= \frac{\partial x^\alpha}{\partial x'^\mu} \omega_\alpha \frac{\partial x'^\mu}{\partial x^\nu} e^\nu \\
&= \frac{\partial x^\alpha}{\partial x'^\mu} \frac{\partial x'^\mu}{\partial x^\nu} \omega_\alpha e^\nu \\
&= \omega_\alpha e^\nu \delta^\alpha_\nu \\
&= \omega_\alpha e^\alpha = \omega.
\end{aligned}$$

This transformation law for the components ω_ν is the same one that will be used to *define* a dual vector in Eq. (3.29). By a similar proof, vector components V^μ may be shown to have the transformation law (3.31); see Problem 3.21.

In Sections 3.7–3.10 such transformation laws will be offered as a *definition* of tensors. In the index-free picture currently under discussion tensors are defined instead as linear maps of some number of vectors and dual vectors to the real numbers, with the transformation laws and associated tensor calculus that will be discussed in Sections 3.7–3.10 following as a *consequence* of that definition.

3.6.3 Tensors of Higher Rank

Higher-rank tensors may be constructed by taking *tensor products* (denoted by $\otimes$) of lower-rank tensors, and a basis for higher-rank tensors may be constructed from tensor products of vector and dual vector spaces (see Problem 3.26).[4] Schematically, a mixed tensor T of rank (p, q) may be expressed as

$$T = T^{\mu_1 \mu_2 \cdots \mu_p}{}_{\nu_1 \nu_2 \cdots \nu_q} e_{\mu_1} \otimes e_{\mu_2} \otimes \cdots \otimes e_{\mu_p} \otimes e^{\nu_1} \otimes e^{\nu_2} \otimes \cdots \otimes e^{\nu_q}, \quad (3.20)$$

where $\{e_\mu\}$ is a vector basis and $\{e^\nu\}$ is a dual vector basis. The components of a higher-rank tensor may be evaluated by inserting basis vectors in its input slots, in a generalization of Eq. (3.19). For example, consider a rank-2 tensor T. Its covariant, contravariant, and mixed components are given by

$$T(e_\mu, e_\nu) = T_{\mu\nu} \quad T(e^\mu, e^\nu) = T^{\mu\nu} \quad T(e_\mu, e^\nu) = T_\mu{}^\nu \quad T(e^\mu, e_\nu) = T^\mu{}_\nu, \quad (3.21)$$

from which it follows that for vectors A and B,

$$T(A, B) = T(A^\mu e_\mu, B^\nu e_\nu) = T(e_\mu, e_\nu) A^\mu B^\nu = T_{\mu\nu} A^\mu B^\nu, \quad (3.22)$$

and similarly for contravariant and mixed components. As another example, the tensor product $V \otimes T$ of a vector V and a rank-2 tensor T is a rank-3 tensor with one possible set of components $(V \otimes T)^\mu{}_{\alpha\beta} = (V \otimes T)(e^\mu, e_\alpha, e_\beta) = V(e^\mu) T(e_\alpha, e_\beta) = V^\mu T_{\alpha\beta}$. Finally,

---

[4] The *tensor product* of two finite-dimensional vector spaces $\mathbb{U}$ and $\mathbb{V}$ produces a new vector space $\mathbb{U} \otimes \mathbb{V}$ having a dimensionality that is the product of those for $\mathbb{U}$ and $\mathbb{V}$. If $\mathbb{U}$ has a basis $\{u_1, u_2, \ldots\}$ and $\mathbb{V}$ has a basis $\{v_1, v_2, \ldots\}$, then $\mathbb{U} \otimes \mathbb{V}$ is spanned by a basis consisting of all ordered pairs (u_i, v_j). In general the tensor product does not commute, $S \otimes T \ne T \otimes S$. Note that the tensor product symbol is sometimes omitted and $S \otimes T$ written simply as ST, if the meaning is clear from the context.

consider the scalar product $A \cdot B = g_{\mu\nu} A^\mu B^\nu$ of Eq. (3.9). From this expression $g(A, B)$ is an operator that takes two vectors as input and returns their scalar product, which is a real number,

$$g(A, B) = g(A^\mu e_\mu, B^\nu e_\nu) = g(e_\mu, e_\nu) A^\mu B^\nu$$
$$= g_{\mu\nu} A^\mu B^\nu = A^\mu B_\mu \in \mathbb{R}. \quad (3.23)$$

Since it takes two vectors as input and acts linearly on both of them to return a number (this is an example of a *multilinear mapping*), $g_{\mu\nu}$ represents a rank-2 tensor of type $(0, 2)$.

Example 3.6 As noted in Example 3.3, for quantum mechanics in Dirac notation a ket $|i\rangle$ is a vector and a bra $\langle f|$ is a dual vector in Hilbert space. A matrix element of an operator $\hat{Q}$ is of the form $\langle f|\hat{Q}|i\rangle$ in Dirac notation. Because a matrix element is a scalar, the operator $\hat{Q}$ is a rank-2 tensor of type $(1, 1)$, since it takes one vector and one dual vector as input and produces a scalar. (In quantum mechanics a matrix element may be a complex rather than real number, but that distinction is not important in the present discussion.)

Let us now greatly simplify keeping track of the distinction between upper indices and lower indices on tensors by demonstrating that the metric tensor map (3.23) may be used to establish a one-to-one relationship between a vector in the tangent space and a corresponding dual vector in the cotangent space.

3.6.4 Identification of Vectors and Dual Vectors

Consider the metric tensor in Eq. (3.23), viewed as a rank-2 covariant tensor that accepts two vector inputs and acts on them (multi-)linearly to give a real number. Schematically, this may be written as the operator $g(\cdot, \cdot)$, where the dots indicate the input slots for the two vectors. Suppose that a vector V is inserted into only *one* of the slots, giving $g(V, \cdot)$. What is this object? It has one open slot that can accept a vector, on which it will act linearly to return a real number. But that should sound familiar: it is the definition of a dual vector! Because it is associated directly with the vector V, let us call this dual vector $\tilde{V} \equiv g(V, \cdot)$. The components of this dual vector may be evaluated by inserting a basis vector as argument in the usual way,

$$\begin{aligned} V_\mu \equiv \tilde{V}(e_\mu) &= g(V, e_\mu) \\ &= g(V^\nu e_\nu, e_\mu) \\ &= V^\nu g(e_\nu, e_\mu) \\ &= g_{\mu\nu} V^\nu, \end{aligned} \quad (3.24)$$

Likewise, using that $g_{\mu\nu}$ and $g^{\mu\nu}$ are matrix inverses, $V^\mu = g^{\mu\nu} V_\nu$. Summing over repeated indices in tensor products is called *contraction*. Thus, the properties of the metric tensor allow vectors and dual vectors to be treated effectively as if they were both vectors, one with an upper index and one with a lower index, with the two related by contraction with the metric tensor,

$$V_\mu = g_{\mu\nu} V^\nu \qquad V^\mu = g^{\mu\nu} V_\nu. \qquad (3.25)$$

This is of great practical importance since it allows the same symbol to be used for a vector and its corresponding dual vector, and it reduces the handling of vectors and dual vectors to keeping proper track of the vertical position of indices in the Einstein summation convention. This identification works only for manifolds with metric tensors but that is no limitation for general relativity, which deals *only* with metric spaces.

Once the operations of raising and lowering indices by contraction with the metric tensor are established through Eq. (3.25), the scalar product between two vectors U and V can be calculated as the *complete contraction* $U_\alpha V^\alpha$ of one of the vectors with the dual vector associated with the other vector:

$$g(U, V) = g_{\mu\nu} U^\mu V^\nu = U_\nu V^\nu. \qquad (3.26)$$

The scalar product has no indices left after contraction and is said to be *fully contracted*. Because tensors of higher rank are products of vectors and dual vectors, the preceding discussion is easily generalized and contraction with the metric tensor can be used to raise or lower any index for a tensor of any rank. For example,

$$A^{\mu\nu} = g^{\mu\alpha} g^{\nu\beta} A_{\alpha\beta} \qquad A_{\mu\nu\lambda\sigma} = g_{\mu\rho} A^\rho{}_{\nu\lambda\sigma}.$$

Since indices can be raised or lowered at will by a metric, tensors may be thought of as objects of a particular tensorial rank, irrespective of their particular vertical arrangement of indices (for example, the identification of vectors and dual vectors discussed above). Of course this is true only in the abstract; index placement matters when tensors are evaluated in a basis.

3.6.5 Index-free versus Component Transformations

The material in this section has been a brief introduction to the index-free formulation of tensors favored by mathematicians and more mathematically minded physicists. Its importance for us will be mostly conceptual since for practical reasons our usage of tensors in the solution of actual problems will be primarily in terms of the transformation properties of tensor components summarized below. The discussion above and that in Section 3.7 indicates that the index-free formalism leads to the same transformation laws for tensors, and even in the index-free formalism the solution of real problems is often easiest in a specific basis. Thus the practical outcome will be the same with application of either approach to the issues addressed in this book, but index-free concepts provide a more solid mathematical foundation for physical results while often the component transformation approach affords a more direct path to obtaining them.

3.7 Tensors Specified by Transformation Laws

In the preceding discussion tensors have been introduced at a fundamental level through linear maps from vectors and dual vectors to the real numbers, but it was shown also that these linear maps imply that when tensors of a given type are expressed in an arbitrary basis their components obey well-defined transformation laws under change of coordinates. This view of tensors as groups of quantities obeying particular transformation laws is often the most practical for physical applications because: (1) it is less abstract and requires less new mathematics for the novice, (2) a physical interpretation often requires expression of the problem in a well-chosen basis anyway, and (3) the component index formalism has a built-in error checking mechanism of great practical utility in solving problems: failure of indices to balance on the two sides of an equation is a sure sign of an error.

This section summarizes the use of tensors to formulate invariant equations by exploiting the transformation properties of their components. Tensors may be viewed as generalizing the idea of scalars and vectors, so let's begin with these more familiar quantities. The following discussion is an adaptation to 4-dimensional (possibly curved) spacetime of many concepts discussed already in Chapter 2 for simpler spaces; you are urged to review that material if any conceptual difficulties are encountered in the following material.

3.7.1 Scalar Transformation Law

Consider the behavior of fields under the coordinate transformations introduced in Section 3.2. The simplest possibility is that the field has a single component at each point of the manifold, with a value that is unchanged by the transformation (3.2),

$$\varphi'(x') = \varphi(x). \tag{3.27}$$

Quantities such as $\varphi(x)$ that are unchanged under the coordinate transformation are called *scalars*. As the notation indicates, scalar quantities are generally functions of the coordinates but their value at a given point does not change if the coordinate system changes.[5] An example is temperature displayed for various reporting stations on a weather map. The temperature may vary from point to point but the value of the temperature at any of those points does not depend on the coordinate system that is used to locate points on the surface of the Earth. Thus the field of temperatures is a scalar field. Scalar quantities may be expected to play a central role in physical theories because measurable observables must be scalars if the goal of formulating physical laws such that they do not depend on the coordinate labels is to be fulfilled.

---

[5] This may be contrasted with the behavior of the components of a vector under coordinate transformation, which will be discussed shortly. Geometrically the components of a vector are projections of the vector on particular coordinate axes. Thus, if the coordinate system is changed the components of a vector at a given point typically change their values, but the length of the vector, which is a scalar, does not.

3.7.2 Dual Vector Transformation Law

In parallel with the discussion of vectors defined in the tangent and dual bases for euclidean space in Section 2.1, it is useful to define two kinds of spacetime vectors having distinct transformation laws. Both will be termed vectors because sets of each type separately obey the axioms to form a *vector space,* as discussed in Box 3.3, and because of the duality discussed in Section 3.6.2. By the argument in Section 3.6.4, the same symbol will be used for both but they will be distinguished by vertical placement of indices. By the rules of ordinary partial differentiation the gradient of a scalar field obeys

$$\frac{\partial \varphi(x)}{\partial x'^\mu} = \frac{\partial \varphi(x)}{\partial x^\nu} \frac{\partial x^\nu}{\partial x'^\mu}. \tag{3.28}$$

Consider a vector having a transformation law mimicking that of the scalar field gradient (3.28),

$$A'_\mu(x') = \frac{\partial x^\nu}{\partial x'^\mu} A_\nu(x) \quad \text{(dual vector)}. \tag{3.29}$$

A quantity transforming in this way is termed a *dual vector* (it also goes by the names *one-form, covariant vector,* or *covector*). Understand clearly that in (3.29) the two sides of the equation refer to the *same point* in spacetime. Thus, the argument is x' on the left side and x on the right side, and these label the same spacetime point in two different coordinate systems. Before continuing we stop to note that even the relatively simple expressions that have been introduced so far emphasize the importance of the compact notation that we are using. For example, Eq. (3.29) really means four equations:

$$A'_\mu = \frac{\partial x^0}{\partial x'^\mu} A_0 + \frac{\partial x^1}{\partial x'^\mu} A_1 + \frac{\partial x^2}{\partial x'^\mu} A_2 + \frac{\partial x^3}{\partial x'^\mu} A_3 \quad (\mu = 0, 1, 2, 3)$$

each containing four terms. That is, Eq. (3.29) is a concise way to write the matrix equation

$$\begin{pmatrix} A'_0 \\ A'_1 \\ A'_2 \\ A'_3 \end{pmatrix} = \begin{pmatrix} \partial x^0/\partial x'^0 & \partial x^1/\partial x'^0 & \partial x^2/\partial x'^0 & \partial x^3/\partial x'^0 \\ \partial x^0/\partial x'^1 & \partial x^1/\partial x'^1 & \partial x^2/\partial x'^1 & \partial x^3/\partial x'^1 \\ \partial x^0/\partial x'^2 & \partial x^1/\partial x'^2 & \partial x^2/\partial x'^2 & \partial x^3/\partial x'^2 \\ \partial x^0/\partial x'^3 & \partial x^1/\partial x'^3 & \partial x^2/\partial x'^3 & \partial x^3/\partial x'^3 \end{pmatrix} \begin{pmatrix} A_0 \\ A_1 \\ A_2 \\ A_3 \end{pmatrix}.$$

Also, the partial derivatives appearing in these transformation equations *depend on spacetime coordinates* in the general case and all partial derivatives are understood implicitly to be evaluated at a specific point P labeled by x in one coordinate system and x' in the other. In the short term it may be useful to write out some of these expressions in their full glory to be certain that you understand the formalism but the shorthand notation being introduced here will be essential to maintaining tractability in more complex expressions that will be encountered later.

3.7.3 Vector Transformation Law

Now consider application of the rules of partial differentiation to transformation of the differential,

$$dx'^\mu = \frac{\partial x'^\mu}{\partial x^\nu} dx^\nu. \tag{3.30}$$

This suggests a second vector transformation rule,

$$A'^{\mu}(x') = \frac{\partial x'^{\mu}}{\partial x^{\nu}} A^{\nu}(x) \qquad \text{(vector)}. \qquad (3.31)$$

A quantity behaving in this way is termed a *vector*.[6] Most physical quantities that are thought of loosely as "vectors" (displacement or velocity, for example) are vectors in the restricted sense defined by the transformation law (3.31).

Box 3.4 — **Co-varying and Contra-varying Components**

The use of *covariant* to refer to lower indices and *contravariant* to upper indices arises from the fundamental idea that tensors are invariants that are independent of coordinate systems but their components and basis vectors are not.

Transformations of Vectors and Dual Vectors

In terms of the Jacobian matrix U and the inverse Jacobian matrix $\hat{U}$ that are defined in Example 3.7,

$$U^{\mu}_{\nu} = \partial x'^{\mu}/\partial x^{\nu} \qquad \hat{U}^{\nu}_{\mu} = \partial x^{\nu}/\partial x'^{\mu} \qquad U^{\mu}_{\nu} \hat{U}^{\alpha}_{\mu} = \delta^{\alpha}_{\nu},$$

a vector V transforms as $V'^{\mu} = U^{\mu}_{\nu} V^{\nu}$ and a dual vector ω as $\omega'_{\mu} = \hat{U}^{\nu}_{\mu} \omega_{\nu}$ under the coordinate transformation $x \to x'$, with a corresponding change of basis since the basis depends on the coordinates. The vector V is invariant, which means that the basis vectors must transform in just such a way to cancel the change in the components. Specifically, since invariance of V requires

$$V = V^{\mu} e_{\mu} = V'^{\mu} e'_{\mu},$$

the basis vectors must transform as $e'_{\mu} = \hat{U}^{\nu}_{\mu} e_{\nu}$ so that

$$V'^{\mu} e'_{\mu} = U^{\mu}_{\nu} V^{\nu} \hat{U}^{\alpha}_{\mu} e_{\alpha} = U^{\mu}_{\nu} \hat{U}^{\alpha}_{\mu} V^{\nu} e_{\alpha} = V^{\mu} e_{\mu},$$

where $U^{\mu}_{\nu} \hat{U}^{\alpha}_{\mu} = \delta^{\alpha}_{\nu}$ was used. By a similar proof the invariance of $\omega = \omega_{\mu} e^{\mu}$ requires that the basis dual vectors transform as $e'^{\mu} = U^{\mu}_{\nu} e^{\nu}$.

Covariance and Contravariance

The preceding relations show that lower-index components transform with the inverse Jacobian matrix $\hat{U}$, just as the components of the vector basis e_{μ} transform. Thus, lower-index components *co-vary* with the vector basis and are termed *covariant components*. By a similar token, upper-index components transform with the Jacobian matrix U and thus "opposite" to the transformation of the basis vectors e_{μ}; they *contra-vary* and thus they are termed *contravariant components*.

---

[6] Some authors call vectors *contravariant vectors*, indicating explicitly that the component index is in the upper position, and call dual vectors *covariant vectors*, indicating explicitly that the component index is in the lower position. The origin of this terminology is discussed in Box 3.4.

Example 3.7 Equation (3.29) and (3.31) may be viewed as matrix equations,

$$A'_\mu(x') = \hat{U}_\mu^\nu A_\nu(x) \qquad A'^\mu(x') = U_\nu^\mu A^\nu(x), \tag{3.32}$$

with the matrices $U = \partial x'/\partial x$ and $\hat{U} = \partial x/\partial x'$ obeying $\hat{U}U = I$, where I is the unit matrix; see Problem 3.8. In these transformations the matrix U is called the *Jacobian matrix* and the matrix $\hat{U}$ is called the *inverse Jacobian matrix*.

3.7.4 Duality of Vectors and Dual Vectors

The preceding discussion distinguishes two kinds of "vectors": dual vectors, which carry a lower index and transform like (3.29), and vectors, which carry an upper index and transform like (3.31). This distinction is analogous to that introduced in Section 2.1 for vectors and dual vectors in euclidean spaces. In particular instances vectors and dual vectors may be considered equivalent as a practical matter (albeit with some loss of mathematical rigor; see Section 3.6.2), but generally they are not. However, the duality discussed in Section 3.6.4 allows us to use the same symbol for vectors and dual vectors, with the distinction between them residing in upper and lower positioning of indices in the summation convention.

3.8 Scalar Product of Vectors

Just as was found in Section 2.1.4 for euclidean spaces, the introduction of vectors and dual vectors, and their relationship through the metric tensor, allows a natural definition of a scalar product

$$A \cdot B \equiv A_\mu B^\mu, \tag{3.33}$$

where the dual vector A_μ and the corresponding vector A^μ are related through the metric tensor according to $A_\mu = g_{\mu\nu} A^\nu$. This product transforms as a scalar because from Eqs. (3.29) and (3.31),

$$A' \cdot B' = A'_\mu B'^\mu = \frac{\partial x^\nu}{\partial x'^\mu} A_\nu \frac{\partial x'^\mu}{\partial x^\alpha} B^\alpha = \frac{\partial x^\nu}{\partial x'^\mu} \frac{\partial x'^\mu}{\partial x^\alpha} A_\nu B^\alpha$$

$$= \frac{\partial x^\nu}{\partial x^\alpha} A_\nu B^\alpha = \delta^\nu_\alpha A_\nu B^\alpha = A_\alpha B^\alpha = A \cdot B, \tag{3.34}$$

where the Kronecker delta δ^ν_μ is given by

$$\delta^\mu_\nu = \frac{\partial x'^\mu}{\partial x'^\nu} = \frac{\partial x^\mu}{\partial x^\nu} = \begin{cases} 1 & \mu = \nu \\ 0 & \mu \neq \nu \end{cases}, \tag{3.35}$$

which is a rank-2 tensor with the unusual property that its components take the same value in all coordinate systems.

Table 3.1 Some tensor transformation laws

Tensor	Transformation law
Scalar	$\varphi' = \varphi$
Dual vector	$A'_\mu = \dfrac{\partial x^\nu}{\partial x'^\mu} A_\nu$
Vector	$A'^\mu = \dfrac{\partial x'^\mu}{\partial x^\nu} A^\nu$
Covariant rank-2	$T'_{\mu\nu} = \dfrac{\partial x^\alpha}{\partial x'^\mu} \dfrac{\partial x^\beta}{\partial x'^\nu} T_{\alpha\beta}$
Contravariant rank-2	$T'^{\mu\nu} = \dfrac{\partial x'^\mu}{\partial x^\alpha} \dfrac{\partial x'^\nu}{\partial x^\beta} T^{\alpha\beta}$
Mixed rank-2	$T'^\nu{}_\mu = \dfrac{\partial x^\alpha}{\partial x'^\mu} \dfrac{\partial x'^\nu}{\partial x^\beta} T^\beta{}_\alpha$

3.9 Tensors of Higher Rank

Three types of rank-2 tensors (0, 2), (1, 1), and (2, 0) may be distinguished; they have the transformation laws

$$T'_{\mu\nu} = \frac{\partial x^\alpha}{\partial x'^\mu} \frac{\partial x^\beta}{\partial x'^\nu} T_{\alpha\beta}, \tag{3.36}$$

$$T'^\nu{}_\mu = \frac{\partial x^\alpha}{\partial x'^\mu} \frac{\partial x'^\nu}{\partial x^\beta} T^\beta{}_\alpha, \tag{3.37}$$

$$T'^{\mu\nu} = \frac{\partial x'^\mu}{\partial x^\alpha} \frac{\partial x'^\nu}{\partial x^\beta} T^{\alpha\beta}. \tag{3.38}$$

These transformation rules may be generalized easily to tensors of any rank. Each upper index on the left side requires a right-side "factor" of the form $\partial x'^\mu / \partial x^\nu$ (prime in the numerator), and each lower index on the left side requires a right-side "factor" of the form $\partial x^\mu / \partial x'^\nu$ (prime in the denominator). This "position of the left-side index equals position of the right-side primed coordinate" rule for the partial derivative factors is a useful aid in remembering the forms of the tensor transformation equations. Transformation laws for tensors through rank 2 are summarized in Table 3.1.

It is sometimes useful to employ tensors of rank higher than two. For example, consider the 4-index quantity $\epsilon_{\alpha\beta\gamma\delta}$. Require that $\epsilon_{0123} = 1$ and that $\epsilon_{\alpha\beta\gamma\delta}$ be completely antisymmetric in the exchange of any two indices; thus it must vanish if any two indices are the same. For example, $\epsilon_{0132} = -1$ and $\epsilon_{1123} = 0$. The quantity $\epsilon_{\alpha\beta\gamma\delta}$ is called the *Levi-Civita symbol*. It is not a tensor (it is a *tensor density*).[7] However,

---

[7] The Levi-Civita symbol $\epsilon_{\alpha\beta\gamma\delta}$ is a *tensor density*, which is a generalization of a tensor. A tensor density behaves as a tensor under transformations, except that it is multiplied by a power J^n of the Jacobian determinant J for the coordinate transformation (see Section 3.13.1). The exponent n is called the *weight* of the tensor density. Because $\eta_{\alpha\beta\gamma\delta} \equiv J\epsilon_{\alpha\beta\gamma\delta}$ transforms as a rank-4 tensor, the Levi-Civita symbol $\epsilon_{\alpha\beta\gamma\delta}$ is a tensor density of weight $+1$. (Note: some authors define the weight of a tensor density with sign opposite that used here.)

$\eta_{\alpha\beta\gamma\delta} \equiv J\epsilon_{\alpha\beta\gamma\delta} \equiv |g|^{1/2}\epsilon_{\alpha\beta\gamma\delta}$, where J is the Jacobian determinant of the transformation and g is the determinant of the metric tensor (see Section 3.13.1), does transform as a rank-4 covariant tensor that is termed the *completely antisymmetric 4th-rank tensor*. An even more important rank-4 tensor, the *Riemann curvature tensor* $R_{\mu\nu\gamma\delta}$, will be introduced in Chapter 8. It will define the curvature of 4-dimensional spacetime, and therefore the gravitational field.

3.10 The Metric Tensor

By analogy with the discussion in Section 2.1.7, a rank-2 tensor of special importance is the metric tensor $g_{\mu\nu}$, because it is associated with the line element

$$ds^2 = g_{\mu\nu} dx^\mu dx^\nu \tag{3.39}$$

that determines the *geometry of the manifold*. The metric tensor is symmetric ($g_{\mu\nu} = g_{\nu\mu}$) and satisfies the $(0, 2)$ tensor transformation law

$$g'_{\mu\nu} = \frac{\partial x^\alpha}{\partial x'^\mu} \frac{\partial x^\beta}{\partial x'^\nu} g_{\alpha\beta}. \tag{3.40}$$

The contravariant form of the metric tensor $g^{\mu\nu}$ is defined by the requirement

$$g_{\mu\alpha} g^{\alpha\nu} = \delta^\nu_\mu, \tag{3.41}$$

so $g_{\mu\nu}$ and $g^{\mu\nu}$ are *matrix inverses*. Contractions with the metric tensor may be used to raise and lower (any number of) tensor indices; for example,

$$A^\mu = g^{\mu\nu} A_\nu \qquad A_\mu = g_{\mu\nu} A^\nu \qquad T^\mu_{\ \nu} = g_{\nu\alpha} T^{\mu\alpha} \qquad T^\alpha_{\ \beta\gamma} = g^{\alpha\mu} g_{\gamma\epsilon} T_{\mu\beta}^{\ \ \epsilon}. \tag{3.42}$$

Thus, the scalar product of vectors may be expressed as

$$A \cdot B = g_{\mu\nu} A^\mu B^\nu \equiv A_\nu B^\nu = g^{\mu\nu} A_\mu B_\nu \equiv A^\nu B_\nu. \tag{3.43}$$

Because such products are scalars, they are unchanged by coordinate-system transformations. A specific example of such a conserved quantity is an invariant length such as the line element of Eq. (3.39), as shown in Problem 3.16.

In mixed-tensor expressions like the third or fourth ones in Eq. (3.42) the relative horizontal order of upper and lower indices can be important. For example, in $T^\mu_{\ \nu} = g_{\nu\alpha} T^{\mu\alpha}$ the notation indicates that the mixed tensor on the left side of the equation was obtained by lowering the rightmost index of $T^{\mu\alpha}$ on the right side (since in $T^\mu_{\ \nu}$ the lower index ν is to the right of the upper index μ). This distinction is immaterial if the tensor is symmetric under exchange of indices but which index is lowered or raised by contraction matters for tensors that are antisymmetric under index exchange (see Section 3.11): $T^\mu_{\ \nu} = g_{\nu\alpha} T^{\mu\alpha}$ and $T_\nu^{\ \mu} = g_{\nu\alpha} T^{\alpha\mu}$ are equivalent if T is symmetric, but different if T is antisymmetric.

Vectors and dual vectors are distinct entities that are defined in different spaces (see Section 3.3). However, Eqs. (3.41)–(3.43) and the discussion in Sections 2.1.5 and 3.6.4 make it clear that for the *special case of a manifold with metric,* indices on any tensor may be raised or lowered at will by contraction with the metric tensor, as in Eq. (3.42). Defining a metric establishes a relationship that permits vectors and dual vectors to be treated as if they were (in effect) different representations of the same vector. Our discussion will usually proceed as if A^μ and A_μ are different forms of the same vector that are related by contraction with the metric tensor, while secretly remembering that they really are different, and that it is only for metric spaces that this conflation is not likely to land us in trouble.

The preceding discussion makes clear why the convention in much of nonrelativistic physics to ignore the mathematical distinction between vectors and dual vectors causes few problems. For example, the gradient operator is commonly termed a vector in elementary physics but Section 3.7.2 shows that it is in truth a dual vector. However, physics assumes metric spaces, so vectors and dual vectors are related trivially through the metric tensor and no lasting harm is done by calling the gradient a vector if the bookkeeping is done correctly in equations. Specifically, the gradient dual vector may be converted to the corresponding vector by contracting with the metric tensor to raise the index. The issue is particularly simple if the problem is formulated in euclidean space using cartesian coordinates, in which case the metric tensor is just the unit matrix and the components of the vector and dual vector are the same. The mathematician is required to be more circumspect because the definition of a manifold does not automatically imply existence of a metric to enable this identification.

3.11 Symmetric and Antisymmetric Tensors

The symmetry of tensors under exchanging pairs of indices is often important. An arbitrary rank-2 tensor can always be decomposed into a symmetric and antisymmetric part according to the identity

$$T_{\alpha\beta} = \tfrac{1}{2}(T_{\alpha\beta} + T_{\beta\alpha}) + \tfrac{1}{2}(T_{\alpha\beta} - T_{\beta\alpha}), \qquad (3.44)$$

where the first term is clearly symmetric and the second term antisymmetric under exchange of indices. For completely symmetric and completely antisymmetric rank-2 tensors

$$T_{\alpha\beta} = \pm T_{\beta\alpha} \qquad T^{\alpha\beta} = \pm T^{\beta\alpha}$$

where the plus sign holds if the tensor is symmetric and the minus sign if it is antisymmetric. More generally, a tensor of rank two or higher is said to be *symmetric* in any two of its indices if exchanging those indices leaves the tensor invariant and *antisymmetric* (or *skew-symmetric*) in any two indices if it changes sign upon switching those indices.

Symmetrizing and antisymmetrizing operations on tensor indices are often denoted by a bracket notation in which ( ) indicates symmetrization and [ ] indicates antisymmetrization over indices included in the brackets. For example, symmetrization over all indices for a rank-n covariant tensor $T_{\alpha,\beta,\ldots\omega}$ corresponds to

$$T_{(\alpha,\beta,\ldots\omega)} \equiv \frac{1}{n!} \text{ (Sum over permutations on indices } \alpha, \beta, \ldots \omega\text{)} \quad (3.45)$$

and antisymmetrization over all indices of $T_{\alpha,\beta,\ldots\omega}$ corresponds to

$$T_{[\alpha,\beta,\ldots\omega]} \equiv \frac{1}{n!} (\pm \text{Sum over permutations on indices } \alpha, \beta, \ldots \omega), \quad (3.46)$$

where the notation $\pm$ indicates that the terms of the sum have a plus sign if the permutation corresponds to an even number of index exchanges and a negative sign if it corresponds to an odd number of index exchanges. As an example,

$$T_{[\alpha\beta\gamma]} = \tfrac{1}{6}\left(T_{\alpha\beta\gamma} + T_{\beta\gamma\alpha} + T_{\gamma\alpha\beta} - T_{\beta\alpha\gamma} - T_{\alpha\gamma\beta} - T_{\gamma\beta\alpha}\right)$$

is the totally antisymmetric part of a rank-3 covariant tensor. Analogous rules apply for upper indices. Thus, for example

$$T^{(\alpha\beta)} = \tfrac{1}{2}(T^{\alpha\beta} + T^{\beta\alpha}) \qquad T^{[\alpha\beta]} = \tfrac{1}{2}(T^{\alpha\beta} - T^{\beta\alpha}).$$

Sometimes it is necessary to symmetrize or antisymmetrize over only a subset of indices. If the indices are contiguous the above notation suffices with only the indices to be symmetrized or antisymmetrized included in the brackets. If indices to be symmetrized or antisymmetrized are not adjacent to each other, the preceding notation may be extended by using vertical brackets to exclude indices from the symmetrization or antisymmetrization. For example, $T_{\alpha[\beta|\gamma|\delta]} = \tfrac{1}{2}(T_{\alpha\beta\gamma\delta} - T_{\alpha\delta\gamma\beta})$ corresponds to a rank-4 covariant tensor that has been antisymmetrized in its second and fourth indices only.

3.12 Summary of Algebraic Tensor Operations

Various algebraic operations are permitted for tensors in equations. These valid operations include:

1. *Multiplication by a scalar*: A tensor may be multiplied by a scalar (meaning that each component is multiplied by the scalar) to produce a tensor of the same rank. For example, $aA^{\mu\nu} = B^{\mu\nu}$, where a is a scalar and $A^{\mu\nu}$ and $B^{\mu\nu}$ are rank-2 contravariant tensors.
2. *Addition or subtraction*: Two tensors of the same type may be added or subtracted (meaning that their components are added or subtracted) to produce a new tensor of the same type. For example, $A^\mu - B^\mu = C^\mu$, where A^μ, B^μ, and C^μ are vectors.
3. *Multiplication*: Two or more tensors may be multiplied by forming products of their components (see Section 3.6.3). The rank of the resultant tensor will be the product of the ranks of the tensor factors. For example, $A_{\mu\nu} = U_\mu V_\nu$, where $A_{\mu\nu}$ is a rank-2 covariant tensor and U_μ and V_ν are dual vectors.

4. *Contraction*: For a tensor or tensor product with covariant rank n and contravariant rank m, a tensor of covariant rank $n-1$ and contravariant rank $m-1$ may be formed by setting one upper and one lower index equal and taking the implied sum. For example, $A = A^\mu{}_\mu$, where A is a scalar and $A^\mu{}_\nu$ is a mixed rank-2 tensor, or $A_\mu = g_{\mu\nu} A^\nu$, where the metric $g_{\mu\nu}$ is a rank-2 covariant tensor, A^ν is a vector, and A_μ is a dual vector.

In addition to these algebraic manipulations, it will often be necessary to integrate or differentiate in tensor expressions. This *tensor calculus* is addressed in the following section.

3.13 Tensor Calculus on Curved Manifolds

To formulate physical theories in terms of tensors requires the ability to manipulate tensors mathematically. In addition to the algebraic rules for tensors described in preceding sections, we must formulate a prescription to integrate tensor equations and one to differentiate them. Tensor calculus is mostly a straightforward generalization of normal calculus but additional complexity arises for two reasons:

1. It must be ensured that integration and differentiation preserve the symmetries and invariances embodied in the tensor equations.
2. To preserve the utility of the tensor formalism, it must be ensured that the results of these operations on tensor quantities are themselves tensor quantities.

As will now be shown, the first requirement implies a simple modification of the rules for ordinary integration while the second implies a less-simple modification with far-reaching mathematical and physical implications for the rules of partial differentiation.

3.13.1 Invariant Integration

Under a change of coordinates the volume element for integration over the spacetime coordinates changes according to

$$d^4x' = \det\left(\frac{\partial x'}{\partial x}\right) d^4x = J d^4x, \qquad (3.47)$$

where $d^4x \equiv dx^0 dx^1 dx^2 dx^3$ and $J \equiv \det(\partial x'/\partial x)$ is the Jacobian determinant (determinant of the 4×4 matrix of partial derivatives relating the x and x' coordinates; see Example 3.7). Notice that the right side of Eq. (3.40) may be viewed as a triple matrix product and recall that the determinant of a matrix product is the product of determinants. Thus Eq. (3.40) implies that the determinant of the metric tensor $g \equiv \det g_{\mu\nu}$ transforms as $g' = J^{-2} g$. Hence $J = \sqrt{|g|}/\sqrt{|g'|}$, where the absolute value signs are necessary because the determinant of $g_{\mu\nu}$ is negative in 4D spacetime with Lorentzian metric signature [see Eq. (4.6)], and this result may be substituted into Eq. (3.47) to give $\sqrt{|g'|}\, d^4x' = \sqrt{|g|}\, d^4x$. This implies that an *invariant volume element*,

$$dV = \sqrt{|g|}\,d^4x, \qquad (3.48)$$

must be employed in tensor integration to ensure that the results are invariant under a change of coordinate system. A simple demonstration of using invariant integration to determine an area for a 2-dimensional curved manifold is given in Example 3.8.

Example 3.8 The metric for a 2-dimensional spherical surface (the 2-sphere S^2) is specified by the line element $d\ell^2 = R^2 d\theta^2 + R^2 \sin^2\theta d\varphi^2$, which may be written as the matrix equation

$$d\ell^2 = (d\theta \quad d\varphi) \begin{pmatrix} R^2 & 0 \\ 0 & R^2 \sin^2\theta \end{pmatrix} \begin{pmatrix} d\theta \\ d\varphi \end{pmatrix}.$$

The area of the 2-sphere may then be calculated as

$$A = \int_0^{2\pi} d\varphi \int_0^\pi \sqrt{\det g_{ij}}\, d\theta$$
$$= \int_0^{2\pi} d\varphi \int_0^\pi R^2 \sin\theta\, d\theta = 4\pi R^2,$$

where the metric tensor g_{ij} is the 2×2 matrix in the first equation for the line element. In this 2-dimensional example the sign of the determinant is positive, so no absolute value is required under the radical in Eq. (3.48).

Thus, the extension of ordinary integration to integration over tensor fields requires only that the volume element be made invariant according to the prescription in Eq. (3.48). What about the derivatives of tensor quantities? This is a more complicated issue that we must now address.

3.13.2 Partial Derivatives

Before proceeding it will be useful to introduce a more compact way to write partial derivatives. Two shorthand notations are in common use, as illustrated by the following examples.[8]

$$\partial_\mu \varphi = \varphi_{,\mu} \equiv \frac{\partial \varphi(x)}{\partial x^\mu} \qquad \partial'_\mu \varphi' = \varphi'_{,\mu} \equiv \frac{\partial \varphi'(x')}{\partial x'^\mu} \qquad \partial^\mu \varphi = \varphi^{,\mu} \equiv \frac{\partial \varphi(x)}{\partial x_\mu}. \qquad (3.49)$$

All three of the notations exhibited for partial derivatives in these equations will be used at various places in this book. Let us now consider the covariance of the partial derivative operation applied to tensors.

The transformation law for the derivative of a scalar is given by Eq. (3.28), which is just the transformation law (3.29); therefore the derivative of a scalar is a dual vector and scalars and their first derivatives have well-defined tensorial properties. So far, so good, but now consider the derivative of a dual vector,

---

[8] Higher-order derivatives can be denoted by additional subscripts. For example $\partial^2 \varphi / \partial x^\mu \partial x^\nu = \varphi_{,\mu\nu}$.

$$A'_{\mu,\nu} \equiv \frac{\partial A'_\mu}{\partial x'^\nu} = \frac{\partial}{\partial x'^\nu}\left(A_\alpha \frac{\partial x^\alpha}{\partial x'^\mu}\right)$$

$$= \frac{\partial A_\alpha}{\partial x'^\nu}\frac{\partial x^\alpha}{\partial x'^\mu} + A_\alpha \frac{\partial^2 x^\alpha}{\partial x'^\nu \partial x'^\mu}$$

$$= \frac{\partial A_\alpha}{\partial x^\beta}\frac{\partial x^\beta}{\partial x'^\nu}\frac{\partial x^\alpha}{\partial x'^\mu} + A_\alpha \frac{\partial^2 x^\alpha}{\partial x'^\nu \partial x'^\mu}$$

$$= A_{\alpha,\beta}\frac{\partial x^\beta}{\partial x'^\nu}\frac{\partial x^\alpha}{\partial x'^\mu} + A_\alpha \frac{\partial^2 x^\alpha}{\partial x'^\nu \partial x'^\mu}. \tag{3.50}$$

The first term in the last line transforms as a (rank-2) tensor but the second term does not since it involves second derivatives, ultimately because the partial-derivative matrix implementing the transformation is *position dependent* in curved spacetime. In flat space it is possible to choose coordinates where the second term vanishes but in curved spacetime it cannot be transformed away globally. By a similar procedure it is found that the partial derivatives of vectors and all higher-order tensors exhibit a similar pathology:

> With the exception of derivatives of scalars, ordinary partial differentiation of tensors is *not a covariant operation in curved spacetime* because it fails to preserve the tensor structure of equations.

This will complicate the formalism immensely because the utility of the tensor framework rests on the preservation of tensor structure under transformations. It is highly desirable to define a new covariant derivative operation that does this automatically. In general the terms that violate the tensor transformation laws for partial derivatives of tensors will involve second derivatives, as in Eq. (3.50). The offending non-tensorial contributions can be eliminated systematically by employing additional fields on the manifold. There is more than one way to do this, each leading to a different form of covariant differentiation. The present discussion will address three: (1) *covariant derivatives* and (2) *absolute derivatives,* which use derivatives of the metric tensor field to cancel non-tensorial character, and (3) *Lie derivatives,* which use derivatives of an auxiliary vector field defined on the manifold to the same end. Covariant derivatives will be introduced in Section 3.13.3, absolute derivatives in Section 3.13.4, and Lie derivatives in Section 3.13.5.

3.13.3 Covariant Derivatives

For the manifolds important in general relativity, the most common approach to converting partial differentiation into an operation that preserves tensor structure is to use particular linear combinations of metric-tensor derivatives to create new non-tensorial terms that *exactly cancel* the non-tensorial terms arising from taking the partial

3.13 Tensor Calculus on Curved Manifolds

derivative. Notice that if the *Christoffel symbols* $\Gamma^{\lambda}_{\alpha\beta}$ are introduced and required to obey a transformation law [9]

$$\Gamma'^{\lambda}_{\alpha\beta} = \Gamma^{\kappa}_{\mu\nu} \frac{\partial x^{\mu}}{\partial x'^{\alpha}} \frac{\partial x^{\nu}}{\partial x'^{\beta}} \frac{\partial x'^{\lambda}}{\partial x^{\kappa}} + \frac{\partial^2 x^{\mu}}{\partial x'^{\alpha} \partial x'^{\beta}} \frac{\partial x'^{\lambda}}{\partial x^{\mu}}, \tag{3.51}$$

it may be shown that (see Problem 3.2)

$$\left(A'_{\mu,\nu} - \Gamma'^{\lambda}_{\mu\nu} A'_{\lambda}\right) = \left(A_{\alpha,\beta} - \Gamma^{\kappa}_{\alpha\beta} A_{\kappa}\right) \frac{\partial x^{\alpha}}{\partial x'^{\mu}} \frac{\partial x^{\beta}}{\partial x'^{\nu}}. \tag{3.52}$$

Comparing with Eq. (3.36), the quantity in brackets is seen to transform as a rank-2 covariant tensor. This suggests the utility of introducing a new derivative operation: the *covariant derivative* of a dual vector is defined to be

$$A_{\mu;\nu} \equiv A_{\mu,\nu} - \Gamma^{\lambda}_{\mu\nu} A_{\lambda}, \tag{3.53}$$

where now a subscript comma denotes ordinary partial differentiation and a subscript semicolon denotes covariant differentiation with respect to the variables following it. It will be useful to introduce also an alternative notation for the covariant derivative,

$$\nabla_{\nu} A_{\mu} = A_{\mu;\nu} \equiv \partial_{\nu} A_{\mu} - \Gamma^{\lambda}_{\mu\nu} A_{\lambda}. \tag{3.54}$$

This covariant derivative of a dual vector then transforms as a covariant tensor of rank 2: neither of its terms is a tensor but their *difference* is. Likewise, the covariant derivative of a vector can be introduced in either of the notations

$$A^{\lambda}_{\;;\mu} = A^{\lambda}_{\;,\mu} + \Gamma^{\lambda}_{\alpha\mu} A^{\alpha} \qquad \nabla_{\mu} A^{\lambda} = \partial_{\mu} A^{\lambda} + \Gamma^{\lambda}_{\alpha\mu} A^{\alpha}, \tag{3.55}$$

where the result of (3.55) is a mixed rank-2 tensor, and the covariant derivatives of the three possible rank-2 tensors through

$$A_{\mu\nu;\lambda} = A_{\mu\nu,\lambda} - \Gamma^{\alpha}_{\mu\lambda} A_{\alpha\nu} - \Gamma^{\alpha}_{\nu\lambda} A_{\mu\alpha}, \tag{3.56}$$

$$A^{\mu}_{\;\nu;\lambda} = A^{\mu}_{\;\nu,\lambda} + \Gamma^{\mu}_{\alpha\lambda} A^{\alpha}_{\;\nu} - \Gamma^{\alpha}_{\nu\lambda} A^{\mu}_{\;\alpha}, \tag{3.57}$$

$$A^{\mu\nu}_{\;\;;\lambda} = A^{\mu\nu}_{\;\;,\lambda} + \Gamma^{\mu}_{\alpha\lambda} A^{\alpha\nu} + \Gamma^{\nu}_{\alpha\lambda} A^{\mu\alpha}, \tag{3.58}$$

or in alternative notation

$$\nabla_{\lambda} A_{\mu\nu} = \partial_{\lambda} A_{\mu\nu} - \Gamma^{\alpha}_{\mu\lambda} A_{\alpha\nu} - \Gamma^{\alpha}_{\nu\lambda} A_{\mu\alpha}, \tag{3.59}$$

$$\nabla_{\lambda} A^{\mu}_{\;\nu} = \partial_{\lambda} A^{\mu}_{\;\nu} + \Gamma^{\mu}_{\alpha\lambda} A^{\alpha}_{\;\nu} - \Gamma^{\alpha}_{\nu\lambda} A^{\mu}_{\;\alpha}, \tag{3.60}$$

$$\nabla_{\lambda} A^{\mu\nu} = \partial_{\lambda} A^{\mu\nu} + \Gamma^{\mu}_{\alpha\lambda} A^{\alpha\nu} + \Gamma^{\nu}_{\alpha\lambda} A^{\mu\alpha}, \tag{3.61}$$

where the derivatives in Eqs. (3.56)–(3.61) define rank-3 tensors. In the general case the covariant derivative of a tensor is a tensor of one rank higher than the tensor being

---

[9] To be precise, $\Gamma^{\lambda}_{\alpha\beta}$ is a Christoffel symbol of the *second kind*. The contraction $\Gamma_{\lambda\alpha\beta} = g_{\lambda\gamma} \Gamma^{\gamma}_{\alpha\beta}$ defines a Christoffel symbol of the *first kind*, which has all lower indices. Only Christoffel symbols of the second kind will be used here and they will be termed simply *Christoffel symbols* (or *connection coefficients*; see Section 7.4.2). Indices notwithstanding, $\Gamma^{\lambda}_{\alpha\beta}$ is *not* a tensor because the last term in Eq. (3.51) involves second derivatives. For general relativity the Christoffel symbol is always symmetric under exchange of its lower indices, so it will be written simply as $\Gamma^{\lambda}_{\alpha\beta}$, rather than with horizontal staggering of indices like $\Gamma^{\lambda}_{\;\alpha\beta}$.

differentiated and the heuristic for constructing it is to form the ordinary partial derivative and add one Christoffel symbol term having the sign and form for a dual vector for each lower index, and one having the sign and form for a vector for each upper index of the tensor (see Example 3.9).

In Section 7.4.2 it will be asserted that in general relativity the Christoffel symbols are equivalent to *connection coefficients* (hence the same notation will be employed for both), which have a geometrical significance on the manifold and can be defined in terms of the derivatives of the metric tensor as [see Eq. (7.30)]

$$\Gamma^\sigma_{\lambda\mu} = \tfrac{1}{2} g^{\nu\sigma} \left(\frac{\partial g_{\mu\nu}}{\partial x^\lambda} + \frac{\partial g_{\lambda\nu}}{\partial x^\mu} - \frac{\partial g_{\mu\lambda}}{\partial x^\nu} \right).$$

Thus the correction terms in Eqs. (3.53)–(3.61) that cancel non-tensorial character are indeed composed of derivatives of the metric tensor, as promised above.

Rules for covariant differentiation: Most of the rules for partial differentiation carry over with suitable generalization for covariant differentiation. For instance, the ordinary (Leibniz) rule for differentiating a product applies also to covariant differentiation,

$$(A_\mu B_\nu)_{;\lambda} = A_{\mu;\lambda} B_\nu + A_\mu B_{\nu;\lambda}. \tag{3.62}$$

The most important exception concerns the results of successive covariant differentiations. Partial derivative operators normally commute: if two are applied successively, the outcome does not depend on the order in which they are applied. However, covariant derivative operators generally do not commute and two successive covariant differentiations may give results that depend on the order in which they are applied. In Section 7.4 it will be shown that covariant derivatives, and their non-commuting nature, arise naturally from a prescription for parallel transport of vectors in curved spaces.

Example 3.9 The Leibniz differentiation rule for the product of two vectors may be used to derive the expressions (3.56)–(3.58) for the covariant derivatives of rank-2 tensors from those given for vectors in Eqs. (3.53) and (3.55). For example,

$$\begin{aligned}(U^\alpha V^\beta)_{;\gamma} &= U^\alpha V^\beta_{;\gamma} + U^\alpha_{;\gamma} V^\beta \\ &= U^\alpha V^\beta_{,\gamma} + U^\alpha(\Gamma^\beta_{\rho\gamma} V^\rho) + U^\alpha_{,\gamma} V^\beta + (\Gamma^\alpha_{\rho\gamma} U^\rho) V^\beta \\ &= (U^\alpha V^\beta)_{,\gamma} + \Gamma^\beta_{\rho\gamma}(U^\alpha V^\rho) + \Gamma^\alpha_{\rho\gamma}(U^\rho V^\beta),\end{aligned}$$

where Eq. (3.55) and $U^\alpha V^\beta_{,\gamma} + U^\alpha_{,\gamma} V^\beta = (U^\alpha V^\beta)_{,\gamma}$ were used in the second and third lines, respectively. This is equivalent to

$$A^{\alpha\beta}_{;\gamma} = A^{\alpha\beta}_{,\gamma} + \Gamma^\beta_{\rho\gamma} A^{\alpha\rho} + \Gamma^\alpha_{\rho\gamma} A^{\rho\beta},$$

where $A^{\alpha\beta} \equiv U^\alpha V^\beta$, which is Eq. (3.58) for the covariant derivative of a contravariant rank-2 tensor.

Carrying out similar manipulations as in this example for the rank-2 tensors $A_{\mu\nu}$ and $A^\nu{}_\mu$ suggests the rule given after Eq. (3.61): differentiate in the normal way and add one Christoffel symbol term having the sign and form for a dual vector for each lower index, and one having the sign and form for a vector for each upper index of the tensor.

Implications of covariant differentiation: One rather important consequence of requiring that differentiation preserve tensor structure in the manner described above is that the covariant derivative of the metric tensor vanishes at all points of a manifold,

$$\nabla_\alpha g_{\mu\nu} = g_{\mu\nu;\alpha} = 0, \qquad (3.63)$$

as you will be asked to prove in Problem 7.2.[10] This has several implications. One is practical: in manipulating tensor equations the operation of covariant differentiation commutes with the operation (3.42) of raising or lowering an index by contraction with the metric tensor. For example, $g_{\alpha\beta} \nabla_\gamma V^\beta = \nabla_\gamma (g_{\alpha\beta} V^\beta) = \nabla_\gamma V_\alpha$. Thus the order of covariant differentiation and contraction with the metric tensor can be interchanged without altering the result. Another consequence having profound physical significance will be introduced in Section 18.9.5. There it will be found that Eq. (3.63) allows the Einstein field equations to have a *vacuum energy term* that may be responsible for the most surprising discovery of modern observational cosmology: the expansion of the Universe appears to be *accelerating,* implying that space is permeated by a mysterious *dark energy* having properties quite unlike any substance ever studied under laboratory conditions.

3.13.4 Absolute Derivatives

Absolute derivatives (also termed *intrinsic derivatives*) are closely related to covariant derivatives. Whereas covariant derivatives are defined over an entire manifold in terms of ordinary partial derivatives plus correction terms to cancel non-tensorial character, absolute derivatives are defined only along constrained paths of the manifold in terms of ordinary derivatives plus correction terms to cancel non-tensorial behavior. Using $D/D\sigma$ to denote the absolute derivative along a path parameterized by σ (see Problem 7.19),

$$\frac{DA_\alpha}{D\sigma} = \frac{dA_\alpha}{d\sigma} - \Gamma^\beta_{\alpha\gamma} A_\beta \frac{dx^\gamma}{d\sigma} \quad \text{(Dual vectors)}, \qquad (3.64)$$

$$\frac{DA^\alpha}{D\sigma} = \frac{dA^\alpha}{d\sigma} + \Gamma^\alpha_{\beta\gamma} A^\beta \frac{dx^\gamma}{d\sigma} \quad \text{(Vectors)}, \qquad (3.65)$$

with generalizations for higher-order tensors similar to that discussed earlier for covariant derivatives. The essential utility of both covariant and absolute derivatives is that when they are applied to tensor fields they produce tensor fields, whereas the ordinary derivative operators applied to tensors do not yield tensors except in the special case of flat spaces

---

[10] The validity of Eq. (3.63) is a consequence of particular assumptions concerning the nature of parallel transport in curved spacetime that will be discussed further in Section 7.8.

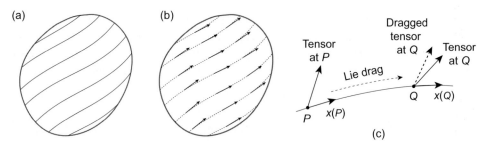

Fig. 3.4 (a) A congruence for a manifold. (b) Tangent vector field for the curves of the congruence. (c) *Lie dragging* using the congruence to compare tensors at two nearby points.

or when the tensor is a scalar, and that they provide a prescription for parallel transport in curved spaces that will be discussed in Section 7.4.3. As for the covariant derivative and the Lie derivative, the absolute derivative obeys the usual Leibniz rule for derivatives of products.

3.13.5 Lie Derivatives

In Section 3.13.3 the covariant derivative was introduced as a modification of partial differentiation that respects tensor structure. An alternative way to modify partial differentiation so that when applied to tensors it yields tensors is through the *Lie derivative*. Although we shall use primarily the covariant derivative to implement differentiation in curved spaces, it will be important in various contexts to have at least a conceptual understanding of the Lie derivative. A primary reason is that symmetries important in general relativity (and for a much broader range of physical applications) have an intimate connection to Lie derivatives. In contrast to the covariant derivative, which uses derivatives of the metric tensor to cancel non-tensorial terms arising in partial differentiation, the Lie derivative uses derivatives of an auxiliary vector field to cancel those same terms.

The basic idea: The starting point is the idea of a *congruence* for a manifold, which is a set of non-intersecting curves that are space-filling in the sense that for each point in the manifold exactly one curve of the congruence passes through it. As an example, for the sphere S^2 curves corresponding to lines of latitude define a congruence since for every point on S^2 exactly one curve of latitude passes through it. Figure 3.4(a) illustrates schematically a congruence for a general manifold. For every point in the manifold a basis vector may be defined by a tangent to the curve at that point, as illustrated in Fig. 3.4(b). Thus a congruence defines naturally a vector field on the manifold. The converse is also true. For any vector field X^μ defined on a manifold, a congruence can be generated by finding the curves for which the vectors of the vector field are tangent at each point of the curve. These curves are solutions of $dx^\mu/d\lambda = X^\mu[x(\lambda)]$. Such congruences are familiar to the physicist in the guise of streamlines connecting magnetic field vectors or illustrating fluid velocity fields. In the present more general context, a curve of the congruence for a vector field X^μ is termed an *orbit* and the vectors X^μ are said to generate the orbit.

Figure 3.4(c) illustrates the essential idea of the Lie derivative. For a tensor field defined on the manifold, a tensor at the point P (here illustrated by a vector) is "dragged" by a prescription that will be given below from the point P to the nearby point Q along a curve of the congruence. This process is called *Lie dragging*. Then the tensor defined naturally by the tensor field at Q, and the tensor dragged from the point P, may be subjected to the usual limiting difference procedure to define a derivative since they are located at the same spacetime point. The resulting derivative is called the *Lie derivative* of the tensor field. That is the basic idea; now some details, using as a guide d'Inverno [76].

Constructing Lie derivatives: Suppose that we have a manifold with a vector field X^μ and a corresponding congruence. Let some tensor field be defined on the manifold. To be definite for this example it will be taken to be a contravariant rank-2 tensor, $T^{\mu\nu}(x)$. Now consider the transformation

$$x'^\mu = x^\mu + X^\mu(x)\delta u, \qquad (3.66)$$

for small δu. This will be regarded as an *active transformation* in which the point P at x^μ is sent to the point Q at $x^\mu + X^\mu(x)\delta u$, with both points labeled in the *same coordinate system*. By construction the point Q lies on the congruence through P that is generated by X^μ. Now consider the tensor with components $T^{\mu\nu}(x)$ at P, which by the transformation (3.66) is mapped to the tensor with components $T'^{\mu\nu}(x')$ at Q. That is, the tensor is *Lie-dragged* from P to Q by the transformation, as illustrated in Fig. 3.4(c). By the usual tensor transformation law,

$$\begin{aligned} T'^{\mu\nu}(x') &= \frac{\partial x'^\mu}{\partial x^\alpha} \frac{\partial x'^\nu}{\partial x^\beta} T^{\alpha\beta}(x) \\ &= T^{\mu\nu}(x) + \left[\partial_\beta X^\nu T^{\mu\beta} + \partial_\alpha X^\mu T^{\alpha\nu}\right]\delta u, \end{aligned} \qquad (3.67)$$

where Eq. (3.66) was used (see Problem 3.17). This gives the tensor $T^{\mu\nu}$ Lie-dragged from P to Q. The Lie derivative of $T^{\mu\nu}$ with respect to the congruence of the vector field X is defined by

$$\mathscr{L}_X T^{\mu\nu} \equiv \lim_{\delta u \to 0} \left(\frac{T^{\mu\nu}(x') - T'^{\mu\nu}(x')}{\delta u}\right). \qquad (3.68)$$

The tensor $T'^{\mu\nu}(x')$ is given by Eq. (3.67) and the tensor $T^{\mu\nu}(x')$ can be determined by expanding in a Taylor series around x,

$$T^{\mu\nu}(x') = T^{\mu\nu}(x) + (\delta u)\partial_\alpha T^{\mu\nu} X^\alpha + \mathscr{O}\left(\delta u^2\right). \qquad (3.69)$$

Then substituting (3.67) and (3.69) into (3.68) gives for the Lie derivative of $T^{\mu\nu}$,

$$\mathscr{L}_X T^{\mu\nu} = X^\alpha \partial_\alpha T^{\mu\nu} - T^{\mu\alpha}\partial_\alpha X^\nu - T^{\alpha\nu}\partial_\alpha X^\mu. \qquad (3.70)$$

By a similar procedure Lie derivatives for tensors of other ranks can be determined. For example, the Lie derivatives of a scalar field φ, vector field A^μ, dual vector field A_μ, rank-2 covariant tensor $A_{\mu\nu}$, and rank-2 contravariant tensor $A^{\mu\nu}$ are

$$\mathscr{L}_X \varphi = X\varphi = X^\alpha \partial_\alpha \varphi, \tag{3.71}$$

$$\mathscr{L}_X A^\mu = X^\alpha \partial_\alpha A^\mu - A^\alpha \partial_\alpha X^\mu, \tag{3.72}$$

$$\mathscr{L}_X A_\mu = X^\alpha \partial_\alpha A_\mu + A_\alpha \partial_\mu X^\alpha, \tag{3.73}$$

$$\mathscr{L}_X A_{\mu\nu} = X^\alpha \partial_\alpha A_{\mu\nu} + A_{\mu\alpha} \partial_\nu X^\alpha + A_{\alpha\nu} \partial_\mu X^\alpha, \tag{3.74}$$

$$\mathscr{L}_X A^{\mu\nu} = X^\alpha \partial_\alpha A^{\mu\nu} - A^{\mu\alpha} \partial_\alpha X^\nu - A^{\alpha\nu} \partial_\alpha X^\mu, \tag{3.75}$$

respectively. Generalizations to tensors of any rank are given in Appendix B of Carrol [63].

Mathematically the Lie derivative is a more primitive concept than the covariant derivative because it requires less added structure on the manifold. For example, in the preceding derivations no appeal was made to either a metric or to connection (Christoffel) coefficients. For the manifolds of interest for general relativity (specifically, those having a metric and a torsion-free connection – see Sections 7.4.2 and 7.8), all partial derivatives ∂_μ may be replaced by covariant derivatives ∇_μ in the Lie derivative because all terms involving connnection coefficients cancel identically under this substitution. For example,

$$\mathscr{L}_X A_{\mu\nu} = X^\alpha \nabla_\alpha A_{\mu\nu} + A_{\mu\alpha} \nabla_\nu X^\alpha + A_{\alpha\nu} \nabla_\mu X^\alpha \tag{3.76}$$

and Eq. (3.74) are of equal validity if the manifold has a torsion-free connection. This demonstrates explicitly that even though only partial derivatives were used in its construction, the Lie derivative of a tensor is itself a tensor.

Lie transport and isometries: A tensor T is said to be *Lie-transported* along a curve of the congruence associated with the vector field V if the Lie derivative vanishes, $\mathscr{L}_V T = 0$. An interesting question for a manifold with metric is whether there exists a vector field K such that the Lie derivative of the metric vanishes, $\mathscr{L}_K g_{\mu\nu} = 0$. We shall take this question up in Section 5.6, where it will be shown that if $\mathscr{L}_K$ applied to the metric tensor gives zero, then K is a *Killing field* and the corresponding *Killing vectors* indicate directions in which the metric is invariant. These symmetries of the metric are called *isometries*.

3.14 Invariant Equations

The properties of tensors elaborated above ensure that any equation will be form-invariant under general coordinate transformations if it equates tensor components having the same upper and lower indices. For example, if the quantities $A^\mu{}_\nu$ and $B^\mu{}_\nu$ each transform as mixed rank-2 tensors according to Eq. (3.37) and $A^\mu{}_\nu = B^\mu{}_\nu$ in the x coordinate system, then in the x' coordinate system $A'^\mu{}_\nu = B'^\mu{}_\nu$ (see Problem 3.12). Likewise, an equation that equates any tensor to zero (that is, sets all its components to zero) in some coordinate system is covariant under general coordinate transformations, implying that the tensor is equal to zero in all coordinate systems. However, equations such as $A^\nu{}_\mu = 10$ or $A^\mu = B_\mu$ might hold in particular coordinate systems but generally are not valid in all coordinate systems because they equate tensors of different kinds (a mixed rank-2 tensor with a scalar in the first example and a dual vector with a vector in the second).

Example 3.10 That a tensor equal to zero in one coordinate system vanishes in any coordinate system can be used to give an alternative proof that the Christoffel symbols $\Gamma^\lambda_{\alpha\beta}$ are not tensors. It will be shown later that they are a measure of curvature and can be made to vanish in a freely falling coordinate system, but not in one where gravity is present and spacetime is curved. But if the $\Gamma^\lambda_{\alpha\beta}$ were tensors they would vanish in *all* coordinate systems by virtue of all their components being equal to zero in a particular one. Thus they are not tensors, as was already deduced from the transformation law (3.51).

The preceding discussion suggests that covariance of a theory under general coordinate transformations will be guaranteed by carrying out the following steps.

1. Formulate all quantities in terms of tensors, with tensor types matching on the two sides of any equation, and with all algebraic manipulations corresponding to valid tensor operations (addition, multiplication, contraction, ...).
2. Redefine any integration to be invariant integration, as discussed in Section 3.13.1.
3. Replace all partial derivatives with the covariant derivatives introduced in Section 3.13.3.
4. Take care to remember that a covariant differentiation generally does not commute with a second covariant differentiation, so the order matters.

As will be demonstrated in subsequent chapters, this prescription in terms of tensors will provide a sleek and powerful formalism for dealing with mathematical relations that would be much more formidable in standard notation.[11]

Background and Further Reading

At the introductory level, see See Hartle [111]; Hobson, Efstathiou, and Lasenby [122]; Cheng [65]; d'Inverno [76]; Peebles [180]; and Zee [258]. For more advanced discussions, see Rindler [204]; Weinberg [246]; Schutz [220, 221]; Carroll [63]; Ryder [212]; and Misner, Thorne, and Wheeler [156]. The discussion of Lie derivatives is based on material in Refs. [76, 94, 176, 199].

Problems

3.1 Show that the products

$$T^{\mu\nu} \equiv V^\mu V^\nu \qquad T_{\mu\nu} \equiv V_\mu V_\nu \qquad T_\mu{}^\nu \equiv V_\mu V^\nu$$

transform as tensors if V_μ and V_ν transform as dual vectors.

---

[11] Strictly, this prescription may fail to yield a unique theory if equations contain products of derivatives, in which case an additional ordering procedure may be needed since covariant derivatives do not commute. Such technicalities will be ignored for the present discussion.

3.2 Demonstrate that the covariant derivative transforms as in Eq. (3.52) if the Christoffel symbol $\Gamma^\kappa_{\alpha\beta}$ obeys the transformation law given by Eq. (3.51). ***

3.3 (a) Given a metric tensor $g_{\mu\nu}$, define its inverse by the requirement $g_{\mu\alpha}g^{\alpha\nu} = \delta^\nu_\mu$ and prove that $g^{\mu\nu}$ is a tensor if $g_{\mu\nu}$ is a tensor. (b) It was asserted that the metric tensor is symmetric. Show that even if it were assumed that $g_{\mu\nu}$ had an antisymmetric part, it would make no contribution to the line element (3.39). ***

3.4 Show that if index raising and lowering operations are *defined* by expressions like

$$T^\mu{}_\nu = g_{\nu\alpha}T^{\mu\alpha} \qquad T_{\mu\nu} = g_{\mu\alpha}g_{\nu\beta}T^{\alpha\beta},$$

then these are valid tensor operations. That is, show that if these relations are true in one coordinate system they are true in all coordinate systems.

3.5 Show that setting an upper and lower index equal on a tensor and summing on that index yields a tensor with those indices removed. Prove that a linear combination of tensors with the same upper and lower indices is itself a tensor, with those same upper and lower indices.

3.6 Write the metric for flat spacetime in spherical coordinates. Use this metric to write the invariant volume element dV for a sphere of radius R and show that integration of the invariant volume element gives the usual formula for the volume of a sphere.

3.7 Derive the transformation law for a dual vector A_μ from that for a vector B^μ by using that $A \cdot B$ transforms as a scalar.

3.8 Cartesian (x, y, z), spherical (r, θ, φ), and cylindrical coordinates (ρ, φ, z) are related by the equations

$$x = r\sin\theta\cos\varphi = \rho\cos\varphi,$$
$$y = r\sin\theta\sin\varphi = \rho\sin\varphi,$$
$$z = r\cos\theta.$$

Consider the 3D dual vector and vector transformations between spherical and cylindrical coordinates written in matrix form, $A'_i(u') = U^j_i A_j(u)$ and $A'^i(u') = \hat{U}^i_j A^j(u)$, treating $u = (r, \theta, \varphi)$ as the unprimed coordinates and $u' = (\rho, \varphi, z)$ as the primed coordinates. Evaluate the entries for the 3×3 matrices U and $\hat{U}$ and show explicitly that $\hat{U}U = I$, where I is the unit matrix. ***

3.9 Write out the terms for a covariant rank-4 tensor antisymmetrized in its first two indices and symmetrized in its last two indices: in the notation of Section 3.11, $T_{[\alpha\beta](\gamma\delta)}$.

3.10 (a) Show that if $A^{\mu\nu}$ is antisymmetric and $B_{\mu\nu}$ is symmetric under index exchange, then $A^{\mu\nu}B_{\mu\nu} = 0$. (b) Show that if a tensor is symmetric or antisymmetric under exchange of two indices it retains that property under coordinate transformation.

3.11 By contracting δ^μ_ν with the components V^ν of an arbitrary vector and using the index raising and lowering properties of the metric tensor, show that $g^{\mu\alpha}g_{\alpha\nu} = \delta^\mu_\nu$.

3.12 Prove that if $T_{\mu\nu}$ and $U_{\mu\nu}$ are rank-2 tensors and $T_{\mu\nu} = U_{\mu\nu}$ in some coordinate system labeled by coordinates x, then $T'_{\mu\nu} = U'_{\mu\nu}$ in an arbitrary new coordinate system labeled by coordinates x'. ***

3.13 A theorem useful in various contexts states that if a set of quantities produces a tensor when contracted with a tensor, then that set of quantities is necessarily a tensor. This is called the *quotient theorem*. Use the definition of the scalar product of vectors and the quotient theorem to argue that the metric $g_{\mu\nu}$ is a tensor.

3.14 As another example of the quotient theorem introduced in Problem 3.13, show explicitly that if for an arbitrary vector V^γ

$$T^\alpha{}_{\beta\gamma} V^\gamma = \frac{\partial x'^\alpha}{\partial x^\delta} \frac{\partial x^\epsilon}{\partial x'^\beta} T^\delta{}_{\epsilon\varphi} V^\varphi,$$

then $T^\alpha{}_{\beta\gamma}$ is necessarily a tensor of type (1, 2).

3.15 (a) Demonstrate that the Kronecker delta δ^ν_μ is a mixed tensor of rank 2. (b) Prove that δ^ν_μ is unusual in that its *components* have the same value in all coordinate systems, which is not true in general for tensors.

3.16 Show that the line element $ds^2 = g_{\mu\nu} dx^\mu dx^\nu$ given in Eq. (3.39) is invariant under transformation of coordinate system. ***

3.17 Prove Eq. (3.67). *Hint*: Use Eq. (3.66) to evaluate the derivatives and keep only terms of order δu. ***

3.18 Use the methods illustrated in Section 3.13.5 and Problem 3.17 to prove that the Lie derivative of a vector is given by Eq. (3.73).

3.19 The Lie derivative obeys the standard Leibniz rule for the derivative of a product, $\mathscr{L}_X(AB) = A\mathscr{L}_X B + (\mathscr{L}_X A)B$, where A and B are tensors. This can be used to construct the Lie derivative of higher-order tensors from those of lower-order tensors. Factor the tensor $A_{\mu\nu}$ as $A_{\mu\nu} = U_\mu V_\nu$, where U and V are dual vectors. Use the formula for the Lie derivative of a dual vector in Eq. (3.73) and the Leibniz rule to derive the Lie derivative of a covariant rank-2 tensor given in Eq. (3.74).

3.20 The Lie derivative discussed in Section 3.13.5 is a generalization of the *Lie bracket* of vector fields A and B, which is defined by the *commutator*, $[A, B] \equiv AB - BA$. Show that the Lie bracket $[A, B]$ of vector fields A and B yields a new vector field that is equivalent to the Lie derivative of a vector field $\mathscr{L}_A B^\mu$ given in Eq. (3.72). *Hint*: Operate with $C \equiv [A, B]$ on an arbitrary function f and expand the vectors in a basis ∂_ν.

3.21 (a) Verify Eqs. (3.19) for the components of vectors and dual vectors. (b) Use the methods of Section 3.6 to derive the vector transformation law given by Eq. (3.31). *Hint*: See Example 3.5. ***

3.22 The Lie bracket or commutator $[A, B] \equiv AB - BA$ of vector fields A and B was introduced in Problem 3.20. It may be shown that for a coordinate (holonomic) basis the Lie bracket of basis vectors vanishes, $[e_\mu, e_\nu] = 0$, but for a non-coordinate (anholonomic) basis $[e_\mu, e_\nu] \neq 0$. Consider the 2D plane with either cartesian coordinates (x, y) or plane polar coordinates (r, θ). Show that the two bases

$$(e_1, e_2) = \left(\frac{\partial}{\partial x}, \frac{\partial}{\partial y}\right) \qquad (e_1, e_2) = \left(\frac{\partial}{\partial r}, \frac{\partial}{\partial \theta}\right)$$

are coordinate bases, but that the normalized basis

$$(\hat{e}_1, \hat{e}_2) = \left(\frac{\partial}{\partial r}, \frac{1}{r}\frac{\partial}{\partial \theta}\right)$$

is a non-coordinate basis.

3.23 It was asserted in Section 3.13.5 that for a manifold with a torsion-free connection (the connection coefficient $\Gamma^\mu_{\alpha\beta}$ is symmetric under exchange of its lower indices), the Lie derivative is unchanged by replacing all partial derivatives by covariant derivatives. Show explicitly that this is true for the Lie derivative of a vector field A^μ.

3.24 Show that the basis vectors e_μ and basis dual vectors e^μ behave as

$$e'_\mu = \frac{\partial x^\nu}{\partial x'^\mu} e_\nu \qquad e'^\mu = \frac{\partial x'^\mu}{\partial x^\nu} e^\nu$$

under coordinate transformations *Hint*: The infinitesimal displacement vector ds must be invariant under the coordinate transformation. ***

3.25 Show that for a vector expanded in the basis (3.6), the transformation law is given by Eq. (3.31). *Hint*: Require that an arbitrary vector be invariant under a general coordinate transformation.

3.26 For two vectors U and V the tensor product $U \otimes V$ yields a rank-2 tensor defined as the operator with two input slots that outputs a real number according to

$$(U \otimes V)(\alpha, \beta) = U(\alpha)V(\beta) \in \mathbb{R},$$

where α and β are arbitrary input vectors, $U(\alpha) = \langle U, \alpha \rangle = U_\mu \alpha^\mu$, and $V(\beta) = \langle V, \beta \rangle = V_\mu \beta^\mu$.

(a) Show that the covariant components of the resulting rank-2 tensor are given by $(U \otimes V)_{\mu\nu} = U_\mu V_\nu$.

(b) Generalize the results of (a) to define a rank-4 tensor S by the tensor product $S = U \otimes V \otimes \Omega \otimes W$, where U, V, W are vectors and Ω is a dual vector. Show that the mixed-tensor components $S^{\mu\nu}{}_\lambda{}^\epsilon$ are given by $S^{\mu\nu}{}_\lambda{}^\epsilon = U^\mu V^\nu \Omega_\lambda W^\epsilon$.

(c) Let T be a rank-2 tensor and V a rank-1 tensor. Show that the mixed tensor components $S^\mu{}_{\nu\gamma}$ of the tensor product $S = T \otimes V$ are given by $S^\mu{}_{\nu\gamma} = T^\mu{}_\nu V_\gamma$.

(d) Show that the rank-2 tensor product of two basis vectors e_μ and e_ν acting on two basis dual vectors gives

$$(e_\mu \otimes e_\nu)(e^\alpha, e^\beta) = \delta^\alpha_\mu \delta^\beta_\nu,$$

and thus that the sum

$$T^{\mu\nu}(e_\mu \otimes e_\nu)(e^\alpha, e^\beta) = T^{\alpha\beta}$$

corresponds to the contravariant components of T. Therefore T may be expanded as $T = T^{\mu\nu}(e_\mu \otimes e_\nu)$ and the tensor product $e_\mu \otimes e_\nu$ is a basis for $T = U \otimes V$.

Hint: Evaluate tensor components by inserting basis vectors as arguments. ***

4 Lorentz Covariance and Special Relativity

To go beyond the Newtonian theory of gravity requires considering the intimate relationship between gravity and the curvature of space and time that was first envisioned by Einstein. Mathematically, this extension is bound inextricably to the geometry of spacetime, and in particular to the aspect of geometry that permits quantitative measurement of distances. The tensor methods that we introduced in Chapter 3 will be critical in this formulation. In Chapter 7 a covariant formalism suitable for curved spacetime will be developed but in the present chapter the idea and approach will be introduced by considering the simpler problem of covariance with respect to Lorentz transformations in flat spacetime. This will allow an elegant and powerful statement of the special theory of relativity in terms of tensor equations. The discussion that follows in this chapter will assume a basic familiarity with the concepts of special relativity (for example, Lorentz transformations, space contraction, time dilation, and relativity of simultaneity), as discussed in the modern physics sections of a typical introductory physics textbook. It is not meant to be a comprehensive introduction to special relativity, but rather to remind the reader of some essential concepts, and to introduce a *geometrical view* of spacetime that is the most elegant way to view special relativity, and that will be an important foundation for the general theory of relativity to follow in later chapters.

4.1 Minkowski Space

A manifold equipped with a prescription for measuring distances is termed a *metric space* and the mathematical function that specifies distances is termed the *metric* for the space. Some familiar examples of metric spaces were introduced in Chapters 2 and 3. In this section those ideas are applied to flat 4-dimensional spacetime, which is commonly termed *Minkowski space.* Although many concepts will be similar to those introduced in Chapter 2, fundamentally new features will enter. Many of these new features are associated with the *indefinite metric* of Minkowski space.

4.1.1 The Indefinite Metric of Spacetime

Although Minkowski space is flat it is not euclidean, for it does not possess a euclidean metric. All of the metrics employed in Chapter 2 could be put into a diagonal form in which the signs of the diagonal entries could all be chosen positive. Such a metric is termed *positive definite.* In contrast, we will see that the Minkowski metric can be put into diagonal form but it is an essential property of Minkowski spacetime that the diagonal entries cannot all be chosen positive. Such a metric is termed *indefinite,* and it leads to properties of Minkowski space differing fundamentally from those of euclidean spaces.

4.1.2 Scalar Products and the Metric Tensor

In a particular inertial frame we may introduce unit vectors e_0, e_1, e_2, and e_3 that point along the t, x, y, and z axes, respectively, such that any 4-vector A may be expressed in the form,

$$A = A^0 e_0 + A^1 e_1 + A^2 e_2 + A^3 e_3. \tag{4.1}$$

Thus (A^0, A^1, A^2, A^3) are the (contravariant) components of the 4-vector A.[1] The scalar product of 4-vectors is given by

$$A \cdot B = B \cdot A = (A^\mu e_\mu) \cdot (B^\nu e_\nu) = e_\mu \cdot e_\nu A^\mu B^\nu. \tag{4.2}$$

Introducing the definition

$$\eta_{\mu\nu} \equiv e_\mu \cdot e_\nu, \tag{4.3}$$

we may express the scalar product as

$$A \cdot B = \eta_{\mu\nu} A^\mu B^\nu. \tag{4.4}$$

The metric tensor $\eta_{\mu\nu}$ is just a special case of the general metric tensor $g_{\mu\nu}$ discussed in Section 3.10; however, in flat spacetime $g_{\mu\nu}$ is a constant matrix independent of the coordinates and it is common to denote it by the special symbol $\eta_{\mu\nu}$.

4.1.3 The Line Element

The line element ds^2 in Minkowski space measures the square of the distance between points with infinitesimal separation and is given by

$$ds^2 = -c^2 d\tau^2 = \eta_{\mu\nu} dx^\mu dx^\nu = -c^2 dt^2 + dx^2 + dy^2 + dz^2, \tag{4.5}$$

where τ is the proper time (the time measured by a clock carried in an inertial frame; thus, it is the time measured between events that are at the same spatial point), and where the metric tensor of flat spacetime may be represented by the constant diagonal matrix

$$\eta_{\mu\nu} = \begin{pmatrix} -1 & 0 & 0 & 0 \\ 0 & 1 & 0 & 0 \\ 0 & 0 & 1 & 0 \\ 0 & 0 & 0 & 1 \end{pmatrix} \equiv \text{diag}(-1, 1, 1, 1). \tag{4.6}$$

Then Eq. (4.5) for the line element may be written as the matrix equation

---

[1] Recall that in our notation non-bold symbols are being used to denote 4-vectors and bold symbols are reserved for 3-vectors, and that a notation such as A^μ may stand either for the full 4-vector, or for a component of it, depending on the context.

$$ds^2 = (cdt \ dx \ dy \ dz) \begin{pmatrix} -1 & 0 & 0 & 0 \\ 0 & 1 & 0 & 0 \\ 0 & 0 & 1 & 0 \\ 0 & 0 & 0 & 1 \end{pmatrix} \begin{pmatrix} cdt \\ dx \\ dy \\ dz \end{pmatrix}, \quad (4.7)$$

where ds^2 represents the square of the spacetime interval between x and $x + dx$ with

$$x = (x^0, x^1, x^2, x^3) = (ct, x^1, x^2, x^3). \quad (4.8)$$

A point in Minkowski space defines an *event* and the path followed by an object in spacetime is termed the *worldline* for the object. This 4-dimensional spacetime with indefinite metric is termed a *Lorentzian manifold* (or sometimes a *pseudo-euclidean manifold*).

Example 4.1 Given a Minkowski vector with components (A^0, A^1, A^2, A^3), what are the components of the corresponding dual vector? From Eq. (3.42) with $\eta_{\mu\nu}$ substituted for the metric tensor, the indices may be lowered through the contraction $A_\mu = \eta_{\mu\nu} A^\nu$. Therefore, using the metric tensor (4.6) the elements of the corresponding dual vector are $A_\mu = (-A^0, A^1, A^2, A^3)$. This illustrates explicitly that vectors and dual vectors generally are not equivalent in non-euclidean manifolds, but that they are in one-to-one correspondence though contraction with the metric tensor.

The metric (4.6) is diagonal, with relative sign of the diagonal terms $(-+++)$. This sign pattern is termed the *signature of the metric*. (Some authors instead define the signature to be an *integer* equal to the difference of the number of positive signs and number of negative signs.) It is also common in the literature to see the opposite signature, corresponding to the pattern $(+---)$ that results from multiplying the metric (4.6) by -1. This choice is purely conventional but it is an essential property of Minkowski space that it is not possible for all terms in the signature of the metric to have the same sign. That is, Minkowski space has an *indefinite metric,* in contrast to the positive definite metric characteristic of euclidean spaces.

4.1.4 Invariance of the Spacetime Interval

Special relativity follows from two assumptions: (1) the speed of light is constant for all observers, and (2) the laws of physics cannot depend on spacetime coordinates. The postulate that the speed of light is a constant is equivalent to a statement that the spacetime interval ds^2 of Eq. (4.5) is an invariant that is unchanged by transformations between inertial systems (the Lorentz transformations to be discussed below; see Problem 4.1). This is not true for the euclidean spatial interval $dx^2 + dy^2 + dz^2$, nor is it true for the time interval $c^2 dt^2$; it is true only for the particular combination of spatial and time intervals defined by Eq. (4.5). Because of this invariance, Minkowski space is the natural manifold for the formulation of special relativity.

Example 4.2 The metric can be used to determine the relationship between the time coordinate t and the proper time τ. From Eq. (4.5)

$$d\tau^2 = \frac{-ds^2}{c^2} = \frac{1}{c^2}(c^2 dt^2 - dx^2 - dy^2 - dz^2)$$

$$= dt^2 \left\{ 1 - \frac{1}{c^2} \left[\left(\frac{dx}{dt}\right)^2 + \left(\frac{dy}{dt}\right)^2 + \left(\frac{dz}{dt}\right)^2 \right] \right\}$$

$$= \left(1 - \frac{v^2}{c^2}\right) dt^2, \tag{4.9}$$

where v is the magnitude of the velocity. Therefore, the proper time that elapses between coordinate times t_1 and t_2 is

$$\tau_{12} = \int_{t_1}^{t_2} \left(1 - \frac{v^2}{c^2}\right)^{1/2} dt. \tag{4.10}$$

The proper time interval τ_{12} is shorter than the coordinate time interval $t_2 - t_1$ because the square root in the integrand of Eq. (4.10) is always less than one. This is the special-relativistic *time dilation effect,* stated in general form. For the special case of constant velocity, (4.10) yields

$$\Delta\tau = \left(1 - \frac{v^2}{c^2}\right)^{1/2} \Delta t, \tag{4.11}$$

which is the formulation of special relativistic time dilation that is found commonly in textbooks. From this example it is clear that the origin of time dilation in special relativity lies in the geometry of spacetime, specifically in the indefinite nature of the Minkowski metric.

The first postulate of special relativity (constant speed of light for all observers) is ensured by the invariance of the interval (4.5) under transformations between inertial frames. As was suggested by the discussion in Chapter 3, the second postulate (coordinate invariance of physical law) can be ensured by formulating the equations of special relativity in terms of tensors defined in Minkowski space, which we now address.

4.2 Tensors in Minkowski space

In Minkowski space the transformations between coordinate systems are particularly simple because they are *independent of spacetime coordinates.* Therefore, the derivatives appearing in the general definitions of Table 3.1 for tensors are constants and the transformation of a coordinate vector x^μ may be expressed as $x'^\mu = \Lambda^\mu{}_\nu x^\nu$, where the matrix $\Lambda^\mu{}_\nu$ does not depend on the spacetime coordinates. Hence, for flat spacetime the tensor transformation laws simplify to

$$\varphi' = \varphi \qquad \text{Scalar}$$
$$A'^{\mu} = \Lambda^{\mu}{}_{\nu} A^{\nu} \qquad \text{Vector}$$
$$A'_{\mu} = \Lambda_{\mu}{}^{\nu} A_{\nu} \qquad \text{Dual vector} \qquad (4.12)$$
$$T'^{\mu\nu} = \Lambda^{\mu}{}_{\gamma} \Lambda^{\nu}{}_{\delta} T^{\gamma\delta} \qquad \text{Contravariant rank-2 tensor}$$
$$T'_{\mu\nu} = \Lambda_{\mu}{}^{\gamma} \Lambda^{\delta}{}_{\nu} T_{\gamma\delta} \qquad \text{Covariant rank-2 tensor}$$
$$T'^{\mu}{}_{\nu} = \Lambda^{\mu}{}_{\gamma} \Lambda^{\delta}{}_{\nu} T^{\gamma}{}_{\delta} \qquad \text{Mixed rank-2 tensor}$$

and so on. In addition, for flat spacetime it is possible to choose a coordinate system for which non-tensorial terms like the second term of Eq. (3.50) can be transformed away so *covariant derivatives are equivalent to partial derivatives* in Minkowski space. In the transformation laws (4.12) the $\Lambda^{\mu}{}_{\nu}$ are elements of *Lorentz transformations* that we will now discuss in more detail.

4.3 Lorentz Transformations

Inertial frames enjoy a privileged role in Newtonian mechanics. Newton's first law is unchanged in special relativity and inertial frames can be constructed in the same way as for Newtonian mechanics. What is different about the inertial frames of special relativity is that because of the requirements imposed by the constant speed of light and principle of relativity postulates, the transformations between inertial frames are no longer the Galilean transformations of Newtonian mechanics but rather the *Lorentz transformations.* Hence, the inertial frames of special relativity are often termed *Lorentz frames.* Rotations are an important class of transformations in euclidean space because they *change the direction but preserve the length* of an arbitrary 3-vector. It is desirable to generalize this idea to investigate abstract rotations in the 4-dimensional Minkowski space that change the direction but preserve the length of 4-vectors. As we will now demonstrate, such rotations in Minkowski space are just the Lorentz transformations alluded to above.

4.3.1 Rotations in Euclidean Space

First we consider a rotation of the coordinate system in euclidean space, as illustrated in Fig. 2.5. The condition that the length of an arbitrary vector be unchanged by this transformation corresponds to the requirement that the transformation matrix R implementing the rotation [see Eq. (2.42)] act on the metric tensor g_{ij} in the following way

$$R g_{ij} R^{\mathrm{T}} = g_{ij}, \qquad (4.13)$$

where R^{T} denotes the transpose of R. For euclidean space the metric tensor is just the unit matrix so the requirement (4.13) reduces to $R R^{\mathrm{T}} = 1$, which is the condition that R be an orthogonal matrix. Thus, we have obtained the well-known result that rotations in euclidean space are implemented by orthogonal matrices in a rather pedantic manner.

But Eq. (4.13) is valid generally, not just for euclidean spaces. Therefore, it may be used as guidance for constructing more general rotations in Minkowski space.

4.3.2 Generalized 4D Minkowski Rotations

By analogy with the above discussion of rotations in euclidean space, which left the length of 3-vectors invariant, let us now seek a set of transformations that leave the length of a 4-vector invariant in the Minkowski space. The coordinate transformation may be written in matrix form,

$$dx'^{\mu} = \Lambda^{\mu}{}_{\nu} dx^{\nu}, \qquad (4.14)$$

where the transformation matrix $\Lambda^{\mu}{}_{\nu}$ is expected to satisfy the analog of Eq. (4.13) for the Minkowski metric $\eta_{\mu\nu}$,

$$\Lambda \eta_{\mu\nu} \Lambda^{T} = \eta_{\mu\nu}, \qquad (4.15)$$

or explicitly in terms of components, $\Lambda_{\mu}{}^{\rho} \Lambda^{\sigma}{}_{\nu} \eta_{\rho\sigma} = \eta_{\mu\nu}$.[2] This property may now be used to construct the elements of the transformation matrix $\Lambda^{\mu}{}_{\nu}$. The possible transformations include rotations about the spatial axes (corresponding to rotations within inertial systems) and transformations between inertial systems moving at different constant velocities that are termed Lorentz boosts.[3] Consider first the simple case of rotations about the z axis.

4.3.3 Lorentz Spatial Rotations

Rotations about a single spatial axis in Minkowski space correspond to a 2-dimensional problem with euclidean metric, so the condition (4.13) may be written as

$$\begin{pmatrix} a & b \\ c & d \end{pmatrix} \begin{pmatrix} 1 & 0 \\ 0 & 1 \end{pmatrix} \begin{pmatrix} a & c \\ b & d \end{pmatrix} = \begin{pmatrix} 1 & 0 \\ 0 & 1 \end{pmatrix}, \qquad (4.16)$$

where a, b, c, and d parameterize the transformation matrices. Carrying out the matrix multiplications explicitly on the left side gives

$$\begin{pmatrix} a^2 + b^2 & ac + bd \\ ac + bd & c^2 + d^2 \end{pmatrix} = \begin{pmatrix} 1 & 0 \\ 0 & 1 \end{pmatrix},$$

and comparison of the two sides of the equation implies the conditions

$$a^2 + b^2 = 1 \qquad c^2 + d^2 = 1 \qquad ac + bd = 0.$$

---

[2] Note that in this discussion we are using the (common) convention that $\eta_{\mu\nu}$ is either a symbol standing for the full tensor or a specific component of the tensor distinguished by indices μ and ν, depending on context. In a matrix equation like (4.15) the order of the factors matters because matrices don't generally commute, but when the matrix equation is written out in terms of sums over component products as in $\Lambda_{\mu}{}^{\rho} \Lambda^{\sigma}{}_{\nu} \eta_{\rho\sigma} = \eta_{\mu\nu}$ the order of factors can be rearranged at will, since the components of the matrices are just numbers that commute with each other.

[3] Two inertial frames may differ in displacement, rotational orientation, and uniform velocity. This corresponds to 10 possible transformations between inertial frames: three velocity boosts along the spatial axes, three rotations about the three spatial axes, and four translations in the space and time directions. These 10 transformations form a group called the *Poincaré group* (see Box 7.1). The six Lorentz transformations correspond to the velocity boosts and spatial rotations, and they form a group called the *Lorentz group* that is a subgroup of the Poincaré group, also discussed in Box 7.1.

Obviously one choice of parameters that satisfies these conditions is

$$a = \cos \varphi \qquad b = \sin \varphi \qquad c = -\sin \varphi \qquad d = \cos \varphi.$$

This leads to the standard result

$$\begin{pmatrix} x'^1 \\ x'^2 \end{pmatrix} = R \begin{pmatrix} x^1 \\ x^2 \end{pmatrix} = \begin{pmatrix} \cos \varphi & \sin \varphi \\ -\sin \varphi & \cos \varphi \end{pmatrix} \begin{pmatrix} x^1 \\ x^2 \end{pmatrix}, \quad (4.17)$$

which is Eq. (2.41) restricted to rotations about a single axis. Now we shall apply this same technique to determine the elements of a Lorentz boost transformation.

4.3.4 Lorentz Boost Transformations

Consider a boost from one inertial system to a second one moving in the positive direction at uniform velocity along the x axis, as illustrated in Fig. 4.1. Since the y and z coordinates will not be affected, this is a 2-dimensional problem in the time coordinate t and the spatial coordinate x. The transformation is of the general form

$$\begin{pmatrix} cdt' \\ dx' \end{pmatrix} = \begin{pmatrix} a & b \\ c & d \end{pmatrix} \begin{pmatrix} cdt \\ dx \end{pmatrix}, \quad (4.18)$$

and the condition (4.15) can be written out explicitly as

$$\begin{pmatrix} a & b \\ c & d \end{pmatrix} \begin{pmatrix} -1 & 0 \\ 0 & 1 \end{pmatrix} \begin{pmatrix} a & c \\ b & d \end{pmatrix} = \begin{pmatrix} -1 & 0 \\ 0 & 1 \end{pmatrix}, \quad (4.19)$$

which is identical in form to the rotation case, except for the indefinite metric. Multiplying the matrices on the left side and comparing with the right side gives the conditions

$$a^2 - b^2 = 1 \qquad -c^2 + d^2 = 1 \qquad -ac + bd = 0,$$

which clearly are satisfied by the parameterization

$$a = \cosh \xi \qquad b = \sinh \xi \qquad c = \sinh \xi \qquad d = \cosh \xi,$$

where ξ is a hyperbolic variable taking the values $-\infty \leq \xi \leq \infty$. Therefore, we may write the boost transformation as

$$\begin{pmatrix} cdt' \\ dx' \end{pmatrix} = \begin{pmatrix} \cosh \xi & \sinh \xi \\ \sinh \xi & \cosh \xi \end{pmatrix} \begin{pmatrix} cdt \\ dx \end{pmatrix}. \quad (4.20)$$

A geometrical interpretation of this result is discussed in Box 4.1.

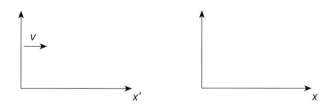

Fig. 4.1 A Lorentz boost along the positive x axis by a velocity v.

> **Box 4.1** **Minkowski Rotations**
>
> The respective derivations make clear that the appearance of hyperbolic functions in Eq. (4.20) instead of trigonometric functions as in Eq. (4.17) traces to the role of the indefinite metric diag $(-1, 1)$ in Eq. (4.19) relative to the positive-definite metric diag $(1, 1)$ in Eq. (4.16). The boost transformations are in a sense "rotations" in Minkowski space, but these rotations have unusual properties relative to normal rotations in euclidean space since they mix space and time, and may be viewed as rotations through imaginary angles (see Problem 4.6). These properties follow from the metric because the invariant interval is neither the length of vectors in space nor the length of time intervals, but rather the specific mixture of space and time intervals implied by the Minkowski line element (4.5) with indefinite metric (4.6). This is quite different from Newtonian mechanics, where time is a universal parameter common to all observers and the Galilean transformations conserve only the *space interval*.

The Lorentz boost transformation (4.20) can be put into a more familiar form by exchanging the boost parameter ξ for the boost velocity. For convenience, let's replace the differentials in Eq. (4.20) with finite space and time intervals ($dt \to t$ and $dx \to x$). The velocity of the boosted system is $v = x/t$. From Eq. (4.20), the origin ($x' = 0$) of the boosted system is given by

$$x' = ct \sinh \xi + x \cosh \xi = 0.$$

Therefore, $x/t = -c \sinh \xi / \cosh \xi$, from which it may be concluded that

$$\beta \equiv \frac{v}{c} = \frac{x}{ct} = -\frac{\sinh \xi}{\cosh \xi} = -\tanh \xi. \qquad (4.21)$$

This relationship between ξ and β is plotted in Fig. 4.2. Utilizing the identity $1 = \cosh^2 \xi - \sinh^2 \xi$ and the definition

$$\gamma \equiv \left(1 - \frac{v^2}{c^2}\right)^{-1/2} \qquad (4.22)$$

of the *Lorentz γ-factor*, we may write

$$\cosh \xi = \sqrt{\frac{\cosh^2 \xi}{1}} = \frac{1}{\sqrt{1 - \sinh^2 \xi / \cosh^2 \xi}} = \frac{1}{\sqrt{1 - v^2/c^2}} = \gamma, \qquad (4.23)$$

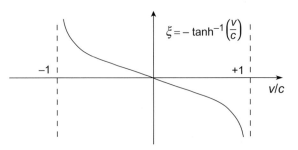

Fig. 4.2 Dependence of the Lorentz parameter ξ on $\beta = v/c$.

and from this result and (4.21),

$$\sinh \xi = -\beta \cosh \xi = -\beta \gamma. \tag{4.24}$$

Inserting (4.23)–(4.24) into (4.20) for finite intervals gives

$$\begin{pmatrix} ct' \\ x' \end{pmatrix} = \gamma \begin{pmatrix} 1 & -\beta \\ -\beta & 1 \end{pmatrix} \begin{pmatrix} ct \\ x \end{pmatrix} \tag{4.25}$$

and writing this matrix expression out explicitly gives the Lorentz boost equations (for the specific case of a positive boost along the x axis) in standard textbook form,

$$t' = \gamma \left(t - \frac{vx}{c^2} \right) \qquad x' = \gamma(x - vt) \qquad y' = y \qquad z' = z, \tag{4.26}$$

with the inverse transformation corresponding to the replacement $v \to -v$. By inspection, these reduce to the Galilean boost equations (2.43) if $v/c \to 0$ and it is easy to verify (Problem 4.1) that the Lorentz transformations leave invariant the spacetime interval ds^2.

4.4 Lightcone Diagrams

The line element (4.5) defines a cone, implying that Minkowski spacetime may be classified according to the *lightcone diagram* exhibited in Fig. 4.3. The *lightcone* is a 3-dimensional surface in the 4-dimensional spacetime and intervals in spacetime may be characterized according to whether they are inside of, outside of, or on the lightcone. Assuming the metric signature $(-+++)$ employed here, the standard terminology is

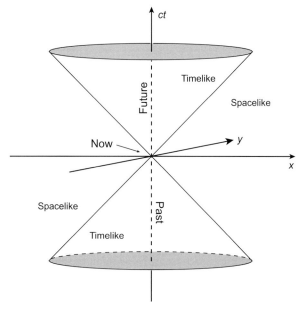

Fig. 4.3 Lightcone diagram for flat spacetime in two space and one time dimensions. The future lightcone is the surface swept out by a spherical light pulse emitted from the origin.

- If $ds^2 < 0$ the interval is termed *timelike*.
- If $ds^2 > 0$ the interval is termed *spacelike*.
- If $ds^2 = 0$ the interval is called *lightlike* (or *null*).

These regions for flat spacetime are labeled in Fig. 4.3. This classification can be extended also to surfaces. For example, a *spacelike surface* is a collection of points for which any pair of points has a spacelike separation and a *lightlike surface* is a collection of points for which any pair of points has a lightlike separation. The lightcone classification makes clear the distinction between Minkowski spacetime and a mere 4-dimensional euclidean space in that two points in the Minkowski spacetime may be separated by a distance that when squared could be positive, negative, or zero. This is not possible in a euclidean metric. Notice in particular that for lightlike particles, which have worldlines confined to the lightcone, the square of the spacetime interval between any two points on a worldline is *zero*.

Example 4.3 The Minkowski line element (4.5) in one space and one time dimension [which often is termed (1+1) dimensions] is given by $ds^2 = -c^2 dt^2 + dx^2$. Thus, if the spacetime interval is lightlike, $ds^2 = 0$ and

$$-c^2 dt^2 + dx^2 = 0 \quad \longrightarrow \quad \left(\frac{dx}{dt}\right)^2 = c^2 \quad \longrightarrow \quad v = \pm c.$$

This result can be generalized easily to the full space, leading to the conclusion that events in Minkowski space separated by a null interval ($ds^2 = 0$) are connected by signals moving at light velocity, $v = c$. If the time axis (ct) and space axes have the same scales, this means that the worldline of a freely propagating photon (or any massless particle, necessarily moving at light velocity) always makes $\pm 45°$ angles in the lightcone diagram. By similar arguments, events at timelike separations (inside the lightcone) are connected by signals with $v < c$, and those with spacelike separations (outside the lightcone) could be connected only by causality-violating signals with $v > c$.

The lightcone illustrated in Fig. 4.3 was placed at the origin for illustration but each point in the spacetime has its own lightcone, as illustrated in Fig. 4.4. From Example 4.3, the tangent to the worldline of any particle at a point defines the local velocity of the particle at that point and constant velocity implies straight worldlines. Therefore, as illustrated in Fig. 4.5(b), light travels in a straight line in flat spacetime and always on the lightcone because it has constant local velocity, $v = c$, while the worldline for any massive particle must lie inside the local lightcone because it must always have $v \leq c$ (in the jargon, a worldline for a massive particle is always *timelike*). The worldline for the massive particle in Fig. 4.5(a) is curved, indicating an acceleration since the velocity is changing with time. For non-accelerated massive particles the worldline would be straight, but always within the local lightcone.

In the Galilean relativity of Newtonian mechanics an event picks out a hyperplane of simultaneity in the spacetime diagram consisting of all events occurring at the same time as the event. All observers agree on what constitutes this set of simultaneous events because in Galilean relativity simultaneity is independent of the observer. In Einstein's relativity, *simultaneity depends on the observer* and hyperplanes of constant coordinate time have no

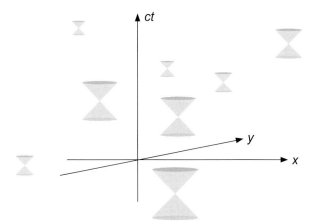

Fig. 4.4 Lightcones are local concepts. Each point of spacetime should be imagined to have its own local lightcone.

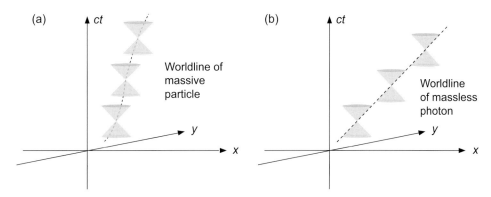

Fig. 4.5 Worldlines for (a) massive (timelike) particles and (b) massless (lightlike) particles. For massive particles the trajectory must always lie inside the local lightcone at each point; for massless particles it must always lie on the lightcone at each point.

invariant meaning. However, *all observers agree on the lightcones associated with events*, because the speed of light is an invariant for all observers. Thus, the *lightcones define an invariant spacetime structure* permitting unambiguous causal classification of events.

4.5 The Causal Structure of Spacetime

As noted above, the causal properties of Minkowski spacetime are encoded in its invariant lightcone structure. Each point in spacetime may be viewed as lying at the apex of a lightcone ("Now" in Fig. 4.3), as illustrated in Fig. 4.4. From Example 4.3, the lightcone defines a locus of points that are connected to the origin by signals moving at light velocity, events inside the lightcone may be connected to the origin by signals moving at less than light velocity, and events outside the lightcone may be connected to the origin only by signals having $v > c$. Therefore, the event at the origin of a local lightcone may influence any event within its forward lightcone (the "Future" in Fig. 4.3) through signals propagating at $v < c$. Likewise, the event at the origin of the lightcone may be influenced by events within its backward lightcone (the "Past" in Fig. 4.3) through signals with speeds less than that of light. Conversely, events at spacelike separations may not be influenced, or have an influence on, the event at the origin except by signals that require $v > c$. Finally, events on the lightcone are connected by signals that travel exactly at c. Thus, events at spacelike separations are causally disconnected and the lightcone is a surface separating the knowable from the unknowable for an observer at the apex of the lightcone.

This lightcone structure of spacetime ensures that all velocities obey locally the constraint $v \leq c$. Since velocities are defined and measured locally, covariant field theories in either flat or curved spacetime are guaranteed to respect the speed limit $v \leq c$, irrespective of whether velocities appear to exceed c globally. For example, in the Hubble expansion of the Universe galaxies beyond a certain distance (the horizon) appear to recede at velocities in excess of c because of the expansion of space, and light coming to us from near the horizon is stretched in wavelength and takes longer to propagate to us than it would in a flat, non-expanding spacetime. However, all local measurements in that expanding, possibly curved spacetime would determine the velocity of light to be c, in accordance with the axioms of special relativity and the associated lightcone structure of spacetime.

Example 4.4 Time machines are of enduring interest in science fiction and in the public imagination. The local lightcones of Minkowski spacetime embody the causal structure of special relativity, and it will be seen later that general relativity inherits the local lightcone structure of Minkowski space. Therefore, as explored further in Box 4.2, lightcone diagrams provide a simple way to answer the question of whether special or general relativity allow going back in time to prevent your own birth? Which in turns prevents you from going back in time to prevent your own birth, . . .

From the preceding discussion we may conclude that the axioms of special relativity are fundamentally at odds with the Newtonian concept of absolute simultaneity, since the demand that light have the same speed for all observers necessarily means that the apparent temporal order of two events depends upon the observer. However, the abolishment of absolute simultaneity introduces *no causal ambiguity* because all observers agree on the lightcone structure of spacetime and hence all observers will agree that event A can cause event B only if A lies in the past lightcone of B, for example.

4.6 Lorentz Transformations in Spacetime Diagrams

It is instructive to examine the action of Lorentz transformations in the spacetime (lightcone) diagram. If consideration is restricted to boosts only in the x direction, the relevant part of the spacetime diagram in some inertial frame corresponds to a plot with axes ct and x, as illustrated in Fig. 4.6.

> **Box 4.2** — **Time Machines and Causality Paradoxes**
>
> When time travel comes up it is usually about going *backward* in time. Traveling *forward* in time requires no special talent and it is easy to arrange various scenarios consistent with relativity where a person could travel into a future time even faster than normal (at least as thought experiments; I leave procurement and engineering details to you!) [111]. For example, in the twin paradox of Example 4.5 it is possible to arrange for the traveler (whether twin or not) to arrive back at Earth centuries in the future relative to clocks that remain on Earth — a kind of time travel. Similar options exist using the gravitational time dilation to be described in Section 9.2. (You'll need a black hole and an unlimited fuel budget, or a planet-size batch of incredibly strong and thin material; but again I leave procurement and engineering to you!) However, the real question is, could you go *back in time* to explore your earlier history?
>
> No! Not according to current understanding. To bend a forward-going timelike worldline continuously into a backward-going one requires going outside the local lightcone, requiring that $v > c$. If *closed timelike loops* were permitted, travel to earlier times might be possible. However, they are forbidden if there are no negative energy densities (see Box 7.3) and the Universe has the topology in evidence. Thus, the determined time traveler has two options: find some negative energy, or find structures with an exotic spacetime topology allowing closed timelike loops. However, negative energy is probably forbidden in classical gravity (it is unclear whether quantum mechanics could provide any loopholes), and there is no evidence at present for exotic spacetime topologies with closed timelike loops. Hence, I would advise against taking a strong investment position in (if you will) time-travel futures! These statements are based entirely on classical gravity considerations; it is unknown at present whether they could be modified by some future understanding of quantum gravity.

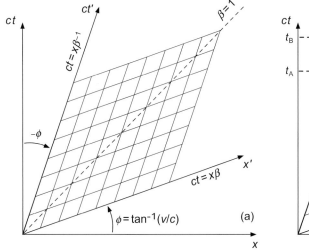
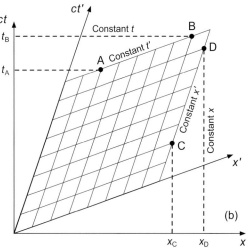

Fig. 4.6 (a) Lorentz boost transformation in a spacetime diagram. (b) Comparison of events in boosted and unboosted reference frames.

4.6.1 Lorentz Boosts and the Lightcone

What happens to the axes in Fig. 4.6 under a Lorentz boost? From the first two of Eqs. (4.26),

$$ct' = c\gamma \left(t - \frac{vx}{c^2}\right) \qquad x' = \gamma(x - vt). \qquad (4.27)$$

The t' axis corresponds to $x' = 0$, which implies from the second of Eqs. (4.27) that $ct = x\beta^{-1}$, with $\beta = v/c$, is the equation of the t' axis in the (ct, x) coordinate system. Likewise, the x' axis corresponds to $t' = 0$, which implies from the first of Eqs. (4.27) that $ct = x\beta$. The x' and ct' axes for the boosted system are also shown in Fig. 4.6(a) for a boost corresponding to a positive value of β. The time and space axes are rotated by the same angle, but in *opposite directions* by the boost (a consequence of the indefinite Minkowski metric). The angle of rotation is related to the boost velocity through $\tan\varphi = v/c$.

Many characteristic features of special relativity are apparent from Fig. 4.6. For example, relativity of simultaneity follows directly, as illustrated in Fig. 4.6(b). Points A and B lie on the same t' line, so they are simultaneous in the boosted frame. But from the dashed projections on the ct axis, in the unboosted frame event A occurs before event B. Likewise, points C and D lie at the same value of x' in the boosted frame and so are spatially congruent, but in the unboosted frame $x_C \neq x_D$. Relativistic time dilation and space contraction effects follow rather directly from these observations, as illustrated in the examples below.

Example 4.5 The time registered by a clock moving between two points in spacetime depends on the path followed, as suggested by Eq. (4.9). That this is true even if the path returns to the initial spatial position is the source of the *twin paradox* of special relativity. Assume twins, initially at rest in the same inertial frame. Twin 2 travels at near lightspeed to a distant star, turns around, and then returns at the same speed to the starting point; Twin 1 remains at the starting point for the entire period. The corresponding spacetime paths are illustrated in the following figure

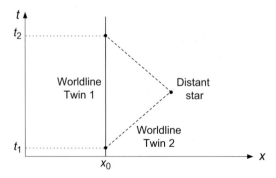

(assuming rapid accelerations at the distant star and at Earth for Twin 2). The elapsed clock time for Twin 1 is $t_2 - t_1$. The elapsed time on the clock carried by Twin 2 is always smaller because of the square root factor in (4.10). The (seeming) paradox arises if things are described from the point of view of Twin 2, who sees Twin 1 move away and then back. This seems to be symmetric with the case of Twin 1 watching Twin 2 move away and then back. But it isn't: the twins travel *different worldlines* (for example, the worldlines differ because Twin 2 experiences accelerations that Twin 1 does not experience), and different distances along these worldlines. Their clocks record the proper time on their respective worldlines and thus differ when they are rejoined, indicating unambiguously that Twin 2 is younger at the end of the journey.

Example 4.6 Consider the following schematic representation of the spacetime diagram illustrated in Fig. 4.6, where a rod of length L_0, as measured in its own rest frame (t, x), is oriented along the x axis.

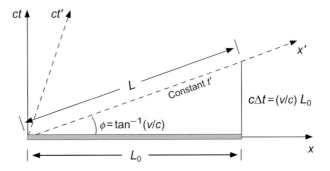

The frame (t', x') is boosted by a velocity v along the $+x$ axis relative to the (t, x) frame. Therefore, in the primed frame the rod will have a velocity v in the negative x' direction. Determining the length L observed in the primed frame requires that the positions of the ends of the rod be measured *simultaneously in that frame*. The axis labeled x' corresponds to constant t' [see Fig. 4.6(b)], so the distance marked as L is the length in the primed frame. This distance *seems* longer than L_0, but this is deceiving because a slice of Minkowski spacetime is being represented on a piece of euclidean paper. (The printer was fresh out of Minkowski-space paper!) Much as a Mercator projection of the globe onto a euclidean sheet of paper gives misleading distance information (Greenland isn't really larger than Brazil), the metric must be trusted to determine the correct distance in a space. From the Minkowski indefinite metric and the triangle in the figure above, $L^2 = L_0^2 - (c\Delta t)^2$. But from Eq. (4.27) it was found that the equation for the x' axis is $c\Delta t = (v/c)L_0$, from which it follows that

$$L = (L_0^2 - (c\Delta t)^2)^{1/2} = \left(L_0^2 - \left(\frac{v}{c}L_0\right)^2\right)^{1/2} = L_0(1 - v^2/c^2)^{1/2},$$

which is the familiar length-contraction formula of special relativity: L is *shorter* than L_0, even though it appears to be longer in the figure above.

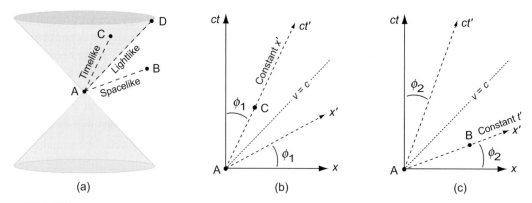

Fig. 4.7 (a) Timelike, lightlike (null), and spacelike separations for events. (b) A Lorentz transformation that brings the timelike separated points A and C of (a) into spatial congruence (they lie along a line of constant x' in the primed coordinate system). (c) A Lorentz transformation that brings the spacelike separated points A and B of (a) into coincidence in time (they lie along a line of constant t' in the primed coordinate system).

4.6.2 Spacelike and Timelike Intervals

As noted previously, the spacetime interval between any two events may be classified in a relativistically invariant way as timelike, lightlike, or spacelike by constructing the lightcone at one of the points, as illustrated in Fig. 4.7(a). This geometry and that of Fig. 4.6 then suggest another important distinction between events at spacelike separations [the line AB in Fig. 4.7(a)] and timelike separations [the line AC in Fig. 4.7(a)]:

(i) If two events have a timelike separation, a Lorentz transformation exists that can bring them into spatial congruence. Figure 4.7(b) illustrates geometrically a coordinate system (ct', x'), related to the original system by an x-axis Lorentz boost of $v/c = \tan \varphi_1$, in which A and C have the same coordinate x'.

(ii) If two events have a spacelike separation, a Lorentz transformation exists that can synchronize the events. Figure 4.7(c) illustrates an x-axis Lorentz boost by $v/c = \tan \varphi_2$ to a system in which A and B have the same time t'.

Maximum values of φ_1 and φ_2 are limited by the $v = c$ line. The Lorentz transformation to bring A into spatial congruence with C exists only if C lies to the left of $v = c$ (C separated by a timelike interval from A). Likewise, the Lorentz transformation to synchronize A with B exists only if B lies to the right of $v = c$ (B separated by a spacelike interval from A).

4.7 Lorentz Covariance of the Maxwell Equations

We conclude this chapter by examining the Lorentz invariance of the Maxwell equations that describe classical electromagnetism. There are several motivations. First, it provides a nice example of how useful Lorentz invariance and Lorentz tensors can be. Second, the properties of the Maxwell equations influenced Einstein strongly in his development of the special theory of relativity. Finally, there are many useful parallels between general relativity and the Maxwell theory, particularly for weak gravity where the Einstein field equations may be linearized. Understanding covariance of the Maxwell equations will prove particularly important when gravitational waves are discussed beginning in Chapter 22.

4.7.1 Maxwell Equations in Noncovariant Form

In free space, using Heaviside–Lorentz, $c = 1$ units (see Appendix B for a discussion of $c = 1$ units), the Maxwell equations may be written as

$$\nabla \cdot \boldsymbol{E} = \rho \tag{4.28}$$

$$\frac{\partial \boldsymbol{B}}{\partial t} + \nabla \times \boldsymbol{E} = 0 \tag{4.29}$$

$$\nabla \cdot \boldsymbol{B} = 0 \tag{4.30}$$

$$\nabla \times \boldsymbol{B} - \frac{\partial \boldsymbol{E}}{\partial t} = \boldsymbol{j}, \tag{4.31}$$

where $\boldsymbol{E}$ is the electric field, $\boldsymbol{B}$ is the magnetic field, ρ is the charge density, and $\boldsymbol{j}$ is the current vector, with the density and current required to satisfy the equation of continuity

$$\frac{\partial \rho}{\partial t} + \nabla \cdot \boldsymbol{j} = 0. \tag{4.32}$$

It is well known that the Maxwell equations are consistent with the special theory of relativity. However, in the form (4.28)–(4.32) this covariance is *not manifest,* since these equations are formulated in terms of 3-vectors and separate derivatives with respect to space and time, not in terms of Minkowski tensors. It proves useful in a number of contexts to reformulate the Maxwell equations in a manner that is manifestly covariant with respect to Lorentz transformations. The usual route to accomplishing this begins by replacing the electric and magnetic fields by new variables.

4.7.2 Scalar and Vector Potentials

The electric and magnetic fields appearing in the Maxwell equations may be eliminated in favor of a vector potential $\boldsymbol{A}$ and a scalar potential φ through the definitions

$$\boldsymbol{B} \equiv \nabla \times \boldsymbol{A} \qquad \boldsymbol{E} \equiv -\nabla \varphi - \frac{\partial \boldsymbol{A}}{\partial t}. \tag{4.33}$$

The vector identities

$$\nabla \cdot (\nabla \times \boldsymbol{B}) = 0 \qquad \nabla \times \nabla \varphi = 0, \tag{4.34}$$

may then be used to show that the second and third Maxwell equations are satisfied identically, and the identity

$$\nabla \times (\nabla \times \boldsymbol{A}) = \nabla(\nabla \cdot \boldsymbol{A}) - \nabla^2 \boldsymbol{A}, \tag{4.35}$$

may be used to write the remaining two Maxwell equations as the coupled second-order equations

$$\nabla^2 \varphi + \frac{\partial}{\partial t} \nabla \cdot A = -\rho \qquad (4.36)$$

$$\nabla^2 A - \frac{\partial^2 A}{\partial t^2} - \nabla \left(\nabla \cdot A + \frac{\partial \varphi}{\partial t} \right) = -j. \qquad (4.37)$$

These equations may then be decoupled by exploiting a fundamental symmetry of electromagnetism termed *gauge invariance*.

4.7.3 Gauge Transformations

Because of the identity $\nabla \times \nabla \varphi = 0$, the simultaneous transformations

$$A \rightarrow A + \nabla \chi \qquad \varphi \rightarrow \varphi - \frac{\partial \chi}{\partial t} \qquad (4.38)$$

for an arbitrary scalar function χ do not change the E and B fields; thus, they leave the Maxwell equations invariant. The transformations (4.38) are termed (classical) *gauge transformations*. This freedom of gauge transformation may be used to decouple Eqs. (4.36)–(4.37). For example, if a set of potentials (A, φ) that satisfy

$$\nabla \cdot A + \frac{\partial \varphi}{\partial t} = 0, \qquad (4.39)$$

is chosen, the equations decouple to yield

$$\nabla^2 \varphi - \frac{\partial^2 \varphi}{\partial t^2} = -\rho \qquad \nabla^2 A - \frac{\partial^2 A}{\partial t^2} = -j, \qquad (4.40)$$

which may be solved independently for A and φ.

A constraint of the sort (4.39) is termed a *gauge condition* and the imposition of such a constraint is termed *fixing the gauge*. This particular choice of gauge that leads to the decoupled equations (4.40) is termed the *Lorentz gauge*. Another common gauge is the *Coulomb gauge*, with a gauge-fixing condition

$$\nabla \cdot A = 0, \qquad (4.41)$$

which leads to the equations

$$\nabla^2 \varphi = -\rho \qquad \nabla^2 A - \frac{\partial^2 A}{\partial^2 t} = \nabla \frac{\partial \varphi}{\partial t} - j. \qquad (4.42)$$

Let us utilize the shorthand notation for derivatives introduced in Eq. (3.49):

$$\partial^\mu \equiv \frac{\partial}{\partial x_\mu} = (\partial^0, \partial^1, \partial^2, \partial^3) = \left(-\frac{\partial}{\partial x^0}, \nabla \right),$$

$$\partial_\mu \equiv \frac{\partial}{\partial x^\mu} = (\partial_0, \partial_1, \partial_2, \partial_3) = \left(\frac{\partial}{\partial x^0}, \nabla \right), \qquad (4.43)$$

where, for example, $\partial^1 = \partial/\partial x_1$ and

$$\nabla \equiv (\partial^1, \partial^2, \partial^3) \qquad (4.44)$$

is the 3-divergence. A compact and covariant formalism then results from introducing the 4-vector potential A^μ, the 4-current j^μ, and the d'Alembertian operator $\Box$ through

$$A^\mu \equiv (\varphi, \boldsymbol{A}) = (A^0, \boldsymbol{A}) \qquad j^\mu \equiv (\rho, \boldsymbol{j}) \qquad \Box \equiv \partial_\mu \partial^\mu. \tag{4.45}$$

Then a gauge transformation takes the form

$$A^\mu \to A^\mu - \partial^\mu \chi \equiv A'^\mu \tag{4.46}$$

and the preceding examples of gauge-fixing constraints become[4]

$$\partial_\mu A^\mu = 0 \quad \text{(Lorentz gauge)} \qquad \nabla \cdot \boldsymbol{A} = 0 \quad \text{(Coulomb gauge)}. \tag{4.47}$$

The operator $\Box$ is Lorentz invariant since

$$\Box' = \partial'_\mu \partial'^\mu = \Lambda^\nu{}_\mu \Lambda_\lambda{}^\mu \partial_\nu \partial^\lambda = \partial_\mu \partial^\mu = \Box.$$

Thus, the Lorentz-gauge wave equation may be expressed in the manifestly covariant form

$$\Box A^\mu = j^\mu \tag{4.48}$$

and the continuity equation (4.32) becomes

$$\partial_\mu j^\mu = 0. \tag{4.49}$$

The covariance of the Maxwell wave equation (4.48) in the Lorentz gauge, coupled with the gauge invariance of electromagnetism, ensures that the Maxwell equations are covariant in all gauges. However, as was seen in the example of the Coulomb gauge, the covariance may not be manifest for a particular choice of gauge.

4.7.4 Maxwell Equations in Manifestly Covariant Form

The Maxwell equations may be cast in a form that is manifestly covariant by appealing to Eqs. (4.33) to construct the components of the electric and magnetic fields in terms of the potentials [Problem 4.15(b)]. Proceeding in this manner, we find that the six independent components of the 3-vectors $\boldsymbol{E}$ and $\boldsymbol{B}$ are elements of an antisymmetric rank-2 *electromagnetic field tensor*

$$F^{\mu\nu} = -F^{\nu\mu} = \partial^\mu A^\nu - \partial^\nu A^\mu, \tag{4.50}$$

which may be expressed in matrix form as

$$F^{\mu\nu} = \begin{pmatrix} 0 & -E^1 & -E^2 & -E^3 \\ E^1 & 0 & -B^3 & B^2 \\ E^2 & B^3 & 0 & -B^1 \\ E^3 & -B^2 & B^1 & 0 \end{pmatrix}. \tag{4.51}$$

That is, the electric field $\boldsymbol{E}$ and the magnetic field $\boldsymbol{B}$ are *vectors* in 3-dimensional euclidean space but in Minkowski space their six components together form an antisymmetric

---

[4] This notation shows explicitly that the Lorentz condition is a covariant constraint because it is formulated in terms of 4-vectors; however, the Coulomb gauge condition is not covariant because it is formulated in terms of only three of the components of a 4-vector.

rank-2 tensor. Now let us employ the Levi-Civita symbol $\epsilon_{\alpha\beta\gamma\delta}$ introduced in Section 3.9, which has the value $+1$ for $\alpha\beta\gamma\delta = 0123$ and cyclic permutations, -1 for odd permutations, and zero if any two indices are equal, and which further satisfies $\epsilon_{\alpha\beta\gamma\delta} = -\epsilon^{\alpha\beta\gamma\delta}$. If the *dual field tensor* $\mathscr{F}^{\mu\nu}$ is then defined by

$$\mathscr{F}^{\mu\nu} = \tfrac{1}{2}\epsilon^{\mu\nu\gamma\delta} F_{\gamma\delta} = \begin{pmatrix} 0 & -B^1 & -B^2 & -B^3 \\ B^1 & 0 & E^3 & -E^2 \\ B^2 & -E^3 & 0 & E^1 \\ B^3 & E^2 & -E^1 & 0 \end{pmatrix}, \qquad (4.52)$$

the Maxwell equations (4.28) and (4.31) may be written

$$\partial_\mu F^{\mu\nu} = j^\nu, \qquad (4.53)$$

and the Maxwell equations (4.29) and (4.30) may be written as

$$\partial_\mu \mathscr{F}^{\mu\nu} = 0. \qquad (4.54)$$

The Maxwell equations in this form are manifestly covariant because they are formulated exclusively in terms of Lorentz tensors.

Background and Further Reading

At the introductory level, see Hartle [111], Cheng [65], Kogut [134], Lambourne [142], and d'Inverno [76]. Kogut [134] has an extensive discussion of the effect of Lorentz transformations on spacetime diagrams. For more advanced discussions of the material in this chapter, see Hobson, Efstathiou and Lasenby [122]; Weinberg [246]; and Misner, Thorne, and Wheeler [156]. An introduction to gauge invariance for the Maxwell equations may be found in Cheng [65] or Guidry [103].

Problems

4.1 Verify explicitly that the Lorentz transformation of Eq. (4.20)

$$cdt' = c\cosh\xi\, dt + \sinh\xi\, dx$$
$$dx' = c\sinh\xi\, dt + \cosh\xi\, dx$$
$$dy' = dy$$
$$dz' = dz$$

leaves invariant the Minkowski line element ds^2. ***

4.2 Consider a simple clock consisting of two mirrors that reflect a light pulse back and forth, with the period of each "tick" equal to the time for light to travel between the mirrors. In the rest frame of the clock the situation is as illustrated on the left side of the following diagram:

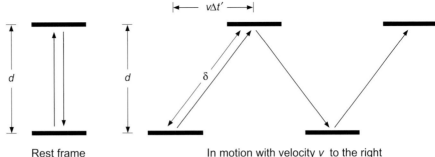

| Rest frame | In motion with velocity v to the right |

with the period of a tick equal to $\Delta t = d/c$. Now suppose the clock is in uniform motion with respect to a stationary observer, who observes the situation shown schematically on the right side of the preceding diagram. Derive the time dilation formula of special relativity from this information.

4.3 Use the Lorentz transformations (4.26) expressed in differential form to obtain the velocity transformation rules consistent with Lorentz invariance. Show that for the special case of two inertial frames moving along the x axis with relative velocity v, the velocity transformation law is

$$u' = \frac{u-v}{1 - uv/c^2},$$

and that this embodies the constancy of the speed of light in all inertial frames, but reduces to the result expected from Galilean invariance for small v.

4.4 Consider the following spacetime triangle:

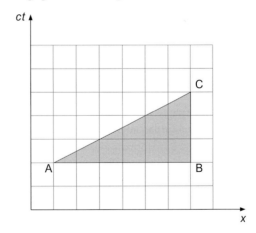

Which side is longest and which is shortest? Give your reasoning. How far is it from A to C by the paths A → C and A → B → C, in units of the grid spacing?

4.5 Show that a lightlike (null) vector in Minkowski space must be orthogonal to itself. *Hint*: What condition indicates that two vectors are orthogonal?

4.6 What is the inverse of the transformation given by Eq. (4.20)? Show that this transformation is equivalent to rotation through an imaginary angle. ***

4.7 Use Lorentz invariance to derive a Doppler-shift equation valid for relativistic velocities of the source.

4.8 Consider the following diagram

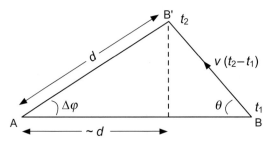

where a distant source at point B is moving at a velocity $v \sim c$ toward the point B'. At time t_1 the source at point B emits a light signal that is detected at time t_1' by an observer at point A. When the source reaches the point B' at time t_2, it emits another light signal that is detected by observer A at time t_2'. You may assume the distance from A to B to be very large compared with the distances associated with transverse motion.

(i) Derive an expression for the time observed at A for the source to appear to move from B to B' and an expression for the apparent transverse velocity β_T (apparent velocity perpendicular to the line of sight) for the source observed at A in terms of the angle θ and the true velocity $\beta = v/c$.

(ii) Show that this apparent velocity is maximized for an angle $\theta_{max} = \cos^{-1}\beta$, find the expression for the maximal apparent velocity, and show that it has no upper bound, even though the actual velocity satisfies $\beta < 1$ [you may find $\sin(\cos^{-1}\beta) = (1 - \beta^2)^{1/2}$ to be useful in this regard].

(iii) Calculate the apparent transverse velocity if $\theta = 10°$ and the actual velocity from B to B' is $\beta = v/c = 0.995$.

This optical illusion of an apparent transverse velocity exceeding that of light is observed frequently in radio astronomy where it is called *superluminal motion*. This is discussed further in Box 15.1. ***

4.9 If a Lorentz transformation is denoted by $\Lambda^\mu{}_\nu$, prove that the Lorentz transformation given by $\Lambda_\mu{}^\nu = \eta_{\mu\alpha}\eta^{\nu\beta}\Lambda^\alpha{}_\beta$ is its inverse.

4.10 Sketch the spacetime diagram corresponding to Fig. 4.6(a), but for the axes x and t plotted in the coordinate system (x', t').

4.11 The primed axes (x', t') plotted in the coordinate system of the unprimed axes in Fig. 4.6 do not appear to be orthogonal, but they must be since the unprimed axes are orthogonal (their scalar product vanishes) and the scalar product is preserved under Lorentz transformations. Show generally that two vectors in a spacetime diagram are orthogonal if each makes the same angle with respect to the lightcone. Use this result to show that a lightlike vector is necessarily orthogonal to itself.

4.12 Two bodies have the same temperature T initially and are found to have identical masses (of order grams). If one body is heated by a small amount to a temperature

$T + \Delta T$ and then an identical force that is small in magnitude is applied to each body, which body accelerates faster and by how much?

4.13 A coordinate system S' moves with constant velocity along the x axis of a second coordinate system S. If, as measured in S, two events occur at the same place and are separated in time by Δt, what is the spatial separation of these two events as measured in S' if they are observed to be separated by a time $\Delta t'$ in the S' frame? *Hint*: Use the invariance of the Minkowski interval.

4.14 Two events in Minkowski space have a timelike separation. Show that the time between the events as measured by any inertial observer is always greater than or equal to the proper time between the events.

4.15 (a) Prove that the Maxwell field tensor $F^{\mu\nu}$ is invariant under a gauge transformation of the 4-vector potential A^{μ}. (b) By appealing to the definitions (4.33), show that the Maxwell field tensor $F^{\mu\nu}$ has components given by Eqs. (4.50) and (4.51). (c) Show that the Maxwell equations written in the covariant form (4.53) and (4.54) are equivalent to the noncovariant form of the Maxwell equations given by Eqs. (4.28)–(4.31). ***

5 Lorentz-Invariant Dynamics

In Chapter 4 the Minkowski metric was introduced and covariance with respect to Lorentz transformations between inertial systems was explored. The basic properties of special relativity: time dilation, space contraction, and relativity of simultaneity, appeared naturally as a direct consequence of the metric structure of Minkowski space, in particular through the requirement that the line element be an invariant. This chapter continues that discussion for flat Minkowski space by considering general properties of worldlines for particles and for light in the Minkowski spacetime. We begin by introducing a set of units that will prove useful in the kinds of problems to be addressed in this and later chapters. These are called *natural units* because they are suggested by the specific nature of the problem being addressed.

5.1 A Natural Set of Units

The speed of light c appears often in equations for flat spacetime and in curved spacetime the gravitational constant G will appear routinely in addition to c. It is convenient to introduce a natural set of units called *geometrized* or $c = G = 1$ units in which c and G take unit value and so do not appear explicitly in equations (or just $c = 1$ units if, as in this chapter, gravity is not considered).[1] Geometrized units are explained in Box 5.1, Examples 5.1–5.3, and Appendix B. They will be used unless restoration of c and G factors is desirable for clarity or emphasis, or it is necessary to calculate a physical quantity in standard units.

Example 5.1 In geometrized units all explicit instances of G and c are dropped in the equations. Calculating quantities in standard units then requires reinserting appropriate combinations of c and G to give the right physical dimensions for each term. For example, it will be shown in Chapter 11 that in geometrized units the Schwarzschild radius defining the event horizon for a spherical black hole is $r_S = 2M$, where M is the mass, so both sides of this equation have dimensions of length in geometrized units. What is the Schwarzschild radius of the Sun? The result may be obtained by inspection since from Box 5.1 the mass of

---

[1] If quantum mechanics is important it may be useful to also set $\hbar = 1$ and in statistical applications the Boltzmann constant is sometimes set to $k_B = 1$. These more general sets of natural units are discussed further in Appendix B.

> **Box 5.1** **Geometrized Units**
>
> Assuming $c = G = 1$ and setting
>
> $$1 = c = 2.9979 \times 10^{10} \text{ cm s}^{-1} \qquad 1 = G = 6.6741 \times 10^{-8} \text{ cm}^3 \text{ g}^{-1} \text{ s}^{-2}$$
>
> permits solving for standard units like seconds in terms of these new units. For example, from the first equation $1\text{s} = 2.9979 \times 10^{10}$ cm and from the second $1\text{g} = 6.6741 \times 10^{-8} \text{ cm}^3 \text{ s}^{-2}$, so
>
> $$1\text{g} = 6.6741 \times 10^{-8} \text{ cm}^3 \left(\frac{1}{2.9979 \times 10^{10} \text{ cm}}\right)^2 = 7.4260 \times 10^{-29} \text{ cm}$$
>
> and both time and mass have the dimension of length in geometrized units. Likewise, the above relations may be used to derive
>
> $$1 \text{ erg} = 1 \text{ g cm}^2 \text{ s}^{-2} = 8.2625 \times 10^{-50} \text{ cm}$$
> $$1 \text{ g cm}^{-3} = 7.4260 \times 10^{-29} \text{ cm}^{-2} \qquad 1 M_\odot = 1.4766 \text{ km},$$
>
> and so on. Velocity is dimensionless in these units (v is measured in units of v/c). See also Examples 5.1–5.3 and Appendix B.

the Sun is 1.4766 km in geometrized units. Thus, for the Sun $r_S = 2M_\odot = 2 \times 1.4766$ km $= 2.95$ km. Alternatively, to convert this equation to CGS units note that $r_S = 2M$ implies that the right side must be multiplied by a combination of G and c having the units of cm g^{-1} to make it dimensionally correct in the CGS system. Clearly this requires the combination G/c^2, so in CGS units the Schwarzschild radius is $r_S = 2GM/c^2$. Table B.1 of Appendix B also may be used to read off the same result: from the table, conversion of $r_s = 2M$ from natural to standard units requires the replacements $r_s \to r_s$ and $M \to GM/c^2$, which gives $r_s = 2GM/c^2$. Finally as a check, if the problem is worked directly in CGS units:

$$r_S^\odot = \frac{2(6.674 \times 10^{-8} \text{ cm}^3 \text{ g}^{-1} \text{ s}^{-2})(1.989 \times 10^{33} \text{ g})}{(3 \times 10^{10} \text{ cm s}^{-1})^2} = 2.95 \times 10^5 \text{ cm},$$

which is the same result as obtained above.

Example 5.2 In Section 9.5 it will be shown that in geometrized units the escape velocity from the radial coordinate R outside a spherical black hole is $v_{\text{esc}} = (2M/R)^{1/2}$. Velocity in the CGS system has units of cm s^{-1}, so to convert to the CGS system the right side of the above equation must be multiplied by a combination of G and c to give this dimensionality. The required factor is clearly $G^{1/2}$ and $v_{\text{esc}} = (2GM/R)^{1/2}$ in CGS units. Equivalently, from Table B.1 conversion of $v_{\text{esc}} = (2M/R)^{1/2}$ from geometrized to standard units requires the replacements $v \to v/c$, $R \to R$, and $M \to GM/c^2$, which gives the same result.

Example 5.3 The world record for the 100 meter dash as of 2017 was a little less than 9.6 seconds (set by Usain Bolt of Jamaica in 2009). What is this in $c = 1$ natural units? We have $c = 1 = 3 \times 10^8$ ms^{-1}. Therefore, 1 second is equal to 3×10^8 meters and 9.6 seconds is $9.6 \times (3 \times 10^8) = 2.88 \times 10^9$ meters. The meaning is perfectly sensible when c is restored: the world record in the 100 meter dash is equal to the time for light to travel 2.88×10^9 meters, which is 9.6 seconds. But be warned! As useful as these units are in a scientific context, your friends may consider you to be rather eccentric, or even a bit dangerous, if you insist on discussing sporting events in such units.

5.2 Velocity and Momentum for Massive Particles

Particles with finite mass follow timelike worldlines. The worldline for a particle is conveniently parameterized in terms of a variable that changes continuously along its worldline. For timelike trajectories the natural choice is the proper time τ, which is the time that would be measured by a clock carried along the worldline. The equation of the worldline may then be expressed as $x^\mu = x^\mu(\tau)$, and a velocity 4-vector (the *4-velocity*) may be defined by

$$u^\mu = \left(\frac{dx^0}{d\tau}, \frac{dx^1}{d\tau}, \frac{dx^2}{d\tau}, \frac{dx^3}{d\tau} \right), \tag{5.1}$$

where the proper time interval $d\tau$ is related to the spacetime interval ds through

$$d\tau^2 = -ds^2, \tag{5.2}$$

and the coordinate time interval dt and the proper time interval $d\tau$ are related through (see Example 4.2)

$$d\tau = dt \left(1 - v^2\right)^{1/2}, \tag{5.3}$$

where $\boldsymbol{v}$ is the 3-velocity, $v^i = dx^i/dt$. (Remember: units chosen so that $c = 1$ are being employed! This would read $d\tau = dt(1 - v^2/c^2)^{1/2}$ in standard "engineering" units.)[2]

The 4-velocity is tangent to the worldline of a particle at any point and lies within the forward lightcone, since the displacement is given by $\Delta x^\mu = u^\mu \Delta \tau$. Figure 5.1 illustrates. Then,

$$u^0 = \frac{dx^0}{d\tau} = \frac{dt}{d\tau} = \left(1 - v^2\right)^{-1/2} \qquad u^i = \frac{dx^i}{d\tau} = \frac{dx^i}{dt}\frac{dt}{d\tau} = v_i \left(1 - v^2\right)^{-1/2}$$

so that for the components of the 4-velocity

$$u^\mu = (\gamma, \gamma \boldsymbol{v}), \tag{5.4}$$

where the Lorentz γ-factor is defined by

$$\gamma = \left(1 - v^2\right)^{-1/2} \tag{5.5}$$

in $c = 1$ units. The scalar product of u with itself gives the normalization

$$u \cdot u = \eta_{\mu\nu} u^\mu u^\nu = \eta_{\mu\nu} \frac{dx^\mu}{d\tau} \frac{dx^\nu}{d\tau} = -1, \tag{5.6}$$

---

[2] For 3-space quantities like $\boldsymbol{v}$ it is possible to define an orthogonal coordinate system in which the distinction between upper and lower indices is not important. Thus, it is not uncommon in the literature to see the 3-velocity components written as v_i rather than v^i.

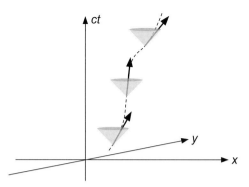

Fig. 5.1 The 4-velocity along a timelike worldline.

where Eqs. (4.5) and (5.2) were used. A corresponding 4-momentum may be defined by

$$p^\mu = mu^\mu, \tag{5.7}$$

where m is the rest mass. The equation of motion is then

$$m\frac{du}{d\tau} = \frac{dp}{d\tau} = f,$$

where $du/d\tau$ is the 4-acceleration and f is the 4-force. From (5.7) and (5.6), the normalization of the 4-momentum is

$$p^2 \equiv p \cdot p = m^2 u \cdot u = -m^2, \tag{5.8}$$

and by virtue of Eq. (5.4) the components of the 4-momentum are

$$p^\mu = (E, \boldsymbol{p}) = (m\gamma, m\gamma \boldsymbol{v}). \tag{5.9}$$

Since $p_\mu = (-E, \boldsymbol{p})$, Eq. (5.8) can be expressed as

$$E = \sqrt{\boldsymbol{p}^2 + m^2}, \tag{5.10}$$

which is just the familiar Einstein relation written in units for which $c = 1$.

5.3 Geodesics and a Variational Principle

A metric allows the definition of *geodesics*: paths that represent the shortest distance between any two points. In flat spaces one has the aphorism that "the shortest distance between two points is a straight line." Thus, the geodesics in euclidean space are given by

$$\frac{d^2\boldsymbol{r}}{dt^2} = 0, \quad \text{(Newton's first law)} \tag{5.11}$$

and those of Minkowski space are given by

$$\frac{d^2 t}{d\tau^2} = 0 \qquad \frac{d^2\boldsymbol{r}}{d\tau^2} = 0. \tag{5.12}$$

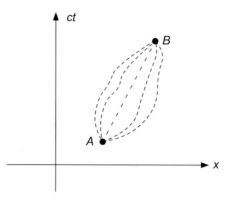

Fig. 5.2 *Principle of Extremal Proper Time*: of all the possible classical paths between two points in spacetime, the one taken by a free particle extremizes the proper time.

In both cases, the geodesics are straight lines and correspond to the motion expected for free particles with no forces acting on them. Thus Minkowski space is *not euclidean* because its indefinite metric isn't euclidean but it is *flat*, just as euclidean space is. However, gravitation is associated with non-euclidean curved spacetime and everyday experience living on the 2-dimensional surface of a spherical planet suggests that the geodesics in curved spaces generally will not be straight lines. We shall see later that geodesics also can be defined as the "straightest possible lines" within a curved space. Since the squares of spacetime intervals can be positive, negative, or zero, this is sometimes a more convenient specification than one defined in terms of shortest distances.

We shall now show that the motion of free particles in spacetime is governed by a simply stated principle:[3]

> **Principle of Extremal Proper Time:** The worldline for free particles between points separated by timelike intervals extremizes the proper time.

This principle is illustrated in Fig. 5.2, and a simple example where the optimal path is one of least time rather than least distance is given in Fig. 5.3. The proper time between the points A and B in Fig. 5.2 can be computed from Eq. (4.5) as,

$$\tau_{AB} = \int_A^B (dt^2 - dx^2 - dy^2 - dz^2)^{1/2}. \tag{5.13}$$

The path may be parameterized by a variable σ that varies continuously from 0 to 1 as the particle moves from A to B. Thus, the path is specified by $x^\mu = x^\mu(\sigma)$ and

$$\tau_{AB} = \int_0^1 \left[\left(\frac{dt}{d\sigma}\right)^2 - \left(\frac{dx}{d\sigma}\right)^2 - \left(\frac{dy}{d\sigma}\right)^2 - \left(\frac{dz}{d\sigma}\right)^2 \right]^{1/2} d\sigma. \tag{5.14}$$

Finding the path that extremizes this proper time is a standard variational problem with a standard solution that is illustrated in Box 5.2. The condition for an extremum is that

---

[3] Or, in a delightfully concise phrasing, "Spacetime shouts 'Go straight!' The free stone obeys... The stone's wristwatch verifies that its path is straight" [231].

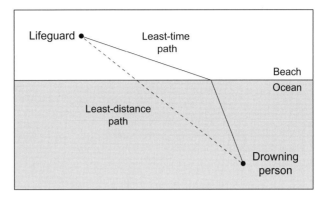

Fig. 5.3 Thought exercise, due to Richard Feynman, for which it is optimal to extremize time rather than distance. If the lifeguard can run faster on the beach than swim in the water, the optimal path is not a straight line but rather one like the solid line that accounts for the trade-off between distance and differing speeds on the beach and in the water. To save the swimmer in distress it is crucial to minimize the time rather than the distance traveled.

$$\delta \int d\tau = 0. \tag{5.15}$$

Defining a *Lagrangian* L as

$$L = \left(-g_{\mu\nu}\frac{dx^\mu}{d\sigma}\frac{dx^\nu}{d\sigma}\right)^{1/2}, \tag{5.16}$$

where $g_{\mu\nu}$ is the metric tensor for the (not necessarily flat) space, so that

$$\tau_{AB} = \int_0^1 L\, d\sigma, \tag{5.17}$$

then leads to the *Euler–Lagrange equation of motion*

$$-\frac{d}{d\sigma}\left(\frac{\partial L}{\partial(dx^\mu/d\sigma)}\right) + \frac{\partial L}{\partial x^\mu} = 0. \tag{5.18}$$

Therefore, satisfying the variational condition (5.15) is equivalent to requiring that the Lagrangian L defined in Eq. (5.16) obey the differential equation (5.18).

Example 5.4 Consider $x^\mu = x^1$ in Minkowski space. For constant $g_{\mu\nu} = \eta_{\mu\nu}$ the Lagrangian L does not depend on x^1 and the last term in Eq. (5.18) vanishes, giving for the Euler–Lagrange equation

$$\frac{d}{d\sigma}\left(\frac{1}{L}\frac{dx^1}{d\sigma}\right) = 0.$$

Utilizing Eq. (5.17) in differential form and multiplying through by $d\sigma/d\tau$ gives $d^2 x^1/d\tau^2 = 0$. Applying similar steps to the other terms then gives (Problem 5.1)

$$\frac{d^2 x^\mu}{d\tau^2} = 0, \tag{5.19}$$

which is Eq. (5.12). Therefore, the variational principle (5.15) implies that geodesics in Minkowski space are straight lines.

> **Box 5.2**   **The Euler–Lagrange Equations**
>
> Consider the paths in spacetime between the fixed points labeled A and B in Fig. 5.2. For a function $L(x^\mu(\sigma), \dot{x}^\mu(\sigma))$, where σ parameterizes the path and $\dot{x}^\mu \equiv dx^\mu/d\sigma$, define the path integral
>
> $$S = \int_A^B L\left(x^\mu(\sigma), \dot{x}^\mu(\sigma)\right) d\sigma.$$
>
> For an arbitrary small variation in the path $x^\mu(\sigma) \to x^\mu(\sigma) + \delta x^\mu(\sigma)$, the corresponding variation in the value of the integral is
>
> $$\delta S \equiv \int_A^B \delta L \, d\sigma = \int_A^B \left(\frac{\partial L}{\partial \dot{x}^\mu(\sigma)} \delta \dot{x}^\mu(\sigma) + \frac{\partial L}{\partial x^\mu(\sigma)} \delta x^\mu(\sigma) \right) d\sigma.$$
>
> The first term can be integrated by parts, giving
>
> $$\delta S = \frac{\partial L}{\partial \dot{x}^\mu(\sigma)} \delta x^\mu(\sigma) \bigg|_A^B + \int_A^B \left[-\frac{d}{d\sigma}\left(\frac{\partial L}{\partial \dot{x}^\mu(\sigma)}\right) + \frac{\partial L}{\partial x^\mu(\sigma)} \right] \delta x^\mu(\sigma) d\sigma,$$
>
> but the variation $\delta x^\mu(\sigma)$ vanishes at the fixed endpoints so the first term is zero and
>
> $$\delta S = \int_A^B \left[-\frac{d}{d\sigma}\left(\frac{\partial L}{\partial \dot{x}^\mu(\sigma)}\right) + \frac{\partial L}{\partial x^\mu(\sigma)} \right] \delta x^\mu(\sigma) d\sigma.$$
>
> For paths that extremize the integral $\delta S = 0$ and for arbitrary variations $\delta x^\mu(\sigma)$ this will be true only if the expression inside the square brackets of the above integrand vanishes. Thus
>
> $$-\frac{d}{d\sigma}\left(\frac{\partial L}{\partial \dot{x}^\mu(\sigma)}\right) + \frac{\partial L}{\partial x^\mu(\sigma)} = 0,$$
>
> which is the *Euler–Lagrange equation* (5.18). Satisfying the variational condition $\delta S = 0$ characterizing an extremal path is equivalent to requiring that the function L satisfy the Euler–Lagrange equation.

5.4 Light and other Massless Particles

For particles moving at lightspeed the rest mass is identically zero. They move on the lightcone with the proper time between two points given by $d\tau^2 = -ds^2 = 0$. Thus, τ is not a useful parameter for the worldline of photons and other massless particles and the 4-velocity (5.1) is undefined for such particles.[4]

5.4.1 Affine Parameters

The proper time is not an appropriate parameter for photons but the curve $x = t$ (corresponding to $v = c$ in $c = 1$ units) may be written parametrically as $x^\mu = u^\mu \lambda$,

---

[4] A vector tangent to the worldline for a photon can be defined easily enough; the complication is to define one of unit length since points on the lightcone are separated by a null distance. Ultimately this follows because in special relativity there is no reference frame in which the photon is at rest [221].

where $u^\mu = (1, 1, 0, 0)$ is a tangent 4-vector, $u^\mu = dx^\mu/d\lambda$, and λ is a parameter [compare Eq. (5.1)]. But u^μ lies on the lightcone so it is a null vector ($ds^2 = 0$). Thus, by steps analogous to those leading to (5.6), it must satisfy

$$u \cdot u = 0. \tag{5.20}$$

With this choice of parameterization the equation of motion for the light ray may be put into the same form as that for a massive particle

$$\frac{du}{d\lambda} = 0, \tag{5.21}$$

which is analogous to Newton's first law. Parameters for which this is true are termed *affine parameters*. Affine parameters are convenient for light rays because they lead to equations of motion that mimic those for timelike particle trajectories. The primary differences between the motion of massive particles and of massless particles will be associated with the difference between Eqs. (5.6) and (5.20).

5.4.2 Energy and Momentum

The energy E and 3-momentum $\boldsymbol{p}$ of photons are given by

$$E = \hbar\omega \qquad \boldsymbol{p} = \hbar\boldsymbol{k}, \tag{5.22}$$

where ω is the frequency and $\boldsymbol{k}$ is the wavevector. Thus,

$$p^\mu = (E, \boldsymbol{p}) = \hbar(\omega, \boldsymbol{k}) = \hbar k^\mu. \tag{5.23}$$

Since photons are massless, the 4-momentum obeys

$$p \cdot p = 0. \tag{5.24}$$

The equations of motion for photons (5.21) may also be expressed in terms of the 4-momentum, $dp/d\lambda = 0$, where λ is an affine parameter.

5.5 Observers

To test theory against data it is convenient to introduce idealized *observers*. Traditionally these observers are armed with measuring rods and clocks as the instruments of their trade. In Minkowski space for a laboratory assumed to be in an inertial frame measurements may be referenced to a set of axes defined for the laboratory, and this set may be taken to be global (one set of axes can be chosen that applies to the entire spacetime). However, as will be discussed in Section 6.7, inertial frames can be defined *locally* around any spacetime point but *curved spacetime precludes the existence of global inertial frames*. Hence it is important to define observers and their observations in a manner that does not assume the existence of global inertial frames.

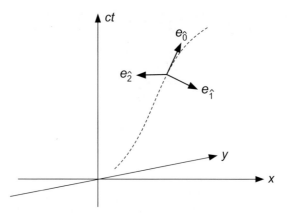

Fig. 5.4 Unit vectors of a local coordinate system at a point on an observer's worldline for two space and one time dimension. It is convenient to choose an orthonormal rather than our usual coordinate basis for the laboratory, as indicated explicitly by the hats on the basis vector indices (see Section 3.4.1).

An observer moving through spacetime may be thought of as occupying a local laboratory moving on a (timelike) worldline. The laboratory carries four orthogonal unit vectors $e_{\hat{0}}$, $e_{\hat{1}}$, $e_{\hat{2}}$, and $e_{\hat{3}}$ that specify a local, orthonormal coordinate system. This defines (locally) a time direction and three space directions to which the observer can reference all measurements; Fig. 5.4 illustrates. The timelike component $e_{\hat{0}}$ will be tangent to the observer's worldline (the observer's clock is moving in that direction if it is at rest in the laboratory). Since the 4-velocity u of the observer is a unit tangent vector ($u \cdot u = -1$), the unit vector in the time direction is $e_{\hat{0}} = u$. Any mutually orthogonal set of three unit spatial vectors that also are orthogonal to e_0 may be chosen to complete the set. Observers measure components of 4-vectors in the laboratory coordinate system, which correspond to scalar products with the basis 4-vectors. For example, the 4-momentum of a particle is $p = p^{\hat{\mu}} e_{\hat{\mu}}$, so the energy measured for the particle by an observer with 4-velocity $u = e_{\hat{0}}$ is

$$E = p^0 = -p \cdot e_{\hat{0}}, \qquad (5.25)$$

which is valid even for accelerated observers (see Problem 5.13).

5.6 Isometries and Killing Vectors

In differential geometry *Killing vectors,* which are associated with the German mathematician Wilhelm Killing (1847–1923), are a standard tool for analyzing symmetries such as those that arise as conservation laws (constants of motion) in the usual Lagrangian or Hamiltonian formulations of mechanics. An introduction to Killing vector fields may be found in Box 5.3.

> **Box 5.3** **Killing Vector Fields**
>
> As discussed in Section 5.6.1, a metric is said to have a symmetry (an isometry) if some transformation leaves it unchanged, and this implies that some quantity is conserved. This box and Section 5.6.2 deal with identifying and exploiting that symmetry. Consider the transformation law (3.36) applied to the metric tensor with primed and unprimed coordinates interchanged,
>
> $$g_{\mu\nu}(x) = \frac{\partial x'^{\alpha}}{\partial x^{\mu}} \frac{\partial x'^{\beta}}{\partial x^{\nu}} g'_{\alpha\beta}(x').$$
>
> Assume an infinitesimal transformation in the direction of symmetry
>
> $$x^{\mu} \to x'^{\mu} = x^{\mu} + \epsilon K^{\mu},$$
>
> where ϵ is small and K^{μ} is a vector field. As the transformation is in a direction of symmetry the metric retains the same functional form so $g'_{\mu\nu}(x') = g_{\mu\nu}(x')$. Making this replacement, substituting the second equation into the first, and expanding $g_{\alpha\beta}(x')$ in a Taylor series around $g_{\alpha\beta}(x)$ gives (see Problem 5.11)
>
> $$g_{\mu\nu}(x) = g_{\mu\nu}(x) + \epsilon \left[K^{\gamma} \partial_{\gamma} g_{\mu\nu}(x) + \partial_{\mu} K^{\beta} g_{\beta\nu} + \partial_{\nu} K^{\beta} g_{\mu\beta} + \mathcal{O}\left(\epsilon^2\right) \right].$$
>
> Since ϵ is arbitrary, this can hold generally only if the quantity in square brackets vanishes, which leads to
>
> $$\partial_{\nu} K_{\mu} + \partial_{\mu} K_{\nu} + K^{\gamma} \partial_{\gamma} g_{\mu\nu} = 0.$$
>
> But (as you are asked to show in Problem 5.12) from the formula for the Lie derivative given by Eq. (3.74)
>
> $$\mathscr{L}_K g_{\mu\nu} = \partial_{\nu} K_{\mu} + \partial_{\mu} K_{\nu} + K^{\gamma} \partial_{\gamma} g_{\mu\nu}.$$
>
> This equation, the basic properties of Lie derivatives and covariant derivatives, and the preceding equation lead to
>
> $$\nabla_{\nu} K_{\mu} + \nabla_{\mu} K_{\nu} = \partial_{\nu} K_{\mu} + \partial_{\mu} K_{\nu} = 0,$$
>
> which is known as *Killing's equation*. The vector fields K that solve it are the Killing vector fields associated with this isometry. Generally for such a vector field a basis may be introduced such that vectors may be put in the form (5.27) corresponding to a unit vector in a direction of isometry.

5.6.1 Symmetries of the Metric

In all spacetimes, whether flat or not, one constant of motion for timelike worldlines may be deduced from the normalization (5.6) of the 4-velocity $u^{\mu} = dx^{\mu}/d\tau$,

$$u \cdot u = g_{\mu\nu} u^{\mu} u^{\nu} = -1. \tag{5.26}$$

For a general metric (not necessarily restricted to flat spacetime) this will be the only constant of motion, corresponding to the preservation of $u \cdot u$. If there are additional constants of motion, they must arise from specific symmetries in the problem.

> **Box 5.4** **Noether's Theorem**
>
> That continuous symmetries imply conservation laws is a qualitative statement of *Noether's theorem,* proposed in 1918 by German mathematician Emmy Noether (1882–1935). Noether's theorem – as for many transformative ideas – is deceptively simple, but it lies at the heart of many of the advances in modern quantum field theory that involve symmetries (that is, most of them).
>
> In Erlangen in the late 1800s women could not register for university math classes but Noether's father was a math professor who let her audit his lectures. She took an exam on which her score was so high that she was awarded an undergraduate degree permitting her to enter graduate school. She earned her doctorate in three years in 1907, one of the first women to attain a mathematics doctorate in Germany. Attempts to hire her in a teaching position were blocked because she was female but she worked as an unpaid research assistant and published papers, including her famous theorem in 1918. She was Jewish and was forced to migrate from Germany to the United States in 1933, where she became a professor at Bryn Mawr College. She died unexpectedly at age 53 from surgery complications.
>
> Publications on quantum field theory are replete with references to Noether's theorem, and Einstein regarded Noether as the most important woman mathematician ever. Certainly few mathematicians, male or female, have had as much influence on the development of modern physics. Although famous in physics for her theorem, in mathematics Noether is best known for her work in abstract algebra.

By virtue of *Noether's theorem* (see Box 5.4), in ordinary classical and quantum mechanics continuous symmetries imply conservation laws. For example, energy conservation is implied by a force derivable from a potential (a conservative force) that is independent of time, conservation of linear momentum in the direction x is implied by the potential being constant in that direction, and conservation of angular momentum follows from a potential that is spherically symmetric. If a spacetime metric has a symmetry (a transformation that leaves the metric invariant), that too will generally imply that some quantity is conserved. As an example, suppose that the metric is independent of one of the spacetime coordinates, say x^0, such that $x^0 \to x^0 + constant$ leaves the metric unchanged. For such a symmetry (called an *isometry*),[5] a unit vector may be defined that points along the direction in which the metric is constant, in this case,

$$K^\mu = (1,\ 0,\ 0,\ 0), \qquad (5.27)$$

which is termed the *Killing vector* associated with the symmetry. For example, in flat space described by cartesian coordinates, $ds^2 = dx^2 + dy^2 + dz^2$ and conservation of the components of linear momentum is associated with the three Killing vectors $(1, 0, 0)$, $(0, 1, 0)$, and $(0, 0, 1)$ indicating invariance under translations in the x, y, and z directions, respectively. In most of the relatively simple examples to be discussed here the symmetry

---

[5] Mathematically, an isometry group (the theory of groups is discussed in Box 7.1) is the set of one-to-one, distance-preserving maps from a metric space onto itself. *Example:* the isometry group of the sphere S^2 is the orthogonal group O(3) of 3D rotations and reflections, since the sphere transforms into itself under those operations.

and direction of symmetry will be obvious intuitively, but in Box 5.3 a more systematic approach to identifying the Killing vectors for a spacetime is described.

5.6.2 Quantities Conserved along Geodesics

The symmetry implied by a spacetime Killing vector means that *some quantity is conserved along a geodesic*. As noted in Section 5.3, the timelike geodesics in a spacetime may be found by extremizing the proper time according to Eq. (5.15), which implies the Euler–Lagrange equation (5.18) with a Lagrangian given by Eq. (5.16). Suppose the metric to be independent of the coordinate x^1. Then $\partial L/\partial x^1 = 0$ and utilizing $L d\sigma = d\tau$ [see Eq. (5.17)], Eqs. (5.16) and (5.1), and that $g_{\mu\nu}$ is symmetric under exchange of its indices [111],

$$\frac{\partial L}{\partial (dx^1/d\sigma)} = -\frac{g_{1\mu}}{L}\frac{dx^\mu}{d\sigma} = -g_{\alpha\mu}K^\alpha u^\mu = -K \cdot u, \qquad (5.28)$$

where $K^\alpha = (0, 1, 0, 0)$ is the Killing vector associated with the metric being independent of the coordinate x^1. Then the condition implied by Eq. (5.18) reduces to

$$\frac{d}{d\sigma}(K \cdot u) = 0 \qquad (5.29)$$

and $K \cdot u$ is conserved along a geodesic if K is a Killing vector associated with an isometry of the spacetime. As will be demonstrated in later examples, this observation can be used to greatly simplify problems if the metric has a high degree of spacetime symmetry.

Background and Further Reading

See Hartle [111]; Cheng [65]; and d'Inverno [76] for introductions. For more advanced discussions, see Schutz [221]; Weinberg [246]; and Misner, Thorne, and Wheeler [156].

Problems

5.1 Prove, by extremizing the distance interval along the worldline $\delta \int ds = 0$, that in special relativity free particles move with uniform velocity on geodesics that are straight lines. ***

5.2 Consider a particle of rest mass m at rest in an inertial frame. An observer moves through the inertial frame with velocity v and the worldlines of the particle and observer intersect. What energy does the observer detect for the particle? (You can probably guess the answer, but work it out formally to gain experience with calculating observable quantities.)

5.3 (a) Convert 1 Joule to $c = 1$ units. (b) What is a pressure of 1 atmosphere (10^5 Nm^{-2}) in $c = 1$ units?

5.4 (a) Convert an acceleration of 1 m^{-1} in $c = 1$ units to SI units. (b) What is an energy density of 2 kgm^{-3} in $c = 1$ units equal to in SI units?

5.5 Supply the details in the derivation of Eq. (5.29).

5.6 Verify that $m\gamma$ in Eq. (5.9) behaves like the total energy at low velocity.

5.7 Consider a set of units where the gravitational constant G, the speed of light c, and Boltzmann's constant k_B are all chosen to be equal to one (this is a common choice in cosmology). (a) In astrophysics pressure is commonly quoted in units of dyne cm^{-2}. What is 1 dyne cm^{-2} in the current units? (b) In nuclear physics, elementary particle physics, and some branches of astrophysics a common unit of energy is MeV (million electron volts). Use the current set of units to express 1 MeV in ergs, in grams, in centimeters, and in kelvin (degrees).

5.8 In Section 23.2.1 it will be shown that the the gravitational wave luminosity L produced by revolution of a binary star system having components with masses m_1 and m_2, and period P can be written in geometrized units as

$$L = \frac{128}{5} 4^{2/3} M^{4/3} \mu^2 \left(\frac{\pi}{P}\right)^{10/3},$$

where $M = m_1 + m_2$ and $\mu = m_1 m_2/(m_1 + m_2)$. Supply the factors of c and G to convert this expression into standard (CGS) units with the luminosity in ergs^{-1}. ***

5.9 Assume a Lagrangian $L(x, \dot{x}) = \frac{1}{2} m \dot{x}^2 - V(x)$, where $V(x)$ is the potential energy. Show that extremizing the classical action $S = \int L dt$ for this Lagrangian leads to Newton's second law of motion.

5.10 Assume an accelerated worldline in 2-dimensional Minkowski spacetime (t, x) with the parameterization

$$x(\sigma) = a \cosh \sigma \qquad t(\sigma) = a \sinh \sigma,$$

where a is a constant and the parameter σ varies from $-\infty$ to $+\infty$. Plot the worldline in the x–t plane. Reparameterize the equations using the proper time τ rather than σ, assuming that $\tau = 0$ when $\sigma = 0$. Write expressions for the 4-velocity components as a function of τ. Show that the resulting 4-velocity has the proper normalization given by (5.6), and that the spatial component of the velocity approaches c asymptotically but never reaches it.

5.11 In support of the discussion in Box 5.3, use

$$g_{\mu\nu}(x) = \frac{\partial x'^{\alpha}}{\partial x^{\mu}} \frac{\partial x'^{\beta}}{\partial x^{\nu}} g_{\alpha\beta}(x')$$

and the transformation $x^{\mu} \to x'^{\mu} = x^{\mu} + \epsilon K^{\mu}$ to show that to first order in the small parameter ϵ,

$$g_{\mu\nu}(x) = \left(\delta^{\alpha}_{\mu} + \epsilon \partial_{\mu} K^{\alpha}\right) \left(\delta^{\beta}_{\nu} + \epsilon \partial_{\nu} K^{\beta}\right) g_{\alpha\beta}(x').$$

Then, by expanding $g_{\alpha\beta}(x')$ in a Taylor series around $g_{\alpha\beta}(x)$, show that to $\mathcal{O}(\epsilon)$ the above equation implies that

$$\partial_{\nu} K_{\mu} + \partial_{\mu} K_{\nu} + K^{\gamma} \partial_{\gamma} g_{\mu\nu} = 0$$

for the vector field K^{α}. ***

5.12 Show that the equation

$$\partial_\nu K_\mu + \partial_\mu K_\nu + K^\gamma \partial_\gamma g_{\mu\nu} = 0$$

in Box 5.3 is equivalent to Killing's equation

$$\nabla_\nu K_\mu + \nabla_\mu K_\nu = 0.$$

Hint: Take the Lie derivative of the metric tensor $g_{\mu\nu}$ (see Section 3.13.5) and use that for metric spaces (1) the covariant derivative of the metric tensor vanishes, and (2) partial derivatives and covariant derivatives are equivalent in the expression for the Lie derivative. ***

5.13 Show that for an accelerated observer in Minkowski space following the worldline described in Problem 5.10, a photon emitted by a star at frequency ω_0 in the star's rest frame will be observed to have a frequency $\omega = \omega_0 e^{-\tau/a}$, where τ is the proper time. *Hint*: This problem will be easiest if you solve Problem 5.10 first, work in the rest frame of the star, and use Eq. (5.25). ***

6 The Principle of Equivalence

The general theory of relativity rests upon two principles that are in fact related: the *principle of equivalence* and the *principle of general covariance*. The principle of equivalence asserts that under suitable conditions gravitational acceleration is indistinguishable from acceleration by any other force, while general covariance requires that the laws of physics retain their form under the most general coordinate transformations. This chapter introduces the principle of equivalence and Chapter 7 takes up the principle of general covariance in curved spacetime. Then, in Chapter 8 the principle of equivalence will be used to enforce the principle of general covariance for gravitational fields, which will lead to the general theory of relativity.

6.1 Einstein and Equivalence

As we have introduced already in the discussion of Section 1.8, Einstein came to realize the significance of what we now call the equivalence principle when the thought came to him in 1907 that if he were in free fall in a gravitational field he would not be able to feel his own weight. Einstein said that this was the happiest thought of his life, for he saw in it the idea that would eventually allow him to formulate a theory of gravity that is consistent with the principles of relativity.

Until that time, Einstein had struggled with understanding how to modify special relativity to create a covariant theory of gravity, with a major stumbling block being that special relativity is formulated in terms of Lorentz transformations between inertial frames that can be defined globally, but a gravitational field does not permit the notion of global inertial frames. In the equivalence principle Einstein saw a possible way around this impasse. He did not yet have sufficient understanding of the required mathematics so he groped his way – not always forward – for eight years (often with the aid of his friend, the mathematician Marcel Grossmann), with the equivalence principle serving often as a guide, until finally in late 1915 he was able to forge his intuitive ideas into a mathematically consistent theory of gravity.

Historians debate the importance of the equivalence principle and whether Einstein could have formulated general relativity without its aid. It seems fair to say that the equivalence principle at least hastened the formulation of general relativity, both because of the intuitive guidance that it offered, and because Einstein was able to derive some results of general relativity (the deflection and redshift of light in a weak gravitational field, for example) using equivalence arguments even before he was able to construct the field equations of general relativity.

6.2 Inertial and Gravitational Mass

The principle of equivalence originates in the observation that in Newtonian physics there are two ways in which the mass enters dynamical equations:

1. The *inertial mass* m_i is defined through Newton's second law of motion: $m_i = F/a$, where F is the magnitude of the force and a is the magnitude of the acceleration.
2. The *gravitational mass* m_g is defined through Newton's law of gravitation: $m_g = r^2 F/GM$, where r is the separation of the mass from a gravitating sphere like the Earth, F is the magnitude of the gravitational force, M is the mass of the gravitating sphere, and G is the gravitational constant.

These definitions are highly asymmetric in scope. The inertial mass is defined in terms of a response to *any* force, while the gravitational mass is defined in terms of response only to a particular force, gravity. The quantitative relationship between the inertial and gravitational masses for a given object was suggested by Galileo's experiments with inclined planes showing that different objects fall at the same rate in a gravitational field, but was first established to high precision in the Eötvös experiment of 1893, which is illustrated in Fig. 6.1.

Two equal weights composed of different material A and B are suspended from a sensitive torsion balance. Except at the Earth's equator and the poles, if the inertial and gravitational masses differ a couple will be produced by the action on the inertial mass of the centrifugal force associated with Earth's rotation. The results of such tests of the equivalence principle are often reported in terms of the *Eötvös parameter* η, which is defined by the difference in acceleration Δa relative to the average gravitational acceleration a_g for spheres made from two substances A and B,

$$\eta(A, B) = \frac{\Delta a}{a_g} = \frac{\left(m_g/m_i\right)_A - \left(m_g/m_i\right)_B}{\frac{1}{2}\left[\left(m_g/m_i\right)_A + \left(m_g/m_i\right)_B\right]}. \tag{6.1}$$

The Eötvös experiments established that the inertial and gravitational masses are proportional to each other with a measured $\eta \sim 10^{-9}$. More recently this equivalence has been extended to a precision $\eta \sim 10^{-13}$ by improved torsion balance experiments [217], and

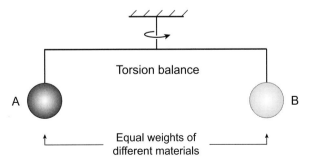

Fig. 6.1 The Eötvös experiment to measure the difference between gravitational and inertial mass.

by using laser rangefinding to track precisely the distance between the Earth and Moon as they are accelerated in the Sun's gravitational field [165, 254]. By a suitable choice of units this proportionality becomes an equality and we arrive at what is sometimes termed the *weak principle of equivalence*:

$$m_i = m_g. \tag{6.2}$$

An equally valid statement of weak equivalence is that all particles experience the same acceleration in a gravitational field. The gravitational acceleration is given by $g = \nabla\varphi$, where φ is the gravitational potential. By Newton's gravitational law and second law then

$$\boldsymbol{F}_g = -m_g \nabla\varphi = m_i \boldsymbol{a},$$

so if $m_g = m_i$ the acceleration is

$$\boldsymbol{a} = -\left(\frac{m_g}{m_i}\right)\nabla\varphi = -\nabla\varphi, \tag{6.3}$$

which does not depend on the mass (or any other intrinsic property) of the object that is falling in the gravitational field.

6.3 The Strong Equivalence Principle

Einstein extended the weak equivalence principle to the modern equivalence principle, sometimes called the *strong principle of equivalence*.

6.3.1 Elevators, Gravity, and Acceleration

The strong principle of equivalence is suggested by thought experiments using the elevator illustrated in Fig. 6.2. The elevator is assumed to be an idealized device in which the occupant is unable to see or hear anything going on outside, so that her perception of the world depends solely on the observations that she can make inside the elevator. By considering various experiments with such an elevator, it may be concluded that the occupant is unable to distinguish an acceleration of the elevator

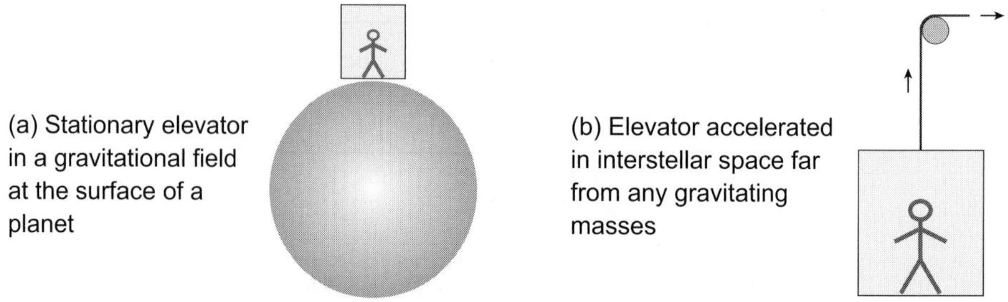

(a) Stationary elevator in a gravitational field at the surface of a planet

(b) Elevator accelerated in interstellar space far from any gravitating masses

Fig. 6.2 The Einstein elevator. (a) Acceleration produced by gravity. (b) Acceleration produced by non-gravitational forces. The two are indistinguishable if the extent of the elevator is small.

at some point in space where no gravitational fields act from the effect of a gravitational field in a non-accelerating elevator: *Physics in a non-accelerating reference frame with gravitational acceleration* **g** *is indistinguishable through any possible measurement from physics in a reference frame with no gravity but accelerating with* **a** $= -$**g**. In either case the elevator occupant feels a force pressing her against the bottom of the elevator. This may be formalized as the *Strong Equivalence Principle:*

> **The Strong Equivalence Principle:** For an observer in free fall in a gravitational field, the results of all local experiments are independent of the magnitude of the gravitational field.

Notice that the equivalence of gravitational and inertial mass is central to these arguments because if two different test masses fell with different accelerations this could be used to detect the presence of a gravitational field.

Thus, Einstein realized that the gravitational "force" is a fictitious force having only a relative existence that is created by observations in a non-inertial frame, much like the Coriolis force is a pseudoforce necessitated by formulation of the rotational problem in a non-inertial frame. An observer standing at the surface of the Earth feels a gravitational acceleration and corresponding gravitational force because the observer is in a non-inertial frame. But an observer in free fall in that same gravitational field feels no gravitational force because he is then observing from a local inertial frame.[1]

6.3.2 Alternative Statements of the Equivalence Principle

Henceforth "equivalence" will be understood to mean "strong equivalence". There are some alternative statements of the equivalence principle that will prove useful in various contexts, so we collect them here:

1. All *local*, freely falling, non-rotating laboratories are fully equivalent for the performance of physical experiments. Such a laboratory is called a *local inertial frame (LIF)* or *Lorentz frame*.
2. In any *sufficiently local* region of spacetime the effect of gravity can be transformed away.
3. In any *sufficiently local* region of spacetime a *local inertial system* may be constructed in which the special theory of relativity is valid, even in a very strong gravitational field.
4. All forms of mass and energy contribute equivalent quantities of gravitational and inertial mass.

---

[1] It was Einstein's remarkable physical intuition, not primarily his mathematical ability (though one suspects that his mathematical abilities exceeded that of most of us), that led to general relativity. The famous mathematician David Hilbert is said to have remarked "Every schoolboy in the streets of Göttingen understands more about four-dimensional geometry than Einstein. Yet ... Einstein did the work and not the mathematicians" [231]. Hilbert's remarks are hyperbolic (surely not *every* schoolboy, even in Göttingen!), but they convey succinctly that Einstein's deep physical intuition allowed him to go where those with greater mathematical ability had not even contemplated going.

In these statements a practical test of whether a laboratory is freely falling and therefore a local inertial frame is *weightlessness*: experiments would reveal any object in the laboratory to be weightless.

6.3.3 Equivalence and the Path to General Relativity

Einstein used the equivalence principle to obtain two important results of general relativity even before he could formulate and solve the corresponding field equations. The first was the deflection of light in a gravitational field (though his initial value for the amount was too small by a factor of two because his early formulations of general relativity failed to account completely for the curvature of spacetime). The second was the gravitational redshift and the gravitational time dilation that it implies. Let us now understand how Einstein was led to these two results by the principle of equivalence.

In both cases the general strategy will be similar. An elevator will fall freely in a gravitational field with one observer in the elevator, falling with it, and one fixed observer external to the elevator, watching it fall. For the observer in the elevator the equivalence principle implies that gravity can be transformed away and special relativity is valid. For the external observer the manifest downward acceleration of the elevator implies that the elevator and its contents are subject to a gravitational field that must be accounted for in analyzing the situation. The important conclusions then will follow when we require that these two points of view be compatible with a single set of physical laws for the Universe.

6.4 Deflection of Light in a Gravitational Field

The effect of a gravitational field on the direction of light propagation may be examined by considering the thought experiment illustrated in Fig. 6.3.

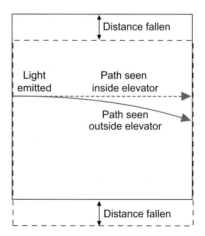

Fig. 6.3   The equivalence principle and deflection of light in a gravitational field.

6.4.1 A Thought Experiment

An elevator hangs suspended in a gravitational field. A light source affixed to one wall of the elevator emits a ray of light directed horizontally across the elevator and at the same instant the cable supporting the elevator is cut, allowing the elevator to fall freely in the gravitational field. Now let us consider the path of the light as viewed from the interior of the elevator and from the outside.

Internal and external observers: In the interior, if the extent of the elevator is sufficiently small (this requirement will be quantified later) the principle of equivalence implies that the effect of gravity can be transformed away. Thus, an observer in the interior is unaware of any gravitational field, special relativity is valid, and the light travels in a straight line, striking the other side of the elevator at the same distance above the floor as it was emitted. If the same experiment is observed from the outside at a stationary point the external observer is aware of the gravitational field and sees the elevator falling in it. The external observer finds that the spot at which the light strikes the right wall of the elevator has fallen by the same amount as the elevator has fallen in the time for the light to traverse the interior of the elevator.

Reconciling the observations: Since the internal and external observations must lead to a consistent physical interpretation, as viewed from the outside where the effect of gravity is manifest the light must follow a *curved path* across the elevator, exactly as if it had mass and fell with the elevator in the gravitational field (see Problem 6.2). Thus, *the equivalence principle requires that the path of light be deflected by a gravitational field.*

6.4.2 Curvature Radius and the Strength of Gravity

The amount of light deflection in the gravitational field may be quantified through a *radius of gravitational curvature* (obtained by fitting a circle to the local curved path) given by

$$r_c = \frac{c^2}{g}, \qquad (6.4)$$

where g is the local gravitational acceleration and c is the speed of light (see Problem 6.1). A stronger gravitational field thus leads to a smaller curvature radius. A natural measure for the strength of gravity at the surface of a gravitating object such as a star may then be formed by taking the ratio of the actual radius of the object to the gravitational radius of curvature (6.4),

$$\epsilon \equiv \frac{R}{r_c} = \frac{GM}{Rc^2} = \frac{\text{Actual radius}}{\text{Light curvature radius}}, \qquad (6.5)$$

where $g = GM/R^2$ has been used.[2] If $\epsilon \ll 1$ the field may be characterized as weak. At the other extreme, notice that if this ratio approaches unity the gravitational field is capable of putting light into orbit around the object. This point will be revisited when black hole solutions to the general theory of relativity are discussed in Chapter 11.

---

[2] Note that ϵ is the only dimensionless quantity that can be formed from a combination of M, r, G, and c.

Equation (6.5) also may be written in the form

$$\epsilon = \frac{GMm/R}{mc^2} = \frac{E_g}{E_0} = \frac{\text{Gravitational energy}}{\text{Rest mass energy}}, \quad (6.6)$$

where E_g is the gravitational energy and E_0 is the rest mass energy. Thus, the weak-field gravitational condition $\epsilon \ll 1$, derived above in terms of the curvature radius, is equivalent to a condition that the gravitational energy of a test particle be much less than its rest mass energy. If the gravitational field is weak Newtonian gravity is approximately valid, with only small general relativity corrections. On the other hand, if the weak-field conditions are not satisfied, *strong gravity* prevails and deviations from Newtonian predictions become substantial, necessitating a general relativistic description of gravity.

As illustrated in Problem 6.1, most gravitational fields are very weak on the scale set by (6.5) or (6.6). For example, the white dwarf Sirius B has an average density of approximately 10^6 g cm^{-3}, which gives $\epsilon \simeq 10^{-4}$. Even at the surface of a white dwarf the gravitational field is weak on the natural scale set by light curvature (though it is enormous by Earth standards) and Newtonian gravity is still a rather good approximation. But for gravity generated by even more compact objects, such as that near the surface of a neutron star or the event horizon of a black hole, the gravitational curvature radius and actual radius will be of comparable size ($\epsilon \sim 1$) and a correct description of gravity requires general relativity. More generally, it will be convenient to use $\epsilon = GM/Rc^2$, with R a characteristic distance (not necessarily a radius) over which gravity generated by the source M is acting, as a measure of the intrinsic strength of gravity in various contexts.

6.5 The Gravitational Redshift

The equivalence principle implies that light follows a curved path in a gravitational field, as if it possessed an effective mass. Therefore, it is not surprising that a gravitational field has nontrivial consequences for the vertical propagation of light and that the equivalence principle can be used to deduce them.

6.5.1 A Second Thought Experiment

Consider the elevator of the preceding example but now arrange things so that a monochromatic source of light is attached to the floor of the elevator and points directly upward at a detector attached to the ceiling that measures the frequency of the light (Fig. 6.4). At the instant that the light source is turned on the cable is cut, allowing the elevator to fall freely. As before, let us analyze the situation both with respect to an observer inside the elevator and one outside watching the elevator fall in the gravitational field.

Internal and external observers: For the observer in the elevator the extent of the elevator is assumed small enough that its interior can be considered a local inertial frame in which the effects of gravity may be transformed away. Thus, the internal observer is unaware of any gravitational field and the frequency of the light detected at the ceiling of the elevator

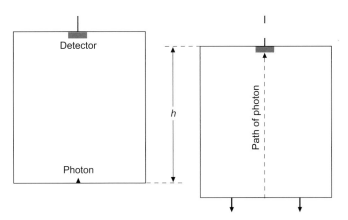

Fig. 6.4 The principle of equivalence and the gravitational redshift.

is just ν_0, the same as the emitted frequency. Now consider the external observer who, unlike the internal observer, is aware of the gravitational field and must account for the effect of gravitational forces in analyzing the experiment. The elevator begins to fall as the light begins propagating upward. When the light reaches the ceiling a time $t = h/c$ has elapsed and the elevator has accelerated from rest to a velocity $v = gh/c$, where g is the local gravitational acceleration. The detector is in motion toward the light source when it receives the signal so the detected light will be Doppler shifted to higher frequency. The time of fall is assumed to be short so $v \ll c$ and the nonrelativistic Doppler formula may be employed to calculate an expected frequency shift for the light of

$$\frac{\Delta \nu}{\nu} = \frac{v}{c} = \frac{gh/c}{c} = \frac{GMh}{R^2 c^2}, \tag{6.7}$$

where G is the gravitational constant, M is the mass of the Earth, and R is its radius. Thus, the external observer should see the detected light blueshifted by this amount.

Reconciling the observations: Both the internal and external observers must agree on the detected frequency since that is a physical observation. To avoid a contradiction it must be assumed that there is an *additional redshift* produced by the gravitational field – which is the single different ingredient in the external observer's analysis – that is exactly equal in magnitude to the blueshift, so it exactly cancels it. (For an alternative argument that reaches the same conclusion, see Problem 6.2.) Hence photons propagating upward in a gravitational field for a short distance experience a redshift with magnitude given by Eq. (6.7). By similar arguments, light propagating downward in a gravitational field from a source to a detector will experience a *blueshift* to higher frequencies.

6.5.2 The Total Redshift in a Gravitational Field

We have just shown that gravity acts on vertically propagating light as if the light had an effective mass and alters its energy and hence its frequency. The total redshift for light leaving a gravitational field may be found rather simply if gravity can be considered weak in the sense defined by Eq. (6.5). In that case the light may be viewed as propagating through a sequence of approximate inertial frames and from Eq. (6.7) the integrated redshift for light leaving the gravitational field and detected at a radius s is given by

$$\int_{\nu_0}^{\nu_s} \frac{d\nu}{\nu} = -\int_R^s \frac{GM}{r^2 c^2} dr, \qquad (6.8)$$

which is to be solved for the frequency ν_s. Integrating, exponentiating both sides, and then expanding the right-side exponential by virtue of the weak-field assumption gives for the total frequency shift observed at the radius s,

$$\frac{\nu_s}{\nu_0} \simeq 1 - \frac{GM}{c^2}\left(\frac{1}{R} - \frac{1}{s}\right). \qquad (6.9)$$

For a distant observer $s \to \infty$, yielding an asymptotic frequency shift

$$\frac{\nu_\infty}{\nu_0} \simeq 1 - \frac{GM}{Rc^2}. \qquad (6.10)$$

The corresponding *gravitational redshift* z for the weak-field limit is

$$z \equiv \frac{\lambda_\infty - \lambda_0}{\lambda_0} = \frac{\nu_0}{\nu_\infty} - 1. \qquad (6.11)$$

Equation (6.10) gives $\nu_0/\nu_\infty \simeq 1 + GM/Rc^2$ and thus the gravitational redshift is

$$z \simeq \frac{GM}{Rc^2} \quad \text{(weak-field limit)}, \qquad (6.12)$$

where R is the radius of the gravitating spherical object and M is its mass. The exact result of general relativity for a spherical gravitational field is

$$1 + z = \left(1 - \frac{2GM}{Rc^2}\right)^{-1/2}, \qquad (6.13)$$

which will be derived in §9.2. This reduces to Eq. (6.12) in the weak-field limit where $GM/Rc^2 \ll 1$.

Example 6.1 Consider the gravitational redshift for the white dwarf Sirius B, assuming that $R = 5.85 \times 10^8$ cm and $M = 1.95 \times 10^{33}$ g. Inserting these values into Eq. (6.12) gives $z \simeq 2.5 \times 10^{-4}$ [which justifies the weak field approximation used to derive (6.12)]. This theoretical value may be compared with an observational redshift of $z = 2.7 \pm 0.2 \times 10^{-4}$ determined from displacement of spectral lines corrected for motion of Sirius B relative to the Earth [39]. The checkered history of measuring gravitational redshifts for white dwarfs is recounted in Box 6.1.

Even the gravitational redshift caused by Earth's extremely weak field has been measured and found to agree with Eq. (6.12). Problem 6.8 and Problem 6.3 are of interest in this regard.

> **Box 6.1** **Measuring Gravitational Redshifts for White Dwarfs**
>
> Measuring the gravitational redshift is difficult with normal stars because the effect is very small and in first order cannot be disentangled from the kinematic Doppler shift of spectral lines caused by relative motion (see Problem 6.10). Well-observed binary systems containing a compact object such as Sirius A+B represent the best option for isolating the gravitational redshift from Doppler effects.
>
> **The Redshift of Sirius B**
>
> The gravitational redshift of Sirius B was first measured in 1925 by Walter Adams and shown to be in good agreement with calculations of Arthur Eddington. For some time this agreement was touted as one of the observational proofs of the correctness of general relativity (alongside the gravitational deflection of light by the Sun and the advance of the perihelion of Mercury), and as support for Eddington's partially erroneous ideas about the structure of white dwarfs. However, in retrospect the reported value was a factor of four smaller than that for better modern measurements (probably because of contamination by scattered light from Sirius A in the measurements of Adams), and there was a corresponding error in the calculations because Eddington's flawed ideas about white dwarfs led him to assume incorrect temperatures and radii for Sirius B. Thus the apparent agreement of theory and observation in 1925 was meaningless [123]. The situation for the Sirius B redshift was resolved conclusively only in later decades, through better observations and better understanding of white dwarf structure.
>
> **Gravitational Redshifts and the History of Science**
>
> It has been argued that the saga of the redshift for Sirius B did not have a substantial effect on acceptance of general relativity, but it hindered progress in understanding white dwarfs because the initial erroneous measured redshift of Sirius B was used by Eddington to argue against Chandrasekhar's relativistic model of white dwarfs (now known to be correct), which predicted more compact white dwarfs with much larger redshifts (also correct). This caused a substantial delay in the acceptance of Chandrasekhar's fundamental contributions to the modern understanding of dense matter [123].

6.5.3 Gravitational Time Dilation

The redshift of light caused by a gravitational field may be viewed equivalently as a gravitational dilation of time. Simply imagine the frequency of a light wave to define a clock, with each cycle of the wave corresponding to one tick. Thus, if a time Δt_0 passes at the surface R of a gravitating sphere the corresponding time period Δt_∞ measured at a large distance is given by

$$\frac{\Delta t_0}{\Delta t_\infty} \simeq \frac{\nu_\infty}{\nu_0} \simeq 1 - \frac{GM}{Rc^2} \qquad \text{(weak-field limit)}, \qquad (6.14)$$

where again weak gravity is assumed. As shown in Section 9.2, the corresponding exact result is

$$\frac{\Delta t_0}{\Delta t_\infty} = \left(1 - \frac{2GM}{Rc^2}\right)^{1/2}, \qquad (6.15)$$

which reduces to (6.14) if GM/Rc^2 is small compared with unity.

Example 6.2 For Sirius B the field is weak by the standards of general relativity and Eq. (6.14) gives that $\Delta t_0/\Delta t_\infty \simeq 1 - GM/Rc^2 \simeq 0.99975$. A (sturdy!) clock placed at the surface of Sirius B would run slow by about one second per hour for a distant observer because of gravitational time dilation.

The time dilation described above is *purely gravitational in origin,* independent of any additional special relativistic time dilation caused by relative motion between source and observer. See Problem 6.3 for an example of everyday practical importance where both special relativistic and general relativistic time dilation effects are significant.

6.6 Equivalence and Riemannian Manifolds

We now introduce a mathematical digression that will prove crucial to the discussion of the equivalence principle and its relationship to the full theory of general relativity. It is obvious that a global cartesian coordinate system cannot be set up on a curved surface. However, using two dimensions to illustrate, if the metric takes the form

$$ds^2 = a(x,y)dx^2 + 2b(x,y)dxdy + c(x,y)dy^2, \qquad (6.16)$$

where a, b, and c are some functions of (x, y), then the corresponding space is *locally euclidean.* That is, in the neighborhood of any point in the space a cartesian coordinate system that is valid locally may be constructed. A metric that can be expressed in the quadratic form (6.16) is termed a *Riemannian metric* and a continuous, differentiable manifold having such a metric is termed a *Riemannian manifold.*[3] This may not seem to be very profound because spaces of interest to physicists usually have Riemannian metrics, but mathematically it is quite possible to invent manifolds for which this is not the case. In a Riemannian space the metric is locally euclidean so the geometry is locally euclidean. For example, in such a space it is found that

- the circumference of a circle is $2\pi R$ plus higher-order terms, and
- the sum of the angles of a triangle is π plus higher-order terms,

---

[3] Strictly, such a manifold is termed *Riemannian* if $ds^2 > 0$ and *pseudo-Riemannian* if ds^2 can be positive, negative, or zero (as is the case for special and general relativity). However, it is common in physics to use "Riemannian" to refer to both cases. *Differentiable manifolds* have a precise mathematical specification discussed in Box 3.1. However, the spaces of interest here will all meet that specification and our discussion will often use "space" as shorthand for differentiable, pseudo-Riemannian manifold.

with the magnitude of higher-order corrections decreasing smoothly as circles and triangles are made smaller. These mathematical considerations hint at a close relationship between the principle of equivalence and Riemannian geometry. Since equivalence will be key to developing a relativistic theory of gravity, this implies further an intimate connection between Riemannian geometry and gravitation. This relationship was foreshadowed already in the work of Gauss and Riemann during the nineteenth century, and is summarized in Box 6.2.

You will recall that the metric structure introduced in Section 3.10 was of the quadratic form (6.16) [compare Eq. (3.39)]. Therefore, spacetime in the tensor formulation of Chapter 3 corresponds to a (pseudo-)Riemannian manifold that exhibits euclidean geometry locally, implying that the tensor formalism introduced in Chapter 3 is the relevant mathematical framework with which to explore implications of the equivalence principle for a covariant theory of gravity. The story of how Einstein came to this realization is summarized in Box 6.3.

Box 6.2 — **Gauss, Riemann, and the Inner Properties of Curved Spaces**

The properties of a manifold may be divided into those that are *inner* or *intrinsic*, determined solely from measurements in the manifold itself, and those that are *outer* or *extrinsic*, which concern embedding of the manifold in higher dimensions.

Gaussian Curvature

Karl Friedrich Gauss (1777–1855) showed that all inner properties of a curved space are described by the derivatives $\partial \xi^\alpha / \partial x^\mu$ of the functions $\xi^\alpha(x)$ defining the transformation between a general coordinate system x^μ and a local cartesian coordinate system $\xi^\alpha(x)$ that is valid only in a localized region of a (Riemannian) curved space. For a 2-dimensional space, this reduces to the Gaussian curvature parameter that will be introduced in Section 7.2.2.[a] For spaces with more dimensions the expressions are more complicated, as will be discussed in Section 8.4, but for those spaces too the appropriate generalization of Gaussian curvature will describe the inner properties of the space, independent of any embedding in a higher dimension.

The Equivalence Principle and Riemannian Geometry

It will be seen also that, because of the equivalence principle, all effects of a gravitational field may be described in terms of the derivatives $\partial \xi^\alpha / \partial x^\mu$ of the functions $\xi^\alpha(x)$ defining the transformation between a general coordinate system x^μ and a local inertial coordinate system that may (by equivalence) be constructed at any spacetime point, and in which (freely falling) coordinate system special relativity is valid. This will imply that the natural mathematical expression of the equivalence principle lies in Riemannian geometry, which mathematicians had studied for more than half a century before Einstein conceived the principle of equivalence. The history of Einstein choosing to base general relativity on Riemannian geometry is given in Box 6.3.

[a] That Gaussian curvature is an *intrinsic invariant* is known as the "Remarkable Theorem" *(Theorema Egregium* in latin*)*. Proof of this theorem by Gauss in 1828 laid the foundations of differential geometry. Bernhard Riemann (1826–1866) extended these ideas to higher-dimensional manifolds, where the generalization of Gaussian curvature is now known as *Riemannian curvature*. Much of the rest of this book is either explicitly or implicitly about Riemannian curvature of the spacetime manifold.

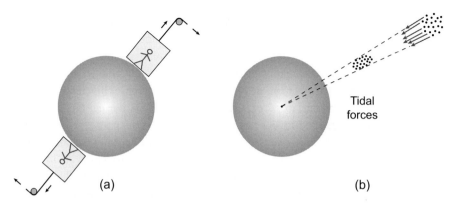

Fig. 6.5 (a) The Einstein elevator in two different local inertial frames. (b) Tidal (differential gravitational) forces; an object experiences tidal forces if the gravitational force (magnitude or direction) is not the same for different parts of the object.

Box 6.3   **Einstein, Grossmann, and Differential Geometry**

By the time Einstein returned to Zurich from Prague in 1912, he understood intuitively that gravity was associated with curved spacetime and hence with non-euclidean geometry. However, he had insufficient mathematical background to formulate a corresponding theory. Einstein had often skipped geometry classes while a university student (earning the epithet "lazy dog" from his teacher Minkowski), and had depended on notes from his friend Marcel Grossmann to prepare for tests. As a result, he had scored 4.5 out of 6 in his geometry course – a passing student, but no Einstein! On the other hand, Grossmann had scored 6 of 6, wrote a thesis and published various papers on non-euclidean geometry, and had become a professor of mathematics in Zurich.

Einstein went to his old friend and asked him how to implement his ideas about gravity and curved spacetime. Grossmann decided that Riemannian geometry, a challenging subject for mathematicians and one virtually unknown to physicists, was the required approach. Thus began a collaboration in which Grossmann taught Einstein tensors (by his own account Einstein struggled at this) and he and Einstein developed sufficient Riemannian geometry in a series of papers for Einstein to unveil, in late 1915 after several false starts, the field equations of general relativity.

6.7 Local Inertial Frames and Inertial Observers

The term *local* appears prominently in discussion of the equivalence principle. The elevator experiments illustrated in Fig. 6.5(a) indicate that a precise meaning is required for this adjective. According to the principle of equivalence, elevator occupants on opposite sides of the Earth may replace the gravitational field by a local acceleration. This is not a contradiction because the equivalence principle applies *locally.* Clearly the two elevator occupants in this case cannot be in the same local inertial frame.

6.7.1 Locality and Tidal Forces

We may give an operational definition of local by requiring tidal effects like those of Fig. 6.5(b) to be negligible. This will be true if the laboratory for observations is sufficiently small. Thus, in the elevator thought experiments used to illustrate the equivalence principle, if the spatial extent of the elevator were large enough two particles released at rest near the walls would be observed to drift inward gradually, since they are falling on non-parallel radial lines toward the Earth's center [Fig. 6.5(b)]. Likewise, two particles released at different heights near the center of the elevator will drift apart vertically because they feel somewhat different strengths of the gravitational field. These tidal forces are never zero but over short times for small enough spatial regions they may be ignored and the falling elevator approximates a local inertial system in which special relativity is valid. Later, this condition will be expressed quantitatively in terms of spacetime curvature.

6.7.2 Inertial Observers

In subsequent discussion the idea of an *inertial observer* will be a central concept. We may define an inertial observer as a laboratory that records the spacetime location of events using an associated local coordinate system satisfying the following conditions [221]:

1. The distance between any two spatial points in the laboratory does not change with time.
2. Clocks may be imagined to sit at each spatial point that are synchronized and that all run at the same rate.
3. At any time the geometry of the space within the laboratory is euclidean.

This definition does not address directly whether inertial observers can exist. Strictly, from the preceding discussion inertial observers *cannot* exist in a gravitational field, but it may be imagined that in a sufficiently localized region of space and time the conditions are well-enough fulfilled (tidal forces are sufficiently small) that no serious error is made by assuming an inertial observer. The coordinate system of the inertial observer then may be caricatured as a rigid lattice of rods, with an array of synchronized clocks located at the intersections of the rods. The rigid rods and the synchronized clocks then allow the spatial location and time of any event to be recorded without ambiguity by the inertial observer.

6.7.3 Definition of Local Inertial Frames

The preceding ideas may be expressed in more precise language by specifying exactly what is meant by a *local inertial frame or LIF*. The equivalence principle asserts that at *each point P* of a curved spacetime manifold described by a metric $g_{\mu\nu}(x)$ it is possible to transform to a new basis where the metric becomes the constant Minkowski metric:

$$g'_{\mu\nu}(x'_P) = \eta_{\mu\nu} = \text{diag}(-1, 1, 1, 1). \tag{6.17}$$

Specifically, since $g_{\mu\nu}(x)$ at a point $x = x_P$ is a real symmetric matrix there exists an orthogonal transformation that will diagonalize it, and once diagonalized each coordinate can be rescaled if needed to give the metric tensor $\eta_{\mu\nu} = \text{diag}(-1, 1, 1, 1)$. Notice two important things, however: (1) Such a transformation cannot change the signature of the metric. (2) This transformation is *local to the point* P; generally a *different transformation* is required to diagonalize the metric at a different point P' if the space is curved.

Although, Eq. (6.17) is valid only at a point, it is possible to choose the transformation so that the *first derivatives* of the metric tensor evaluated at x_P vanish also:

$$g'_{\mu\nu}(x'_P) = \eta_{\mu\nu} = \text{diag}(-1, 1, 1, 1) \qquad \left.\frac{\partial g'_{\mu\nu}}{\partial x'^\alpha}\right|_{x=x_P} = 0. \qquad (6.18)$$

A coordinate system satisfying Eq. (6.18) is called a *local inertial frame* (LIF). However, no transformation satisfying Eq. (6.18) makes all *second derivatives* at x_P vanish. Thus, the adjective "local" in the formulation of the equivalence principle means that the observer occupies a sufficiently small laboratory (in both space and time) that effects depending on second derivatives of the metric are negligible. Then up to first order the space of the freely falling laboratory is flat Minkowski space. Since second derivatives of the metric are associated with tidal forces, an alternative statement is that tidal forces are negligible in a LIF. Because of Eq. (6.18) the gravitational effects of spacetime curvature are expected to appear first in the second derivatives of the metric. This will indeed turn out to be the case.

The preceding discussion means that gravity cannot be identified with the Newtonian gravitational force, since by the equivalence principle that can be transformed to zero in a suitable reference frame. Newtonian gravity is an inertial force caused by observation in a non-inertial frame, analogous to fictitious centrifugal or coriolis forces that appear in rotating frames but can be eliminated by going to a non-rotating frame. What cannot be transformed away are the tidal forces arising from the non-uniformity of the gravitational field, so these represent the true effects of gravity in general relativity.

In summary, by the equivalence principle there *are* inertial frames in the spacetime of general relativity but they are *local, freely falling frames*. In the presence of a non-uniform gravitational field these local inertial frames at different points have accelerations differing in both magnitude and direction, so they cannot be glued together to form a global inertial frame unless the gravitational field vanishes, in which case Minkowski space is recovered globally, not just in separate local frames defined at each point.

6.8 Lightcones in Curved Spacetime

Because a local inertial frame can be defined at each point of a curved spacetime, *general relativity inherits the local lightcone structure of special relativity*. This structure is a coordinate-invariant statement that the speed limit is c, so in general relativity velocities (which are defined locally) satisfy $v \leq c$ and the *local causal classification* of events given in Section 4.4 into timelike, spacelike, and null carries over in curved spacetime. However, the *global organization* of lightcones in curved spacetime can lead to causal structure that does not exist for unaccelerated observers in Minkowski space. In flat spacetime the metric can be chosen constant over all of spacetime and local lightcones are rigid in the sense that the lightcone at one point can be obtained from one at another point by translation without change in geometry of the lightcone. In curved spacetime this is no longer true since the metric depends on location and the global causal structure can become complicated. An extreme example will be encountered when event horizons of black holes are considered in later chapters.

6.9 The Road to General Relativity

If the equivalence principle is used to replace curved spacetime locally with flat spacetime, global spacetime must be viewed as a patchwork of such locally flat frames meshed together smoothly to describe a curved space, as illustrated schematically in Fig. 6.6. According to Einstein then, the inhomogeneity of the gravitational field is caused by the inhomogeneity of gravitating matter, implying that the spacetime is curved, with the curvature related to the distribution of matter.

The key to finding the relationship between the curvature and the distribution of gravitating matter is the previously noted connection between the equivalence principle and Riemannian geometry. The presence of gravitating matter precludes the existence of extended inertial frames, implying that spacetime cannot be globally flat (not the Minkowski pseudo-euclidean manifold). However, by the equivalence principle the space must be *locally* Minkowski, and thus spacetime must generally correspond to a Riemannian manifold with the metric determined by a nonlinear relation among the mass, curvature, and density. The close connection of the physical principle of equivalence with the mathematical assumption of Riemann geometry for spacetime means that in a sufficiently local region particle trajectories are straight lines, but *globally* these lines are not straight if the manifold is curved. This idea is familiar from our experience on the curved surface of the Earth. A sufficiently local region looks flat and lines of longitude at the equator appear to be straight and parallel locally, but globally lines of longitude are not parallel, eventually crossing at the poles. The goal of the next two chapters will be to deduce exactly

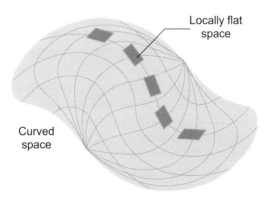

Fig. 6.6   Curved space and various flat local inertial systems that differ globally.

what nonlinear relationship among the mass, curvature, and density is required to knit a patchwork of locally flat spaces into a smooth Riemannian manifold describing curved 4-dimensional spacetime.

Background and Further Reading

For elementary discussions see Hartle [111], Cheng [65], Kogut [134], Carrol and Ostlie [62], Roos [207], and d'Inverno [76]. For more advanced discussions, see Schutz [221]; Raychaudhuri, Banerji, and Banerjee [199]; Rindler [204]; Weinberg [246]; and Misner, Thorne, and Wheeler [156]. The Global Positioning System is an excellent practical illustration of the principle of equivalence, as illustrated in Problem 6.3. A summary of the nontrivial role played by relativity in the functioning of the GPS may be found in Ashby [31].

Problems

6.1 In the falling elevator experiment illustrating the deflection of light in a gravitational field, approximate the curved path of the light as a segment of a circle and find the radius of this circle (assume the angle of deflection to be small in your derivation). This is termed the radius of curvature r_c for the gravitational field at that point. Evaluate the resulting expression for the surface of the Earth, for the surface of the white dwarf having a radius of 5.5×10^8 cm and a mass of 2.1×10^{33} g, and for the surface of a neutron star, which you may take to have a radius of 10 km and a mass of one solar mass. Assuming the elevator to have a width of 2 meters, how far does the light "fall" (as seen by the outside observer) in traveling from one side of the elevator to the other for each of these cases? Finally, calculate the ratio ϵ of Eq. (6.5) characterizing the strength of the gravitational field for the surface of the Earth, Sirius B, and a neutron star. ***

6.2 Consider the following thought experiment, originally proposed by Einstein. Assume a uniform gravitational field and a photon that propagates vertically in the field from a point z_1 to a higher point z_2. A device (the design is an engineering task left to you!) converts the photon at z_2 into an equivalent mass m with 100% efficiency and the mass is then dropped back to z_1, assuming Newtonian gravity to be valid. At z_1, another device converts the mass back to a photon of equivalent energy with 100% efficiency and the photon travels back up to z_2 again, and so on. Show that this hypothetical experiment is a perpetual motion machine that creates energy unless the photon suffers a redshift $g(z_2 - z_1)/c^2$ as it moves vertically in the gravitational field. Thus, show that a gravitational redshift is required by energy conservation. ***

6.3 The Global Positioning System (GPS) consists of 24 satellites in orbits with 12-hour periods (assume the orbits to be circular for this exercise). Timing on signals received from multiple satellites can be used to determine the position of a ground receiver.

In the most precise applications, the system strives for about two-meter accuracy in locating position on the Earth's surface. The atomic clocks on the satellites, synchronized regularly by signals from the ground, keep very accurate time but there are two relativistic sources of correction in relating clocks on the satellites to clocks on the ground. Because the satellites are in motion with respect to the ground, there is a special relativistic time dilation effect, and because the clocks on the ground are in a deeper gravitational potential than those on the satellites, there is also a general relativistic (gravitational) time dilation effect. Use this information to estimate the magnitude of each of these corrections (paying careful attention to their relative signs) and the total relativistic correction for time dilation. If two-meter accuracy is required in position determination, approximately how long can the GPS system operate without corrections for these relativistic effects before position information is compromised? *Hint*: A summary of the essential role played by relativity in the GPS system may be found in Ashby [31]. ***

6.4 By applying the equivalence principle to light moving vertically in a gravitational field, show that photons behave under the influence of gravity as if they had an effective mass $m = h\nu/c^2$. This is useful heuristic insight (roughly: photons have energy given by the quantum formula, energy and mass are equivalent, so photons are influenced by gravity as if they had a mass). However, it is potentially misleading since photons do not have a mass. Note that the thought experiment of Problem 6.2 deals with essentially the same issue but never assumes the photon to be massive.

6.5 The US atomic standard clock at Boulder sits at an elevation of approximately 1650 meters. Assume a second clock identical to the first but at an elevation of 25 meters. How much should these two clocks differ in time over a period of a year because of gravitational time dilation?

6.6 Write the equations involving inertial mass and gravitational mass (considered as separate quantities) for the small-amplitude motion of a pendulum with massless string and for sliding blocks on frictionless inclined planes. Describe how experiments with the blocks and pendulum could be used to test for the equivalence of inertial and gravitational mass.

6.7 Beginning from the blue Doppler shift in Eq. (6.7), supply all steps in the proof that a clock on a satellite a distance r_s above the center of the Earth experiences a shift in frequency relative to an identical clock on the ground given by

$$\frac{\nu_s - \nu_0}{\nu_0} = \frac{GM}{c^2}\left(\frac{1}{R} - \frac{1}{r_s}\right),$$

where R is the radius and M the mass of the Earth, G is the gravitational constant, ν_0 is the clock frequency on the ground, and ν_s is the clock frequency on the satellite.

6.8 The first direct measurement of the redshift caused by Earth's gravitational field was made by Pound, Rebka, and Snider [191, 192, 193], who detected gamma-rays emitted vertically from a radioactive ^{57}Fe source using an ^{57}Fe target located 22.5 meters above the source (giving a gravitational redshift) or below the source (giving a gravitational blueshift). If there is a gravitational frequency shift the photons striking the target will be shifted off resonance in energy and not be absorbed, but by moving the target slowly a Doppler shift can be induced to compensate for the gravitational shift and cause the photons to be shifted back on resonance and absorbed. Normally the very high precision required for the measurement would not be attainable because of thermal noise caused by the emitter and absorber being embedded in vibrating crystal lattices. However, the experiment used the then newly discovered Mössbaur effect, for which certain atoms with a rigid crystalline structure exhibit essentially no recoil because the recoil is transferred to the entire lattice, giving a sufficiently sharp emission energy to make the experiment feasible. (a) For this experiment, estimate the precision required in the frequency measurement to detect the gravitational redshift. (b) If the redshift predicted by the equivalence principle is correct, how large must the relative velocity between emitter and absorber be to exactly compensate for the gravitational redshift so that the photons would be absorbed? ***

6.9 Two masses enter the gravitational force equation. One may be viewed as the source of the field (*active mass* m_a) and one as the "test charge" (passive mass, m_p). Show that active and passive masses are equivalent in appropriately chosen units, so that one can speak of a single gravitational mass m_g. *Hint*: Newton's third law.

6.10 Show that the gravitational redshift causes a fractional shift in spectral lines of order 10^{-6} for the Sun and other main sequence stars. Hence it is very difficult to measure the redshift for them. *Hint*: From observational systematics the ratio of mass to radius is almost constant (within a factor $\sim$ 2–3) for main sequence stars the mass of the Sun or greater. ***

7 Curved Spacetime and General Covariance

Lorentz covariance makes it manifest that the principles of special relativity (constant speed of light and invariance under Lorentz transformations) are obeyed by some set of equations. More generally, the six transformations of the Lorentz group may be enlarged by adding to them the four possible uniform translations in space and time. The corresponding 10-parameter group is called the *inhomogeneous Lorentz group* or the *Poincaré group* (Box 7.1 gives an introduction to the theory of groups). Poincaré invariance is a mathematical statement that physics does not depend on choice of coordinate system origin, orientation, and so on, and leads to conservation laws such as those for energy and angular momentum. In the absence of gravity it is thought that the laws of physics are absolutely invariant under Poincaré transformations. However, even the Poincaré transformations assume flat spacetime, so covariance with respect to Poincaré transformations is insufficient to deal with gravity. Therefore, we shall now seek a more general covariance that embraces the possibility of gravity arising from curved spacetime.

7.1 General Covariance

General covariance will be taken to mean that a physical equation holds in a gravitational field provided that:

- It holds in the absence of gravity (agrees with special relativity in flat spacetime).
- It maintains its form under the most general coordinate transformations $x \to x'$.

As discussed by Weinberg [246], general covariance is in fact just a restatement of the equivalence principle but this restatement will prove to be particularly useful for developing the mathematical formalism of general relativity. In this chapter we shall extend the idea of covariance to curved manifolds. As part of that discussion several topics crucial to the later formulation of general relativity will make their appearance:

1. *Gaussian curvature* and its generalization *Riemannian curvature* as intrinsic measures of curvature for a manifold;
2. the *stress–energy tensor*, which will provide a covariant description of the coupling of gravity to matter, radiation, and pressure;
3. the introduction of additional structure on the manifold associated with *connection coefficients,* which will be necessary to specify a unique prescription for parallel transport;

4. the relationship of *covariant derivatives*, *absolute derivatives*, and *Lie derivatives* to parallel transport on curved manifolds; and
5. the central thesis of general relativity that gravity corresponds to the motion of *free particles* but in a curved manifold, with that motion described mathematically by the *geodesic equation*.

We begin this discussion by considering coordinate systems appropriate for curved spacetime manifolds and a measure of intrinsic curvature in those manifolds.

7.2 Curved Spacetime

The discussion of the equivalence principle and the deflection of light by gravity in Chapter 6 suggests that gravity is associated with curvature of spacetime. Thus, let us leave behind the restriction to flat spaces and consider the more general problem of covariance in curved spacetime. The first issues to address are those of appropriate coordinate systems for curved manifolds and a quantitative measure of curvature in those manifolds.

7.2.1 Coordinate Systems

Physics cannot depend on coordinate choice but the ease of solving particular problems may depend strongly on a judicious choice of coordinates. In a flat spacetime the choice of special coordinate systems suggested by symmetries often can lead to a simplification of problems. For example, in special relativity it is particularly advantageous to formulate problems in inertial frames because of the symmetry associated with the Lorentz transformations that connect inertial frames (see Box 7.1). In contrast, for general relativity in curved spacetime there are particular coordinate system choices that may be advantageous for specific problems with very high symmetry, but there is no special coordinate system that leads to a general simplification. Therefore, there is no particular advantage to formulating general relativity in a specific coordinate system and it is useful to develop methods for working in arbitrary coordinates. We have anticipated this in Section 3.4 and the coordinate (holonomic) bases introduced there will be used almost exclusively for our subsequent development of general relativity.

7.2.2 Gaussian Curvature

Most of us have an intuitive feeling for what curvature is but that intuition is not quantitative; worse, it can be misleading because it often mixes up curvature associated with a manifold (intrinsic curvature) and curvature resulting from embedding that manifold in a higher-dimensional manifold (extrinsic curvature), as discussed further in Box 8.1. In general relativity our primary interest lies in intrinsic curvature, so it is desirable to formulate a quantitative measure of curvature for manifolds that does not depend on any embedding in a higher-dimensional space. The metric tensor determines the intrinsic geometry of a manifold but it is not always manifest from the form of $g_{\mu\nu}$

> **Box 7.1** **Symmetries and Groups**
>
> A group is a set $G = \{x, y, \ldots\}$ for which a binary operation $a \cdot b = c$ called *group multiplication* is defined that has the following properties
>
> (i) *Closure:* If x and y are elements of G, then $x \cdot y$ is an element of G also.
> (ii) *Identity:* An identity element e exists such that $e \cdot x = x \cdot e = x$ for $x \in G$.
> (iii) *Existence of an Inverse:* For every group element x there is an inverse x^{-1} in the set such that $xx^{-1} = e$.
> (iv) *Associativity:* Multiplication is associative: $(x \cdot y) \cdot z = x \cdot (y \cdot z)$ for $x, y, z \in G$.
>
> For groups of transformations multiplication corresponds to applying first one and then the other transformation. The group definition requires associativity but not commutivity. A group consisting of commutative elements only is *abelian*; otherwise, it is *nonabelian*.
>
> **Example: The Lorentz Group**
>
> There are six independent Lorentz transformations: three rotations about the spatial axes parameterized by real angles, and three boosts along the spatial axes parameterized by boost velocities. Because rotation angles and boost velocities can take continuous real values, the set of Lorentz transformations is infinite. The Lorentz transformations form a group:
>
> 1. Two successive transformations are equivalent to some other transformation.
> 2. Every Lorentz transformation has an inverse that is the transformation in the opposite direction (for example, $v \rightarrow -v$ for boosts).
> 3. The identity corresponds to no transformation.
> 4. It doesn't matter how three successive transformations are grouped, so multiplication is associative, but the *order* matters, so the Lorentz group is nonabelian.
>
> An important class of groups corresponds to transformations that are continuous (analytical) in their parameters. These are called *Lie groups*. The Lorentz group is a Lie group. Groups also can be classified according to whether their parameter spaces are closed and bounded (*compact groups*) or not (*noncompact groups*). Because boost velocities can approach c asymptotically but never reach it, the Lorentz-group parameter space is not closed and the group is noncompact.
>
> **Example: The Poincaré Group**
>
> The Poincaré group is formed by adding to the Lorentz transformations the uniform translations along the four spacetime axes. It is a nonabelian, noncompact Lie group, and contains the Lorentz group as a *subgroup* (a subset of group elements that satisfy the same group postulates as the parent group).

whether the manifold is curved. For example, if the metric tensor is constant everywhere then the manifold is necessarily flat but the metric tensor for even a flat manifold can be parameterized to make it position-dependent (see the examples in Problem 7.1). What we need is a measure of curvature that: (1) can be determined by measurements intrinsic to the manifold, and (2) is not obscured by different parameterizations. The required curvature invariant is called *Riemannian curvature* in an arbitrary number of dimensions and

Gaussian curvature when restricted to two dimensions. Gaussian curvature is introduced in this chapter and it is generalized to Riemannian curvature of spacetime in Chapter 8.

To illustrate how to quantify the intrinsic curvature of a manifold, consider first 2-dimensional surfaces such as the 2-sphere defined by

$$x^2 + y^2 + z^2 = S^2, \tag{7.1}$$

where S is a constant. Gauss demonstrated that for 2-surfaces there is a single invariant characterizing the curvature (see Box 6.2). Accordingly, this quantity is termed the *Gaussian curvature*. For a 2-dimensional coordinate system (x^1, x^2) with the metric in diagonal form and non-zero elements g_{11} and g_{22}, the Gaussian curvature K is [65]

$$K = \frac{1}{2g_{11}g_{22}} \left\{ -\frac{\partial^2 g_{22}}{(\partial x^1)^2} - \frac{\partial^2 g_{11}}{(\partial x^2)^2} + \frac{1}{2g_{11}} \left[\frac{\partial g_{11}}{\partial x^1} \frac{\partial g_{22}}{\partial x^1} + \left(\frac{\partial g_{11}}{\partial x^2}\right)^2 \right] \right.$$
$$\left. + \frac{1}{2g_{22}} \left[\frac{\partial g_{11}}{\partial x^2} \frac{\partial g_{22}}{\partial x^2} + \left(\frac{\partial g_{22}}{\partial x^1}\right)^2 \right] \right\}, \tag{7.2}$$

which is position-dependent in the general case and is an *intrinsic quantity* expressed entirely in terms of the metric for the space and its derivatives. For orthogonal coordinates (x, y) the Gaussian curvature K at a point (x_0, y_0) also can be expressed as

$$K(x_0, y_0) = \frac{1}{R_x(x_0) R_y(y_0)}, \tag{7.3}$$

where $R_x(x_0)$ is the radius of curvature[1] in the x direction and $R_y(y_0)$ is the radius of curvature in the y direction. For a 2-sphere $R_x = R_y = R$ for all points and therefore the Gaussian curvature is constant and given by $K = R^{-2}$.

Example 7.1 The Gaussian curvature of a 2D surface may be obtained through the following geometrical argument. Consider Fig. 7.1 and use the circumference of a circle relative to that for flat space to measure deviation from flatness. As shown in Section 2.4, a circle in the 2-dimensional space may be defined by marking a locus of points lying a constant distance S from a reference point, chosen to be the north pole in Fig. 7.1. Then the circumference of the circle is

$$C = 2\pi r = 2\pi R \sin \frac{S}{R} = 2\pi S \left(1 - \frac{S^2}{6R^2} + \cdots \right), \tag{7.4}$$

where the angle subtended by S is S/R and $r = R \sin(S/R)$. If the space were flat $2\pi S$ would be the circumference of the circle, so the higher-order terms measure the curvature. Substituting $R^2 = 1/K$ in Eq. (7.4) and solving for K gives a general expression for the Gaussian curvature,

$$K = \lim_{S \to 0} \frac{3}{\pi} \left(\frac{2\pi S - C}{S^3} \right) = \lim_{S \to 0} \frac{6}{S^2} \left(1 - \frac{C}{2\pi S} \right). \tag{7.5}$$

---

[1] That is, the radius of a circle having an arc that approximates the curvature locally at the point (x_0, y_0).

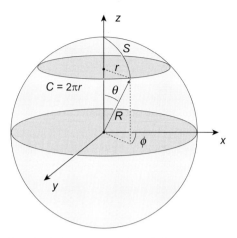

Fig. 7.1 Measuring the circumference of a circle of radius S in a curved manifold.

Thus, the Gaussian curvature for a 2-dimensional surface may be determined by measuring the circumference C of arbitrary small circles on that surface and considering the deviation of C from $2\pi S$ in the limit that the radius S tends to zero. This procedure is illustrated in Box 7.2 and in several problems at the end of this chapter. A sphere was used here for illustration but the procedure in fact valid for any smooth 2D manifold.

In Section 8.4 the Gaussian curvature parameter for a 2-dimensional surface will be generalized to a set of parameters (elements of the Riemann curvature tensor) that describe the curvature of 4-dimensional spacetime.

7.2.3 Distance Intervals

In curved spacetime the interval between two events may be expressed by replacing the flat-space metric tensor $\eta_{\mu\nu}$ in Eq. (4.5) with the general metric tensor $g_{\mu\nu}(x)$ for a manifold that is possibly curved,

$$ds^2 = g_{\mu\nu}(x)dx^\mu dx^\nu. \tag{7.6}$$

The metric tensor $g_{\mu\nu}(x)$ in a curved spacetime has a more complicated form than that for Minkowski space in Eq. (4.7) because it is a function of the spacetime coordinates. A metric tensor that is a function of spacetime coordinates leads naturally to the utility of basis vectors that depend on the coordinates. Coordinate bases having this property were introduced in Section 3.4.

7.3 A Covariant Description of Matter

In the theory of general relativity curved spacetime is responsible for gravity and mass, energy, and pressure are responsible for curving spacetime. Therefore, it is critical in the

> **Box 7.2** **Small Circles and Gaussian Curvature**
>
> The results of Section 7.2.2 were derived by embedding a 2-sphere in 3D euclidean space. However, Eq. (7.5) indicates that intrinsic curvature may be deduced by measurements made entirely *within the space*, without embedding in higher dimensions.
>
> **Intrinsic Determination of Gaussian Curvature**
>
> We may illustrate using the geometry of small circles drawn on the unit 2-sphere, which has a line element in spherical polar coordinates $ds^2 = d\theta^2 + \sin^2\theta \, d\varphi^2$. In the coordinates (θ, φ) a line segment from $(0, 0)$ to $(\lambda, 0)$ has a length
>
> $$S = \int \sqrt{ds} = \int_0^\lambda d\theta = \lambda,$$
>
> since φ is constant so $d\varphi^2 = 0$. The set of points (λ, φ) with φ ranging from 0 to 2π then defines a circle of radius $S = \lambda$, with a circumference C given by
>
> $$C = \int \sqrt{ds} = \int_0^{2\pi} \sin\theta \, d\varphi = \sin\lambda \int_0^{2\pi} d\varphi = 2\pi \sin\lambda.$$
>
> Thus, from Eq. (7.5) the Gaussian curvature is
>
> $$K = \lim_{S \to 0} \frac{6}{S^2}\left(1 - \frac{C}{2\pi S}\right) = \lim_{\lambda \to 0} \frac{6}{\lambda^2}\left(1 - \frac{\sin\lambda}{\lambda}\right).$$
>
> Since our interest is in the limit $\lambda \to 0$, we expand $\sin\lambda$ in a power series,
>
> $$\frac{\sin\lambda}{\lambda} = \frac{1}{\lambda}\left(\lambda - \frac{\lambda^3}{3!} + \cdots\right) \simeq 1 - \frac{\lambda^2}{6}.$$
>
> This gives
>
> $$K = \lim_{\lambda \to 0} \frac{6}{\lambda^2}\left(1 - \left(1 - \frac{\lambda^2}{6}\right)\right) = 1,$$
>
> which is the expected result, since it was noted earlier that the Gaussian curvature of a 2-sphere having radius R is equal to $1/R^2$ and $R = 1$ has been assumed.
>
> **Intrinsic Properties without Embedding**
>
> The Gaussian curvature has been obtained from measurements made entirely within the 2D space. Hence the Gaussian curvature is an *intrinsic property* of the 2D space that can be deduced by local measurements made entirely within the manifold.

formulation of the theory to describe the distribution of these quantities and their coupling to gravity in covariant fashion. To that end, it is convenient to introduce the *stress–energy* (or *energy–momentum*) tensor $T^{\mu\nu}$, which has the components

1. $T^{00} = \epsilon$ (energy density).
2. $T^{ii} = P^i$ (pressure in i direction; equivalently, momentum components per unit area).
3. T^{0i} (energy flux in the direction i).

Box 7.3 Energy Conditions and the Stress–Energy Tensor

It is common to refer generically to the fields giving finite contributions of mass, energy, or pressure to the stress–energy tensor as "matter." The properties of the matter contributing to $T^{\mu\nu}$ must be suitably constrained if physically meaningful solutions are to be obtained. It is standard to impose a set of *energy conditions* on $T^{\mu\nu}$ that are basically *assumptions* about the way that any reasonable form of matter should behave, and that are obeyed by all presently known forms of matter. Three such energy conditions are most commonly used [26],

1. **Weak energy condition:** $T_{\mu\nu}u^\mu u^\nu \geq 0$ for any unit timelike vector u^μ, which means that the energy density seen by any observer may not be negative.
2. **Strong energy condition:** $T_{\mu\nu}u^\mu u^\nu + \frac{1}{2}T^\mu{}_\mu \geq 0$ for any unit timelike vector u^μ, implying physically that the energy density plus the sum of the principle pressures must be non-negative (for a perfect fluid this is equivalent to requiring $\epsilon + 3P \geq 0$, where ϵ is the energy density and P the isotropic pressure).
3. **Null energy condition:** $T_{\mu\nu}k^\mu k^\nu \geq 0$ for any null vector k^μ, which means physically that the sum of the energy density plus any of the principle pressures may not be negative ($\epsilon + P \geq 0$ for a perfect fluid).

The first and second conditions are independent but both imply the null energy condition. These energy conditions are more in the nature of recipes based on classical experience about the way matter should behave rather than physical law, and they can be violated by quantum (and some classical) fields. Nevertheless, they often are invoked in proofs.

4. T^{i0} (momentum density in the direction i).
5. T^{ij} ($i \neq j$) (shear of the pressure component P^i in the j direction).

By physical arguments the tensor $T^{\mu\nu}$ is symmetric, with 10 independent components. Physically meaningful results from general relativity require that the form of $T^{\mu\nu}$ be constrained further by hypotheses that are discussed in Box 7.3. For present purposes it will be sufficient to simplify greatly and restrict attention to *perfect fluids,* which neglect complications arising from energy transport and viscosity that characterize more realistic fluids.

7.3.1 Stress–Energy for Perfect Fluids

For flat spacetime the most general form of the stress–energy tensor for a perfect fluid that is consistent with Lorentz invariance is

$$T^{\mu\nu} = (\epsilon + P)u^\mu u^\nu + P\eta^{\mu\nu} \qquad \text{(flat spacetime)}, \qquad (7.7)$$

where $\eta^{\mu\nu}$ is the Minkowski metric, P is the pressure, $\epsilon = \rho c^2$ is the energy density, $u^\mu = dx^\mu/d\tau$ is the 4-velocity [with $x^\mu(s)$ describing the worldline of a particle in terms of the proper time τ]. Conservation of 4-momentum may then be expressed by the 4-divergence relation

$$\partial_\mu T^{\mu\nu} = 0 \quad \text{(flat spacetime)}. \tag{7.8}$$

The generalization of (7.7) to curved spacetime is

$$T^{\mu\nu} = (\epsilon + P)u^\mu u^\nu + P g^{\mu\nu} \quad \text{(curved spacetime)}, \tag{7.9}$$

where $g^{\mu\nu}$ is the metric. The corresponding generalization of the conservation law (7.8) to curved spacetime is the covariant derivative relation

$$\nabla_\mu T^{\mu\nu} = 0 \quad \text{(curved spacetime)}. \tag{7.10}$$

Equations (7.7)–(7.10) imply a basic difference between general relativistic and Newtonian gravity concerning the source of the gravitational field, as discussed in Box 7.4.

7.3.2 Local Conservation of Energy

As indicated in Box 7.4, there is no well-defined idea of local energy and momentum conservation in general relativity. This is a consequence of the equivalence principle, which requires that all effects of gravity vanish over a small enough region, but it may be viewed more fundamentally in terms of spacetime symmetries [111]. In non-GR physics conservation laws such as those of momentum or energy are consequences of spacetime symmetries: for example, momentum is conserved because of spatial translational invariance (Noether's theorem; see Section 5.6.1). Since all physical systems are expected to be translationally invariant, the law of momentum conservation is always valid in non-GR physics.

But in general relativity spacetime is the *solution,* not a pre-defined stage on which physics plays out. As a consequence, specific GR solutions may have certain symmetries, but there are *no symmetries guaranteed to be common to all GR spacetimes* and hence no associated generic local conservation laws. Thus, the absence of a law of local energy conservation in general relativity is a feature, not a bug. It reflects the shift from viewing gravity as a force acting in spacetime to viewing gravity as dynamical, curved spacetime itself. It does make sense, however, to assume approximate energy conservation in general relativity when averaged over a large enough region of spacetime, and the total energy of a spacetime is a well-defined concept if the spacetime is asymptotically flat (tends to flat Minkowski space on the boundaries).

7.4 Covariant Derivatives and Parallel Transport

As discussed in Section 3.13.2, ordinary differentiation is not a covariant operation in curved manifolds for tensors of rank one and above. In Section 3.13.3 the covariant derivative was introduced to remedy this problem. This section elaborates further on covariant derivatives and shows that they have an alternative geometrical interpretation following from general considerations of how to compare vectors located at two different spacetime points. This issue becomes critical for the calculation of derivatives because, by

> **Box 7.4** **Conceptual Differences between Newtonian and Einstein Gravity**
>
> General relativity and Newtonian gravity differ quantitatively in their predictions for physical observables but they also differ fundamentally in their physical interpretation. This box summarizes a few of the more important conceptual differences, some of which may have large implications for our understanding of the Universe.
>
> **What Couples to Gravity?**
> The form of Eq. (7.9) points to an essential difference between Einstein gravity and the Newtonian theory. All components of the stress–energy tensor must be viewed as contributing to the curvature and thence to gravity, so energy, mass, and pressure are all sources of the gravitational field. In contrast only mass is a source of Newtonian gravity. By smuggling in $E = mc^2$ from special relativity one can (by a stretch) view energy as a source for Newtonian gravity, but not pressure [it does not appear in the Poisson equation (8.1) that determines the gravitational potential].
>
> **Pressure in Stars and Cosmology**
> But (you may object), what about the role of pressure in stabilizing stars against gravitational contraction in Newtonian gravity? In that case the forces opposing gravity are not produced by a pressure but rather by a *pressure gradient*. In contrast, gravity couples directly to the magnitude of the local pressure in general relativity, so there can be forces associated with pressure even if there is no pressure gradient. It follows that in a universe having a finite but constant pressure the existence of the pressure could still be detected by its (general relativistic) gravitational effect. This is precisely the nature of the cosmological *vacuum energy* to be discussed in Chapters 17 and following. That increasing the pressure increases the strength of gravity in general relativity also has implications for the gravitational stability of stars. Specifically, it suggests a limit beyond which even increasing the pressure by an arbitrary amount cannot stop a massive object from collapsing under the influence of its own gravity. This will lead soon to the idea of a *black hole*.
>
> **Energy Conservation in General Relativity**
> Another important difference implied by Einstein gravity concerns Eqs. (7.8) and (7.10). These appear to be similar formally but their physical meanings are different. Energy and momentum are conserved in flat spacetime and Eq. (7.8) expresses a *conservation law* for 4-momentum. In a curved spacetime the constraint (7.10) does not imply a conservation law because *the gravitational energy is not included in the stress–energy tensor.* That is, there is no well-defined concept of local energy and momentum conservation in general relativity, ultimately because it is difficult to construct a sensible local expression for gravitational energy (see also the discussion of gravity and mass for neutron stars in Section 10.3). The best that one can achieve is approximate 4-momentum conservation over a finite volume.

definition, constructing the derivative of a vector requires taking the difference of vectors at two different points within the manifold, and vectors at two different points are defined in different tangent spaces.[2]

---

[2] Do not misconstrue the convention in elementary physics of representing vectors by a directed line segment to mean that vectors connect two spacetime points. A vector is defined *at a single point*, as discussed in Sections 3.3 and 3.4; it does not connect two points separated by finite distance in the manifold.

We would like to construct a new derivative operation that fulfills three requirements: (1) The operation should exhibit the properties expected of a derivative, such as the Leibniz rule for the derivative of a product. (2) The derivative of a tensor should transform as a tensor. (3) The derivative should represent the change of the whole tensor, not just its components. As we will now demonstrate geometrically, the derivative satisfying these requirements corresponds to the covariant derivative already introduced in Section 3.13.3. To be definite, let us illustrate for derivatives of vectors. The formal definition of the derivative that we seek is

$$\nabla_\nu V^\mu u^\nu e_\mu = \lim_{\delta\lambda \to 0} \left(\frac{V_\parallel(\lambda + \delta\lambda) - V(\lambda)}{\delta\lambda} \right), \tag{7.11}$$

where $V_\parallel(\lambda + \delta\lambda)$ represents the vector $V(\lambda)$ parallel transported along the curve parameterized by λ from λ to $\lambda + \delta\lambda$, a basis vector field $e_\mu(\lambda)$ is assumed to be defined in the vicinity of the curve, $u^\nu \equiv dx^\nu/d\lambda$, and we use the symbol ∇_ν in (7.11), anticipating equivalence with the covariant derivative. From Eq. (7.11), we must now understand how to transport a vector from λ to $\lambda + \delta\lambda$ while preserving its intrinsic properties.

7.4.1 Parallel Transport of Vectors

For a flat space the tangent space is congruent to the space itself and there is no conceptual difficulty in comparing two vectors at different points. Just move one vector – keeping its orientation fixed with respect to a global set of coordinate axes – to the position of the other vector and compare. However, as suggested by Fig. 7.2(a), on a curved surface this issue becomes more complicated because the tangent spaces in which the vectors are defined are different at the two points. As Figs. 7.2(b) and 7.2(c) illustrate, a vector transported along a path in a curved manifold such that at each small step it makes the same angle relative to the path (which can serve as an operational definition of parallel transport and defines the "straightest possible path") gets rotated as it is transported. Hence, a comparison of vectors at two points must untangle how much of any difference is an intrinsic difference

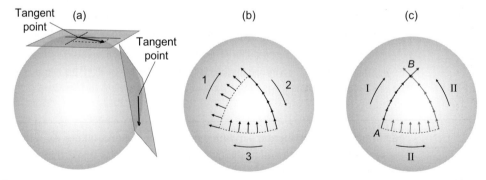

Fig. 7.2 (a) Tangent spaces and vectors in curved spaces (see Fig. 3.1). (b) Parallel transport of a vector in a closed path on a curved surface. The vector rotates by 90° for parallel transport on the closed path 1 → 2 → 3. (c) Dependence of parallel transport on the path taken. Parallel transport from A to B on the direct path labeled I rotates the vector by a different amount than for parallel transport from A to B on the two-segment path labeled II.

between the vectors and how much results from transformation of the coordinate system. To proceed it is desirable to specify more formally the meaning of parallel transport for vectors in curved manifolds.

7.4.2 The Affine Connection and Covariant derivatives

Generalizing the discussion in Section 2.3 to curved spacetime, differentiation of a 4-vector $V = V^\mu e_\mu$ with respect to a parameter λ gives two contributions,

$$\frac{dV}{d\lambda} = \frac{d}{d\lambda}\left(V^\mu e_\mu\right) = \frac{dV^\mu}{d\lambda} e_\mu + \frac{de_\mu}{d\lambda} V^\mu, \tag{7.12}$$

where the first term represents the change in the vector components in a fixed coordinate system defined by the basis e_μ and the second term represents the change in the coordinate system in moving from one point to another. Introducing $u^\nu \equiv dx^\nu/d\lambda$, this may be written

$$\frac{dV}{d\lambda} = \partial_\nu V^\mu u^\nu e_\mu + V^\mu u^\nu \partial_\nu e_\mu. \tag{7.13}$$

It is reasonable to expect that for infinitesimal separation between x and x' the second term will be linear in V^μ, so it can be expanded in the vector basis e_μ as in Eq. (2.36), giving (Problem 7.20)

$$\frac{dV}{d\lambda} = \left(\partial_\nu V^\mu + \Gamma^\mu_{\alpha\nu} V^\alpha\right) u^\nu e_\mu, \tag{7.14}$$

where $\Gamma^\mu_{\alpha\nu}$ is called the *affine connection coefficient* or often just the *connection*. The coefficients $\Gamma^\mu_{\alpha\nu}$ define a *connection* between tangent spaces at different points of the manifold, which permits a vector in the tangent space at one point to be parallel transported (moved, preserving lengths and angles locally) and compared with a vector defined in the (different) tangent space at another point. Physically $\Gamma^\alpha_{\mu\nu}$ may be interpreted as the α component of the rate of change of the basis vector e_μ by a displacement in the direction specified by e_ν.

Use of the same notation for the connection coefficient of Eq. (7.14) as for the Christoffel symbol of Eq. (3.51) is deliberate because *the connection coefficient and the Christoffel symbol may be viewed as equivalent,* as discussed in Box 7.5. Just as for the Christoffel symbol, don't be fooled by the surfeit of indices on the connection coefficient; it has indices aplenty but it does not transform as a tensor. (Which is exactly the point! If it *did* transform as a tensor it would be of no use to us.) It will be shown later that the affine connection can be constructed from the metric tensor and its derivatives, that it vanishes in a space with constant metric, and that the Riemann tensor describing the local intrinsic curvature of the spacetime may be constructed from the affine connection. Thus it is central to developing the theory of general relativity. Assuming equivalence of the Christoffel coefficients and connection coefficients, comparison of Eq. (7.14) with Eq. (3.55) indicates that the quantity in parentheses in Eq. (7.14) corresponds to the *covariant derivative of the vector* V^μ,

$$\nabla_\nu V^\mu = \partial_\nu V^\mu + \Gamma^\mu_{\alpha\nu} V^\alpha. \tag{7.15}$$

The covariant derivative represents the change of the *whole vector V*, not only its components V^μ.

> **Box 7.5** **Connection Coefficients and Christoffel Symbols**
>
> As noted in the text, Christoffel symbols and connection coefficients have been denoted by the same symbols because they may be viewed as equivalent. Specifically, Christoffel symbols are connection coefficients expressed in a coordinate basis. The German mathematician Erwin Bruno Christoffel (1821–1900) introduced what are now called Christoffel symbols and covariant derivatives to enable covariant differentiation of tensor fields. The connection coefficient described here is often termed the *Levi-Civita connection*, after Italian mathematician Tullio Levi-Civita (1873–1941), whose writings and correspondence were instrumental in Einstein's assimilation of tensor calculus. Einstein is said to have replied to a question about what he liked best about Italy, "spaghetti and Levi-Civita", and once wrote in a letter to Levi-Civita, "I admire the elegance of your method of computation; it must be nice to ride through these fields upon the horse of true mathematics while the likes of us have to make our way laboriously on foot."

7.4.3 Absolute Derivatives and Parallel Transport

For a euclidean or pseudo-euclidean (Minkowski) manifold, parallel transport of a vector along a path parameterized by λ means that the length and direction of the vector (as referenced to a universal cartesian coordinate system that can be defined for the flat space) don't change, implying that the components satisfy $dA^\mu/d\lambda = 0$. For a more general Riemannian or pseudo-Riemannian manifold there is no universal cartesian coordinate system and this condition for parallel transport of a vector along a curve parameterized by λ generalizes to

$$\nabla_\nu V^\mu u^\nu = \partial_\nu V^\mu u^\nu + \Gamma^\mu_{\alpha\nu} V^\alpha u^\nu = 0, \tag{7.16}$$

where the vector V was expanded in a basis $V = V^\mu e_\mu$ and $u^\nu = dx^\nu/d\lambda$. But in Eq. (7.16) $\partial_\nu V^\mu u^\nu = dV^\mu/d\lambda$ by the chain rule and comparison with Eq. (3.65) indicates that the condition for parallel transport is equivalent to

$$\frac{DV^\mu}{D\lambda} = 0, \tag{7.17}$$

where $D/D\lambda$ is the absolute derivative along the path parameterized by λ. This implies a fundamental relationship between the covariant derivative evaluated along a path and parallel transport of vectors in curved spaces, and thence with covariant differentiation. For flat spaces covariant differentiation is equivalent to ordinary differentiation and parallel transport may be implemented using ordinary derivatives, but in curved spaces it is the covariant derivative that implements parallel transport. Combining (7.17) with (3.65), parallel transport of the components of a vector along a curve parameterized by λ requires that

$$\frac{dV^\mu}{d\lambda} = -\Gamma^\mu_{\alpha\nu} V^\alpha \frac{dx^\nu}{d\lambda}, \tag{7.18}$$

which represents transport along the curve such that at each infinitesimal step the vector is parallel transported. In a curved space, this is the nearest that one can approximate the euclidean-space prescription for parallel transport of vectors.

7.4.4 Geometry and Covariant Derivatives

A prescription for parallel transport defines a unique map from a vector to vectors that are infinitesimally nearby (the uniqueness of this map will be discussed in Section 7.8), and the covariant derivative then uses this map to determine the infinitesimal change in the vector field under transport. Did you see that coming? The covariant derivative, which was introduced as an *algebraic constraint* in Section 3.13.3 to ensure that taking derivatives preserves tensor structure is now seen to correspond equivalently to a *geometrical constraint* on parallel transport ensuring that the difference of two vectors at different points of the manifold (and thus the derivative) is taken correctly.[3]

From the preceding discussion (see also Section 2.3), for curved spaces the covariant derivative differs from the partial derivative because the coordinate basis vectors change from point to point in the manifold. This makes clear why for scalars partial differentiation is equivalent to covariant differentiation: scalars do not depend on the basis vectors, so the covariant derivative of a scalar is just the ordinary partial derivative. Likewise, for flat spaces the basis vectors can be chosen constant and the covariant derivative is just the partial derivative. But for rank-1 and higher tensors in curved manifolds basis vectors will generally depend on coordinates and covariant derivatives will not be equivalent to partial derivatives.

7.5 Gravity and Curved Spacetime

Now we shall employ the metric and curvature of spacetime, and the principle of equivalence expressed in terms of general covariance, to construct in the remainder of this chapter and in the next the general theory of relativity and the corresponding theory of gravitation. The starting point is to take seriously the suggestion from the equivalence principle that gravity is not a force in the normal sense but rather is a consequence of motion that takes place in curved spacetime.

7.5.1 Free Particles

Consider a particle moving solely under the influence of gravity (in a gravitational field but experiencing no electromagnetic, weak, or strong forces). Such a particle is termed a *free particle* in general relativity because it may be viewed as moving with no force acting on it, but in a curved spacetime. By the principle of equivalence, in a freely falling coordinate system labeled by coordinates ξ^μ the special theory of relativity is valid and the equation of motion is given by the special relativistic generalization of Newton's second law [see Eqs. (5.12) and (4.5)],

---

[3] Many authors choose to introduce covariant differentiation in terms of parallel transport, rather than introducing it algebraically and then later interpreting it in terms of parallel transport, as we have done here.

$$\frac{d^2\xi^\mu}{d\tau^2} = 0 \qquad d\tau^2 = -\eta_{\mu\nu}d\xi^\mu d\xi^\nu, \qquad (7.19)$$

where the metric is the $\eta_{\mu\nu} = \text{diag}(-1, 1, 1, 1)$ of Minkowski space.

7.5.2 The Geodesic Equation

Following the discussion in Weinberg [246], introduce another *arbitrary* coordinate system x^μ that is not necessarily inertial. The freely falling coordinates ξ^μ are functions of the new coordinates, $\xi^\mu = \xi^\mu(x)$, and the chain rule may be used to write

$$\frac{d^2\xi^\alpha}{d\tau^2} = \frac{d}{d\tau}\left(\frac{d\xi^\alpha}{d\tau}\right) = \frac{d}{d\tau}\left(\frac{\partial\xi^\alpha}{\partial x^\mu}\frac{dx^\mu}{d\tau}\right) = 0,$$

which becomes, upon taking the derivative of the product and using the chain rule,

$$\frac{\partial\xi^\alpha}{\partial x^\mu}\frac{d^2x^\mu}{d\tau^2} + \frac{\partial^2\xi^\alpha}{\partial x^\mu \partial x^\nu}\frac{dx^\mu}{d\tau}\frac{dx^\nu}{d\tau} = 0.$$

Multiplying through by $\partial x^\lambda / \partial \xi^\alpha$ and employing

$$\frac{\partial x^\lambda}{\partial \xi^\alpha}\frac{\partial \xi^\alpha}{\partial x^\mu} = \delta^\lambda_\mu \qquad (7.20)$$

yields the *geodesic equation*,

$$\frac{d^2x^\lambda}{d\tau^2} + \Gamma^\lambda_{\mu\nu}\frac{dx^\mu}{d\tau}\frac{dx^\nu}{d\tau} = 0, \qquad (7.21)$$

where the affine connection $\Gamma^\lambda_{\mu\nu}$ is defined by

$$\Gamma^\lambda_{\mu\nu} = \frac{\partial x^\lambda}{\partial \xi^\alpha}\frac{\partial^2 \xi^\alpha}{\partial x^\mu \partial x^\nu}, \qquad (7.22)$$

and is equivalent to the affine connection employed in Eq. (7.14) and to the Christoffel symbol of Eq. (3.51) (see Box 7.5). From Eq. (5.1) the geodesic equation (7.21) may also be expressed in terms of 4-velocities u,

$$\frac{du^\lambda}{d\tau} + \Gamma^\lambda_{\mu\nu}u^\mu u^\nu = 0. \qquad (7.23)$$

The proper time interval in the arbitrary coordinate system takes the form

$$d\tau^2 = -\eta_{\alpha\beta}d\xi^\alpha d\xi^\beta = -\eta_{\alpha\beta}\frac{\partial\xi^\alpha}{\partial x^\mu}dx^\mu \frac{\partial\xi^\beta}{\partial x^\nu}dx^\nu$$

$$= -\eta_{\alpha\beta}\frac{\partial\xi^\alpha}{\partial x^\mu}\frac{\partial\xi^\beta}{\partial x^\nu}dx^\mu dx^\nu = -g_{\mu\nu}dx^\mu dx^\nu, \qquad (7.24)$$

where the metric tensor is defined by

$$g_{\mu\nu} \equiv \frac{\partial\xi^\alpha}{\partial x^\mu}\frac{\partial\xi^\beta}{\partial x^\nu}\eta_{\alpha\beta}, \qquad (7.25)$$

and is clearly modified from its form $\eta_{\mu\nu}$ in the local inertial system.

7.6 The Local Inertial Coordinate System

It will now be demonstrated that the affine connection $\Gamma^\lambda_{\mu\nu}$ and the metric tensor $g_{\mu\nu}$ at a point X in an arbitrary coordinate system x^μ are sufficient to define the local inertial coordinates ξ^α in the neighborhood of X [246]. A differential equation for the local inertial coordinates may be obtained by multiplying Eq. (7.22) through by $\partial \xi^\beta / \partial x^\lambda$ and utilizing Eq. (7.20) to give

$$\frac{\partial^2 \xi^\alpha}{\partial x^\mu \partial x^\nu} = \Gamma^\lambda_{\mu\nu} \frac{\partial \xi^\alpha}{\partial x^\lambda}. \tag{7.26}$$

This has a power series solution that may be expressed near the point X as

$$\xi^\alpha(x) = a^\alpha + b^\alpha{}_\mu (x^\mu - X^\mu) + \tfrac{1}{2} b^\alpha{}_\lambda \Gamma^\lambda_{\mu\nu} (x^\mu - X^\mu)(x^\nu - X^\nu) + \cdots, \tag{7.27}$$

where $a^\alpha \equiv \xi^\alpha(X)$ and $b^\alpha{}_\lambda \equiv \partial \xi^\alpha(X)/\partial X^\lambda$. In addition, from these results the metric tensor (7.25) may be written in the form

$$g_{\mu\nu}(X) = b^\alpha{}_\mu b^\beta{}_\nu \eta_{\alpha\beta}. \tag{7.28}$$

Thus, $\Gamma^\lambda_{\mu\nu}$ and $g_{\mu\nu}$ at the point X determine the local inertial coordinates ξ^α up to order $(x - X)^2$. This is sufficient, since the inertial coordinates are valid only in the vicinity of the point X.[4]

7.7 The Affine Connection and the Metric Tensor

The preceding derivation has shown that: (1) the affine connection $\Gamma^\lambda_{\mu\nu}$ determines the gravitational force and therefore may be viewed as the gravitational field, and (2) the metric tensor determines the properties of the interval $d\tau$. Now it will be shown that, in fact, $g_{\mu\nu}$ determines the *full effect of gravitation* because $\Gamma^\lambda_{\mu\nu}$ can be expressed in terms of the metric tensor and its derivatives, and thus the metric tensor may be viewed as the gravitational potential in general relativity. Continuing to follow Weinberg [246], first differentiate Eq. (7.25) with respect to x^λ:

$$\frac{\partial g_{\mu\nu}}{\partial x^\lambda} = \frac{\partial^2 \xi^\alpha}{\partial x^\lambda \partial x^\mu} \frac{\partial \xi^\beta}{\partial x^\nu} \eta_{\alpha\beta} + \frac{\partial \xi^\alpha}{\partial x^\mu} \frac{\partial^2 \xi^\beta}{\partial x^\lambda \partial x^\nu} \eta_{\alpha\beta}.$$

But the inertial coordinates ξ^α have been shown already to obey the differential equation (7.26). Inserting this in the preceding equation yields

$$\frac{\partial g_{\mu\nu}}{\partial x^\lambda} = \Gamma^\rho_{\lambda\mu} \frac{\partial \xi^\alpha}{\partial x^\rho} \frac{\partial \xi^\beta}{\partial x^\nu} \eta_{\alpha\beta} + \Gamma^\rho_{\lambda\nu} \frac{\partial \xi^\alpha}{\partial x^\mu} \frac{\partial \xi^\beta}{\partial x^\rho} \eta_{\alpha\beta}$$

$$= \Gamma^\rho_{\lambda\mu} g_{\rho\nu} + \Gamma^\rho_{\lambda\nu} g_{\rho\mu}, \tag{7.29}$$

---

[4] Strictly, if ξ^α are local inertial coordinates so are the coordinates $\xi'^\alpha = \Lambda^\alpha{}_\beta \xi^\beta + c^\alpha$ obtained by a Poincaré transformation (see Box 7.1). Thus, the local inertial coordinates are determined only up to an arbitrary Poincaré transformation. Since the laws of physics are absolutely inviolate with respect to Poincaré transformations in inertial systems, no physical observables depend on this ambiguity.

where (7.25) has been used. This equation may be solved for the connection $\Gamma^{\rho}_{\lambda\mu}$ by observing that if the indices $\lambda \leftrightarrow \mu$ and $\lambda \leftrightarrow \nu$ are switched in the above expression one can write

$$\frac{\partial g_{\mu\nu}}{\partial x^{\lambda}} + \frac{\partial g_{\lambda\nu}}{\partial x^{\mu}} - \frac{\partial g_{\mu\lambda}}{\partial x^{\nu}} = g_{\rho\nu}\Gamma^{\rho}_{\lambda\mu} + g_{\rho\mu}\Gamma^{\rho}_{\lambda\nu} + g_{\rho\nu}\Gamma^{\rho}_{\mu\lambda}$$
$$+ g_{\rho\lambda}\Gamma^{\rho}_{\mu\nu} - g_{\rho\lambda}\Gamma^{\rho}_{\nu\mu} - g_{\rho\mu}\Gamma^{\rho}_{\nu\lambda}.$$

But both $\Gamma^{\rho}_{\mu\nu}$ and $g_{\mu\nu}$ are symmetric in their lower indices (see Section 7.8), so this reduces to

$$\frac{\partial g_{\mu\nu}}{\partial x^{\lambda}} + \frac{\partial g_{\lambda\nu}}{\partial x^{\mu}} - \frac{\partial g_{\mu\lambda}}{\partial x^{\nu}} = 2g_{\rho\nu}\Gamma^{\rho}_{\lambda\mu}.$$

Multiplying by $g^{\nu\sigma}$ and using $g^{\nu\sigma}g_{\rho\nu} = \delta^{\sigma}_{\rho}$ gives finally

$$\Gamma^{\sigma}_{\lambda\mu} = \tfrac{1}{2}g^{\nu\sigma}\left(\frac{\partial g_{\mu\nu}}{\partial x^{\lambda}} + \frac{\partial g_{\lambda\nu}}{\partial x^{\mu}} - \frac{\partial g_{\mu\lambda}}{\partial x^{\nu}}\right). \tag{7.30}$$

Thus the connection, and thence the gravitational field, are *determined completely by the metric tensor and its first derivatives.*

Equation (7.30) gives a method for determining the connection coefficients entirely from the properties of the metric that is well-suited to computer implementation. However, in finding connection coefficients by hand a method illustrated in Problem 7.4 that is based on comparing the Euler–Lagrange equation of motion with the geodesic equation often proves more efficient than direct computation from Eq. (7.30). The problems at the end of this chapter give practice in using both methods.

7.8 Uniqueness of the Affine Connection

The affine connection is an *additional feature* imposed on the differential structure of a manifold through a definition that is *not unique*. For example, recall that the result of parallel transport – which is implemented through the connection – depends on the path taken. However, if a manifold has both a metric and a connection defined for it, certain *compatibility demands* are usually imposed that constrain the connection. In particular, the connection and the metric are said to be *compatible* if the inner (scalar) product of two arbitrary vectors is preserved under parallel transport.

In proving Eq. (7.30), symmetry of the connection under exchange of its lower indices has been assumed. The *torsion tensor* $T^{\lambda}_{\mu\nu}$ is defined by

$$T^{\lambda}_{\mu\nu} \equiv \Gamma^{\lambda}_{\mu\nu} - \Gamma^{\lambda}_{\nu\mu}. \tag{7.31}$$

(Note that neither term on the right side of Eq. (7.31) is a tensor but their difference is.) Then, it can be shown that the connection defined on a manifold with metric $g_{\mu\nu}$ is unique and determined completely by the metric if we assume that

1. The connection is torsion-free, $T^\lambda_{\ \mu\nu} = 0$.
2. The covariant derivative (defined in terms of the connection) of the metric tensor vanishes everywhere, $\nabla_\alpha g_{\mu\nu} = g_{\mu\nu;\alpha} = 0$, which is sufficient to guarantee that the scalar product of vectors is preserved under parallel transport (see Problem 7.2 and Problem 7.12).

In this case the connection is said to be a *metric connection* or it is said to be *metric compatible*. Equation (7.30) determines the affine connection uniquely in terms of the metric tensor as a consequence of the *assumptions* (1) and (2) above. These assumptions are then justified after the fact by the correctness of the resulting theory when compared with observations.

Background and Further Reading

At the introductory level, see See Hartle [111], Cheng [65], and d'Inverno [76]. For more advanced discussions, see Weinberg [246]; Schutz [220]; and Misner, Thorne, and Wheeler [156].

Problems

7.1 Consider the general expression for Gaussian curvature K in a 2-dimensional space that is defined in Eq. (7.2).

(a) Show that $K = 0$ for any metric that is globally position-independent. Thus, since a 2-dimensional planar surface can be parameterized globally in cartesian coordinates, its Gaussian curvature is zero.

(b) Show that if the planar surface from part (a) is parameterized instead in polar coordinates (r, θ), the metric becomes position-dependent but again $K = 0$, indicating that the intrinsic surface is flat, even though this parameterization gives a position-dependent metric.

(c) Show that for a spherical surface parameterized with the polar coordinates (S, φ) defined in Fig. 7.1, the Gaussian curvature is $K = 1/R^2$, indicating that this surface has non-zero but constant intrinsic curvature.

(d) Show that the same result as in (c) is obtained if the spherical surface is parameterized instead by the cylindrical coordinates (r, φ) of Fig. 7.1.

These results indicate that a position-independent metric is sufficient to ensure $K = 0$ but that the parameterization may give a position-dependent metric, even if the space is intrinsically flat, and that K is an intrinsic measure of curvature that is independent of the parameterization chosen. ***

7.2 Write an expression for the covariant derivative of the metric tensor $g_{\mu\nu}$. Use general covariance and the principle of equivalence to show that it vanishes in all frames. Verify the correctness of this result by computing the covariant derivative of $g_{\mu\nu}$ explicitly from the connection coefficients. ***

7.3 Write the line element for cylindrical polar coordinates in 3-dimensional euclidean space. Find $g_{\mu\nu}$, $g^{\mu\nu}$, det g, and the geodesic equations for this case.

7.4 Christoffel symbols (equivalently, affine connections) $\Gamma^{\gamma}_{\alpha\beta}$ may be constructed directly from Eq. (7.30), but often a faster way to obtain them is by comparing (term by term) the Euler–Lagrange equations of motion corresponding to Eqs. (5.18) and (5.16) with the geodesic equation of motion (7.21). As a simple illustration of this method, consider a line element $ds^2 = dr^2 + r^2 d\theta^2$ for a 2-dimensional space parameterized in polar coordinates (r, θ). Construct the Euler–Lagrange equations for motion on a 1-dimensional path parameterized by a variable σ ranging from 0 to 1 and deduce the non-zero Christoffel symbols Γ^k_{ij} by comparing with the terms of the geodesic equations (7.21). Check your results by computing the Γ^k_{ij} for this case directly from Eq. (7.30). This is a simple 2-dimensional example but a similar technique can be used in 4-dimensional curved spacetime. ***

7.5 Consider the metric implied by the line element

$$ds^2 = -dt^2 + dr^2 + (\rho^2 + r^2)(d\theta^2 + \sin^2\theta \, d\varphi^2),$$

where ρ is a constant. Find all non-vanishing Christoffel symbols $\Gamma^{\gamma}_{\alpha\beta}$.

7.6 A possible metric describing the large-scale structure of space and time in our Universe is defined by the line element

$$ds^2 = -dt^2 + a(t)^2(dx^2 + dy^2 + dz^2),$$

where $a(t)$ is a *scale factor* that changes smoothly with time. Find the non-vanishing Christoffel symbols $\Gamma^{\gamma}_{\alpha\beta}$ for this metric.

7.7 The angular excess ϵ (deviation of the sum of interior angles for a polygon from that expected for 2-dimensional flat space) is related to the area A of the polygon by $\epsilon = KA$, where K is the Gaussian curvature for the 2-dimensional surface. Starting from the formula for the area of a lune (figure bounded by segments of two great circles), verify this explicitly for the special case of triangles on 2-dimensional spheres.

7.8 Show that the equation of motion for flat space, $d^2x^\mu/d\sigma^2 = 0$, where σ parameterizes the paths, is transformed to the geodesic equation (7.21) for motion in curved space by the prescription of replacing partial derivatives by covariant derivatives. *Hint:* Use the chain rule to rewrite $d^2x^\mu/d\sigma^2$.

7.9 Prove that for a connection coefficient contracted on one upper and one lower index, $\Gamma^\sigma_{\lambda\sigma} = \frac{1}{2} g^{\nu\sigma} g_{\nu\sigma,\lambda}$.

7.10 (a) Prove that the covariant derivative of the Kronecker delta field δ^μ_ν defined on a manifold is zero. (b) Use the result from part (a) to prove that $g^{\mu\nu}$ has vanishing covariant derivative if the connection on the manifold is a metric connection. *Hint:* Use $\delta^\mu_\nu = g^{\mu\epsilon} g_{\epsilon\nu}$.

7.11 Show that the condition (7.17) for parallel transport implies that a geodesic defined by Eq. (7.21) represents the straightest possible path in that it may be constructed by parallel transport of its own tangent vector. *Hint:* Set the general vector in Eq. (3.65) equal to the tangent vector of the parameterized curve.

7.12 Use the absolute derivative defined in Section 3.13.4 to prove that the inner (scalar) product of two vectors is preserved under parallel transport on a path if the connection defined on the manifold is a metric connection. ***

7.13 In the limit of weak, slowly varying gravitational fields it may be assumed that the metric deviates only slightly from that for flat space:

$$g_{\mu\nu} = \eta_{\mu\nu} + h_{\mu\nu}(x),$$

where $\eta_{\mu\nu}$ is the constant Minkowski metric and $h_{\mu\nu}$ represents a small spacetime-dependent correction that is approximately constant in time. Show that the connections coefficients may be approximated by

$$\Gamma^\sigma_{0\mu} \simeq \frac{1}{2}\eta^{\nu\sigma}\left(\frac{\partial h_{0\nu}}{\partial x^\mu} - \frac{\partial h_{0\mu}}{\partial x^\nu}\right) \qquad \Gamma^\sigma_{00} \simeq -\frac{1}{2}\eta^{\nu\sigma}\frac{\partial h_{00}}{\partial x^\nu}$$

in this limit.

7.14 Prove that in a local inertial frame the connection coefficients vanish but their first derivatives do not.

7.15 If a 4-vector V is expanded in a basis e_μ that may be position dependent, $V = V^\mu e_\mu$, the derivative of V is

$$\frac{\partial V}{\partial x^\nu} = \frac{\partial V^\mu}{\partial x^\nu}e_\mu + V^\mu\frac{\partial e_\mu}{\partial x^\nu},$$

where the first term is associated with the change in the components and the second term is associated with the change in the basis. (See Section 2.3.) The partial derivative factors in the second term are themselves vectors that may be expanded in the basis,

$$\frac{\partial e_\mu}{\partial x^\nu} = \Gamma^\lambda_{\mu\nu}e_\lambda,$$

where the expansion coefficients $\Gamma^\lambda_{\mu\nu}$ are just the connection coefficients (Christoffel symbols) discussed in Sections 3.13.3 and 7.4. Consider as a simple example 2D euclidean space, parameterized in (constant) cartesian coordinates (x^1, x^2) or (position-dependent) polar coordinates (r, θ). Express basis vectors for the polar coordinates in terms of the cartesian coordinates and use the expansion in the equation for $\partial e_\mu/\partial x^\nu$ above to evaluate the eight $\Gamma^\lambda_{\mu\nu}$ in polar coordinates. Check your result using the formula given by Eq. (7.30).

7.16 Consider a 2D space with coordinates (φ, θ) and a line element

$$ds^2 = \cos^2\varphi\, d\varphi^2 + \sin^2\varphi\, d\theta^2,$$

with the variable θ ranging from 0 to 2π. By applying Eq. (7.5) to circles drawn around the origin, determine whether this space is flat or curved.

7.17 Consider a 2D space with coordinates (r, θ) and a line element

$$ds^2 = (1 + \alpha r^2)dr^2 + r^2 d\theta^2,$$

with the variable θ ranging from 0 to 2π. Apply Eq. (7.5) to circles drawn around the origin of this space to determine the Gaussian curvature.

7.18 The stress–energy tensor for a perfect fluid is given by $T_{\mu\nu} = (\epsilon + P)u_\mu u_\nu + P g_{\mu\nu}$, where ϵ is the energy density, u is the 4-velocity, P is the pressure, and $g_{\mu\nu}$ is the metric tensor. Show that in a frame where the fluid is at rest, $T^\mu{}_\nu = \text{diag}(-\epsilon, P, P, P)$. *Hint*: Use the requirement of Eq. (5.6) that $u \cdot u = -1$. ***

7.19 Assume that a vector field $V(\lambda)$ is defined along a curve $x^\mu(\lambda)$ parameterized by λ for some manifold. Prove that the covariant derivative of a vector V evaluated along the path parameterized by λ is given by Eq. (3.65), assuming that $V(\lambda) = V^\mu(\lambda) e_\mu(\lambda)$, where $e_\mu(\lambda)$ is a basis vector evaluated at the point on the curve labeled by λ. ***

7.20 (a) Utilizing the arguments as in Section 2.3, derive Eq. (7.14) from Eq. (7.12).
(b) Show that Eq. (7.14) is equivalent to the definition of the covariant derivative for a vector given in Eq. (3.55). ***

8 The General Theory of Relativity

Einstein's theory of gravity is a relationship between mass and curvature that implements the principles of equivalence and general covariance. Most of the tools have been assembled in preceding chapters and we are now ready in this relatively short chapter to construct the *Einstein field equations,* which are the fundamental equations describing a covariant theory of gravity. Our procedure will be to use the success of Newtonian gravity as a guide to what the correct theory should look like in the weak-field limit, and to basically (as Einstein did) use this limit to guess the correct form of the field equations. However, even with the field equations in hand we will be only partway to our goal. To apply general relativity to physical systems it is necessary to *solve* these formidable equations. Hence, we will conclude the chapter by giving a brief introduction to methods that might be used to find solutions of the field equations.

8.1 Weak-Field Limit

Given the success of Newton's theory of gravity in most applications, it may be expected that the weak-field limit of Einstein's theory must reduce to Newton's theory. Therefore, let us begin by considering the limit of Einstein's theory when the gravitational field is very weak as a guide to the structure of the full theory. Because general relativity is a field theory, it will prove convenient to formulate Newtonian gravity also as a field theory. The Newtonian gravitational field may be derived from a scalar potential φ that obeys the *Poisson equation,*

$$\nabla^2 \varphi = 4\pi G \rho, \tag{8.1}$$

where $\nabla \equiv \boldsymbol{i}\dfrac{\partial}{\partial x} + \boldsymbol{j}\dfrac{\partial}{\partial y} + \boldsymbol{k}\dfrac{\partial}{\partial z}$. The Newtonian equation of motion for a particle is then

$$F^i = \frac{d^2 x^i}{dt^2} = -\frac{\partial \varphi}{\partial x^i}, \tag{8.2}$$

where $\boldsymbol{F}$ is the gravitational force and unit mass was assumed. For a point-like source mass M the potential is $\varphi = -GM/r$. Earlier it was shown that the geodesic equation of motion is given by Eq. (7.21). If the space is flat $g_{\mu\nu}(x) = \eta_{\mu\nu} =$ constant, so $\partial g_{\mu\nu}(x)/\partial x^\alpha = 0$ and the affine connection (7.30) vanishes, the covariant derivatives are equivalent to the normal partial derivatives, and the equation of motion (7.21) becomes that of a free particle in Minkowski space. Generally though spacetime is curved by mass and the second term of Eq. (7.21) does not vanish, so let us consider that case for small deviations from a flat metric.

Assuming for the moment gravitational fields that are static and weak, with characteristic velocities well below c,

$$\frac{\partial g_{\mu\nu}}{\partial x^0} = 0 \qquad \frac{dx^i}{d\tau} \ll \frac{dx^0}{d\tau} \simeq 1, \tag{8.3}$$

the equation of motion (7.21) becomes ($c = 1$ units)

$$\frac{d^2 x^\mu}{d\tau^2} + \Gamma^\mu_{00}\left(\frac{dx^0}{d\tau}\right)^2 = 0, \tag{8.4}$$

and the connection (7.30) reduces to

$$\Gamma^\mu_{00} = \frac{1}{2} g^{\mu\alpha}\left(\frac{\partial g_{0\alpha}}{\partial x^0} + \frac{\partial g_{0\alpha}}{\partial x^0} - \frac{\partial g_{00}}{\partial x^\alpha}\right) = -\frac{1}{2} g^{\mu\alpha}\frac{\partial g_{00}}{\partial x^\alpha}. \tag{8.5}$$

The field is assumed weak so the metric can be expanded around the flat metric,

$$g_{\mu\nu} = \eta_{\mu\nu} + h_{\mu\nu} \tag{8.6}$$

where $h_{\mu\nu}$ is small relative to $\eta_{\mu\nu}$. Then, $\partial g_{00}/\partial x^\alpha = \partial h_{00}/\partial x^\alpha$ and to lowest order

$$\Gamma^\mu_{00} = -\frac{1}{2} \eta^{\mu\alpha}\frac{\partial h_{00}}{\partial x^\alpha}. \tag{8.7}$$

Since the metric $\eta^{\mu\nu}$ is diagonal the relevant connection components are

$$\Gamma^0_{00} = \frac{1}{2}\frac{\partial h_{00}}{\partial x^0} = 0 \qquad \Gamma^i_{00} = -\frac{1}{2}\frac{\partial h_{00}}{\partial x^i}, \tag{8.8}$$

which gives for the equations of motion (Problem 8.5)

$$\frac{d^2 x^0}{d\tau^2} = 0 \qquad \frac{d^2 x^i}{dt^2} = \frac{1}{2} c^2 \frac{\partial h_{00}}{\partial x^i}, \tag{8.9}$$

where factors of c have been restored. Consistency with the Newtonian equation (8.2) requires that $h_{00} = -2\varphi/c^2$ and thus that

$$g_{00} = \eta_{00} + h_{00} = -\left(1 + \frac{2\varphi}{c^2}\right), \tag{8.10}$$

implying a scalar-field source for weak gravity having the form

$$\varphi = -\tfrac{1}{2} c^2 (g_{00} + 1). \tag{8.11}$$

Thus the weak-gravity limit yields a strong glimmer of the Einstein conjecture that gravity derives from the geometry of spacetime, with the metric tensor $g_{\mu\nu}$ as its source.

Example 8.1 For the specific case of a weak, static field at the surface of a gravitating sphere of radius R and mass M, the potential is $\varphi = -GM/R$ and Eq. (8.10) may be written as

$$g_{00} = -\left(1 - \frac{2GM}{Rc^2}\right). \tag{8.12}$$

The second term on the right side of Eq. (8.12) measures the deviation of the metric from that for flat space. It is proportional to the radius of an object divided by the radius of curvature associated with the surface gravitational field, or equivalently to the ratio of gravitational energy to rest mass energy for a test particle at the surface, and thus measures the strength of the gravitational field (see Section 6.4.2). This term is of order 10^{-9} for the Earth, 10^{-6} for the Sun, and is still only of order 10^{-4} for a white dwarf.

As the preceding example shows, for objects up to white dwarf density the deviation of the metric from that for flat space is extremely small and Newtonian gravity is an excellent approximation. For a typical neutron star the second term of Eq. (8.12) is an appreciable fraction of one, invalidating the assumptions of the preceding weak-gravity derivation. Neutron star densities and higher signal the onset of significant effects from the curvature of spacetime and non-negligible general relativistic corrections to Newtonian gravity.

8.2 Recipe for Motion in a Gravitational Field

The preceding discussion suggests a recipe for writing the equations of motion in a gravitational field by applying the principles of equivalence and general covariance.

1. Invoke the equivalence principle to justify a local Minkowski coordinate system ξ^μ and formulate the appropriate equations of motion for flat Minkowski spacetime in tensor form in those coordinates.
2. Replace the Minkowski coordinates ξ^μ by general curvilinear coordinates x^μ in all equations. (This is similar to converting from local euclidean coordinates describing a small region on the surface of the Earth to general spherical coordinates that describe the Earth's surface globally.) The constant Minkowski metric $\eta_{\mu\nu}$ is thereby replaced by the general position-dependent metric $g_{\mu\nu}(x)$.
3. Replace all partial derivatives with the corresponding covariant derivatives and all integral volume elements by invariant volumes.

The resulting equations may be expected to describe physics in a gravitational field and, this procedure relates gravity to the curvature of spacetime. This recipe has been suggested by considering weak gravity, but it will now be shown that it is expected to be valid also for strong gravity.

8.3 Towards a Covariant Theory of Gravitation

Combining the Poisson equation (8.1), the density expressed in terms of the time–time component of the stress–energy tensor, $\rho = T_{00}/c^2$, and the weak-gravity scalar field $\varphi = -\frac{1}{2}c^2(g_{00} + 1)$ of Eq. (8.11), leads to

$$\nabla^2 g_{00} = -\frac{8\pi G}{c^4} T_{00}. \tag{8.13}$$

Equation (8.13) isn't covariant since it is expressed in terms of tensor components, not tensors, and it is expected to hold only for weak, slowly varying fields. Nevertheless, it may be used as a guide to guess the correct form of a fully covariant gravitational theory. Let us – with Einstein – conjecture the following concerning the form of a covariant theory of gravity:

1. The right side of (8.13) is not a tensor, but since the Newtonian limit of the equation being sought is proportional to one component of the stress–energy tensor, let us assume that the right side should be modified by the replacement $T_{00} \to T_{\mu\nu}$.
2. Since the right side now transforms as a rank-2 tensor (the constants are scalars), covariance requires that the left side of Eq. (8.13) be replaced by something having the same transformation properties. The following general requirements on the new left side may be surmised:

    (a) Since the weak-field limit is proportional to a curvature $\nabla^2 g_{00}$, the correct left side may be expected to be a covariant measure of spacetime curvature.
    (b) It must be a symmetric covariant tensor of rank 2 to match the tensorial properties of the right side.
    (c) It must be divergenceless with respect to covariant differentiation since $T_{\mu\nu}$ has that property, by virtue of Eq. (7.10).

    Thus, we require quantities that allow construction of a left side that is a symmetric, rank-2, covariant measure of spacetime curvature, with vanishing covariant divergence.
3. The candidate field equations must reduce to the Poisson equation describing Newtonian gravitation in the limit of weak, slowly varying fields and nonrelativistic velocities.

These conjectures and the preceding development will allow us to formulate covariant field equations for general relativity that reduce to Newtonian gravity in the weak-field limit. But before doing that there is one final and extremely important housekeeping task that must be implemented.

8.4 The Riemann Curvature Tensor

By the preceding reasoning we need a covariant measure of curvature in 4D spacetime. The earlier discussion of curvature was limited to Gaussian curvature in 2D space. We are now in a position to generalize that to a covariant description of curvature in 4D spacetime by using the connection coefficients introduced in Section 7.4.2. The required generalization is the *Riemann curvature tensor*, $R^{\sigma}{}_{\mu\nu\lambda}$, which may be defined through (see Problems 8.3 and 8.13)

$$R^{\sigma}{}_{\mu\nu\lambda} = \Gamma^{\sigma}_{\mu\lambda,\nu} - \Gamma^{\sigma}_{\mu\nu,\lambda} + \Gamma^{\sigma}_{\alpha\nu}\Gamma^{\alpha}_{\mu\lambda} - \Gamma^{\sigma}_{\alpha\lambda}\Gamma^{\alpha}_{\mu\nu}. \tag{8.14}$$

This tensor has symmetries under index exchange that are seen most easily after lowering the upper index through contraction with the metric tensor. Then it is found that the Riemann tensor obeys the conditions (see Problem 8.14)

> **Box 8.1** **Curvature and Space Dimensionality**
>
> The curvature tensor has n^4 components in n dimensions but symmetries reduce that to $n^2(n^2 - 1)/12$ independent components. Thus in 4D the curvature tensor has 20 independent components and in 3D it has 6 independent components. In 2D there are 15 symmetry relations on the $2^4 = 16$ components of the Riemann curvature tensor, leaving only 1 independent component of curvature – the Gaussian curvature of Section 7.2.2. In 1D curvature cannot be defined.
>
> **Intrinsic and Extrinsic Curvature**
>
> The assertion that curvature cannot be defined in 1D may not seem right – what about a curved line? But curvature is meant here as an *intrinsic property of a space* and in 1D there can be no intrinsic curvature. What is really meant by a "curved line" is the *embedding of a 1D surface in a higher-dimensional manifold*; the perceived curvature of the line then is a property of the embedding, not an intrinsic property of the 1D space. As discussed further by Carrol [63] in his Appendix D, a manifold embedded in a higher-dimensional space inherits an *induced metric* from the embedding and the apparent curvature of embedded spaces having no intrinsic curvature results from this induced metric. This is termed *extrinsic curvature*.
>
> In a similar vein, a cylinder is a *flat 2-dimensional surface*. (Its Gaussian curvature vanishes and it may be constructed by identifying two opposite edges of a flat surface.) This also may be verified by parallel transporting a vector in a closed rectangular path on the cylinder. Unlike for a sphere [see Fig. 7.2(b)], the vector remains unchanged under parallel transport on the cylinder. The curvature perceived for the cylinder is an artifact of viewing the 2D cylinder embedded in a 3D space. General relativity usually deals with *intrinsic curvature*, which is specified by the Riemann curvature tensor and is independent of any embedding in higher dimensions.
>
> **Gravity, Curvature, and Dimensionality**
>
> A geometric theory ascribing gravity to spatial curvature is self-consistent only in four or more dimensions. Thus, there would be no gravity in a universe of fewer dimensions than ours. Conversely, it has been proposed that the Universe actually has more than four dimensions and that the seeming weakness of gravity in our world is because its strength has "leaked" into unseen dimensions. This idea will be discussed further in Chapter 26.

$$R_{\sigma\mu\nu\lambda} = -R_{\mu\sigma\nu\lambda} = -R_{\sigma\mu\lambda\nu}$$
$$R_{\sigma\mu\nu\lambda} = R_{\nu\lambda\sigma\mu} \qquad R_{\sigma\mu\nu\lambda} + R_{\sigma\lambda\mu\nu} + R_{\sigma\nu\lambda\mu} = 0,$$
(8.15)

(the last relation is called the *cyclic identity*). Because of the symmetries (8.15) only 20 of the $4^4 = 256$ components of the Riemann tensor are independent in 4-dimensional spacetime (see the discussion in Box 8.1). All components of the Riemann tensor vanish in flat spacetime. Conversely, if it is found that if all components of the Riemann tensor vanish the geometry of spacetime is necessarily flat. The independent components of the Riemann curvature tensor thus represent the covariant generalization of the Gaussian curvature describing the intrinsic curvature of 2-dimensional space to a rank-4 tensor describing the intrinsic curvature of 4-dimensional spacetime.

8.5 The Einstein Equations

The curvature is described by the Riemann tensor but it is rank-4 and by the reasoning in Section 8.3 we need the left side of the equation being sought to correspond to a symmetric rank-2 tensor. This suggests that the Riemann tensor must enter in contracted form. The symmetric *Ricci tensor* $R_{\mu\nu}$ may be formed by contraction of the Riemann tensor with the metric tensor,[1]

$$R_{\mu\nu} = R_{\nu\mu} = g^{\lambda\sigma} R_{\lambda\mu\sigma\nu} = R^{\sigma}{}_{\mu\sigma\nu},$$
$$= \Gamma^{\lambda}_{\mu\nu,\lambda} - \Gamma^{\lambda}_{\mu\lambda,\nu} + \Gamma^{\lambda}_{\mu\nu}\Gamma^{\sigma}_{\lambda\sigma} - \Gamma^{\sigma}_{\mu\lambda}\Gamma^{\lambda}_{\nu\sigma}, \qquad (8.16)$$

and the *Ricci scalar*, R, by a further metric-tensor contraction,

$$R = R^{\mu}{}_{\mu} = g^{\mu\nu} R_{\mu\nu}. \qquad (8.17)$$

(It is common practice to use the same symbol R for the Riemann curvature tensor, Ricci tensor, and Ricci scalar, with the number of indices distinguishing among these tensors.) In addition to the algebraic symmetry constraints implied by Eq. (8.15), the curvature tensor obeys an important differential constraint called the *Bianchi identity*:

$$\nabla_{\lambda} R_{\mu\nu\alpha\beta} + \nabla_{\beta} R_{\mu\nu\lambda\alpha} + \nabla_{\alpha} R_{\mu\nu\beta\lambda} = 0. \qquad (8.18)$$

Taking multiple contractions of the Bianchi identity with the metric tensor and using Eqs. (8.15), (8.16), and (8.17) leads to the crucial identity (see Problem 8.15)

$$\nabla_{\mu} G^{\mu\nu} = 0, \qquad (8.19)$$

where the symmetric *Einstein tensor* $G^{\mu\nu}$ is defined by

$$G^{\mu\nu} \equiv R^{\mu\nu} - \tfrac{1}{2} g^{\mu\nu} R. \qquad (8.20)$$

The relation (8.19), which is sometimes called the *contracted Bianchi identity,* indicates that the Einstein tensor $G^{\mu\nu}$ has *zero covariant divergence*. The Einstein tensor is in fact the rank-2, symmetric curvature tensor with vanishing covariant divergence required as a replacement for the left side of Eq. (8.13). Thus, the covariant theory of gravity may be expressed in terms of the *Einstein equations*,[2]

---

[1] The choice to define the Ricci tensor by contraction on the first and third indices is a common one but some authors define $R_{\mu\nu}$ by contraction on a different pair of indices. This amounts to a sign choice in the definition, since you are asked to show in Problem 8.8 that contracting on any other pair of indices gives zero, or differs from the present definition by at most a sign. A summary of sign conventions employed in the equations of this section by various authors is given in Section 8.7.

[2] If the metric signature $+--- $ is employed instead of the $-+++$ used here, there will be an additional factor of -1 on the right side of (8.21) and in the stress–energy tensor equations (7.7) and (7.9) the last term involving the metric tensor will have a minus sign rather than a plus sign (sign conventions are discussed more extensively in Section 8.7).

$$G_{\mu\nu} = R_{\mu\nu} - \tfrac{1}{2}g_{\mu\nu}R = \frac{8\pi G}{c^4}T_{\mu\nu}, \qquad (8.21)$$

or even more compactly in $c = 1$ or geometrized ($c = G = 1$) units

$$G_{\mu\nu} = 8\pi G T_{\mu\nu} = 8\pi T_{\mu\nu}. \qquad (8.22)$$

Because the tensors are symmetric, these deceptively simple expressions actually represent 10 coupled, nonlinear, partial differential equations that must be solved to determine the effect of gravitation (Box 8.2 elaborates on the nonlinear nature of gravity).

For future reference, note that by contraction with the metric tensor the Einstein equation can also be written in the alternative form (see Problem 22.1)

$$R_{\mu\nu} = \frac{8\pi G}{c^4}(T_{\mu\nu} - \tfrac{1}{2}g_{\mu\nu}T^\alpha{}_\alpha), \qquad (8.23)$$

where the full contraction $T^\alpha{}_\alpha$ represents the trace of the stress–energy tensor. This form is particularly useful for the important class of *vacuum solutions* of the Einstein equations, where the stress–energy tensor is assumed to vanish in the region where the solution is valid. Then from Eq. (8.23) the vacuum Einstein equation can be written as $G_{\mu\nu} = R_{\mu\nu} = 0$, so for vacuum solutions it is sufficient to construct the Ricci tensor $R_{\mu\nu}$ rather than the full Einstein tensor.

However, in the *general case* only the full Riemann curvature tensor $R_{\sigma\mu\nu\lambda}$ with its 20 independent components contains the full information about the spacetime curvature. Because they are contracted quantities the Ricci tensor $R_{\mu\nu}$ contains only 10 independent components and the Ricci scalar R only 1. When $R_{\sigma\mu\nu\lambda}$ vanishes for the entire space then so do $R_{\mu\nu}$ and R, but the converse need not hold. For example, a manifold for which $R_{\mu\nu} = 0$ is termed *Ricci flat*, but such a manifold is not necessarily geometrically flat; only the vanishing of the full curvature tensor $R_{\sigma\mu\nu\lambda}$ ensures that.[3] In subsequent chapters, solutions of the Einstein equations will be sought in 4-dimensional spacetime. To fix ideas, Example 8.2 illustrates calculation of the quantities that enter the Einstein tensor in a simple 2-dimensional case with uniform curvature.

Example 8.2 Consider a 2-dimensional spherical surface in 3-dimensional euclidean space. Let's find the components of the metric tensor, the non-zero connection coefficients, the Riemann curvature tensor, the Ricci tensor, and the Ricci scalar curvature [63]. In standard polar coordinates the line element is

$$ds^2 = a^2(d\theta^2 + \sin^2\theta \, d\varphi^2),$$

---

[3] In three or fewer dimensions a vanishing Ricci tensor implies a vanishing Riemann curvature, but in four spacetime dimensions the Riemann curvature tensor can have non-zero components even if $R_{\mu\nu} = 0$. An example is the Schwarzschild metric to be discussed in Chapter 9, which satisfies $R_{\mu\nu} = 0$ but corresponds to a curved spacetime manifold because it has non-vanishing components of the full Riemann curvature tensor. Indeed, the curvature will be found to be so strong in that case that it can lead to a black hole with an event horizon.

Box 8.2 Quantum Field Theory and the Nonlinearity of Gravity

The Einstein equations are nonlinear because the solution for the gravitational field (metric tensor) is also the source of the gravitational field. Quantum field theory provides a useful perspective on this nonlinearity in terms of a pictorial representation of interaction matrix elements (probability amplitudes) called *Feynman diagrams*. These are highly intuitive and therefore extremely useful: given a Feynman diagram one can (with practice) write the corresponding matrix element and given the matrix element one can sketch the Feynman diagram. Here are some examples:

The solid lines represent (fermion) matter fields and the dashed and wiggly lines represent exchanged virtual gauge bosons that mediate the forces. Each diagram can represent several related processes, depending on the direction in which it is read. For example, diagram (a) read from the bottom represents two electrons interacting by exchanging a virtual photon (symbolized by γ) and diagram (b) read from the bottom represents an interaction in which a neutron (n) exchanges a virtual W^- intermediate vector boson with an electron neutrino (ν_e), converting the neutron to a proton and the neutrino to an electron.

The photon and the W^- particle differ fundamentally in that the photon is an *abelian gauge boson* but the W^- is a *nonabelian gauge boson* (signified graphically in the above diagrams by using a dashed line for abelian gauge bosons and a wiggly line for nonabelian gauge bosons). This is a consequence of their symmetry under local gauge transformations. In essence, abelian gauge bosons are associated with a single operator or a set of commuting operators, while nonabelian gauge bosons are associated with a set of operators that do not commute among themselves.

Diagram (c) above illustrates an important difference between abelian and nonabelian theories. It implies that nonabelian gauge bosons can interact directly with themselves (self-coupling), which produces a nonlinear theory. Such diagrams are forbidden for abelian theories like electromagnetism (built on the exchange of abelian photons). The virtual exchange particle that mediates gravity (*graviton*) is expected to be nonabelian. Thus gravitons can couple to themselves, implying that gravity is described by a *nonlinear field theory*, in contrast to the electromagnetic field, which does not self-couple and is described by a *linear field theory*. In fairness it should be noted that this argument is by analogy and it has not been proved that gravity obeys a quantum field theory sharing all the features described above.

where a is the radius of the sphere. This corresponds to a diagonal metric with

$$g_{\theta\theta} = a^2 \qquad g_{\varphi\varphi} = a^2 \sin^2\theta \qquad g_{\theta\varphi} = g_{\varphi\theta} = 0,$$

8.5 The Einstein Equations

and since $g^{\mu\nu}$ is the matrix inverse of $g_{\mu\nu}$,

$$g^{\theta\theta} = \frac{1}{a^2} \qquad g^{\varphi\varphi} = \frac{1}{a^2 \sin^2\theta}.$$

From Eq. (7.30) the connection coefficients with θ as an upper index are

$$\Gamma^\theta_{\varphi\varphi} = -\sin\theta\cos\theta \qquad \Gamma^\theta_{\theta\theta} = \Gamma^\theta_{\theta\varphi} = \Gamma^\theta_{\varphi\theta} = 0$$

and the connection coefficients with φ as an upper index are

$$\Gamma^\varphi_{\theta\varphi} = \Gamma^\varphi_{\varphi\theta} = \cot\theta \qquad \Gamma^\varphi_{\theta\theta} = \Gamma^\varphi_{\varphi\varphi} = 0.$$

The Riemann curvature tensor is given by Eq. (8.14). Since this is a 2-dimensional space there will be only one independent component. Up to symmetries this component may be taken to be

$$R^\theta_{\ \varphi\theta\varphi} = \frac{\partial \Gamma^\theta_{\varphi\varphi}}{\partial \theta} - \frac{\partial \Gamma^\theta_{\varphi\theta}}{\partial \varphi} + \Gamma^\theta_{\alpha\theta}\Gamma^\alpha_{\varphi\varphi} - \Gamma^\theta_{\alpha\varphi}\Gamma^\alpha_{\varphi\theta}$$

$$= \frac{\partial \Gamma^\theta_{\varphi\varphi}}{\partial \theta} - \Gamma^\theta_{\varphi\varphi}\Gamma^\varphi_{\varphi\theta} = \sin^2\theta.$$

The metric tensor may be used to lower the upper index, $R_{\mu\nu\alpha\beta} = g_{\mu\lambda} R^\lambda_{\ \nu\alpha\beta}$, which leads to

$$R_{\theta\varphi\theta\varphi} = R_{\varphi\theta\varphi\theta} = a^2 \sin^2\theta,$$

where Eq. (8.15) has been used. The Ricci tensor is then given by Eq. (8.16). The non-vanishing components are

$$R_{\varphi\varphi} = g^{\theta\theta} R_{\theta\varphi\theta\varphi} = \sin^2\theta \qquad R_{\theta\theta} = g^{\varphi\varphi} R_{\varphi\theta\varphi\theta} = 1.$$

Finally, the Ricci scalar curvature is the full contraction (8.17) of the Ricci tensor with the metric tensor,

$$R = g^{\mu\nu} R_{\mu\nu} = g^{\varphi\varphi} R_{\varphi\varphi} + g^{\theta\theta} R_{\theta\theta} = \frac{2}{a^2},$$

which is, up to a multiplicative factor, just the Gaussian curvature that was discussed in Section 7.2.2.

You are now encouraged to work through Problem 8.2, which does the same thing as Example 8.2 but for a general spherical metric in 4-dimensional spacetime. For future reference, the results of Problem 8.2 are summarized in Appendix C.

8.6 Limiting Behavior of the Einstein Tensor

It is not very difficult to demonstrate that the Einstein tensor exhibits the following limiting behavior:

1. For weak, nonrelativistic fields, $G_{00} \to \nabla^2 g_{00}$, as required in the earlier derivation of the weak-field limit.
2. If spacetime is flat (no curvature), $G_{\mu\nu} \to 0$.
3. If there is no matter, energy, or pressure in the universe, $G_{\mu\nu} \to 0$.

These are exactly the properties to be expected from a theory of gravitation in which curved spacetime is responsible for gravity and mass–energy–pressure is responsible for curving spacetime, and that reduces to Newtonian gravitation in the limit of weak fields and low velocities.

8.7 Sign Conventions

Unfortunately there is no uniform standard for the sign convention that different authors choose for the metric tensor, curvature tensors, and Einstein field equations. The conventions used in a few books and in the present text are summarized in Table 8.1 in terms of overall signs S_1, S_2, and S_3 chosen for the metric, Riemann curvature tensor, Ricci tensor, and Einstein field equation, respectively:

$$\begin{aligned} \eta^{\mu\nu} &= S_1 \times \text{diag}\,(-1,+1,+1,+1), \\ R^{\sigma}_{\ \mu\nu\lambda} &= S_2 \times (\Gamma^{\sigma}_{\mu\lambda,\nu} - \Gamma^{\sigma}_{\mu\nu,\lambda} + \Gamma^{\sigma}_{\alpha\nu}\Gamma^{\alpha}_{\mu\lambda} - \Gamma^{\sigma}_{\alpha\lambda}\Gamma^{\alpha}_{\mu\nu}), \\ R_{\mu\nu} &= S_2 \times S_3 \times R^{\sigma}_{\ \mu\sigma\nu}, \\ G_{\mu\nu} &= S_3 \times 8\pi\, T_{\mu\nu}. \end{aligned} \qquad (8.24)$$

The choice of sign convention is arbitrary as long as the convention is used consistently but care must be taken in comparing relations from different sources since different sign conventions may be in use.

8.8 Solving the Einstein Equations

The foregoing development has produced a set of field equations describing gravity covariantly. However, to apply this formalism systematically to comparison with data

Sign	This text	Refs. [63, 65, 111, 156, 221]	Refs. [76, 122, 204]	Ref. [246]
S_1	+	+	−	+
S_2	+	+	+	−
S_3	+	+	−	−

Table 8.1 Sign conventions in Eq. (8.24) used by various authors

it is necessary to find *solutions* for the field equations subject to appropriate boundary conditions. This is no easy task, given that Eq. (8.21) represents a set of coupled, nonlinear, partial differential equations, and that the appropriate boundary conditions may involve tricky issues, particularly in the limit of strong gravity.[4] In subsequent chapters the task of finding physically meaningful solutions will be pursued. It will be found that two assumptions can be used to find important solutions analytically that appear to represent physically observable objects: assume the field to be weak, or assume the metric describing the field to have a high degree of symmetry.

8.8.1 Solutions in the Limit of Weak Fields

Analytical solutions of the Einstein field equations may be obtained by positing a relatively weak gravitational field, so that it is justified to expand the metric about the Minkowski-space metric. Since *most* gravitational fields are weak, this can be very useful. For example, this approach was used in Section 8.1 to show that Newtonian gravity is the weak-field limit of general relativity, and will be used to predict the existence of gravitational waves in Chapter 22 (though it will be inadequate for strong-field sources of gravitational waves discussed in Chapter 24, which can be studied quantitatively only by using numerical computer solutions of the Einstein equations; see Section 8.8.3 below).

8.8.2 Solutions with a High Degree of Symmetry

A second assumption that can lead to meaningful analytical solutions of the field equations is to idealize the problem by assuming a high degree of symmetry for the metric. Then it is possible to decouple the Einstein equations and reduce the problem to solving a subset of the original equations. In this case it may be possible to obtain solutions without making a weak-field assumption. This will be the approach that we shall take to finding analytical solutions relevant for black holes and for cosmology in later chapters.

---

[4] When Einstein first derived his field equations he was unable to find exact solutions. However, this did not keep him from exploring the implications using the equivalence principle in the weak-field limit even before the field equations were available, as has been discussed in Chapter 6, and solutions to linearized forms of the equations for weak fields, as will be discussed in Chapter 22. Even without exact solutions this led to the prediction of gravitational light deflection, gravitational redshift and time dilation, and gravitational waves. The first exact solution of the field equations was not found by Einstein but rather by Schwarzschild shortly after the field equations were published, as will be discussed in Chapter 9.

8.8.3 Solutions by Numerical Relativity

If gravity is strong without a high degree of symmetry for the source, neither the weak-field nor the high-symmetry assumption is valid and it is necessary to resort to *numerical relativity,* where solutions are obtained from large-scale computer simulations. This was an important but somewhat academic issue as long as no data existed to probe asymmetric strong-gravity problems directly. This changed dramatically with the discovery in 2015 of gravitational waves from the merger of two $\sim 30 M_\odot$ black holes [10], which clearly represents a highly asymmetric problem in the strong-gravity limit (little spatial symmetry is expected – see Figs. 24.6 and 24.8 – and presumably many of the gravitational waves that were detected originated in the strong-gravity environment near the event horizons of the black holes, as discussed in Chapter 24).

A major issue in implementing numerical relativity is that the standard approaches to solution of partial differential equations on a computer require that the usual textbook representation of general relativity be reformulated. Ideally one would like to use a reasonably well-understood model of an object described by general relativity (for example, a binary black hole system near merger) to supply some "initial data," and then evolve these initial data numerically according to the Einstein field equations to some final equilibrium situation (for example the final black hole after the merger of the binary). The problem is that the coordinate independence of the usual formulation of the field equations means that there is no natural notion of "time" built into the equations, which means that the idea of "initial data" has no clearly defined meaning. A typical approach is to reformulate general relativity by splitting spacetime into 3+1 dimensions of space and time, for which Einstein's equations take a form better adapted to solving the initial value problem numerically on a computer. This approach is often termed the *3+1 formalism.*

Full-blown numerical relativity is rather technical and beyond the scope of the material presented here because it involves advanced issues in general relativity, numerical analysis, and computational science. However, given its ascendant standing in the emerging discipline of gravitational wave astronomy, later chapters will provide some qualitative allusions to the subject, along with literature references. For those wishing to go further, comprehensive introductions may be found in the books by Alcubierre [26], and by Baumgarte and Shapiro [40].

Background and Further Reading

See Roos [207]; Hartle [111]; Schutz [221]; Weinberg [246]; Hobson, Efstathiou, and Lasenby [122]; Cheng [65]; Islam [125]; and Misner, Thorne, and Wheeler [156]. A bibliography of some key papers in the development of general relativity that was compiled in celebration of the 100-year anniversary of the theory may be found at the webpage [8]. An introduction to numerical relativity is given in Alcubierre [26]; and in Baumgarte and Shapiro [40].

Problems

8.1 Consider a 2D spacetime (x^0, x^1) with a line element

$$ds^2 = -(dx^0)^2 + f(x^0)^2 (dx^1)^2,$$

where $f(x^0)$ is a function of time. Find the connection coefficients Γ^k_{ij} and the Riemann curvature component $R^1{}_{101}$.

8.2 Find the non-vanishing connection coefficients (Christoffel symbols) for a metric of the form

$$ds^2 = -e^\sigma dt^2 + e^\lambda dr^2 + r^2(d\theta^2 + \sin^2\theta d\varphi^2)$$

where $\sigma = \sigma(r,t)$ and $\lambda = \lambda(r,t)$. Find the nontrivial components of the corresponding Riemann curvature tensor $R_{\alpha\beta\gamma\delta}$, Ricci tensor $R_{\mu\nu}$, Ricci scalar R, and Einstein tensor $G_{\mu\nu}$. ***

8.3 The commutator $[A, B]$ of two objects A and B is defined by $[A, B] = AB - BA$. Show that the action of two successive covariant differentiations with respect to the indices ν and then λ on an arbitrary dual vector V_μ is

$$V_{\mu;\nu\lambda} = \partial_\lambda \partial_\nu V_\mu - (\partial_\lambda \Gamma^\sigma_{\mu\nu})V_\sigma - \Gamma^\sigma_{\mu\nu}\partial_\lambda V_\sigma$$
$$- \Gamma^\alpha_{\mu\lambda}(\partial_\nu V_\alpha - \Gamma^\sigma_{\alpha\nu}V_\sigma) - \Gamma^\alpha_{\nu\lambda}(\partial_\alpha V_\mu - \Gamma^\sigma_{\mu\alpha}V_\sigma),$$

where $\partial_\beta \equiv \partial/\partial x^\beta$. *Hint:* Take the covariant derivative of V_μ and then take the covariant derivative of the result, but remember that the covariant derivative of the dual vector yields a rank-2 tensor, so the second covariant derivative is not that of a dual vector but of a rank-2 tensor. Use this result to show that the Riemann curvature tensor is the commutator of the covariant derivatives. That is, for an arbitrary dual vector V_μ, prove that

$$[\nabla_\nu, \nabla_\lambda]V_\mu = R^\sigma{}_{\mu\nu\lambda}V_\sigma,$$

where ∇_α denotes taking a covariant derivative with respect to x^α and the curvature tensor is defined in terms of the connection coefficients in Eq. (8.14).

8.4 Estimate the deviation of the spacetime metric from the Minkowski metric in weak-gravity approximation at the surface of a proton, the Earth, the Sun, a white dwarf, and a neutron star. Is weak gravity a good approximation for all these objects?

8.5 Demonstrate that the approximate equation of motion (8.9) follows from Equations (8.4) and (8.8). ***

8.6 A clock falls slowly in a weak gravitational field. Show using the equivalence principle, special-relativistic time dilation, and energy conservation that

$$d\tau = 1 + \frac{\varphi}{c^2} = \sqrt{-g_{00}}dx^0,$$

where τ is the proper time, φ is the Newtonian gravitational potential, and $g_{\mu\nu}$ is the metric tensor.

8.7 For a 2-dimensional space the Riemann curvature tensor reduces to a single independent, nontrivial component, which may be taken to be R_{1212}. Show that for a diagonal metric in two dimensions R_{1212} is just the Gaussian curvature (7.2), up to a normalization factor $-\det g^{-1}$.

8.8 Use symmetries to show that, up to signs, the contraction $R_{\mu\nu} = R^\sigma{}_{\mu\sigma\nu}$ on the first and third indices that defines the Ricci tensor in Eq. (8.16) is the only possible independent, non-zero contraction of the Riemann curvature tensor on two indices. That is, show that any other choice for contracting two indices gives $\pm R_{\mu\nu}$ or zero. ***

8.9 Use the symmetries of the Riemann curvature tensor to prove that the Ricci tensor $R_{\mu\nu}$ is symmetric in its indices.

8.10 Geodesics must satisfy the geodesic equation (7.21). Assuming the Earth to be a sphere, use this to prove that the equator is the only circle of latitude that is a geodesic. *Hint*: The required connection coefficients for a sphere were already derived in Example 8.2.

8.11 Prove that the Einstein equations reduce to the Poisson equation (8.1) for Newtonian gravity in the static, low-velocity, weak-field limit. *Hint*: For this problem it is convenient to use the Einstein equations in the form (8.23).

8.12 To solve a general relativistic problem requires finding solutions for the Einstein equations (8.21) or (8.23). The stress–energy tensor obeys Eq. (7.10) so one can choose to solve the equation $T^\mu_{\nu;\mu} = 0$ in place of solving one of the Einstein equations. In many cases this can lead to a faster solution than solving all the Einstein equations directly. As an example of such manipulations, show that for a static metric

$$ds^2 = -e^\sigma dt^2 + e^\lambda dr^2 + r^2(d\theta^2 + \sin^2\theta d\varphi^2),$$

with $\sigma = \sigma(r)$ and $\lambda = \lambda(r)$, and a diagonal stress–energy tensor of the form $T^\mu_\nu = \text{diag}(-\rho, P, P, P)$, the constraint $T^\mu_{\nu;\mu} = 0$ leads to the equation $P' + \frac{1}{2}(P+\rho)\sigma' = 0$, where primes denote partial derivatives with respect to r. *Hint*: The connection coefficients that you will need to evaluate the covariant derivatives are derived in Problem 8.2 and summarized in Appendix C. ***

8.13 Use the results of Problem 8.3 and the quotient theorem illustrated in Problem 3.13 to argue that the Riemann curvature tensor $R^\sigma{}_{\mu\nu\lambda}$ is indeed a tensor.

8.14 Obtain an expression for the Riemann tensor $R_{\sigma\mu\lambda\nu}$ for the special case of a local inertial (freely falling) frame. (*Hint*: See Problem 7.14.) Since $R_{\sigma\mu\lambda\nu}$ is a tensor (see Problem 8.13), conclusions derived from this equation in this particular frame are valid in all frames. Use this simplified equation to verify the symmetry relations

$$R_{\sigma\mu\nu\lambda} = -R_{\mu\sigma\nu\lambda} = -R_{\sigma\mu\lambda\nu} \qquad R_{\sigma\mu\nu\lambda} = R_{\nu\lambda\sigma\mu}$$

given in Eq. (8.15).

8.15 Prove that the Bianchi identity implies that the Einstein tensor $G^{\mu\nu}$ has vanishing covariant 4-divergence. *Hint*: Use multiple contractions with the metric tensor on the Bianchi identity but be mindful of signs according to the footnote preceding Eq. (8.16) and the results of Problem 8.8. ***

9 The Schwarzschild Spacetime

One of the simplest solutions of the Einstein field equations corresponds to a metric that describes the spacetime around a static, spherical, uncharged mass without angular momentum and isolated from all other mass. The requisite metric was discovered by Karl Schwarzschild (1873–1916) in 1916, before Einstein himself was able to find any solutions to Eq. (8.21). For simplicity in the following, the object to be described by this metric will be referred to often as a star, though many of its most interesting applications are to things that are not stars in the normal sense. The Schwarzschild solution is a *vacuum solution* of the Einstein equations, meaning that it is valid in the absence of matter (the stress–energy tensor vanishes). As discussed in Section 8.5, a vacuum solution is equivalent to solving only $R_{\mu\nu} = 0$ where $R_{\mu\nu}$ is the Ricci tensor, instead of the full set of Einstein equations (8.21). Because the Schwarzschild solution is a vacuum solution it is approximately valid outside the Sun or a neutron star if their angular momentum is neglected and they are assumed to be spherical, but the interior of a star or neutron star will be described by a different metric.

9.1 The Form of the Metric

In spherical coordinates (r, θ, φ) the angular part of the metric may be expected to be unchanged from its form in empty space because of the spherical symmetry but the parts of the metric involving dt and dr are expected to be modified by functions that depend on the radial coordinate r. Therefore, let us write

$$ds^2 = -B(r)dt^2 + A(r)dr^2 + r^2 d\theta^2 + r^2 \sin^2\theta d\varphi^2, \qquad (9.1)$$

where $A(r)$ and $B(r)$ are unknown (positive) functions that may depend only on r because of the assumptions of spherical symmetry and no time dependence for the solution (later it will be found that these assumptions are not independent). The functions $A(r)$ and $B(r)$ may be determined by requiring that this metric be consistent with the vacuum Einstein field equations for gravity.

9.1.1 The Schwarzschild Solution

The solution, found by Schwarzschild in 1916 by substituting the metric implied by Eq. (9.1) into (8.21) is remarkably simple (Problem 9.1),

$$B(r) = 1 - \frac{2M}{r} \qquad A(r) = B(r)^{-1}, \qquad (9.2)$$

where M is the single parameter, entering mathematically as an integration constant but which will be interpreted below as the mass of the star. At large r,

$$\lim_{r \to \infty} A(r) = \lim_{r \to \infty} B(r) = 1, \qquad (9.3)$$

as expected since far from the star gravity becomes weak and the flat Minkowski metric should be recovered asymptotically. The Schwarzschild line element may be expressed in the form[1]

$$ds^2 = -\left(1 - \frac{2M}{r}\right) dt^2 + \left(1 - \frac{2M}{r}\right)^{-1} dr^2 + r^2 d\theta^2 + r^2 \sin^2\theta d\varphi^2, \qquad (9.4)$$

where $d\tau^2 = -ds^2$. The corresponding metric tensor is

$$g_{\mu\nu} = \begin{pmatrix} -\left(1 - \frac{2M}{r}\right) & 0 & 0 & 0 \\ 0 & \left(1 - \frac{2M}{r}\right)^{-1} & 0 & 0 \\ 0 & 0 & r^2 & 0 \\ 0 & 0 & 0 & r^2 \sin^2\theta \end{pmatrix}. \qquad (9.5)$$

The radial coordinate r entering this equation may be interpreted in terms of a spherical surface area, as described in Box 9.1 and Section 9.1.3.

9.1.2 The Schwarzschild Radius

Comparing the coefficient g_{00} in Eq. (9.4) with the corresponding result for Newtonian gravity, the parameter M appearing in the Schwarzschild metric may be identified with the total mass that is the source of the gravitational curvature. This total mass includes the rest mass and any contributions from the energy densities of fields, pressure, and the spacetime curvature itself. Outside a spherical source, the line element (9.4) indicates that the geometry of spacetime is determined completely by the total mass–energy M and not by any details of its internal distribution. This is the analog of the Newtonian result that the external gravitational field for a spherical mass distribution is determined by the mass interior to a given radius. The quantity

$$r_S \equiv 2M \qquad (9.6)$$

appearing prominently in the preceding equations is called the *Schwarzschild radius*. It plays a central role in the description of the Schwarzschild spacetime geometry.

Notice that the line element (9.4) appears to contain two singularities, one for $r = 0$ and one for the Schwarzschild radius $r = r_S = 2M$. The singularity at $r = 2M$ is a *coordinate singularity*; that is, it can be made to disappear by an appropriate change of variables, which indicates that it is a consequence of the current parameterization and not a physical singularity. However, the singularity at $r = 0$ is physical. These singularities of

---

[1] A metric having no components that depend on time is termed *stationary*. If in addition the metric has no cross terms like $dt d\varphi$ coupling timelike and spacelike components, the metric is termed *static*. The form (9.4) for the Schwarzschild line element implies a static and stationary metric.

> **Box 9.1** **The Schwarzschild Radial Coordinate**
>
> The Schwarzschild coordinate r may be interpreted in terms of the surface area A of spheres at fixed r and t through $A = 4\pi r^2$. This avoids attempting to interpret it as a distance from a center, which is problematic because
>
> 1. distances are measured by the line element, not directly by coordinates,
> 2. the Schwarzschild metric is a vacuum solution and thus not valid within the central mass that is the gravitational source,
> 3. there may be an event horizon preventing fundamentally even being able to see the end of a measuring rod extended toward the origin, and (most exotically)
> 4. a space could have spheres without centers.
>
> As an example of the final point, consider the 2D surface displayed in the following figure,
>
>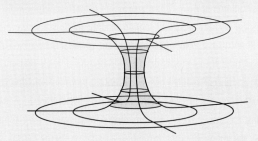
>
> where circles are drawn that have no centers within the 2D space (the throat is not part of the space). General relativity is a local theory that does not specify the global structure of spacetime. There is no evidence that spacetime has an exotic topology like that sketched above, but the possibility cannot be excluded out of hand.

the Schwarzschild metric are discussed further in Box 9.2 and Example 9.1 illustrates how a simple transformation of coordinates can produce a coordinate singularity for a perfectly well-behaved manifold.

Example 9.1 Consider the line element $ds^2 = dr^2 + r^2 d\varphi^2$ in 2D polar coordinates and introduce the transformation $r = a^2/\rho$, where a is a constant. Then the line element expressed in the new coordinates is given by

$$ds^2 = \frac{a^4}{\rho^4}\left(d\rho^2 + \rho^2 d\varphi^2\right).$$

This is singular at $\rho = 0$ but there is nothing pathological about the 2D plane at that point, suggesting that this is a coordinate problem. The reason is not very difficult to fathom. The transformation has mapped all points at infinity into $\rho = 0$ and introduced a coordinate singularity because the geometry is not correctly represented by the (ρ, φ) coordinates at that point.

Often no single set of coordinates describes correctly the geometry of a complete manifold and overlapping coordinate patches are required to give a comprehensive description.

> **Box 9.2** **Singularities of the Schwarzschild Metric**
>
> The singularities of the Schwarzschild metric may be classified as either *coordinate singularities* or *physical singularities*.
>
> **Coordinate Singularities**
>
> A coordinate singularity is a place where a chosen set of coordinates fails to describe the geometry properly. For example, at the north pole of a spherical coordinate system the azimuthal angle φ takes a continuum of values 0–2π, all corresponding to a single point. However, on the surface of a uniform, non-rotating sphere there is no physical difference between the north pole and any other point. As discussed further in Section 9.1.2, the singularity at the Schwarzschild radius r_S is a coordinate singularity that can be removed by choosing new coordinates.
>
> **Physical Singularities**
>
> The singularity at $r = 0$ is physical and cannot be removed by a coordinate transformation. Its meaning is unclear and probably will remain so until there is a viable theory of quantum gravity. In 1935 Einstein and Rosen [81] attempted to interpret the physical Schwarzschild singularity as a portal to another region of spacetime, leading to an idea that has become a staple of science fiction, spacetime *wormholes* [147]. Later it was established that any attempt to enter a wormhole would collapse it too rapidly to permit even a photon to go through [95, 141, 156]. More recently the possibility has been raised that a wormhole could be stabilized against collapse, but only by threading it with exotic material having a stress–energy tensor that may be physically impossible (one exhibiting a negative energy density in the frame of a particle moving through the wormhole). Moreover, it was argued that a stabilized wormhole could be converted into a time machine [159, 234].
>
> Hence, wormholes are not precluded by general relativity, provided that a mechanism exists to generate the required spacetime topology: since the Einstein equations are differential equations they are *local* and do not determine the topology of spacetime, which is a *global* property associated with connectedness of the manifold. But it is unclear whether it would be possible to travel through a wormhole, even if they did exist. For a broader discussion, see Thorne [234].

9.1.3 Measuring Distance and Time

What is the physical meaning of the Schwarzschild coordinates (t, r, θ, φ)? As suggested in Box 9.1, a practical definition may be assigned to the radial coordinate r by enclosing the origin of the Schwarzschild spacetime in a series of concentric spheres, measuring for each sphere a surface area (conceptually by laying measuring rods end to end), and then assigning a radial coordinate r to that sphere using Area $= 4\pi r^2$ for the sphere.[2] Then distances and trigonometry can be used to define the angular coordinate variables θ and φ, and finally a coordinate time t can be defined in terms of clocks attached to the

---

[2] Thus r may be interpreted as what would be measured directly for the radius of the sphere *if* the spacetime that it contained were euclidean. Of course generally it *isn't*, but this is still a perfectly sensible – and quite useful – *definition* of a coordinate.

concentric spheres. For Newtonian theory with its implicit assumption that events occur on a passive background of euclidean space and constantly flowing time, this would be the whole story. But by now we know enough to be highly suspicious of assigning such direct physical significance to the coordinates of a theory.

Coordinate distance and proper distance: The Schwarzschild spacetime becomes flat at large distance from the gravitational source and the coordinates (t, r, θ, φ) provide a global reference frame *for an observer at infinity*. However, physical quantities measured by arbitrary observers are not specified directly by the coordinates but rather must be computed from the metric. Consider a radial distance interval. Setting $dt = d\theta = d\varphi = 0$ in Eq. (9.4) gives an interval of radial distance

$$ds = \frac{dr}{\sqrt{1 - 2GM/rc^2}}, \qquad (9.7)$$

where factors of G and c have been restored. In Eq. (9.7) ds is termed the *proper distance* and dr is termed the *coordinate distance*. The physical radial interval measured by a local observer is the proper distance ds, *not* the coordinate distance dr. Because GM/rc^2 is a measure of the intrinsic strength of gravity (see Section 6.4.2), the proper distance and coordinate distance are equivalent only if gravity is negligibly weak, either because the source mass M is small, or because of being at a very large coordinate distance r from the source.

The relationship between the coordinate distance interval dr and the proper distance interval ds is illustrated further in Fig. 9.1. The circles C_1 and C_3 represent spheres having radius r in euclidean space, while the circles C_2 and C_4 represent spheres having an infinitesimally larger radius $r + dr$ in euclidean space. In euclidean space the distance that would be measured between the spheres is dr but in the curved space the measured distance between the spheres is ds, which is larger than dr, by virtue of Eq. (9.7). However, at large distance from the source of the gravitational field the Schwarzschild spacetime becomes flat asymptotically and in this limit $dr \sim ds$. Thus, for an observer at infinity the Schwarzschild coordinate distance and the proper distance are equivalent.

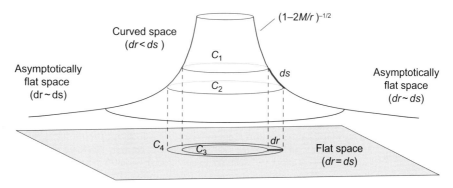

Fig. 9.1 The relationship between coordinate radial interval dr and proper radial interval ds in Schwarzschild spacetime (adapted from Ref. [89]).

Coordinate time and proper time: Likewise, to measure a time interval for a stationary clock at r we may set $dr = d\theta = d\varphi = 0$ in the line element and use $ds^2 = -d\tau^2 c^2$ to obtain

$$d\tau = \sqrt{1 - \frac{2GM}{rc^2}}\, dt. \tag{9.8}$$

In this expression $d\tau$ is termed the *proper time* and dt is termed the *coordinate time*. The physical time interval measured by a local observer is given by the proper time $d\tau$, not by the coordinate time dt. Just as for the relationship between coordinate distance and proper distance, dt and $d\tau$ coincide only if the effect of the gravitational field vanishes. Thus the Schwarzschild coordinate time t is the time measured by an observer at infinity.

To summarize, for the gravitational field outside a spherical mass distribution the coordinates r and t may be interpreted as corresponding directly to physical distance and time in Newtonian gravity, but in general relativity the physical (proper) distances and times must be computed from the metric and are not given directly by the coordinates. Only in those regions of spacetime where the gravitational field is very weak is the Newtonian interpretation recovered. All this is as it should be, since the goal of relativity is to make the laws of physics independent of the coordinate system in which they are formulated.[3]

9.1.4 Embedding Diagrams

It is sometimes useful to form a mental image of the structure for a curved space by embedding the space or a subset of it in 3D euclidean space. Two dimensions of Schwarzschild spacetime can be embedded in a 3D euclidean space. Let's illustrate by choosing $\theta = \frac{\pi}{2}$ and $t = 0$, to give a 2D metric

$$d\ell^2 = \left(1 - \frac{2M}{r}\right)^{-1} dr^2 + r^2 d\varphi^2. \tag{9.9}$$

The metric of the 3D embedding space is conveniently represented in cylindrical coordinates as $d\ell^2 = dz^2 + dr^2 + r^2 d\varphi^2$, which can be written as

$$d\ell^2 = \left(\frac{dz}{dr}\right)^2 dr^2 + dr^2 + r^2 d\varphi^2 = \left[1 + \left(\frac{dz}{dr}\right)^2\right] dr^2 + r^2 d\varphi^2 \tag{9.10}$$

on $z = z(r)$. Comparing Eq. (9.9) with Eq. (9.10) implies that

$$z(r) = 2\sqrt{2M(r - 2M)}, \tag{9.11}$$

---

[3] Foster and Nightingale [89] give a nice analogy. The coordinates in a physical theory are like street numbers. They (supplemented by an additional coordinate varying along a street) provide a labeling for points in a space, but knowing the street numbers is not sufficient to find distances. The question of whether the distance between 36th Street and 37th Street is the same as the distance between 40th Street and 41st Street can't be answered until it is known whether the streets are equally spaced: distances must be computed from a *metric* that gives a distance-measuring prescription. Streets that are always equally spaced correspond to a "flat" space, but streets with irregular spacing correspond to a position-dependent metric in a "curved" space. For the flat space the difference in street number corresponds directly (up to a scale) to a physical distance, but in the more general (curved) case it does not until supplemented by a position-dependent metric.

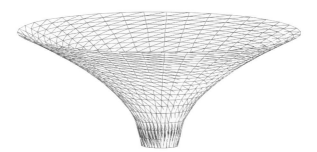

Fig. 9.2 An embedding diagram for the Schwarzschild $(r - \varphi)$ plane. The embedding is defined by Eq. (9.11) and the figure shows $z(r)$ from r_S out to $8r_S$. In such embedding diagrams *only points on the embedded surface are relevant*. Points in the embedding space not on the embedded surface have no physical meaning.

which defines an embedding surface $z(r)$ having the same geometry as the Schwarzschild metric in the $(r - \varphi)$ plane. This embedding is illustrated in Fig. 9.2. Although Fig. 9.2 certainly is not what the Schwarzschild spacetime "looks like," it still is a striking and useful visualization of the increasingly large curvature of space near the Schwarzschild radius and the asymptotic flattening of the Schwarzschild geometry at large r. For this reason such embedding diagrams are a standard representation of curved spacetime in popular-level discussion.

9.2 The Gravitational Redshift

With the Schwarzschild solution in hand, let us revisit the gravitational redshift and time dilation effects introduced in Section 6.5. Consider the emission of light from a Schwarzschild radial coordinate R_1 that is then detected by a stationary observer at a Schwarzschild radial coordinate r, with $r \gg R_1$. Figure 9.3 illustrates in a spacetime diagram. The frequency shift was estimated in Eq. (6.10) assuming weak fields and the validity of the equivalence principle. Let us apply the Schwarzschild solution to the calculation of gravitational redshifts, without invoking the weak field assumption. For an observer with 4-velocity u, the energy measured for a photon with 4-momentum p is from Eq. (5.25),

$$E = \hbar\omega = -p \cdot u, \tag{9.12}$$

where ω is the frequency, the 4-velocity is $u^\mu = dx^\mu/d\tau$, and from Eq. (5.26) the 4-velocity obeys

$$u \cdot u = g_{\mu\nu}(x)\frac{dx^\mu}{d\tau}\frac{dx^\nu}{d\tau} = -1, \tag{9.13}$$

where τ is the proper time. Observers are assumed to be stationary (in space, not in time!), so $u^i(r) = 0$. Thus only the timelike component contributes to the implied summation in Eq. (9.13) and

$$g_{00}(x)u^0(r)u^0(r) = -1. \tag{9.14}$$

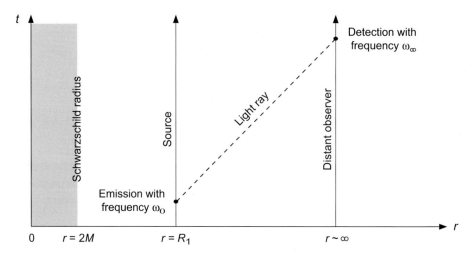

Fig. 9.3 A spacetime diagram for gravitational redshift in the Schwarzschild metric.

From the Schwarzschild metric (9.5), the g_{00} component is

$$g_{00}(r) = -\left(1 - \frac{2M}{r}\right) \tag{9.15}$$

and combining (9.14) and (9.15) gives

$$u^0(r) = \left(1 - \frac{2M}{r}\right)^{-1/2}. \tag{9.16}$$

The relationship between the frequency at emission and the frequency observed at infinity can be found easily from this result by utilizing a symmetry associated with the time-independence of the metric.

9.2.1 Exploiting a Symmetry of the Metric

The Schwarzschild metric (9.5) does not depend on time. As discussed in Section 5.6, this implies the existence of a timelike Killing vector

$$K^\mu = (t, r, \theta, \varphi) = (1, 0, 0, 0) \tag{9.17}$$

associated with symmetry of the metric under time displacement. Thus, for a stationary observer at a distance r,

$$u^\mu(r) = \left(\left(1 - \frac{2M}{r}\right)^{-1/2}, 0, 0, 0\right) = \left(1 - \frac{2M}{r}\right)^{-1/2} K^\mu. \tag{9.18}$$

The energy of the photon measured by a stationary observer at r is then

$$\hbar\omega(r) = -p \cdot u = -\left(1 - \frac{2M}{r}\right)^{-1/2} (K \cdot p)_r, \tag{9.19}$$

where the notation indicates that the scalar product is to be evaluated at r. But as discussed in Section 5.6.2, $(K \cdot p)$ is *conserved along the photon geodesic* if K is a Killing vector, so $(K \cdot p)$ is in fact independent of r and thus

$$\hbar\omega_0 \equiv \hbar\omega(r = R_1) = -\left(1 - \frac{2M}{R_1}\right)^{-1/2}(K \cdot p) \qquad \hbar\omega_\infty \equiv \hbar\omega(r \to \infty) = -(K \cdot p),$$

from which a gravitational redshift follows immediately,

$$\omega_\infty = \omega_0 \left(1 - \frac{2M}{R_1}\right)^{1/2}. \tag{9.20}$$

As discussed in Section 6.5.3, by viewing the photon frequency as defining clock ticks this result can be interpreted alternatively as a gravitational time dilation.

9.2.2 Recovering the Weak-Field Limit

No weak-field assumptions have been used in the preceding derivation so Eq. (9.20) is expected to be valid for weak and strong fields. For the special case of weak fields $2M/R_1$ is small, the square root in Eq. (9.20) can be expanded in a binomial series, and the factors involving G and c restored to give

$$\omega_\infty \simeq \omega_0 \left(1 - \frac{GM}{R_1 c^2}\right) \qquad \text{(weak-field limit)}, \tag{9.21}$$

which is the result derived previously in Eq. (6.10) by assuming weak fields and using the equivalence principle.

9.3 Particle Orbits in the Schwarzschild Metric

Let us now find the appropriate equations of motion for free particles in the Schwarzschild spacetime, first for massive particles and then for photons. This task is simplified by exploiting the symmetries inherent in the problem, in a generalization of the approach taken above to determine the gravitational redshift. Our discussion will adapt often from the excellent treatment of the Schwarzschild metric in Hartle [111].

9.3.1 Conserved Quantities

The Schwarzschild metric is independent of time and spherically symmetric. Because of the time independence there is a Killing vector with components $K_t = (1, 0, 0, 0)$ associated with symmetry under time displacement, which was invoked already in Eq. (9.17). Because the metric does not depend on the angle φ there is another Killing vector with components $K_\varphi = (0, 0, 0, 1)$ that is associated with symmetry under rotations about the z axis. (There are additional Killing vectors associated with the full symmetry of the spacetime but they will not be needed here.) As discussed in Section 5.6, each of these

Killing vectors has associated with it a quantity conserved along a geodesic, corresponding to the scalar products $-K_t \cdot u$ and $K_\varphi \cdot u$. Denoting these by ϵ and ℓ, respectively,

$$\epsilon \equiv -K_t \cdot u = \left(1 - \frac{2M}{r}\right)\frac{dt}{d\tau}, \tag{9.22}$$

$$\ell \equiv K_\varphi \cdot u = r^2 \sin^2\theta \frac{d\varphi}{d\tau}. \tag{9.23}$$

From Eq. (9.23), at low velocities the constant of motion ℓ may be interpreted as the orbital angular momentum per unit rest mass. Furthermore, for large r

$$\lim_{r\to\infty} \epsilon = \frac{dt}{d\tau} = \frac{E}{m}, \tag{9.24}$$

where $E = p^0 = mu^0 = m\, dt/d\tau$ has been used (see Section 5.2). Thus, at large distance ϵ may be interpreted as the energy per unit rest mass. In addition to the constants of motion ϵ and ℓ, one has always the constraint

$$u \cdot u = g_{\mu\nu} u^\mu u^\nu = -1 \tag{9.25}$$

following from the normalization of the velocity for a timelike particle [see Eq. (5.26)].

9.3.2 Equation of Motion

Conservation of angular momentum confines the particle motion to a plane, which conveniently may be chosen as the equatorial plane with $\theta = \frac{\pi}{2}$, implying that $u^2 = u^\theta = 0$. Writing the constraint (9.25) out in the Schwarzschild metric gives then

$$-\left(1 - \frac{2M}{r}\right)(u^0)^2 + \left(1 - \frac{2M}{r}\right)^{-1}(u^1)^2 + r^2(u^3)^2 = -1. \tag{9.26}$$

Utilizing (5.1), (9.22), and (9.23), this may be expressed as

$$\frac{\epsilon^2 - 1}{2} = \frac{1}{2}\left(\frac{dr}{d\tau}\right)^2 + \frac{1}{2}\left[\left(1 - \frac{2M}{r}\right)\left(\frac{\ell^2}{r^2} + 1\right) - 1\right] \tag{9.27}$$

and finally as

$$E = \frac{1}{2}\left(\frac{dr}{d\tau}\right)^2 + V_{\text{eff}}(r), \tag{9.28}$$

with the definition

$$E \equiv \frac{1}{2}(\epsilon^2 - 1) \tag{9.29}$$

and with an effective potential

$$V_{\text{eff}}(r) = \frac{1}{2}\left[\left(1 - \frac{2M}{r}\right)\left(\frac{\ell^2}{r^2} + 1\right) - 1\right] = -\frac{M}{r} + \frac{\ell^2}{2r^2} - \frac{M\ell^2}{r^3}. \tag{9.30}$$

This is analogous to the energy integral of Newtonian mechanics with an effective potential V_{eff}. The form for V_{eff} differs from that of Newtonian mechanics only in the $-M\ell^2/r^3$ term (and in the physical interpretation of the coordinates), but this will have significant implications.

9.3.3 Classification of Orbits

Figure 9.4 illustrates schematically the effective potential (9.30) (left) and the kinds of orbits that can occur (right), while Fig. 9.5 illustrates the Schwarzschild effective potential for several values of ℓ/M and also compares the Schwarzschild potential with a Newtonian

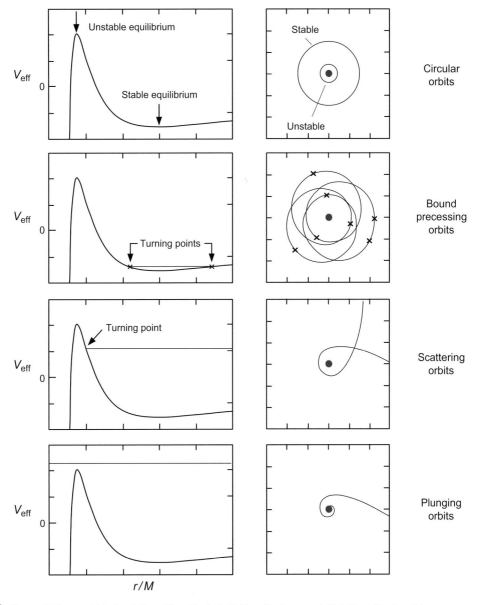

Fig. 9.4 Classes of Schwarzschild orbits (adapted from Hartle [111]). The effective potential is indicated by the solid curves. The radial turning points occur when E from Eq. (9.28) is equal to V_{eff}. The dashed horizontal lines (and the point denoted by the arrows in the top panel) indicate a typical value of E for a given class of orbits.

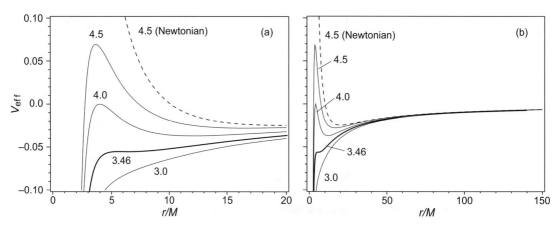

Fig. 9.5 (a) Effective potentials (9.30) for timelike particles in the Schwarzschild geometry, with curves labeled by values of ℓ/M. The Schwarzschild curves with $\ell/M > \sqrt{12} \simeq 3.46$ have one maximum and one minimum. The innermost stable circular orbit corresponding to $\ell/M = \sqrt{12} \simeq 3.46$ is indicated by a heavier curve. For the $\ell/M = 4.5$ case both the Schwarzschild potential (solid line) and the corresponding Newtonian potential (dotted line) are displayed. (b) Behavior of the solutions in (a) at large distances.

potential for one case. The Schwarzschild potential generally has one maximum and one minimum if $\ell^2/M^2 > 12$ (see Problem 9.2).

Stable and unstable circular orbits: The top row of Fig. 9.4 illustrates two circular orbits (positions of the arrows) that can occur when the energy E is exactly equal to an extremum of V_{eff}. The inner orbit corresponds to a maximum of the potential and is unstable; the outer orbit corresponds to a minimum of the potential and is stable. Stable and marginally stable circular orbits are important in astrophysics and will be discussed further in Section 9.3.4.

Bound precessing orbits: The second row of Fig. 9.4 illustrates bound orbits that move between two radial turning points (indicated by arrows on the left side and x's on the right side). These orbits do not close as Kepler ellipses would but instead precess in the potential (see Fig. 9.6). A quantitative description of this precession, which is a precise test of general relativity, will be given in Section 9.4.

Scattering orbits: The third row of Fig. 9.4 illustrates a scattering trajectory having only a single radial turning point. Because of the relativistic corrections, this scattering orbit differs substantially from the parabola expected from Newtonian gravitational scattering.

Plunge orbits: The fourth row of Fig. 9.4 illustrates a "plunge orbit," for which E is greater than the effective potential for any r and the particle spirals to $r = 0$. As illustrated in Fig. 9.5, such orbits do not occur in Newtonian gravity because the centrifugal potential diverges to infinity as r approaches zero, causing any trajectory with finite orbital angular momentum to miss the origin. In the Schwarzschild problem the $-M\ell^2/n^3$ term causes the potential to decrease at short distances, so particles with sufficient energy for a given ℓ surmount the inner barrier of the Schwarzschild effective gravitational potential and plunge to the origin. The effective potentials in Fig. 9.5 illustrate a general result that will be found

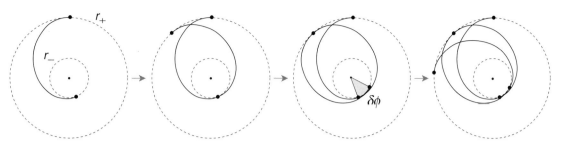

Fig. 9.6 Precession of orbits in a Schwarzschild metric (highly exaggerated). The radial coordinate of the inner turning point r_- and of the outer turning point r_+ are represented by dashed circles. Dots on the inner circle indicate perihelion for each orbit and dots on the outer circle indicate the corresponding aphelion for each orbit. The quantity $\delta\varphi$ indicates the shift in angle of the perihelion for one orbital period.

to recur in various contexts: gravity derived from the full theory of general relativity is characteristically more attractive than the gravity predicted by Newtonian theory for the same situation.

9.3.4 Stable Circular Orbits

As will be discussed in Section 15.2, accretion onto compact objects is a major energy source for various astrophysical phenomena. Accretion typically occurs through an accretion disk, and tidal forces on the particles in an accretion disk tend to circularize orbits. Therefore, the stable and marginally stable circular orbits for a spacetime are of particular interest. The top panel of Fig. 9.4 indicates that a circular orbit occurs in the Schwarzschild metric when E from Eq. (9.28) is equal to a minimum or maximum of the effective potential (indicated by the arrows). The radial coordinates of these orbits satisfy (Problem 9.2)

$$r = \frac{\ell^2}{2M} \pm \frac{1}{2}\sqrt{\frac{\ell^4}{M^2} - 12\ell^2}, \qquad (9.31)$$

where the plus sign corresponds to the minimum and the minus sign to the maximum of the potential. Equation (9.31) has two real solutions if $\ell^2/M^2 > 12$. The requirement that $E = V_{\text{eff}}$ at this value of r implies that [Problem 9.15(a)]

$$\epsilon^2 = \left(1 - \frac{2M}{r}\right)\left(1 + \frac{\ell^2}{r^2}\right). \qquad (9.32)$$

The angular velocity Ω of a particle as seen by a distant observer in a $\theta = \frac{\pi}{2}$ Schwarzschild orbit is given by [Problem 9.15(b)]

$$\Omega = \frac{d\varphi}{dt} = \frac{1}{r^2}\left(1 - \frac{2M}{r}\right)\left(\frac{\ell}{\epsilon}\right), \qquad (9.33)$$

and from Eqs. (9.31) and (9.32) [111]

$$\frac{\ell}{\epsilon} = \sqrt{Mr}\left(1 - \frac{2M}{r}\right)^{-1},$$

which gives from Eq. (9.33)

$$\Omega = \sqrt{\frac{M}{r^3}}. \quad (9.34)$$

Since the period is given by $P = 2\pi/\Omega$, Eq. (9.34) is equivalent formally to Kepler's third law, but is expressed in terms of r and t Schwarzschild coordinates rather than proper distances and times.

Example 9.2 It will be useful for later applications to write out explicit components of the velocity 4-vector for a circular Schwarzschild orbit in the equatorial plane. The components are

$$u = (u^t, 0, 0, \Omega u^t), \quad (9.35)$$

where the relation

$$\Omega = \frac{d\varphi}{d\tau}\frac{d\tau}{dt} = \frac{1}{u^t}\frac{d\varphi}{d\tau}$$

was used. The timelike component $u^t = dt/d\tau$ may be evaluated explicitly using the normalization Eq. (9.25), which gives

$$u^t = \left(1 - \frac{3M}{r}\right)^{-1/2}, \quad (9.36)$$

as you are asked to show in Problem 9.15(c).

Stable circular orbits do not exist at arbitrarily small values of the radial coordinate in the Schwarzschild spacetime [see Fig. 9.5(a)]. The expression (9.31) has a minimum possible value for a stable orbit that occurs for $\ell^2/M^2 = 12$. The corresponding radius for the *innermost stable circular orbit (ISCO)* is then

$$R_{\text{ISCO}} = 6M. \quad (9.37)$$

The innermost stable circular orbit is important in determining how much gravitational energy can be extracted from matter accreting onto neutron stars and black holes in X-ray binary systems, or in the central engines of active galaxies, as will be discussed further in Chapter 15.

9.4 Precession of Orbits

For periodic orbital motion in the equatorial plane (corresponding to $\theta = \frac{\pi}{2}$ in our chosen coordinate system) an orbit closes if the angle φ sweeps out exactly 2π in the passage between two successive inner or two successive outer radial turning points. In Newtonian gravitation the central potential is of the form r^{-1} and this leads to closed, elliptical Kepler orbits in the absence of perturbations from other bodies. From Eq. (9.30) the Schwarzschild effective potential deviates from r^{-1}. As a consequence the orbits *precess:* the change in the angle φ between successive radial turning points is generally larger than 2π, as illustrated in Fig. 9.6. The inner and outer turning points of radial motion in Fig. 9.6 are denoted by r_- and r_+, respectively. From left to right, the figures show motion between successive turning points, with the precession angle between two successive inner turning points denoted by $\delta\varphi$. The amount of precession in this figure is greatly exaggerated for clarity. In realistic systems the precession for each orbit is usually tiny. As an example, the Binary Pulsar has an orbital period of 7.75 hours and the orbital precession angle for it accumulates to a shift of a little over four degrees in a year (see the discussion in Section 10.4.2). Precession angles for most other objects are much smaller.

9.4.1 The Change in Perihelion Angle

Quantitative investigation of this precession requires an expression for the rate of change $d\varphi/dr$ as the object moves on its orbit. From the energy equation (9.28) and the conservation equation (9.23) for ℓ,

$$\frac{dr}{d\tau} = \pm\sqrt{2(E - V_{\text{eff}}(r))} \qquad \frac{d\varphi}{d\tau} = \frac{\ell}{r^2 \sin^2\theta}.$$

Combining these and recalling that the orbital plane corresponds to $\theta = \frac{\pi}{2}$,

$$\frac{d\varphi}{dr} = \frac{d\varphi/d\tau}{dr/d\tau} = \pm\frac{\ell}{r^2\sqrt{2(E - V_{\text{eff}}(r))}}$$

$$= \pm\frac{\ell}{r^2}\left[2E - \left(1 - \frac{2M}{r}\right)\left(1 + \frac{\ell^2}{r^2}\right) + 1\right]^{-1/2}$$

$$= \pm\frac{\ell}{r^2}\left[\epsilon^2 - \left(1 - \frac{2M}{r}\right)\left(1 + \frac{\ell^2}{r^2}\right)\right]^{-1/2}, \qquad (9.38)$$

where $E = \frac{1}{2}(\epsilon^2 - 1)$ from Eq. (9.29) has been used. The change in φ per orbit, $\Delta\varphi$, can be obtained by integrating Eq. (9.38) over one orbit,

$$\Delta\varphi = \int_{r_-}^{r_+} \frac{d\varphi}{dr}\, dr + \int_{r_+}^{r_-} \frac{d\varphi}{dr}\, dr = 2\int_{r_-}^{r_+} \frac{d\varphi}{dr}\, dr$$

$$= 2\ell \int_{r_-}^{r_+} \frac{dr}{r^2}\left[\epsilon^2 - \left(1 - \frac{2M}{r}\right)\left(1 + \frac{\ell^2}{r^2}\right)\right]^{-1/2}. \qquad (9.39)$$

Evaluation of the integral requires some care. From Eq. (9.27)

$$\frac{dr}{d\tau} = \pm\left[\epsilon^2 - \left(1 - \frac{2M}{r}\right)\left(1 + \frac{l^2}{r^2}\right)\right]^{1/2}, \qquad (9.40)$$

which is a factor in the denominator of Eq. (9.39). But the limits are turning points of the radial motion and $dr/d\tau = 0$ at r_+ or r_-. Thus, the denominator of the integrand in Eq. (9.39) vanishes at the upper and lower limits of the integral. Restoring the constants G and c that have been suppressed in geometrized units requires the substitutions $M \to GM/c^2$ and $\ell \to \ell/c$, and Eq. (9.39) becomes in standard units

$$\Delta\varphi = 2\ell \int_{r_-}^{r_+} \frac{dr}{r^2} \left(c^2(\epsilon^2 - 1) + \frac{2GM}{r} - \frac{\ell^2}{r^2} + \frac{2GM\ell^2}{c^2 r^3} \right)^{-1/2}. \quad (9.41)$$

The first three terms in parentheses give the Newtonian contribution and the last term (proportional to r^{-3}) gives the relativistic correction. In the Solar System and most other applications the perihelion shifts are very small. Expanding the integrand of Eq. (9.41) to first order in $1/c^2$ and evaluating the integral with due care given to the preceding warnings yields [111]

$$\delta\varphi \equiv \Delta\varphi - 2\pi \simeq 6\pi \left(\frac{GM}{c\ell} \right)^2. \quad (9.42)$$

As you are asked to show in Problem 9.5, this may be rewritten in terms of more familiar classical orbital parameters as

$$\delta\varphi = \frac{6\pi GM}{ac^2(1-e^2)}, \quad (9.43)$$

where e is the eccentricity of the orbit and a is the semimajor axis. This form demonstrates explicitly that general relativistic precession for a binary pair is favored by (1) large mass M, (2) compact orbits (small value of a), and (3) large eccentricities e.

9.4.2 Testing the Prediction

The precession observed for known objects is much less than that indicated schematically in Fig. 9.6. Once contributions from other sources have been subtracted, the precession of Mercury's orbit in the Sun's gravitational field because of general relativistic effects is observed to be 43 arcseconds per century (the angular precession of the inner turning point for a planet is termed the *precession of the perihelion*).[4] The orbit of the Binary Pulsar (Section 10.4.1) precesses by about 4.2° per year (the angular motion of the inner turning point for a binary star orbit is termed *precession of the periastron*), and that of the Double Pulsar (Section 10.4.4) by about 17° per year. The precise agreement of these observed precessions with the predictions of general relativity is a strong test of the theory.

9.5 Escape Velocity

Consider a stationary observer at a Schwarzschild radial coordinate R, who launches a projectile radially with a velocity v such that the projectile reaches infinity with exactly zero velocity. This defines the escape velocity in the Schwarzschild metric. No forces act

---

[4] Once precession of the Earth's rotation axis with respect to stars on the celestial sphere is accounted for, a discrepancy of 574.10 ± 0.65 arcseconds per century remains. About 531.6 arcseconds per century come from Newtonian gravitational perturbations on Mercury's orbit that are generated by other planets. The remaining ~ 43 arcseconds/century is attributable to general relativistic precession, except for a tiny effect from a possible quadrupole moment for the Sun.

on the projectile so it follows a radial geodesic. The energy per unit rest mass is ϵ and it is conserved by virtue of Eq. (9.22). At infinity $\epsilon = 1$, since then the particle is at rest and the energy is entirely rest mass energy. If u_{obs} is the 4-velocity of the stationary observer, the energy measured by the observer is [see Eq. (5.25)]

$$E = -p \cdot u_{\text{obs}} = -mu \cdot u_{\text{obs}} = -mg_{\mu\nu}u^\mu u^\nu_{\text{obs}} = -mg_{00}u^0 u^0_{\text{obs}}, \qquad (9.44)$$

where $p^\mu = mu^\mu$ with p^μ the 4-momentum and m the rest mass, and the last step follows because the observer is stationary. But in the Schwarzschild spacetime

$$g_{00} = -\left(1 - \frac{2M}{r}\right) \qquad u^0_{\text{obs}} = \left(1 - \frac{2M}{R}\right)^{-1/2} \qquad u^0 = \left(1 - \frac{2M}{r}\right)^{-1},$$

where the first expression follows from the metric (9.5), the second from Eq. (9.18), and the last from $u^0 = dt/d\tau$, Eq. (9.22), and that $\epsilon = 1$ at infinity and thus at all values of r, since it is conserved along the radial geodesic. Therefore, Eq. (9.44) reduces to

$$E = m\left(1 - \frac{2M}{R}\right)^{-1/2}. \qquad (9.45)$$

But from Eqs. (5.9) and (5.5) the energy and velocity in the observer's rest frame are related by $E = m(1 - v^2)^{-1/2}$, which implies upon comparison with Eq. (9.45) that

$$v_{\text{esc}} = \sqrt{\frac{2M}{R}}, \qquad (9.46)$$

which happens to be the same result obtained for Newtonian theory. Notice that at the Schwarzschild radius $R = 2M$ and $v = 1$ (in $c = 1$ units), implying an escape velocity equal to the velocity of light. The meaning of this will be addressed in Chapter 11.

9.6 Radial Fall of a Test Particle

It will be instructive for later discussion to analyze the particular case of a radial plunge orbit that starts from infinity with zero kinetic energy ($\epsilon = 1$) and zero angular momentum ($\ell = 0$). First, let us find an expression for the proper time as a function of the coordinate r. Setting $\ell = 0$ and $\epsilon = 1$ in Eqs. (9.28)–(9.30) for particle motion in the Schwarzschild geometry gives

$$r^{1/2}dr = \pm(2M)^{1/2}d\tau. \qquad (9.47)$$

Taking the negative sign for an infalling orbit, integrating both sides, and choosing an integration constant $\tau(r = 0) = 0$ gives

$$\frac{\tau}{2M} = -\frac{2}{3}\frac{r^{3/2}}{(2M)^{3/2}} \qquad (9.48)$$

for the proper time τ as a function of the Schwarzschild coordinate r. To find an expression for the coordinate time t as a function of r, note that $\epsilon = 1$ and is conserved. Then from Eq. (9.22) and Eq. (9.47)

$$\frac{dt}{dr} = \frac{dt/d\tau}{dr/d\tau} = -\left(1 - \frac{2M}{r}\right)^{-1}\left(\frac{2M}{r}\right)^{-1/2}, \qquad (9.49)$$

which may be integrated to give

$$t = t_0 - \int \left(1 - \frac{2M}{r}\right)^{-1}\left(\frac{2M}{r}\right)^{-1/2} dr = t_0 - \frac{1}{\sqrt{2M}}\int \frac{r^{3/2}dr}{r - 2M}$$

$$= t_0 - 2M\left(\frac{2}{3}\left(\frac{r}{2M}\right)^{3/2} + 2\left(\frac{r}{2M}\right)^{1/2} + \ln\left|\frac{(r/2M)^{1/2} - 1}{(r/2M)^{1/2} + 1}\right|\right), \qquad (9.50)$$

where t_0 is an integration constant. Equations (9.48) and (9.50) may be inverted to find r as a function of τ or t, respectively, and an expression for $t = t(\tau)$ may be obtained by solving (9.48) for r and substituting in (9.50).

From Eq. (9.48), the fall to the horizon at $r = 2M$ and then to the singularity at $r = 0$ takes finite proper time (the time measured by a clock carried by the falling particle). But from Eq. (9.50), as $r \to 2M$ the last term tends to infinity, so it takes an *infinite* amount of Schwarzschild coordinate time to reach the horizon and the particle never reaches $r = 0$. Figure 9.7 illustrates the different dependence of the radial coordinate on proper time and coordinate time. The smooth behavior of the $\tau(r)$ curve as it passes through $r = 2M$ suggests that there is nothing special about this point. This is further confirmed by examining the spacetime curvature in the vicinity of $r = 2M$. As shown in §31.2 of Misner, Thorne, and Wheeler [156], the components of the Riemann curvature tensor are all finite at the horizon, with the non-zero components typically being of order $1/M^2$. Thus, tidal forces at $r = 2M$ may be large if M is small, but they *remain finite*. In contrast, a similar analysis at $r = 0$ finds that in every local Lorentz frame at least one component of the curvature tensor becomes infinite, implying infinite tidal forces at $r = 0$. These considerations suggest that the Schwarzschild singularity at $r = 0$ is a true physical singularity but the one at $r = 2M$ is an artifact of the coordinate system and its failure to properly describe the geometry using the chosen coordinates in this region.

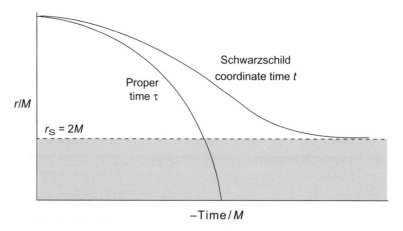

Fig. 9.7 Comparison of dependence of the radial coordinate on proper time and on Schwarzschild coordinate time for a free particle falling radially in a Schwarzschild spacetime.

Later alternative coordinates will be introduced that confirm this idea by explicitly removing the singularity at $r = 2M$ (but not at $r = 0$).[5]

9.7 Orbits for Light Rays

The calculation of orbits for light rays (or any massless particles) in the Schwarzschild metric can be set up to largely parallel that of particle orbits. The primary difference is that timelike particle orbits must be replaced by null photon orbits defined on the lightcone. The immediate impact is through Eq. (5.20):

$$u \cdot u = g_{\mu\nu} \frac{dx^\mu}{d\lambda} \frac{dx^\nu}{d\lambda} = 0, \tag{9.51}$$

where λ is an affine parameter. That is, the tangent vector, which was normalized to -1 for timelike trajectories in Eq. (5.6), is null for lightlike trajectories. Taking the motion to occur in the equatorial plane as before ($\theta = \frac{\pi}{2}$), Eq. (9.51) written out explicitly is

$$-\left(1 - \frac{2M}{r}\right)\left(\frac{dt}{d\lambda}\right)^2 + \left(1 - \frac{2M}{r}\right)^{-1}\left(\frac{dr}{d\lambda}\right)^2 + r^2\left(\frac{d\varphi}{d\lambda}\right)^2 = 0. \tag{9.52}$$

By analogy with the corresponding arguments for particle motion in Eqs. (9.22)–(9.23),

$$\epsilon \equiv -K_t \cdot u = \left(1 - \frac{2M}{r}\right)\frac{dt}{d\lambda}, \tag{9.53}$$

$$\ell = K_\varphi \cdot u = r^2 \sin^2\theta \frac{d\varphi}{d\lambda}, \tag{9.54}$$

are conserved along a geodesic for light. By proper choice of normalization for the affine parameter λ, the conserved quantity ϵ may be interpreted as the photon energy and the conserved quantity ℓ as its angular momentum at infinity.

Equations (9.53) and (9.54) may be used to eliminate $dt/d\lambda$ and $d\varphi/d\lambda$ from Eq. (9.52), which can be written after some rearrangement as [111]

$$\frac{1}{b^2} = \frac{1}{\ell^2}\left(\frac{dr}{d\lambda}\right)^2 + V_{\text{eff}}(r) \qquad V_{\text{eff}}(r) \equiv \frac{1}{r^2}\left(1 - \frac{2M}{r}\right) \qquad b^2 \equiv \frac{\ell^2}{\epsilon^2}. \tag{9.55}$$

The sign of ℓ determines the direction in which light orbits and $b \equiv |\ell/\epsilon|$ can be interpreted as the impact parameter of the photon (see Problem 9.7). The effective potential for photons and various classes of light-ray orbits are illustrated in Fig. 9.8. The top portion of the figure shows a plunging orbit that spirals to the origin, the middle portion shows an unstable

---

[5] It took more than a decade after the Schwarzschild solution was published for a consensus to begin developing that the singularity at the Schwarzschild radius was a coordinate and not physical singularity, but the isssue was only completely resolved in the 1960s. Although the singularity at $r = 2M$ is unphysical, the behavior in this region is nevertheless rather unusual by ordinary standards, as will be shown later. The central singularity is physical in the Schwarzschild solution, but there is the issue of whether general relativity is valid in the vicinity of the origin. Presumably it breaks down on a 4-momentum scale relevant for quantum gravity, for which there is not yet an adequate theory (see Sections 12.6 and 26.3).

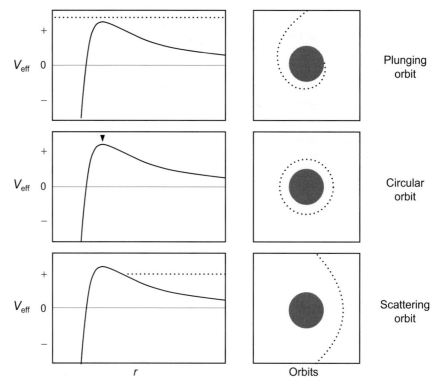

Fig. 9.8 Effective potential for photons (left) and light ray orbits (right) in a Schwarzschild metric (adapted from Hartle [111]).

circular orbit, and the bottom portion illustrates a scattering orbit. The dotted lines on the left side indicate the value of $1/b^2$ for each orbit.

9.8 Deflection of Light in the Gravitational Field

The deflection $d\varphi/dr$ for light in the Schwarzschild metric may be calculated in a manner similar to the determination of the precession angle for orbits of massive objects in Section 9.4. From Eq. (9.54) and the expression for $1/b^2$ in Eq. (9.55),

$$\frac{d\varphi}{dr} = \frac{d\varphi/d\lambda}{dr/d\lambda} = \pm \frac{1}{r^2}\left(\frac{1}{b^2} - V_{\text{eff}}(r)\right)^{-1/2}, \qquad (9.56)$$

where, as before, the sign determines the direction of the orbit around the scattering center. Let r_1 be the radial turning point, as illustrated in Fig. 9.9. Then, using the expression for $V_{\text{eff}}(r)$ in Eq. (9.55),

$$\Delta\varphi = 2\int_{r_1}^{\infty}\frac{dr}{r^2}\left[\frac{1}{b^2} - \frac{1}{r^2}\left(1 - \frac{2M}{r}\right)\right]^{-1/2}. \qquad (9.57)$$

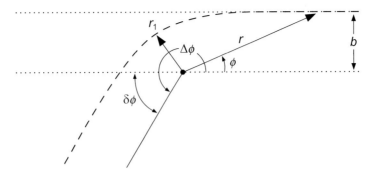

Fig. 9.9 Deflection of light by an angle $\delta\varphi$ in a Schwarzschild metric. Adapted from Hartle [111].

After a change of variables (see Hartle [111], chapter 9), this can be evaluated for small M/b to give $\Delta\varphi \simeq \pi + 4M/b$. From Fig. 9.9 then the deflection angle is

$$\delta\varphi = \Delta\varphi - \pi \simeq \frac{4M}{b}. \tag{9.58}$$

Example 9.3 illustrates the use of this formula for light passing near the Sun.

Example 9.3 Reinstating factors of G and c in Eq. (9.58),

$$\delta\varphi = \frac{4GM}{c^2 b} = 8.477 \times 10^{-6} \left(\frac{M}{M_\odot}\right)\left(\frac{R_\odot}{b}\right) \text{ rad.}$$

For a photon grazing the surface of the Sun, $M = 1\,M_\odot$ and $b = 1\,R_\odot$, which gives $\delta\varphi \simeq 1.75$ arcseconds.

Observation of the deflection estimated in Example 9.3 during a total solar eclipse in 1919 catapulted Einstein to worldwide fame when it was announced in 1920.

9.9 Shapiro Time Delay of Light

Light passing near a gravitating body follows a curved path and the time for light to travel between two points depends on this curvature. The deviation in travel time between that in the curved spacetime and the travel time if there were no curvature is termed the *Shapiro time delay*. As noted in Section 4.5, this does not mean that the speed of light varies. The local speed of light is always c but the observed elapsed time for light to go between two macroscopically separated points in spacetime depends on the metric. Thus, measurement of this time delay is a test of general relativity.

Example 9.4 To determine the time delay of light over a given path it is necessary to evaluate the integral of dt/dr. Proceeding in a similar manner as in Section 9.8, Eqs. (9.53) and (9.55) may be used to write

$$\frac{dt}{dr} = \frac{dt/d\lambda}{dr/d\lambda} = \pm\epsilon\left(1 - \frac{2M}{r}\right)^{-1}\left[\ell^2\left(\frac{1}{b^2} - V_{\text{eff}}\right)\right]^{-1/2}$$

$$= \pm\frac{1}{b}\left(1 - \frac{2M}{r}\right)^{-1}\left(\frac{1}{b^2} - V_{\text{eff}}(r)\right)^{-1/2}.$$

This may be integrated to give the time for light to travel between specified points.

In a typical Shapiro-delay experiment, radar waves are bounced off a planet and the time to go and return is measured for paths that pass very close to the surface of the Sun, or the delay in transmitting signals from space probes to Earth is measured as the signals pass near the Sun. The results of such experiments are consistent with the equation given above, thereby providing further confidence in the validity of general relativity.

9.10 Gyroscopes in Curved Spacetime

In investigating the predictions of general relativity it is often instructive to consider the behavior of gyroscopes in free fall. These will follow a timelike geodesic, with a 4-velocity $u(\tau)$ governed by the geodesic equation (7.23). In addition the gyroscope will have a spacelike *spin 4-vector* $s^\mu = (0, s)$ in this frame. In the freely falling local inertial frame the 4-velocity components of the gyroscope are $u = (1, 0, 0, 0)$, so $s \cdot u = 0$, which is a tensor equation and thus true in all frames. In flat spacetime or in a local inertial frame the spin is constant and $ds^\mu/d\tau = 0$. In curved spacetime the appropriate covariant generalization involves adding a term with the spin and velocity vectors contracted with the connection coefficients, giving an equation analogous to the geodesic equation (7.23),

$$\frac{ds^\mu}{d\tau} + \Gamma^\mu_{\alpha\beta} s^\alpha u^\beta = 0, \tag{9.59}$$

which describes how the components of the gyroscopic spin s^μ change along a geodesic and ensures that the scalar product $s \cdot u$ is preserved on the geodesic. As in classical mechanics the magnitude of the total spin is a constant of motion but the direction of the spin can *precess in angle*. Let us now use Eq. (9.59) to investigate two predictions of general relativity that lead to precession of the spin vector for freely falling gyroscopes in gravitational fields, *geodetic precession* and *dragging of inertial frames*.

9.11 Geodetic Precession

Consider a gyroscope in a circular orbit around a *non-rotating* gravitating sphere of mass M. As will now be demonstrated, an observer comoving with the gyroscope in orbital free fall will find that the gyroscope precesses, even if the source of the field is not rotating.

9.11 Geodetic Precession

This is called *geodetic precession* because the precession is observed along a geodesic.[6] Assume that the spacetime is described by the Schwarzschild metric (9.4), that the radius for the orbit is R, and that the spin points initially in the direction of a distant star. For an observer at rest in the gyroscope's frame the spin has only spatial components and by symmetry it must remain in the same plane, so choosing $\theta = \frac{\pi}{2}$ to define this plane in Schwarzschild coordinates (t, r, θ, φ), any precession occurs in the φ direction. For the 4-velocity $(u^0, u^1, u^2, u^3) \equiv (u^t, u^r, u^\theta, u^\varphi)$ the component in the φ direction is

$$u^\varphi = \frac{d\varphi}{d\tau} = \frac{d\varphi}{dt}\frac{dt}{d\tau} = \Omega u^t \qquad \Omega \equiv \frac{d\varphi}{dt} = \sqrt{\frac{M}{R^3}} \qquad u^t \equiv \frac{dt}{d\tau}, \qquad (9.60)$$

where Ω is the classical orbital angular velocity and Eq. (9.34) has been used. The time evolution for the spin components $(s^t, s^r, s^\theta, s^\varphi)$ will be governed by Eq. (9.59) evaluated in the Schwarzschild basis. Choose initially $s^\theta = 0$. It will remain zero because of symmetry. The component s^t is related to the other components of the spin by the requirement that $s \cdot u = 0$ along the geodesic, which implies that (Problem 9.16)

$$s^t = \frac{R^2 \Omega}{1 - 2M/R} s^\varphi. \qquad (9.61)$$

From Eq. (9.59) the remaining spin components s^r and s^φ require solution of the two equations

$$\frac{ds^r}{d\tau} + \Gamma^r_{\alpha\beta} s^\alpha u^\beta = 0, \qquad (9.62)$$

$$\frac{ds^\varphi}{d\tau} + \Gamma^\varphi_{\alpha\beta} s^\alpha u^\beta = 0. \qquad (9.63)$$

As you are asked to demonstrate in Problem 9.17, the solutions are

$$s^\varphi(t) = -s_0 \sqrt{1 - \frac{2M}{R}} \left(\frac{\Omega}{\omega R}\right) \sin(\omega t) \qquad s^r(t) = s_0 \sqrt{1 - \frac{2M}{R}} \cos(\omega t), \qquad (9.64)$$

where $s_0 = (s \cdot s)^{1/2}$ is the invariant magnitude of the spin and

$$\omega \equiv \left(1 - \frac{3M}{R}\right)^{1/2} \Omega. \qquad (9.65)$$

Imagine a gyroscope on a satellite in a circular Earth orbit and assume that the spin of the gyroscope starts off at $t = 0$ pointing in the radial direction.[7] The world lines for the gyroscope and Earth are illustrated in Fig. 9.10. From Eq. (9.64), after one complete orbit corresponding to a period $t = P = 2\pi/\Omega$,

---

[6] Strictly, "geodetic" is a misnomer because the spin of the gyroscope means that it does not follow an exact geodesic, but this effect is small and will be ignored here. Historically, what is now called geodetic precession was first investigated by Dutch physicist Willem de Sitter (1872–1934) in 1916, by assuming the Earth–Moon system to be a gyroscope in orbital free fall around the Sun. Hence it is also known as *de Sitter curvature precession*. de Sitter predicted that this effect would cause a precession of the entire lunar orbit with respect to the inertial frame of the Solar System by about 19 milliseconds of arc per year. This was not observable at the time of the prediction but has been confirmed by modern laser ranging of the distance to the Moon [205, 254].

[7] Practically the coordinate system may be chosen so that initially the spin vector points toward a distant guide star.

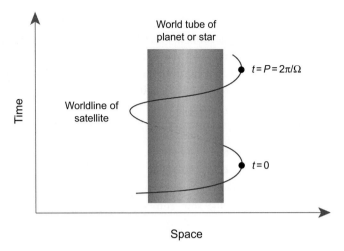

Fig. 9.10 Geodetic worldline for gyroscope in orbit around a spherical mass.

$$s^r(t = P) = s_0(1 - 2M/R)^{1/2} \cos(\omega P)$$
$$= s_0(1 - 2M/R)^{1/2} \cos\left(2\pi \frac{\omega}{\Omega}\right). \qquad (9.66)$$

In the absence of geodetic spin precession (so that $\omega = \Omega$), the angle φ would change by 2π for one orbit, so the additional spin precession angle for each orbit is

$$\Delta\varphi = 2\pi - 2\pi \frac{\omega}{\Omega} = 2\pi \left(1 - \frac{\omega}{\Omega}\right) = 2\pi \left[1 - \left(1 - \frac{3M}{R}\right)^{1/2}\right], \qquad (9.67)$$

in the direction of the orbital motion, where Eq. (9.65) was used in the last step. For any object in the Solar System $M/R = GM/Rc^2$ is very small, so the square root in Eq. (9.67) can be expanded to give

$$\Delta\varphi \simeq \frac{3\pi M}{R} = \frac{3\pi GM}{c^2 R}, \qquad (9.68)$$

for the geodetic precession per orbit for gyroscopes on a satellite in Earth orbit. Since the radial direction is perpendicular to the direction of orbital motion, Eq. (9.68) also gives the precession that would be measured by an observer comoving with the gyroscope in orbit. This geodetic precession effect is small, but cumulative for successive orbits.

Example 9.5 Gravity Probe B (GP-B) tested geodetic precession for gyroscopes aboard a satellite in an almost circular orbit averaging 642 km above the surface of the Earth. Figure 9.11 illustrates. From Problem 9.18, the expected geodetic precession is 1.22×10^{-3} arcsec orbit^{-1}, corresponding to a predicted geodetic precession rate of $\Delta\varphi/\Delta t = 6.6$ arcsec yr^{-1}. As discussed in Box 9.3, the geodetic precession rate measured by GP-B was within 0.07% of this value.

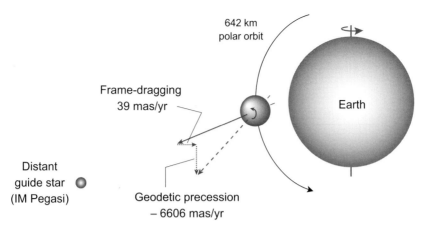

Fig. 9.11 Geodetic precession and frame dragging for a gyroscope on Gravity Probe B (adapted from [85]). Precession angles are greatly exaggerated. The star IM Pegasi was chosen as the directional reference because it was approximately in the desired direction for the gyroscopic spin axis and its proper motion on the celestial sphere was known precisely.

Agreement of the predictions of general relativity with measured geodetic precession for gyroscopes in Earth orbit by Gravity Probe B [85], and with geodetic precession measured for the Moon's orbit by laser rangefinding [205, 254], represent precise confirmation of the theory.

9.12 Gyroscopes in Rotating Spacetimes

The Schwarzschild solution gives a spacetime valid outside any spherical, static, non-rotating body. But astronomical bodies are typically spinning, so it is important physically to ask about solutions of the Einstein equation that are valid in rotating spacetimes. Some objects are spinning slowly, which suggests that their exterior metric might be approximated by an expansion about the Schwarzschild spherical spacetime. On the other hand, in later chapters black holes will be encountered that are spinning with an angular momentum comparable to the maximum allowed by the laws of physics. These cannot in any sense be understood in terms of perturbations of the Schwarzschild metric. In the remainder of this chapter the simpler topic of very slowly rotating spherical spacetimes will be taken up; in Chapter 13 the more complex issue of strongly deformed metrics implying potentially large angular momentum will be addressed.

9.12.1 Slow Rotation in the Schwarzschild Metric

As an example of a slowly rotating astronomical body, consider the Sun. It rotates differentially with a period of about a month, somewhat faster at the equator than at the poles. The spacetime outside the Sun is described by the Schwarzschild solution provided that: (1) it is a vacuum, (2) the gravitational effect of all other bodies in the Universe can

> **Box 9.3** **Testing Geodetic Precession and Frame Dragging**
>
> General relativity predicts that a gyroscope in orbit around Earth will exhibit two effects causing precession relative to a distant inertial frame [216]:
>
> 1. *Geodetic precession* caused by motion of the gyroscope through the spacetime curved by the mass of the Earth. This effect is equivalent to the curvature precession of the Earth–Moon system first described by de Sitter in 1916 [74].
> 2. A *frame dragging* caused by the Earth's rotation, closely related to the dragging of the orbital plane of a satellite around a rotating planet discussed by Lense and Thirring in 1918 [144].
>
> Initial R&D to identify the technologies required to test these predictions was funded in 1963 and it was hoped to launch a satellite experiment by the 1980s. However, the project encountered numerous setbacks, including the 1986 Challenger explosion, which forced the program to a rocket rather than Space Shuttle launch [253].
>
> **Gravity Probe B**
> Finally in 2004 NASA launched *Gravity Probe B (GP-B)*,[a] which used four gyroscopes on a satellite in an almost circular orbit directly over the poles to measure gyroscopic precession, giving direct evidence of both geodetic precession and frame dragging. The following table compares predicted and measured values,
>
> | Effect | GP-B measurement | General relativity |
> |---|---|---|
> | Geodetic precession | -6601.8 ± 18.3 mas yr^{-1} | 6606.1 mas yr^{-1} |
> | Frame dragging | -37.2 ± 7.2 mas yr^{-1} | -39.2 mas yr^{-1} |
>
> where 1 mas = 10^{-3} arcsec. Thus GP-B confirmed predicted geodetic precession within 0.07% and frame dragging within 5% for Earth's rotating gravitational field.
>
> **Summary of the Gravity Probe B Mission**
> The scientific results of Gravity Probe B were presented in Ref. [85] and a special volume of the journal *Classical and Quantum Gravity* published in 2015 contains 22 articles giving a comprehensive overview of the GP-B mission and results [9].
>
> [a] GP-B was launched from Vandenberg Air Force Base with the requisite orbit precision imposing a 1-second launch window. It was the successor to the Gravity Probe A (GP-A) experiment, carried out in a 1976 rocket flight to test the equivalence principle. By comparing timing using hydrogen masers in the rocket and on the ground, GP-A confirmed gravitational time dilation at the 70×10^{-6} level [238].

be neglected, (3) the Sun is static with no spin, and (4) the Sun is spherical. To a very good approximation these conditions are satisfied, so the Schwarzschild metric is almost (but not quite) valid outside the Sun. Let us now parse the "not quite" part of the preceding statement.

Take the exterior of the Sun to be a vacuum and ignore the effect of all other masses, so that deviation from the Schwarzschild metric outside the Sun is caused only by its angular momentum and by its deviation from spherical symmetry. Deviations from spherical symmetry are very small and caused classically by centripetal effects of its rotation. If the Schwarzschild metric were expanded in powers of the angular momentum J, to first order the Sun then would be expected to *remain spherical*, since centripetal forces vary as the angular velocity squared and so come in only at the level of the J^2 term in the expansion.

However, recall that general relativity has many formal similarities with electromagnetism and that electromagnetic forces can arise from charge but also from *motion of charge*. The latter are *magnetic effects*. In general relativity mass acts as the "gravitational charge," which suggests that gravitational "forces" (that is, curvature of spacetime) may arise from *motion of mass* ("mass currents"), in addition to arising from the mass itself. This is indeed the case, and because of the analogy with electromagnetism where magnetic fields are produced by (charge) currents, these are called *gravitomagnetic effects* in general relativity.[8] From Eq. (6.5), the intrinsic strength of gravity near the surface of a gravitating sphere is characterized by GM/Rc^2. In expanding the metric to first order in J it will be found that the resulting gravitomagnetic effects are proportional to $(v/c)(GM/Rc^2)$, or one order lower in an expansion in $(1/c)$ than the leading effects of the gravitational field itself. For low velocities gravitomagnetic effects will be very small but they are potentially measurable and can serve as a test of gravitational theories.

Assuming a slowly rotating spherical field, expansion about the Schwarzschild metric (9.4) to first order in the angular momentum J gives a metric [239]

$$ds^2 = -\left[1 - \frac{2M}{r} + \mathcal{O}\left(\frac{1}{r^2}\right)\right]dt^2 - \left[\frac{4J\sin^2\theta}{r} + \mathcal{O}\left(\frac{1}{r^2}\right)\right]d\varphi\,dt$$
$$+ \left[1 + \frac{2M}{r} + \mathcal{O}\left(\frac{1}{r^2}\right)\right]\left(dr^2 + r^2(d\theta^2 + \sin^2\theta\,d\varphi^2)\right). \tag{9.69}$$

Upon restoring factors of G and c, Eq. (9.69) may be written as [111]

$$ds^2 = ds_0^2 - \frac{4GJ}{c^3 r^2}\sin^2\theta(r\,d\varphi)(c\,dt) + \mathcal{O}\left(J^2\right), \tag{9.70}$$

where ds_0^2 is the contribution from the unperturbed Schwarzschild metric (9.4) and $\mathcal{O}\left(J^2\right)$ indicates that terms of order J^2 and higher have been discarded. For Newtonian theory $J \sim Mrv$, where v is the linear velocity of rotation. Therefore, for the coefficient of the term proportional to J in Eq. (9.70),

$$\frac{GJ}{c^3 r^2} \sim \frac{GMrv}{c^3 r^2} = \frac{v}{c} \times \frac{GM}{c^2 r},$$

which shows that (as expected) the effect of mass motion on spacetime curvature is of order v/c in size relative to the primary effect caused by the mass itself, which is proportional to $GM/c^2 r$. Equations (9.69) and (9.70) are adequate for testing general relativity for slowly spinning objects with near spherical symmetry, such as planets in the Solar System.

---

[8] These effects are caused by rotation of a gravitating body and should be distinguished from the geodetic precession discussed in Section 9.11, which is caused simply by curvature without rotation. A rotating massive body will exhibit both effects, with the geodetic precession generally being much larger than the gravitomagnetic effects for slow rotation.

9.12.2 Dragging of Inertial frames

Consider the following thought experiment. Imagine a spherical body like the Earth in slow rotation, so that the metric Eq. (9.70) is valid approximately in the spacetime surrounding it, and imagine dropping a gyroscope from above the North Pole with an initial spin perpendicular to the rotation axis. Figure 9.12 illustrates. Since spherical coordinates are singular on the z axis, it is convenient to work in cartesian coordinates, as indicated in Fig. 9.12. Upon conversion to cartesian coordinates the metric (9.70) takes the form

$$ds^2 = ds_0^2 \text{ (cartesian)} - \frac{4GJ}{c^3 r^2}(cdt)\left(\frac{xdy - ydx}{r}\right), \qquad (9.71)$$

where ds_0^2 (cartesian) is the unperturbed Schwarzschild metric expressed in cartesian coordinates. The calculation is simplified by noting that only terms up to order $1/c^3$ need be retained. Terms involving the mass M in the Schwarzschild metric will contribute at order $1/c^5$, so for low angular momentum the unperturbed Schwarzschild metric can be replaced by its $M \to 0$ limit, which is just the flat Minkowski metric. Thus, it is sufficient to work with the approximate metric

$$ds^2 = -(cdt)^2 + dx^2 + dy^2 + dz^2 - \frac{4GJ}{c^3 r^2}(cdt)\left(\frac{xdy - ydx}{r}\right). \qquad (9.72)$$

Taking the beginning spin to have only x and y components, initially

$$u^\mu = (u^t, 0, 0, u^z) \qquad s^\mu = (0, s^x, s^y, 0), \qquad (9.73)$$

so clearly $s \cdot u = 0$ initially and by symmetry arguments the spin will evolve with only x and y components. The spin of the gyroscope as it falls freely will be governed by

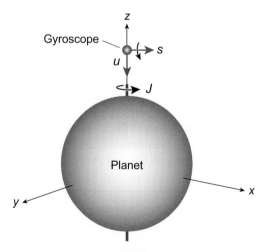

Fig. 9.12 A gyroscope in free fall on the rotation axis of a spherical planet in slow rotation. The initial 4-spin s of the gyroscope is perpendicular to the 4-velocity u, and the angular momentum of the planet is J.

Eq. (9.59). Given the above approximations, to leading order in $1/c$ the only contribution to the summation in the second term of this equation will be from terms with [111]

$$\Gamma^1_{02} \equiv \Gamma^x_{ty} = \frac{2GJ}{c^2 z^3} \qquad \Gamma^2_{01} \equiv \Gamma^y_{tx} = -\frac{2GJ}{c^2 z^3} \qquad (9.74)$$

where $r = z$ since the gyroscope lies on the z axis. For the x and y components of the spin, Eq. (9.59) yields

$$\frac{ds^x}{d\tau} = -\Gamma^x_{yt} s^y u^t = -\frac{2GJ}{c^2 z^3} s^y u^t \qquad \frac{ds^y}{d\tau} = -\Gamma^y_{xt} s^x u^t = \frac{2GJ}{c^2 z^3} s^x u^t.$$

Utilizing $u^t = dt/d\tau$, this may be written

$$\frac{ds^x}{dt} = -\frac{2GJ}{c^2 z^3} s^y \qquad \frac{ds^y}{dt} = \frac{2GJ}{c^2 z^3} s^x, \qquad (9.75)$$

and the corresponding angular rate of precession in the x–y plane Ω_{LT} is given by

$$\Omega_{\text{LT}} = \frac{2GJ}{c^2 z^3}. \qquad (9.76)$$

This result was obtained in a Lorentz frame for which the source of the spherical field is at rest and the gyroscope is falling. However, it is valid also in the frame of the gyroscope because Lorentz boosts along the z axis do not affect the transverse spin components s^x and s^y, and there is no time dilation to leading order in $1/c$.

The gyroscopic precession (9.76) caused by rotation of a gravitational field was pointed out in a 1918 paper by Lense and Thirring [144], and is commonly termed the *Lense–Thirring effect*.[9] It is also called *dragging of inertial frames* or just *frame dragging*, because physically the gyroscopic precession may be viewed as the dragging of local spacetime relative to the distant reference stars by the rotating field. As already noted, the Lense–Thirring effect should be distinguished from the geodetic effect, which occurs even if the mass source for the field is not rotating.

Example 9.6 Equation (9.76) is used in Problem 9.19 to show that for a gyroscope in free fall on the Earth's polar axis the Lense–Thirring precession rate is given by

$$\Omega_{\text{LT}} = 5.65 \times 10^{10} \left(\frac{\text{km}}{z}\right)^3 \text{arcsec yr}^{-1}, \qquad (9.77)$$

where z is the distance from the center of the Earth. For GP-B (Box 9.3), the satellite was in an almost circular polar orbit with a 7027-km semimajor axis. Inserting this for z in Eq. (9.77) gives a precession rate of 0.16 arcsec yr^{-1}, which is much smaller than the geodetic precession rate of more than 6 arcseconds per year. The Lense–Thirring

---

[9] It has been argued that Einstein provided most of the key ideas to Thirring, and that Lense provided only discussion of possible astrophysical applications, so the effect should more properly be termed the *Einstein–Thirring–Lense effect* [118]. But after almost a century the terminology is rather fixed. In the present work the term "frame dragging" will most commonly be used to refer to this effect.

precession rate depends on the latitude of a satellite in polar orbit, so the general relativistic prediction of 0.039 arcsec yr^{-1} shown in Box 9.3 for Gravity Probe B represents an average over the orbit, which is smaller by a factor of about four than our evaluation on the z axis alone.

Confirmation of Lense–Thirring precession by Gravity Probe B (Box 9.3) was the first direct observation of frame dragging, and an important experimental test of general relativity.

Frame dragging has been illustrated here specifically for a slowly rotating Schwarzschild metric but dragging of inertial frames is expected for any metric having a field source that depends on angular momentum. It is a very small effect for Earth's gravitational field but in Chapter 13 extreme frame dragging effects will be discussed that can occur in much stronger, much more rapidly rotating gravitational fields. These may be of large astrophysical importance because frame dragging of spacetime around rotating black holes may help power some of the most energetic events observed in the Universe (see Chapter 15).

Background and Further Reading

Most books dealing with general relativity treat the Schwarzschild metric in some depth. It is one of the simplest possible solutions for the Einstein equations and is of considerable astrophysical importance because it describes the gravitational field in the vacuum external to a spherical mass distribution. The discussions in Hartle [111]; Schutz [221]; and Misner, Thorne, and Wheeler [156] are particularly good and have strongly influenced the presentation in this chapter. Significant parts of the discussion of particle and light orbits, geodetic precession, and dragging of inertial frames in the Schwarzschild metric were adapted from Hartle [111].

Problems

9.1 Use Eq. (9.1) substituted into the Einstein equations (8.21) to obtain the Schwarzschild solution given by Eq. (9.4). ***

9.2 Find the turning points of the radial motion for a free particle in a Schwarzschild spacetime. Show that if $\ell^2/M^2 < 12$, the effective potential is negative for all values of the radial coordinate r. Show that if $\ell^2/M^2 > 12$, the effective potential has one maximum and one minimum. ***

9.3 Demonstrate that the innermost stable circular orbit in the Schwarzschild spacetime is marginally stable [it is at a point of inflection for the potential; see Fig. 9.5(a) for $\ell^2/M^2 \sim 12$].

9.4 Look up the relevant orbital information and estimate to order $1/c^2$ the perihelion precession for Mercury and the periastron precession for the Binary Pulsar PSR

1913+16, assuming their orbits to lie in a Schwarzschild spacetime. (The formulas derived for test particles orbiting large masses are applicable to binary systems having components of comparable mass if the semimajor axis a is replaced by the average separation of the binary pair and mass is replaced by total mass of the binary.) Compare with the observed values of 43 arcseconds per century and about 4.2 degrees per year, respectively. What is the corresponding precession angle per year for Earth's orbit, to order $1/c^2$?

9.5 Assuming Newtonian mechanics for elliptical planetary motion in the Solar System, show that the precession equation (9.42) can be written in the form (9.43) if the gravitational field is weak. *Hint*: For classical orbital motion

$$r^2 \frac{d\varphi}{dt} = \sqrt{1-e^2}\left(\frac{2\pi}{P}\right)a^2,$$

where e is the eccentricity, P the period, and a the semimajor axis of the orbit. ***

9.6 Show that in the Schwarzschild metric the effective potential for motion of light has a single maximum located at $r = 3M$, with height $1/27M^2$. Thus, light can go into an unstable circular orbit of radius $3M$ (that is, $\frac{3}{2}r_S$) in a Schwarzschild spacetime.

9.7 In the following diagram for a scattering photon in the $\theta = \frac{\pi}{2}$ plane of the Schwarzschild metric with $r \gg 2M$,

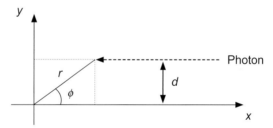

d is the impact parameter. Demonstrate that the parameter b of Eq. (9.55) is equal to d for $r \gg 2M$ and thus may be interpreted as the photon impact parameter. ***

9.8 A rod of length 1 centimeter is oriented radially in the gravitational field at the surface of the Sun. Assuming gravity to be described by a Schwarzschild spacetime, what is the *coordinate* length of the rod? Make a similar estimate for the same rod oriented radially at the surface of a neutron star, again assuming a Schwarzschild metric.

9.9 Introduce a new radial coordinate $\rho = \rho(r)$ and require that the Schwarzschild line element take the form

$$ds^2 = -f(\rho)dt^2 + g(\rho)\left(d\rho^2 + \rho^2 d\theta^2 + \rho^2 \sin^2\theta d\varphi^2\right),$$

where the quantity in parentheses on the right is the line element of euclidean 3-space and $f(\rho)$ and $g(\rho)$ are functions to be determined. Thus show that the Schwarzschild line element can be rewritten in the *isotropic form*

$$ds^2 = -\frac{(1-M/2\rho)^2}{(1+M/2\rho)^2}dt^2 + (1+M/2\rho)^4\left(d\rho^2 + \rho^2(d\theta^2 + \sin^2\theta\, d\varphi^2)\right),$$

where ρ is related to r by $r = \rho(1+M/2\rho)^2$. Hint: Try a solution of the form

$$ds^2 = -(1-2M/r)dt^2 + \alpha(\rho)^2\left(d\rho^2 + \rho^2 d\theta^2 + \rho^2\sin^2\theta\, d\varphi^2\right),$$

and use comparisons of this with the standard form of the Schwarzschild line element to show that ρ and r are related by the differential equation

$$\frac{d\rho}{\rho} = \pm\frac{dr}{\sqrt{r^2 - 2Mr}},$$

(where the positive sign may be chosen by assuming $\rho \to \infty$ for $r \to \infty$). Integrate this with a boundary condition $\rho(r = 2M) = M/2$ (see Problem 9.10 for a justification of this choice) to show that $r = \rho(1+M/2\rho)^2$, use this to find $\alpha(\rho)$, and substitute to give the final form of the isotropic metric after some algebra.

9.10 Use the results of Problem 9.9 to demonstrate that the new radial variable $\rho(r)$ defined for isotropic coordinates has the following properties:

1. For $0 \le r < 2M$, the variable ρ is complex (thus, the region inside the Schwarzschild radius $r_s = 2M$ is not in the real domain of ρ).
2. If $r > 2M$, any value of r gives two real positive values of ρ, so this region of r-space is mapped out *twice* in ρ-space, once in the region $M/2 \le \rho < \infty$ and once in the region $0 \le \rho \le M/2$. [This is the reason for the choice of boundary condition $\rho(r = 2M) = M/2$ in Problem 9.9.]

Hint: Solve the equation $r = \rho(1+M/2\rho)^2$ found in Problem 9.9 for ρ.

9.11 Construct the Schwarzschild line element using the following equivalence principle argument. Imagine a small object in free fall in a spherically symmetric gravitational field. In a local coordinate system falling with the object an observer feels no gravitational field, special relativity is valid, and the line element can be expressed as $ds^2 = -d\bar{t}^2 + d\bar{r}^2$ where $d\bar{r}^2$ is associated with the spatial part of the metric, the coordinates $(\bar{t}, \bar{r})$ are valid only in the freely falling frame, and $c = G = 1$ units are assumed. Now assume a general set of coordinates (T, r) valid for a stationary observer watching the object fall (and who does feel the gravitational field). Assuming that the object has fallen freely from infinity it has a radial velocity $v(r) = -\hat{r}\sqrt{2M/r}$, where M is the mass producing the gravitational field. Show that the two coordinate systems are related by $d\bar{t} = dT$ and $d\bar{r} = dr - vdT$, and that the line element can be expressed in the new coordinates as

$$ds^2 = -\left(1 - \frac{2M}{r}\right)dT^2 - 2\sqrt{\frac{2M}{r}}\,drdT + dr^2,$$

where in spherical coordinates the spatial part of the line element is $d\mathbf{r}^2 = dr^2 + r^2 d\theta^2 + r^2 \sin^2\theta\, d\varphi^2$. This metric is not diagonal because of the cross term in $dr\,dT$. Show that a linear transformation to a new time coordinate t of the form

$$dT = dt - \frac{\sqrt{2M/r}}{1 - 2M/r}dr$$

brings the metric to a diagonal form that corresponds to the standard Schwarzschild line element.

9.12 Derive a general formula for the physical (directly measured) distance between two concentric spherical shells that are at coordinate distances r_1 and r_2 outside the event horizon of a Schwarzschild black hole, assuming that r_1 and r_2 may differ by a non-infinitesimal amount. Calculate the physical distance between shells at coordinate radii 5 km and 6 km if the black hole has a mass of $1 M_\odot$.

9.13 Show that Eq. (9.50) follows from Eq. (9.49). *Hint*: The substitution $r = x^2$ will put the integral into a form where standard tables may be used.

9.14 What is the proper distance for the circumference of a circle in Schwarzschild spacetime that has a coordinate radius r? Interpret your result physically in relation to a circle of radius r drawn in flat spacetime.

9.15 (a) Show that ϵ^2 is given by Eq. (9.32). (b) Prove that the angular velocity with respect to the coordinate time is given in the Schwarzschild spacetime by Eq. (9.33). (c) Show that the timelike component of the 4-velocity for a circular Schwarzschild orbit is given by Eq. (9.36). ***

9.16 Use the requirement that $s \cdot u = 0$ for a gyroscope in orbit to show that the spin component s^t for the gyroscope is given by Eq. (9.61). ***

9.17 Prove that the solutions of the two equations (9.62) and (9.63) are given by Eq. (9.64) using the following steps. (i) Show that Eq. (9.62) can be expressed in terms of s^φ as

$$\frac{ds^r}{dt} - (R - 3m)\Omega s^\varphi = 0.$$

(ii) Then show that Eq. (9.63) can be expressed in terms of s^r as

$$\frac{ds^\varphi}{dt} + \frac{\Omega}{R} s^r = 0.$$

(iii) Solve the above two equations simultaneously to give the solutions for s^r and s^φ given in Eq. (9.64). ***

9.18 Gravity Probe B tested geodetic precession using four gyroscopes on a satellite in an almost circular orbit averaging 642 km above the surface of the Earth (corresponding to a semimajor axis $a = 7027.4$ km for the orbit). What is the prediction of general relativity for the yearly geodetic precession rate that Gravity Probe B should have seen, assuming the orbit of the satellite to be circular? Compare your answer with that actually measured by GP-B, which is discussed in Box 9.3. ***

9.19 Starting from Eq. (9.76), show that for a gyroscope in free fall on the Earth's polar axis the Lense–Thirring precession rate for a gyroscopic spin perpendicular to the rotation axis is given by

$$\Omega_{\text{LT}} = 5.65 \times 10^{10} \left(\frac{1\,\text{km}}{z}\right)^3 \text{ arcsec yr}^{-1},$$

where z is the distance from the center of the Earth. What is the Lense–Thirring precession rate for a gyroscope in a nearly circular polar orbit with semimajor axis 7027.4 km (that of Gravity Probe B in Box 9.3) as it passes over the North Pole? Compare with the yearly geodetic precession rate for a gyroscope on the same satellite. ***

10 Neutron Stars and Pulsars

The existence of the *Chandrasekhar limiting mass* for any object stabilized against gravitational collapse by fermion degeneracy pressure implies that stars cannot shrink to stable white dwarfs if their mass exceeds about 1.4 $M_\odot$ at the end of their lives (see Section 16.2.2 of Ref. [105] for further discussion). Stars exceeding this limit are destined to collapse to objects much more dense even than white dwarfs. If the mass at collapse is more than several solar masses the collapse is likely to produce a black hole, which we will discuss in Chapters 11–14. If the mass is less than this but larger than the white-dwarf Chandrasekhar mass in the final collapse, the star will become a *neutron star*. Neutron stars are of great interest in their own right but they also explain the existence of pulsars and these in turn provide some of the most stringent observational tests of general relativity. In this chapter we consider solutions of the Einstein equations suitable for the description of neutron stars. A crucial new ingredient relative to the Schwarzschild solution of Chapter 9 is that a neutron star requires an interior solution, implying that the full Einstein equation (8.21) must be solved with stress–energy source terms representative of neutron star matter.

10.1 A Qualitative Picture of Neutron Stars

As was discussed in Section 6.4.2, objects of white dwarf density are described reasonably well by Newtonian gravity, with general relativity corrections appearing typically at the $\sim 10^{-4}$ level. For neutron stars the densities are much higher; while Newtonian gravity works *qualitatively* for neutron stars, the gravity is sufficiently strong that a correct *quantitative* description requires general relativity. However, many basic properties can be estimated by employing Newtonian concepts and simple reasoning. For example, in the problems at the end of this chapter the assumption that in a neutron star gravity packs the neutrons down to their hard-core radius of $\sim 10^{-13}$ cm yields immediately that the most massive neutron stars contain about 3×10^{57} baryons (mostly neutrons) within a radius of about 7 km, with a corresponding mass of about 2.3 $M_\odot$. This implies an average density greater than 10^{15} g cm^{-3}, which is several times normal nuclear matter density, and a (gravitational) binding energy of order 100 MeV per nucleon, which is an order of magnitude larger than the binding energy of nucleons in nuclear matter. All of these are reasonable estimates. These problems indicate also that the total gravitational binding energy of a neutron star is within an order of magnitude of the rest mass energy, and that the escape velocity is about 50% of the speed of light, both signaling that general relativistic effects are likely to be significant.

Although general relativity is important for the overall properties of neutron stars, Problem 10.5 makes plausible that over a microscopic scale characteristic of nuclear interactions the metric is essentially constant. This implies that the microphysics (nuclear and elementary particle interactions) of the neutron star can be described by quantum mechanics implemented in flat spacetime (special relativistic quantum field theory). Thus, in studying neutron stars it is possible to largely decouple gravity, which governs the overall structure, from quantum mechanics, which governs the microscopic properties.

10.2 Solutions inside Spherical Mass Distributions

Let's now consider solutions of the Einstein equations for a static, spherical mass distribution. It is already known that outside the mass distribution the Einstein equations take their vacuum form (vanishing stress–energy tensor) and the solution is approximated well by the Schwarzschild metric. Inside the mass distribution the Schwarzschild solution is no longer valid since the stress–energy tensor on the right side of the Einstein equations will have non-zero components. Thus, another solution must be sought for the interior, and then that solution may be matched to the exterior (Schwarzschild) solution at the surface of the spherical mass distribution.

10.2.1 Simplifying Assumptions

In the most general case the solution that we seek could be quite difficult to find since it depends on the internal structure of the star or other spherical object (it will be termed a star for convenience). A relatively simple solution was obtained by Oppenheimer and Volkov [171] by making the following assumptions [99, 199].

1. The matter inside the star is a perfect fluid, with a stress–energy tensor given by Eq. (7.9),
$$T^\mu_{\ \nu} = (\epsilon + P)u^\mu u_\nu + P\delta^\mu_{\ \nu}, \tag{10.1}$$
where $\epsilon = \rho c^2$ is the energy density, P is the pressure, u^μ is the 4-velocity, and for later convenience the rank-2 tensors have been expressed in their mixed form.

2. The system is spherically symmetric, with a line element of the general form[1]
$$ds^2 = -e^{\sigma(r)}dt^2 + e^{\lambda(r)}dr^2 + r^2 d\theta^2 + r^2 \sin^2\theta d\varphi^2, \tag{10.2}$$
implying non-vanishing metric components
$$g_{00}(r) = -e^{\sigma(r)} \quad g_{11}(r) = e^{\lambda(r)} \quad g_{22}(r) = r^2 \quad g_{33}(r,\theta) = r^2 \sin^2\theta.$$

If the neutron star is assumed to be spherical and not vibrating or rotating, the gravitational field outside the mass distribution should correspond to the Schwarzschild

---

[1] Equation (10.2) is just a particular choice for parameterization of Eq. (9.1); see Problem 8.2 and Appendix C. The factors e^σ and e^λ are always positive, so the signature of (10.2) is unambiguous for any σ and λ.

solution and the metric (10.2) must match smoothly to the Schwarzschild metric at the surface of the neutron star.

3. The system is assumed to be in equilibrium so that $\sigma(r)$ and $\lambda(r)$ are functions only of r and not of t, and the 4-velocity has no space components. Thus from $u \cdot u = -1$,

$$u^\mu = (e^{-\sigma/2}, 0, 0, 0). \tag{10.3}$$

Inserting these 4-velocity components in Eq. (10.1), the stress–energy tensor in the rest frame of the fluid takes the diagonal form (see Problem 7.18)

$$T^\nu{}_\mu = \begin{bmatrix} -\epsilon & 0 & 0 & 0 \\ 0 & P & 0 & 0 \\ 0 & 0 & P & 0 \\ 0 & 0 & 0 & P \end{bmatrix}, \tag{10.4}$$

where $\epsilon = \rho$ in $c = 1$ units.

4. For the vacuum Einstein equation only the Ricci tensor is required to construct the Einstein tensor (see Problem 22.1), but in the general non-vacuum case both the Ricci tensor $R_{\mu\nu}$ and the Ricci scalar R will be needed. It is convenient to express the corresponding Einstein equation (8.21) in the mixed tensor form

$$G^\nu{}_\mu \equiv R^\nu{}_\mu - \tfrac{1}{2}\delta^\nu_\mu R = 8\pi T^\nu{}_\mu. \tag{10.5}$$

The stress–energy tensor (10.4) is diagonal, so only diagonal components of $G^\nu{}_\mu$ will be required.

10.2.2 Solving the Einstein Equations

The stress–energy tensor obeys the 4-divergence condition Eq. (7.10). Hence $T^\mu{}_{\nu;\mu} = 0$ can be solved in place of one of the Einstein equations, which can lead to a faster solution than solving all the Einstein equations directly. That strategy will be employed here, using two Einstein equations and the constraint equation $T^\mu{}_{\nu;\mu} = 0$ to obtain a solution.

Solution of the constraint equation: The constraint equation has been solved already in Problem 8.12. There you were asked to show that for a metric of the form (10.2) with a stress–energy tensor given by Eq. (10.4), $T^\mu{}_{\nu;\mu} = 0$ implies that

$$P' + \tfrac{1}{2}(P + \epsilon)\sigma' = 0,$$

with primes denoting partial derivatives with respect to r. Two additional equations are required; the simplest choices are $G^0{}_0 = 8\pi T^0{}_0$ and $G^1{}_1 = 8\pi T^1{}_1$. The Einstein tensors G_{00} and G_{11} were derived in Problem 8.2 and are summarized in Appendix C. The metric is diagonal and contracting with the metric tensor to raise an index gives

$$G^0{}_0 = g^{00} G_{00} = -e^{-\sigma} G_{00} = e^{-\lambda}\left(\frac{1}{r^2} - \frac{\lambda'}{r}\right) - \frac{1}{r^2},$$

$$G^1{}_1 = g^{11} G_{11} = e^{-\lambda} G_{11} = e^{-\lambda}\left(\frac{1}{r^2} + \frac{\sigma'}{r}\right) - \frac{1}{r^2}.$$

From the preceding three equations and Eqs. (10.5) and (10.4),

$$-e^{-\lambda}\left(\frac{1}{r^2} - \frac{\lambda'}{r}\right) + \frac{1}{r^2} = 8\pi\epsilon(r), \tag{10.6}$$

$$e^{-\lambda}\left(\frac{1}{r^2} + \frac{\sigma'}{r}\right) - \frac{1}{r^2} = 8\pi P(r), \tag{10.7}$$

$$P' + \tfrac{1}{2}(P + \epsilon)\sigma' = 0. \tag{10.8}$$

Let us now solve this set of equations.

Solution of first and second Einstein equations: To proceed we observe that the first Einstein equation (10.6) may be rewritten as [156]

$$\frac{2}{r^2}\frac{dm}{dr} = 8\pi\epsilon, \tag{10.9}$$

where a new parameter $m(r)$ has been defined through

$$2m(r) \equiv r(1 - e^{-\lambda}). \tag{10.10}$$

At this point $m(r)$ is only a parameterization of the metric coefficient e^λ since, upon multiplying Eq. (10.10) by e^λ,

$$e^\lambda = \frac{r}{r - 2m(r)} = \left(1 - \frac{2m(r)}{r}\right)^{-1}, \tag{10.11}$$

but $m(r)$ will be interpreted in Section 10.3 below as the total mass–energy enclosed within the radius r. With this interpretation, note that (10.11) is of the Schwarzschild form for r outside the spherical mass distribution of the star. Solving (10.9) for dm gives $dm = 4\pi r^2 \epsilon\, dr$, and thus

$$m(r) = 4\pi \int_0^r \epsilon(r) r^2\, dr, \tag{10.12}$$

where an integration constant $m(0) = 0$ has been chosen on physical grounds. Now consider the second Einstein equation (10.7). Solving it for $\sigma' = d\sigma/dr$ gives

$$\frac{d\sigma}{dr} = e^\lambda\left(8\pi r P(r) + \frac{1}{r}\right) - \frac{1}{r},$$

and substitution of (10.11) for e^λ leads to

$$\frac{d\sigma}{dr} = \frac{8\pi r^3 P(r) + 2m(r)}{r(r - 2m(r))}. \tag{10.13}$$

Therefore Eqs. (10.11) and (10.13) may be used to define the metric coefficients e^σ and e^λ in terms of the parameter $m(r)$ and the pressure $P(r)$.

10.2.3 The Oppenheimer–Volkov Equations

The constraint equation (10.8) may be combined with Eq. (10.13) to give

$$\frac{dP}{dr} = -\frac{(P(r) + \epsilon(r))(4\pi r^3 P(r) + m(r))}{r(r - 2m(r))}.$$

Collecting results gives the *Oppenheimer–Volkov equations* [171] (also called the *Tolman–Oppenheimer–Volkov* or *TOV equations*) for a static, spherical, gravitating perfect fluid

$$\frac{dP}{dr} = -\frac{(P(r) + \epsilon(r))\left(m(r) + 4\pi r^3 P(r)\right)}{r^2 \left(1 - \frac{2m(r)}{r}\right)}, \qquad (10.14)$$

$$m(r) = 4\pi \int_0^r \epsilon(r) r^2 \, dr, \qquad (10.15)$$

where the parameter $m(r)$ will be interpreted in Section 10.3 as the total mass–energy contained within a radius r. Solution of these equations requires specification of an equation of state for the fluid that relates the density to the pressure. They may then be integrated numerically from the origin outward with initial conditions $m(r = 0) = 0$ and an arbitrary choice for the central density $\epsilon(r = 0)$ until the pressure $P(r)$ becomes zero, which defines the surface of the star $r = R$ and the mass of the star $m(R)$. For a given equation of state each choice of $\epsilon(0)$ will give a unique R and $m(R)$ when the equations are integrated. This defines a family of solutions characterized by a specific equation of state and the value of a single parameter (the central density, or a quantity related to it like central pressure) [99].

10.2.4 Interpretation of Oppenheimer–Volkov Equations

Equations (10.14)–(10.15) represent the general relativistic description of hydrostatic equilibrium for a spherical, gravitating perfect fluid.[2] They reduce to the Newtonian description of hydrostatic equilibrium in the limit of weak gravitational fields, but imply significant deviations from the Newtonian description in strong gravitational fields such as those for neutron stars. To exhibit this clearly Eqs. (10.14) and (10.15) may be rewritten in the form

$$4\pi r^2 dP(r) = \frac{-m(r) dm(r)}{r^2}$$
$$\times \left(1 + \frac{P(r)}{\epsilon(r)}\right)\left(1 + \frac{4\pi r^3 P(r)}{m(r)}\right)\left(1 - \frac{2m(r)}{r}\right)^{-1} \qquad (10.16)$$

$$dm(r) = 4\pi r^2 \epsilon(r) dr \qquad (10.17)$$

(see Problem 10.4). For Newtonian gravity, hydrostatic equilibrium requires that

$$4\pi r^2 dP = -\frac{m\, dm}{r^2} \qquad dm = 4\pi r^2 \rho\, dr \qquad (10.18)$$

(in $G = 1$ units), which suggests the following interpretation of Eqs. (10.16)–(10.17):

1. Equation (10.17) gives the mass–energy of a shell lying between radii r and $r + dr$.
2. The left side of Eq. (10.16) is the net force acting outward on this shell because of the pressure gradient.

---

[2] The condition of hydrostatic equilibrium was built into the solution through the assumption (10.3), which constrains the fluid to be static since the 4-velocity has no non-zero space components.

3. The first factor on the right side of Eq. (10.16) is the attractive Newtonian gravity acting on the shell because of the mass interior to it.
4. The three factors on the second line of Eq. (10.16) represent general relativity effects causing deviations from Newtonian gravitation.

In detail, the three factors in Eq. (10.16) giving the general relativistic corrections to Newtonian gravity have the following effect.

1. In lower-mass stars that are described well by Newtonian gravity ϵ is dominated by the rest mass of the baryons and the baryons don't contribute significantly to the pressure, which is dominated by electrons. Thus $P(r)/\epsilon(r) \sim 0$ and $4\pi r^3 P(r)/m(r) \sim 0$, so that the first two correction factors on the second line of Eq. (10.16) are approximately unity in most stars.
2. The first two correction factors tend to exceed unity if the star is more massive. The physical reason is that in both terms, unlike for the Newtonian case, *pressure couples to gravity* (see Section 7.3 and Box 7.4). This also may be seen from the numerator of Eq. (10.14), where the pressure adds linearly to the contribution of both energy density and mass in determining hydrostatic equilibrium.
3. For more massive stars the pressure and the energy density act to steepen the pressure gradient, which compresses the matter and reduces the radius compared with the corresponding case for Newtonian gravity.
4. The final factor on the second line of Eq. (10.16) is approximately equal to one for lower-mass stars described by Newtonian gravity, but it will exceed unity for stars that are more massive.

Because each of the three factors on the second line of Eq. (10.16) is greater than one for more massive stars, in general relativity one finds that gravity is systematically stronger than in the corresponding Newtonian description of the same problem. One of the most important consequences is that the coupling of gravity to pressure in GR ultimately will imply the existence of fundamental limiting masses for strongly-gravitating objects, because if the mass is large enough no amount of pressure will be able to prevent gravitational collapse of the star.

10.3 Interpretation of the Mass Parameter

The parameter $m(r)$ entering the Oppenheimer–Volkov equations through the definition (10.10) has been interpreted provisionally as the total mass–energy enclosed within a radius r. It is time to provide some justification for this interpretation. Outside a star of radius R, the quantity $m(r)$ becomes equal to $m(R)$, which is the mass that would be detected through Kepler's laws for the orbital motion if the star were a component of a well-separated binary system. In the Newtonian limit, it is clear from (10.18) that $m(r)$ can be interpreted unambiguously as the *mass* contained within the radius r. The situation for a star described by general relativity is more complex, as we now discuss.

10.3.1 Total Mass–Energy for a Relativistic Star

For general relativistic stars $m(r)$ may be split into contributions from a rest mass $m_0(r)$, an internal energy $U(r)$, and a gravitational energy $\Omega(r)$,

$$m(r) = m_0(r) + U(r) + \Omega(r), \tag{10.19}$$

as will now be demonstrated [156]. Formally the energy density ϵ can be divided into a contribution from the rest mass and one from internal energy, $\epsilon = \mu_0 n + (\epsilon - \mu_0 n)$, where the first term is the total rest mass of n particles of average mass μ_0 and the second term in parentheses is the contribution of internal energy. The proper volume (see Section 3.13.1) for a spherical shell of thickness dr is

$$\begin{aligned} dV &= 4\pi r^2 \sqrt{g_{11}}\, dr \\ &= 4\pi r^2 \sqrt{e^\lambda}\, dr \\ &= 4\pi r^2 (1 - 2m/r)^{-1/2} dr. \end{aligned}$$

Thus the total rest mass inside the radius r is

$$\begin{aligned} m_0(r) &= \int_0^r \mu_0 n\, dV \\ &= 4\pi \int_0^r r^2 (1 - 2m/r)^{-1/2} \mu_0 n\, dr, \end{aligned}$$

the total internal energy inside r is

$$\begin{aligned} U(r) &= \int_0^r (\epsilon - \mu_0 n)\, dV \\ &= 4\pi \int_0^r r^2 (1 - 2m/r)^{-1/2} (\epsilon - \mu_0 n)\, dr, \end{aligned}$$

and from (10.12) the total mass–energy inside r is

$$m(r) = 4\pi \int_0^r \epsilon(r) r^2\, dr.$$

It follows that the difference

$$\begin{aligned} \Omega(r) &= m(r) - m_0(r) - U(r) \\ &= 4\pi \int_0^r r^2 \epsilon \left(1 - (1 - 2m/r)^{-1/2}\right) dr \end{aligned} \tag{10.20}$$

must be the total gravitational energy inside r. These observations give some confidence that $m(r)$ indeed may be interpreted as the total mass–energy inside the coordinate r.

10.3.2 Gravitational Mass and Baryonic Mass

The integral of Eq. (10.15) is of the same form as that for Newtonian gravity if the mass distribution is given by $\epsilon(r)/c^2$. However, in general relativity $\epsilon(r)$ is not an arbitrary distribution but rather corresponds to a solution P of Eq. (10.14) with an equation of state $\epsilon = \epsilon(P)$. Despite the form of Eq. (10.15), $m(r)$ is the sum of the mass of the star and the gravitational energy, and the mass has no well-defined meaning in isolation from the gravitational energy (see Section 7.3.2). The mass–energy m is termed the *gravitational mass*. It is also useful to ask what the total mass of the nucleons would be if they were dispersed to infinity so that the gravitational energy vanishes. This is termed the *baryonic mass* of the star. The gravitational mass and the baryonic mass are generally not the same, since their difference is the total gravitational binding energy. For typical neutron stars the gravitational mass is about 20% smaller than the baryonic mass.

10.4 Pulsars and Tests of General Relativity

Pulsars are rapidly spinning neutron stars that sweep beamed radiation over the Earth periodically, giving the illusion of pulsation. This apparent pulsation occurs with atomic-clock precision, which means that pulsars offer possibilities for exquisite timing measurements, particularly when they are found as a component of a binary star system. The structure and evolution of pulsars is of considerable intrinsic interest (see Section 16.9 of Ref. [105]), but they also are of great practical importance for general relativity because their superb timing characteristics provide some of the most precise tests available for the theory. In the remainder of this chapter examples will be given of the stringent tests of general relativity afforded by observation of pulsars in close binary systems.

10.4.1 The Binary Pulsar

The *Binary Pulsar* PSR 1913+16 (also known as the *Hulse–Taylor binary* after its discoverers, Russell Hulse and Joseph Taylor, Jr.) is about 6.4 kpc away, near the boundary of the constellations Aquila and Sagitta.[3] This pulsar rotates 17 times a second, giving a pulsation period of 59 milliseconds. The discovery that it is a member of a binary star system followed the observation of a cyclic variation in the arrival time of the pulses, first sooner and then later than expected, that repeats on a 7.75 hour cycle. This would be expected if the pulsar were part of a binary system with that period. The observed pulsar is a rapidly spinning neutron star. Detailed analysis of the system indicates that the binary companion is also a neutron star, but it is not observed as a pulsar.

The Binary Pulsar is a superb laboratory for testing general relativity. Because of the precise repetition frequency of the pulsar, it is a clock of very high quality orbiting in a binary system. Furthermore, because the orbit is highly elliptical, the pulsar velocity and the strength of the gravitational field that it feels changes periodically by significant amounts. Although the structure of the individual neutron stars is strongly influenced by

---

[3] Pulsar names are derived from their position on the celestial sphere. The designation PSR indicates a pulsar, the first part of the number gives the right ascension in hours and minutes, and the second part of the number gives the declination in degrees. For example, PSR 0329+54 is located at right ascension 3 hours and 29 minutes and declination +54 degrees. The more modern convention is to prefix the number with a B if the position is in 1950.0 epoch coordinates and J if it is in 2000.0 epoch coordinates. Thus the Binary Pulsar is referred to variously as PSR 1913+16, PSR B1913+16, and PSR J1915+1606 (where 1606 means 16.06 degrees).

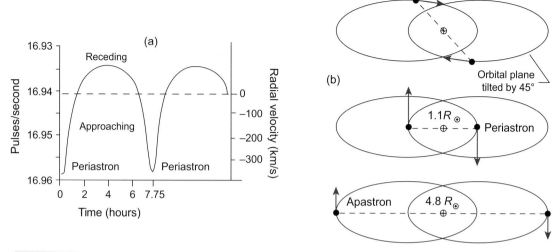

Fig. 10.1 (a) Pulse rate and inferred radial velocity as a function of time for the Binary Pulsar. (b) Orbits for the components of the Binary Pulsar derived from systematic analysis of radio waves emitted from the system [249].

general relativity, they are sufficiently well separated that the orbits in the binary system are largely determined by Kepler's laws, so the precise tests of general relativity are based mostly on observing small deviations from a standard orbital-mechanics analysis.

Because the repetition period for the pulsar is associated with the extremely stable spin of the neutron star, systematic short-term variations in that period as observed from Earth may be ascribed to effects associated with orbital motion of the binary. Thus, we may use these variations to extract very detailed information about the orbit. For example, when the pulsar is moving toward us the repetition rate of the pulses as observed from Earth will be higher than when the pulsar is moving away (Doppler effect), and this can be used to measure the radial velocity. This frequency shift is illustrated in Fig. 10.1(a). Likewise, the pulse arrival times vary as the pulsar moves through its orbit because it takes three seconds longer for the pulses to arrive from the far side of the orbit than from the near side. From this, the Binary Pulsar orbit can be inferred to be about a million kilometers (three light seconds) further away from Earth when on the far side of its orbit than when on the near side.

The orbits determined for the binary system are shown in Fig. 10.1(b). Each neutron star has a mass of about 1.4 solar masses and the orbits are very eccentric (eccentricity parameter $e \sim 0.6$). The minimum separation (periastron) is about 1.1 solar radii, the maximum separation (apastron) is about 4.8 solar radii, and the orbital plane is inclined by about 45 degrees as viewed from Earth. By Kepler's laws, the radial velocity of the pulsar varies substantially as it moves around its elliptical orbit, as illustrated in Fig. 10.1(a). These orbits are not quite closed ellipses because of precession effects associated with general relativity. This causes the location of the periastron to shift a small amount for each revolution. Figure 10.2(a) illustrates this precession (greatly exaggerated), with the points P1, P2, and P3 representing the location of periastron on three successive orbits.

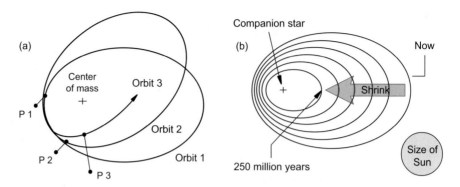

Fig. 10.2 (a) Schematic precession of the periastron for the Binary Pulsar. The amount of precession has been greatly exaggerated in this diagram. (b) Predicted shrinkage of the Binary Pulsar orbit because of gravitational wave emission, based on the current rate of change. Orbital sizes are to scale and are compared with the present diameter of the Sun [249].

10.4.2 Precision Tests of General Relativity

The discovery and detailed study of the Binary Pulsar was of such fundamental importance for testing general relativity that Taylor and Hulse were awarded the 1993 Nobel Prize in Physics for their work (see Box 10.1). The Binary Pulsar provided some of the most stringent tests of general relativity available before the discovery of the Double Pulsar that we shall discuss in Section 10.4.4. Some of these tests are summarized below (see Chapter 5 of Straumann [228] for the detailed procedure used to extract these results from Binary Pulsar data).

Precession of orbits and time dilation: Because space is warped by the gravitational field near the pulsar, the orbit will precess with time, as illustrated in Fig. 10.2(a). This is the same effect as the precession of the perihelion of Mercury discussed in Section 9.4, but it is much larger for the present case. The Binary Pulsar's periastron is observed to advance by 4.2 degrees per year, in accord with the predictions of general relativity. In a single day the orbit of the Binary Pulsar advances by as much as the orbit of Mercury advances in a century! In addition, when the Binary Pulsar is near periastron the gravitational field that it feels is stronger and its velocity is higher and time should run slower. Conversely, near apastron the field is weaker and the velocity lower, so time should run faster. It does both, in the amount predicted by the theory.

Emission of gravitational waves: As will be discussed more extensively in Section 23.2.3, the revolving pair of masses is predicted by general relativity to radiate gravitational waves. Since this takes energy away from the orbital motion, the radius of the orbit must shrink with time if general relativity is correct, as illustrated in Fig. 10.2(b). The shift in the time of periastron can be measured very precisely and is found to correspond to a decrease in the orbital period by 76×10^{-6} s per year (implying a decrease in the size of the orbit by about 3.3 millimeters per revolution). The quantitative decrease in cumulative periastron time is illustrated by the data points in Fig. 23.3.

Because the orbital period of the Binary Pulsar is short, the shift in periastron arrival time accumulates to more than a second earlier every 10 years. The corresponding decay in the

> **Box 10.1** **Relativity and the Nobel Prize**
>
> As this book was being finished it was announced that the 2017 Nobel Prize in Physics had been awarded to Rainer Weiss, Barry Barish, and Kip Thorne of the LIGO collaboration for the discovery of gravitational waves. This was only the second Nobel Prize ever given for either special or general relativity. (The only other was the 1993 Nobel Prize for Physics awarded to Russell Hulse and Joseph Taylor for testing general relativity with the Binary Pulsar.) Remarkably, Einstein won the 1921 Physics Nobel Prize for his work on the quantum theory (most notably the explanation of the photoelectric effect), but never won one for either special or general relativity. The reasons are a subject unto themselves, probably related more to the disciplines of psychology and sociology than to physics.
>
> In essence there were scientists with influence on the Nobel Committee who would not accept the correctness of general relativity and/or special relativity but could not deny Einstein's large impact on physics, so it was decided to give Einstein the 1921 Prize for a "safe" option: his general theoretical work, particularly in quantum theory. But why was Einstein not given a second Nobel for relativity after its correctness was more widely accepted? The awarding of Nobel prizes routinely bypasses highly deserving individuals but surely few inventions in the history of science were more deserving than special and general relativity of at least one Nobel Prize.
>
> A common view is that by the time special and general relativity were more broadly accepted Einstein was seen as diverging from the mainstream of physics with his fixation on a unified field theory of gravity and electromagnetism, and his rejection of the standard probabilistic interpretation of quantum mechanics (though not of quantum mechanics itself, which he helped invent). In this view he was seen as too eccentric to be given an obviously deserved Nobel for relativity. That the period in which Einstein should have won a second Nobel coincided with the rise of virulent antisemitism and irrational rejection of "Jewish science" in Germany likely also was a significant factor.

size of the orbit is accounted for quantitatively by the energy that general relativity predicts should be radiated from the system in the form of gravitational waves. The prediction of the theory for the shift in periastron time is denoted by the dashed curve in Fig. 23.3. Thus, precision measurements on the Binary Pulsar give strong indirect evidence for the correctness of this key prediction of general relativity, as will be discussed further in Section 23.2.3.

10.4.3 Origin and Fate of the Binary Pulsar

The possibility of two neutron stars merging might seem to be a remote one. But *if* an isolated binary neutron star system is formed, merger is inevitable: for any binary the orbital motion will radiate energy as gravitational waves, the orbits will shrink, and eventually the two components *must spiral together*.[4] Formation of a neutron star binary is

---

[4] If the components of the binary are normal stars, or if the binary is interacting with other objects, this outcome might be circumvented. For example, if the system collides with another star, or if one star in the binary explodes as a supernova, the binary might become unbound. But for isolated binary neutron stars no such escape hatches are available and they are destined to merge. The timescale for merger, though, varies enormously

not easy, however. Either a binary (or multiple star system with more than two stars) must form with two stars massive enough to become supernovae, and the neutron stars thus formed must remain bound to each other through the two supernova explosions, or the neutron star binary must result from gravitational capture of one neutron star by another. Although these are improbable events, simulations indicate that they are not impossible, and the existence of the Binary Pulsar (and various similar systems that have been discovered more recently) demonstrates empirically that mechanisms exist for it to happen.

Because of gravitational wave emission and the corresponding shrinkage of the orbit illustrated in Fig. 10.2(b), the two neutron stars are expected to merge in about 3×10^8 years. The sum of the masses of the two neutron stars is likely above the critical mass to form a black hole. Therefore, the probable fate of the Binary Pulsar is for the two neutron stars to merge and collapse to a rotating black hole. As the two neutron stars approach each other they will revolve faster (Kepler's third law). This will cause them to emit gravitational radiation more rapidly, which will in turn cause the orbit to shrink even faster. Thus, near the end the merger of the two neutron stars will proceed rapidly in a positive-feedback runaway and will emit very strong gravitational waves, as will be discussed in Chapter 24. These considerations are valid for any binary star system, not just the Binary Pulsar, but the gravitational wave effects are much more pronounced for binaries involving very compact objects like neutron stars. The death of the Binary Pulsar may produce a gamma-ray burst also, since merging neutron stars are thought to be one mechanism responsible for such events (see Section 15.7).

10.4.4 The Double Pulsar

In 2003 an even more remarkable discovery than that of the Binary Pulsar was made of a binary neutron star system in which *both* neutron stars were observed as pulsars in a very tight, partially eclipsing[5] orbit [60, 139, 148]. This system, consisting of the pulsars denoted PSR J0737-3039A and PSR J0737-3039B, is located in the constellation Puppis at a distance of about a kiloparsec and is known as the *Double Pulsar*. The two neutron stars have masses of $1.3381 \pm 0.0007 M_\odot$ (component A), and $1.2489 \pm 0.0007 M_\odot$ (component B), and spin periods of 22.7 ms (component A) and 2.77 s (component B), respectively. The orbital configuration in January, 2008, is illustrated in Fig. 10.3. The orbit is slightly eccentric ($e = 0.088$) with a mean radius of about 800,000 km (somewhat larger than the radius of the Sun), giving an orbital period of only 147 minutes and a mean orbital velocity of about $10^6 \, \text{km} \, \text{hr}^{-1}$. The binary system also has low transverse velocity relative to Earth.

The very fast orbital period, high orbital velocity and acceleration in the gravitational field, proximity to Earth, nearly edge-on geometry of the orbit, low transverse velocity, and the timing associated with the pulsar clocks have allowed the Double Pulsar to give the most precise tests of general relativity to date [140]. For example, the precession

---

according to the size of the original orbit. For large enough orbits the predicted merger timescale will exceed the present age of the Universe.

[5] The B pulsar can be partially or totally eclipsed by the magnetosphere of the other pulsar. As a result, radio pulses from the B component are not always visible.

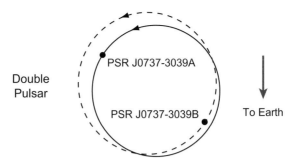

Fig. 10.3 Orbital configuration of the Double Pulsar [139].

of the periastron is about 17° per year, which is more than four times as much as for the Binary Pulsar, and is consistent with the predictions of general relativity. The orbital radius is observed to be decaying by about 7 mm per day, implying a merger in about 85 million years, which is consistent with the predicted energy loss caused by gravitational wave emission from the system. Analysis of the agreement with the predictions of general relativity has been systematized in terms of *post Keplerian (PK) corrections,* which represent the changes in observed system parameters required relative to the assumption of pure Kepler orbits. The measured PK corrections have been found to be within 0.05% of those predicted by general relativity for this system [140].

In addition to the precise tests of general relativity afforded by the Double Pulsar, this system is of intrinsic interest because it likely has an unusual formation history [75, 190]. In one proposed scenario [75] the A component, with a period of milliseconds, is the *older* neutron star and the B component, with a period of seconds, is the *younger* neutron star. This inverts the usual relationship between the age of a pulsar and its spin rate and is thought to result from a mechanism in which a neutron star is spun up by accretion from a companion. That is, after the birth of the first neutron star in a supernova (component A), it was spun up in an accretion phase before the birth of the second neutron star (component B). This mechanism, which is responsible generally for the origin of old but fast *millisecond pulsars,* is discussed further in Section 16.9.5 of Ref. [105].

10.4.5 The Pulsar–White Dwarf Binary PSR J0348+0432

The binary system PSR J0348+0432 is about 2.1 kpc away and contains a 39 ms pulsar of mass 2.01 $M_\odot$ and a white dwarf of mass 0.172 $M_\odot$, with an orbital period of 2.46 hours [30]. An artist's impression of the distorted spacetime produced by PSR J0348+0432 is shown in Fig. 10.4(a), the emission of gravitational waves is illustrated schematically in Fig. 10.4(b), and the geometry of the binary orbit is illustrated in Fig. 10.4(c). The very compact orbit (comparable in width to the diameter of the Sun) suggests that there should be a rapid loss of orbital energy to gravitational waves. Precise radio timing indicates that the orbital period is decreasing by 8.6 μsec year^{-1}, in accord with the prediction of general relativity and implying a time until merger of ∼600 million years.

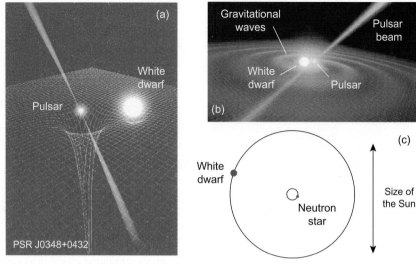

Fig. 10.4 (a) The (highly exaggerated) distortion of spacetime caused by the pulsar–white dwarf binary PSR J0348+0432. The large, compact mass of the neutron star distorts spacetime much more than the smaller, less-compact mass of the white dwarf. Figure from Ref. [30]. Reproduced from Antoniadis *et al.*, A Massive Pulsar in a Compact Relativistic Binary, *Science*, **340**, 6131. Copyright © 2013, American Association for the Advancement of Science. (b) Gravitational wave emission from PSR J0348+0432. Credit: ESO/L. Calçada (c) The binary system PSR J0348+0432. Orbits to scale but object sizes are schematic.

Example 10.1 From Ref. [30] the rate of decrease in orbital period for PSR J0348+0432 because of gravitational wave emission computed from general relativity is $\dot{P}_{\text{GR}} = -2.58 \times 10^{-13}$ s s^{-1} and the measured rate of decrease in the orbital period is

$$\dot{P} = -8.6\,\mu\text{s}\,\text{yr}^{-1} \left(\frac{1\,\text{s}}{10^6\,\mu\text{s}}\right)\left(\frac{1\,\text{yr}}{3.1557 \times 10^7\,\text{s}}\right) = -2.73 \times 10^{-13}\,\text{s}\,\text{s}^{-1}.$$

Therefore $\dot{P}/\dot{P}_{\text{GR}} = 1.05 \pm 0.18$ (where the uncertainty is taken from the error analysis in Ref. [30]), and the orbital period of PSR J0348+0432 is decreasing at a rate accounted for by the rate of gravitational wave emission required by general relativity.

The pulsar PSR J0348+0432 corresponds to one of the most massive neutron stars known, with a gravitational binding energy 60% larger than that of neutron stars in any other binary where gravitational wave damping of the binary orbit has been measured. Thus, it provides a test of general relativity under stronger gravity than for other binary pulsar tests. For example, the orbital period is essentially the same as for the Double Pulsar but PSR J0348+0432 has about twice the gravitational binding energy of each pulsar in the Double Pulsar. Strong-field effects in general relativity depend nonlinearly on the gravitational binding, so the larger binding energy entails a substantially more stringent test of general relativity. Tests in even stronger gravity will be considered in Section 24.4.

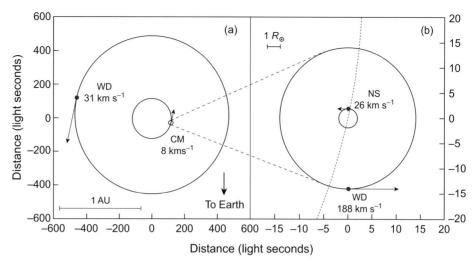

Fig. 10.5 Orbits of PSR J0337+1715 [198]. (a) Orbits of the outer white dwarf (WD) and the center of mass (CM) for the inner white dwarf and neutron star pair. (b) As for left side but scaled up by a factor of 30 to show the orbits for the inner white dwarf and neutron star (NS). Arrows indicate orbital velocities for the center of mass of the inner binary and the individual white dwarfs and neutron star. All orbits lie almost in the same plane, are nearly circular, and have a tilt angle $i \sim 39°$ relative to the line of sight.

10.4.6 The Pulsar–WD–WD Triplet PSR J0337+1715

The triple star system PSR J0337+1715, which contains a millisecond pulsar and two white dwarfs in hierarchical orbits,[6] is illustrated in Fig. 10.5. Timing for the pulsar allows a precise determination of the masses and orbital parameters. The neutron star is found to have a mass of $1.4378 M_\odot$, the inner white dwarf has a mass of $0.1975 M_\odot$, and the outer white dwarf has a mass of $0.4101 M_\odot$. This system has an evolutionary history that is quite out of the ordinary [198, 230], but that isn't of direct relevance here. (It is discussed further within the context of stellar evolution in §16.9.5 of [105].) The primary importance for this discussion is that PSR J0337+1715 should afford a precise test of the general theory of relativity [198].

According to the strong equivalence principle introduced in Section 6.3, objects with different gravitational binding energies should follow the same orbits in a gravitational field. As will be discussed in Section 25.2, this can be tested by tracking the Earth–Moon system using precise laser range finding as it falls gravitationally toward the Sun in its orbit. In PSR J0337+1715 a similar test is possible as the outer white dwarf strongly accelerates the inner binary containing the neutron star and other white dwarf. Gravitational binding energies for the neutron star and white dwarf in the inner binary differ from each other by 4–5 orders of magnitude, and the neutron star binding energy is roughly 10^9 times larger than that of planets or moons in our Solar System (see Problem 10.8). Hence violations

---

[6] A triple-star system is said to be *hierarchical* if two of the stars are close together and the third star is much further away. Such systems are often dynamically more stable than generic triple-star systems.

of the strong equivalence principle should be greatly amplified in J0337+1715 relative to tests possible in the Solar System. This, coupled with the precise timing afforded by the pulsar, should lead to unique new tests of gravitational theory as data are accumulated for this system.

Background and Further Reading

An introduction to special and general relativity with particular emphasis on compact objects like white dwarfs and neutron stars may be found in Glendenning [98, 99], and Shapiro and Teukolsky [223] give a broad introduction to neutron stars, white dwarfs, and black holes. See also Misner, Thorne, and Wheeler [156]; Raychauduri, Banerjii, and Banerjee [199]; Hobson, Efstathiou, and Lasenby [122]; and Walecka [244] for discussions of obtaining the Oppenheimer–Volkov solution from the Einstein equations. A more extensive discussion of pulsars and the pulsar mechanism may be found in Chapter 12 of Ref. [199] and Chapter 10 of Ref. [223]. Neutron stars as endpoints of stellar evolution are discussed in Chapter 16 of Ref. [105]. A more technical discussion of how data from the Binary Pulsar are used to deduce the results discussed here may be found in Chapter 5 of Straumann [228], and an overview of using pulsar timing to test general relativity is given by Stairs [225]. A summary of testing general relativity against observation and experiment may be found in Will [251].

Problems

10.1 Assume that the most massive possible neutron star is spherical and has a radius approximately equal to the Schwarzschild radius for a spherical black hole ($r_s = 2M$ in geometrized units, where M is the mass; see Chapter 11), since if the mass of the neutron star were compressed inside r_s it would collapse to a black hole. If each neutron has an incompressible (hardcore) radius $r_0 \simeq 0.5 \times 10^{-13}$ cm, and gravity packs the neutrons in the neutron star down to this radius (that is, so that the center of each neutron is separated from the center of its nearest neighbors by $2r_0$), estimate the total number of neutrons A, the radius R, the mass M for the neutron star, and the average density $\bar{\rho}$. For this simple estimate you may assume that Newtonian gravity is valid.

10.2 Use the Oppenheimer–Volkov equations to estimate the total gravitational energy of a neutron star (express your answer in ergs, grams, MeV, and solar masses). How does this compare with the total rest mass energy of the neutron star?

10.3 From Eq. (10.11) the metric coefficient e^λ is related to the enclosed mass in the corresponding Newtonian theory. To what quantity in classical gravity is the metric coefficient e^σ related?

10.4 Prove that Eqs. (10.14)–(10.15) describing the hydrostatic equilibrium solution for the Oppenheimer–Volkov equations can be rewritten as Eqs. (10.16)–(10.17). ***

10.5 For the gravity of a neutron star the general relativistic effects are of moderate size but it is usually assumed that the equations governing the nuclear interactions within the star can be solved in flat spacetime. Use the metric corresponding to the Oppenheimer–Volkov solution to justify this approximation. *Hint*: A characteristic range for the relevant nuclear interactions is approximately 10^{-13} cm. ***

10.6 The average binding energy per nucleon in nuclei is about 8 MeV per nucleon. Estimate the corresponding binding energy per nucleon in a neutron star. Explain your result.

10.7 Estimate the escape velocity from the surface of a neutron star. What about from a white dwarf? What does this tell you about the importance of general relativity in these two cases?

10.8 In Section 10.4.6 it was asserted that for the PSR J0337+1715 system the gravitational binding for the neutron star and for the inner white dwarf are much larger, and their difference is much larger, than corresponding quantities for planets and moons in the Solar System. Confirm this by estimating the gravitational strength parameter ϵ defined in Eq. (6.5) for the neutron star and the inner white dwarf in PSR J0337+1715, and for the Earth and Moon. ***

PART II

BLACK HOLES

11 Spherical Black Holes

Perhaps the most spectacular consequences of general relativity is that under certain conditions gravity can become strong enough to trap even light, because space becomes so curved that there are no paths for light to follow from an interior to exterior region. There is extremely strong circumstantial evidence that such *black holes* exist (see Box 11.1 for a brief review of the modern scientific history of the idea). In this chapter the Einstein theory of gravity is applied to black holes using the Schwarzschild solution discussed in Chapter 9. Then Chapter 12 takes a first step in considering how the physics of spherical black holes might be altered by quantum mechanics (*Hawking black holes*) and Chapter 13 considers how the Schwarzschild solution is modified if a black hole is spinning (*Kerr black holes*). Chapters 14 and 15 survey the strong generic evidence for the existence of black holes, and Chapter 24 essentially clinches the case by presenting evidence for gravitational waves emitted from the merger of binary black hole pairs.

11.1 Schwarzschild Black Holes

Let's now consider in more detail the most interesting property of the Schwarzschild solution found in Chapter 9: there can exist an *event horizon* in the Schwarzschild spacetime, and this horizon implies the existence of black holes.

11.1.1 Event Horizons

Imagine an attempt to escape a gravitational field generated by some spherical object of mass M. The condition for escape to a radius r is that the kinetic energy exceed the gravitational potential energy:

$$\tfrac{1}{2}mv^2 \geq \frac{GMm}{r}. \tag{11.1}$$

But the maximal physical velocity for any object is $v = c$; substituting c for v and solving for r,

$$r = \frac{2GMm}{mc^2} = \frac{2GM}{c^2} = r_S, \tag{11.2}$$

where Eq. (9.6) with c and G restored was used. Therefore, r_S is the radius at which the escape velocity becomes equal to the velocity of light. This is just a suggestive result from Newtonian physics supplemented by concepts from special relativity and

> **Box 11.1** **The Modern Idea of Black Holes**
>
> The idea that objects might exist with such strong gravity that they could trap even light dates back to at least the eighteenth century, but the first strong objective evidence that black holes might exist in nature was provided by the work of Robert Oppenheimer (1904–1967) and collaborators [170, 171]. They investigated the structure of neutron stars in a general relativistic context and showed that if neutron stars exceeded a critical mass (estimated now to be about 3 $M_\odot$) they would collapse gravitationally to what would now be called a black hole (a term popularized by John Wheeler in the 1960s). Furthermore, they established the rudiments to be discussed later in this chapter of the observational behavior of light coming from such a collapsing object in the vicinity of what would now be called the event horizon.
>
> Oppenheimer was an enigmatic figure in twentieth-century science. Uniformly agreed to be brilliant by all who knew him and famous as the scientific leader of the Manhattan Project, he won no Nobel Prizes, nor did he leave behind as large a body of oft-cited work as might be expected from one that many regarded as having no scientific peer in his generation. Probably Oppenheimer would have received much broader scientific recognition had he lived long enough for the gravitational-collapse work cited above – which laid the foundation for major portions of modern astronomy and astrophysics – to become mainstream.

some questionable assumptions about the nature of light in Newtonian gravity, but a more rigorous analysis in Section 9.5 comes to exactly the same conclusion: r_S defines a radius where the gravitational curvature is so strong that even light cannot escape. Thus, the Schwarzschild radius r_S may also be termed the *event horizon* of the Schwarzschild solution and, since light is fundamentally trapped inside this radius, the region interior to r_S is termed a *black hole*.

Before proceeding we should clear up a potential source of confusion. Since it has been argued that the Schwarzschild solution is approximately valid outside the Sun or Earth (neglecting their spin and deviation from spherical symmetry), and that $r_s = 2M$ defines an event horizon for a black hole, why aren't the Sun and Earth black holes with corresponding event horizons? The answer is that the Schwarzschild solution is a vacuum solution valid only *outside the mass distribution* producing the gravitational field. In the case of the Sun and Earth it may be verified easily that the Schwarzschild radius lies deep inside both objects, where the Schwarzschild solution is *not valid*.

An event horizon and the associated black-hole properties of the Schwarzschild solution manifest themselves physically only if the mass M responsible for the Schwarzschild gravitational field is contained entirely within the Schwarzschild radius, which implies extremely compact objects of much higher density than planets or normal stars. Thus it is important to distinguish the *Schwarzschild spacetime,* which is approximately valid outside any static spherical mass, and a *Schwarzschild black hole,* which forms in a Schwarzschild spacetime only if the mass M responsible for the gravitational field is contained entirely within its Schwarzschild radius.

The preceding qualitative discussion of event horizons may be placed on firmer grounds by considering a spacecraft approaching the event horizon of a Schwarzschild black hole. For simplicity, the trajectory will be assumed to be radial. Then, as long as the spacecraft does not fire its engines and remains in free fall, the situation is analogous to that of Section 9.6 for the radial fall of a test particle in the Schwarzschild geometry. Let us first consider the view from a point well outside the event horizon, and then consider the view for the (quite doomed) occupants of the spacecraft.

11.1.2 Approaching the Horizon: Outside View

The motion is radial by assumption, so setting $d\theta = d\varphi = 0$ in Eq. (9.4) yields

$$ds^2 = -d\tau^2 = -\left(1 - \frac{2M}{r}\right)dt^2 + \left(1 - \frac{2M}{r}\right)^{-1}dr^2$$
$$= -\left(1 - \frac{r_S}{r}\right)dt^2 + \left(1 - \frac{r_S}{r}\right)^{-1}dr^2. \tag{11.3}$$

As the spacecraft approaches the event horizon its velocity as viewed from a distant fixed frame is $v = dr/dt$. Light signals from the spacecraft travel on the lightcone so $ds^2 = 0$ and thus from (11.3)

$$v = \frac{dr}{dt} = 1 - \frac{r_S}{r}. \tag{11.4}$$

As viewed from a large distance outside r_S, the spacecraft appears to slow as it approaches r_S and eventually stops as $r \to r_S$. Thus, from the exterior a distant observer would never see the spacecraft cross r_S: its image would remain frozen at $r = r_S$ for all eternity.

But let's examine what this means a little more carefully. By rewriting the preceding expression in the form

$$dt = \frac{dr}{1 - r_S/r}, \tag{11.5}$$

it is clear that as $r \to r_S$ the time between successive wave crests for the light coming from the spacecraft tends to infinity. Therefore, for the light waves the wavelength $\lambda \to \infty$ while the energy and frequency tend to zero. The external observer not only sees the spacecraft slow rapidly as it approaches r_S, but the spacecraft image is observed to redshift strongly at the same time. This behavior is just that of the coordinate time already exhibited in Fig. 9.7. Therefore, more properly, the external observation is that the spacecraft approaching r_S slows rapidly and redshifts, until the image fades from view before reaching r_S.

11.1.3 Approaching the Horizon: Spacecraft View

The view is very different from the interior of the spacecraft. The occupants will use their own clocks (measuring proper time) to gauge the passage of time and, starting from a radial position r_0 outside the event horizon, the spacecraft will reach the origin in a proper time τ given by Eq. (9.48):

$$\tau = -\frac{2}{3}\frac{r_0^{3/2}}{(2M)^{1/2}}. \tag{11.6}$$

The spacecraft occupants will notice no spacetime singularity at the horizon. They may be torn apart by tidal forces, but though they may be large any tidal forces at the horizon will remain finite. Assuming that the spacecraft could withstand the increasing tidal forces, it crosses r_S and reaches the center of the black hole in a finite amount of proper time, where it would encounter infinite tidal forces. In fact, the trip from the event horizon to the center is very fast: of order 10^{-4} seconds of proper time for a one solar mass black hole and of order 10 minutes proper time for a billion solar mass black hole – see Problem 11.2.

11.2 Lightcone Description of a Trip to a Black Hole

Curved spacetime inherits the local lightcone structure of Minkowski space and lightcones are invariant for any observer. Thus, it is highly instructive to consider a trip into a Schwarzschild black hole in terms of the local lightcones. Assuming radial light rays ($d\theta = d\varphi = ds^2 = 0$), the lightcone equation at some local coordinate r in the Schwarzschild metric is from Eq. (11.4),

$$\frac{dt}{dr} = \pm \left(1 - \frac{2M}{r}\right)^{-1}, \qquad (11.7)$$

where the plus sign corresponds to outgoing photons (r increasing with time for $r > 2M$) and the minus sign to ingoing photons (r decreasing with time for $r > 2M$). For large r the slope dt/dr becomes equal to ± 1, as for flat spacetime, but as $r \to r_S$ the forward lightcone opening angle tends to zero as $dt/dr \to \infty$. Integrating Eq. (11.7) gives

$$t = \begin{cases} -r - 2M \ln\left|\dfrac{r}{2M} - 1\right| + \text{constant} & \text{(ingoing)} \\ r + 2M \ln\left|\dfrac{r}{2M} - 1\right| + \text{constant} & \text{(outgoing)} \end{cases}. \qquad (11.8)$$

The null geodesics defined by Eq. (11.8) are plotted in Fig. 11.1(a). The tangents at the intersections of the dashed and solid lines define local lightcones corresponding to Eq. (11.7), which are sketched at some representative spacetime points.

11.2.1 Worldline Exterior to the Event Horizon

The corresponding worldline of the spacecraft is illustrated in Fig. 11.1(b), starting at a time t_0 that corresponds to a radius r well exterior to the event horizon of the black hole. The gravitational field there is weak and the lightcone has the usual appearance, with the forward cone opening symmetrically in the direction of increasing time. As illustrated by the dotted line from A, a light signal emitted from the spacecraft at this point can intersect the worldline of an observer remaining at constant distance r_{obs} at a finite time $t_A > t_0$. However, as the spacecraft falls toward the black hole on the worldline indicated the forward lightcone begins to narrow in accordance with Eq. (11.7). Now at point B, say, a light signal from the spacecraft can still intersect the worldline of the external observer, but only at an increasingly distant point in the future (see the arrow on lightcone B). As the spacecraft approaches very closely to r_S, the opening angle of the forward lightcone tends

11.2 Lightcone Description of a Trip to a Black Hole

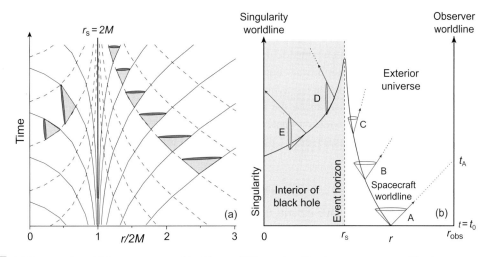

Fig. 11.1 (a) Photon paths and lightcone structure of the Schwarzschild spacetime. The plot assumes constant (θ, φ), so each point should be thought of as representing a 2-sphere. (b) Worldline for a spacecraft falling into a Schwarzschild black hole (adapted from Ref. [156]).

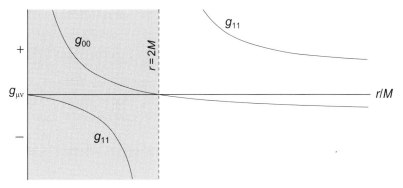

Fig. 11.2 Spacelike and timelike regions for g_{00} and g_{11} in the Schwarzschild metric.

to zero and the length of time for a signal emitted from the spacecraft to reach the external observer's worldline at r_{obs} tends to infinity (see the arrow on lightcone C).

11.2.2 Worldline Interior to the Event Horizon

Now we consider the lightcones inside the event horizon. The structure of the radial and time parts of the Schwarzschild metric (9.4) illustrated in Fig. 11.2 indicate that dr and dt reverse their character at the horizon ($r = 2M$), because the metric coefficients g_{00} and g_{11} *switch signs* at that point. As a consequence, outside the black hole the t direction, $\partial/\partial t$, is timelike ($g_{00} < 0$) and the r direction, $\partial/\partial r$, is spacelike ($g_{11} > 0$). But inside the black hole, $\partial/\partial t$ is spacelike ($g_{00} > 0$) and $\partial/\partial r$ is timelike ($g_{11} < 0$). Thus, a lightcone in the interior takes the rotated appearance exhibited in the left side of Fig. 11.1(b). The worldline of the spacecraft descends inside r_S because the coordinate time decreases (it is now behaving like r) and the decrease in r represents the passage of time, but of course the *proper time* is continuously increasing in this region [see Eq. (11.6)].

Alternatively, outside the event horizon r is a *spacelike coordinate* and application of enough rocket power could reverse the infall and cause r to begin to increase. But inside the event horizon r is a *timelike coordinate* and the radial coordinate of the spacecraft must decrease once inside the horizon for the same reason that time flows into the future in normal experience (whatever that reason is!). Once inside the event horizon it is no more possible to reverse the inward trajectory of the spacecraft than it is possible to reverse the flow of time.

11.2.3 You Can't Get There From Here

Because of these properties of the Schwarzschild spacetime, inside the horizon there are no paths in the forward lightcone of the spacecraft that can reach the external observer at r_0 (the right vertical axis), and no paths in the forward lightcone that can avoid the singularity at the center of the black hole at $r = 0$ (the left vertical axis) – see the lightcones labeled D and E in Fig. 11.1(b). Therefore, all timelike and null paths inside the horizon are bounded by the horizon and must encounter the singularity at $r = 0$.

This discussion illustrates succinctly the real reason that nothing can escape the interior of a black hole. Irrespective of any details concerning the dynamics of the spacecraft (how much fuel it has available and which direction it goes, for example), once inside r_S it cannot escape the black hole because the geometry of spacetime inside the event horizon permits no forward lightcones that intersect exterior regions, and no forward lightcones that do not contain the singularity at the origin. Thus, once inside the event horizon causality implies that there is no escape from the classical Schwarzschild black hole: there are no paths in spacetime that go from the interior to the exterior, and all timelike or null paths within the horizon lead to the $r = 0$ singularity.

11.3 Solution in Eddington–Finkelstein Coordinates

The preceding discussion is illuminating but the interpretation of the results is complicated by the behavior near the singularity at $r = 2M$ in the standard Schwarzschild coordinates. As has been made plausible in earlier discussion, this is a coordinate singularity. In this section and the next two alternative coordinate systems will be discussed for the Schwarzschild geometry that remove the coordinate singularity at the horizon, thus allowing a cleaner interpretation of the Schwarzschild geometry from the horizon inward. Although these two coordinate systems provide advantages for interpreting the interior behavior of the Schwarzschild geometry, the standard coordinates corresponding to the line element (9.4) remain useful for describing the exterior behavior because of their simple asymptotic properties.[1]

---

[1] The *Schwarzschild metric,* which defines a particular geometry of spacetime that could be expressed in many different coordinate systems, should be distinguished from the particular set of coordinates employed in (9.4), which are often referred to as *Schwarzschild coordinates.*

11.3.1 Eddington–Finkelstein Coordinates

In the Eddington–Finkelstein coordinate system, introduced by Eddington originally and later rediscovered by Finkelstein, a new variable v is defined by requiring that [111, 156]

$$t = v - r - 2M \ln \left| \frac{r}{2M} - 1 \right|, \tag{11.9}$$

where r, t, and M have their usual meanings in the Schwarzschild metric and the angular variables θ and φ are assumed to be unchanged from the standard Schwarzschild coordinates. For either $r > 2M$ or $r < 2M$, Eq. (11.9) inserted into the Schwarzschild line element (9.4) gives the line element (Problem 11.1)

$$ds^2 = -\left(1 - \frac{2M}{r}\right)dv^2 + 2dvdr + r^2 d\theta^2 + r^2 \sin^2\theta d\varphi^2. \tag{11.10}$$

The Schwarzschild metric expressed in these new coordinates is manifestly non-singular at $r = 2M$, but the singularity at $r = 0$ remains. This, coupled with the observation that the same metric is obtained starting from either $r < 2M$ or $r > 2M$, suggests that the $r = 0$ singularity is real but the $r = 2M$ singularity is only a coordinate singularity that is unphysical, in that it can be made to disappear with a suitable choice of coordinates.

11.3.2 Behavior of Radial Light Rays

Now consider the behavior of radial light rays expressed in these coordinates. Setting $d\theta = d\varphi = 0$ (radial motion) and $ds^2 = 0$ (of light rays), Eq. (11.10) gives

$$-\left(1 - \frac{2M}{r}\right)dv^2 + 2dvdr = 0. \tag{11.11}$$

This equation has two general solutions and one special solution [111]:

1. *General Solution*: $dv = 0$, implying that $v = $ constant. From Eq. (11.9), at constant v the radial coordinate r must decrease as t increases; thus this solution corresponds to ingoing light rays on constant v trajectories.
2. *General Solution*: If $dv \neq 0$, then

$$\frac{dv}{dr} = 2\left(1 - \frac{2M}{r}\right)^{-1}, \tag{11.12}$$

which yields upon integration

$$v - 2\left(r + 2M \ln \left|\frac{r}{2M} - 1\right|\right) = \text{constant}, \tag{11.13}$$

which changes behavior at $r = 2M$. For $r > 2M$ it is an outgoing solution but for $r < 2M$ it is an *ingoing solution*.
3. *Special Solution*: In the particular case that $r = 2M$, Eq. (11.11) reduces to

$$dvdr = 0, \tag{11.14}$$

which corresponds to stationary light rays at the Schwarzschild radius that can move neither inward nor outward.

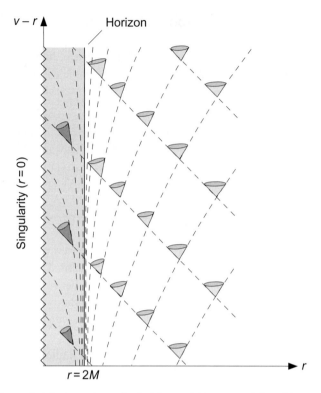

Fig. 11.3 Lightcones in Eddington–Finkelstein coordinates for the Schwarzschild spacetime (adapted from Ref. [156]). Only two coordinates are plotted, so each point represents a 2-sphere of angular coordinates.

These considerations define the lightcone structure of Schwarzschild spacetime geometry, as expressed in Eddington–Finkelstein coordinates. For $r \neq 2M$, solutions 1 and 2 define the left and right sides of the lightcone locally, and for $r = 2M$ light is trapped at the event horizon. Figure 11.3 illustrates this structure in the Eddington–Finkelstein coordinates.

11.3.3 The Event Horizon

The lightcones in Fig. 11.3 at various points are bounded by the two general solutions discussed above and they tilt increasingly inward at smaller radii. The radial light ray bounding the left side of the lightcone at each point moves inward and corresponds to solution 1. The radial light ray bounding the right side of the lightcone corresponds to solution 2; these propagate outward if $r > 2M$ but for $r < 2M$ they change character and *propagate inward*. It follows that for $r < 2M$ the lightcone is tilted toward $r = 0$ sufficiently that no light ray can escape the singularity at $r = 0$. Likewise, since particle trajectories are timelike no particle trajectory can move beyond the horizon or escape the singularity at $r = 0$ once $r < 2M$. Exactly at $r = 2M$, one light ray moves inward (solution 2) and one is trapped at $r = 2M$ (the special solution 3).

11.4 Solution in Kruskal–Szekeres Coordinates

Eddington–Finkelstein coordinates demonstrate explicitly that $r = 2M$ is a coordinate singularity. There is another set of coordinates that is in many contexts even more useful than the Eddington–Finkelstein coordinates and that also exhibits no singularity at $r = 2M$.

11.4.1 Kruskal–Szekeres Coordinates

The *Kruskal–Szekeres coordinate system* introduces variables (v, u, θ, φ), where θ and φ have their usual meaning and the new variables u and v are defined through [111, 156, 221]

$$u = \left(\frac{r}{2M} - 1\right)^{1/2} e^{r/4M} \cosh\left(\frac{t}{4M}\right) \quad (r > 2M) \tag{11.15}$$

$$= \left(1 - \frac{r}{2M}\right)^{1/2} e^{r/4M} \sinh\left(\frac{t}{4M}\right) \quad (r < 2M). \tag{11.16}$$

$$v = \left(\frac{r}{2M} - 1\right)^{1/2} e^{r/4M} \sinh\left(\frac{t}{4M}\right) \quad (r > 2M) \tag{11.17}$$

$$= \left(1 - \frac{r}{2M}\right)^{1/2} e^{r/4M} \cosh\left(\frac{t}{4M}\right) \quad (r < 2M). \tag{11.18}$$

The corresponding line element is

$$ds^2 = \frac{32M^3}{r} e^{-r/2M}(-dv^2 + du^2) + r^2 d\theta^2 + r^2 \sin^2\theta d\varphi^2, \tag{11.19}$$

where $r = r(u, v)$ is defined through

$$\left(\frac{r}{2M} - 1\right) e^{r/2M} = u^2 - v^2. \tag{11.20}$$

As for Eddington–Finkelstein coordinates, the Schwarzschild metric in Kruskal–Szekeres coordinates (11.19) is manifestly non-singular at $r = 2M$.

11.4.2 Kruskal Diagrams

A *Kruskal diagram* corresponds to lines of constant r and t plotted on a $u - v$ grid, as illustrated in Fig. 11.4(a). From Eq. (11.20), curves of constant r are hyperbolas of constant $u^2 - v^2$. If $R > 2M$ the hyperbolas are timelike surfaces (oriented vertically so that any two points have a timelike separation); thus these are permissible trajectories for a massive particle. If $R < 2M$, the hyperbolas are spacelike surfaces (oriented horizontally so that any two points have a spacelike separation); thus objects cannot stay at constant r inside $R = 2M$. The degenerate hyperbola $R = 2M$ is a null line marking the horizon. The two branches of the hyperbola for $r = 0$ correspond to singularities.

Lines of constant t are straight lines with a slope given by $\tanh(t/4M)$ for $r > 2M$ and by $1/\tanh(t/4M)$ for $r < 2M$, since from Eqs. (11.15)–(11.18),

$$v = \begin{cases} u \tanh\left(\dfrac{t}{4M}\right) & (r > 2M) \\ \dfrac{u}{\tanh(t/4M)} & (r < 2M) \end{cases}. \tag{11.21}$$

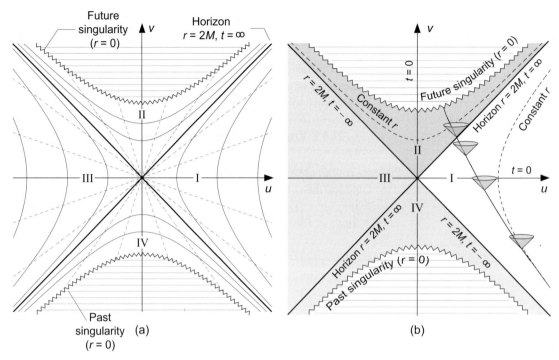

Fig. 11.4 (a) Schwarzschild spacetime in Kruskal–Szekeres coordinates. Only the two coordinates u and v are displayed, so each point is really a 2-sphere corresponding to the variables θ and φ. Spacetime singularities are indicated by jagged curves. The hatched regions above and below the $r = 0$ singularities are not a part of the spacetime. Curves of constant r are hyperbolas and the dashed straight lines are lines of constant t. (b) Worldline of a particle falling into a Schwarzschild black hole in Kruskal–Szekeres coordinates.

For radial light rays in Kruskal–Szekeres coordinates ($d\theta = d\varphi = ds^2 = 0$), Eq. (11.19) yields $dv = \pm du$, so lightcones always make 45-degree angles in the uv parameters, in analogy with flat Minkowski space (but in different coordinates). Furthermore, for the full range of Kruskal–Szekeres coordinates (v, u, θ, φ), the metric component $g_{00} = g_{vv}$ remains negative and $g_{11} = g_{uu}$, $g_{22} = g_{\theta\theta}$, and $g_{33} = g_{\varphi\varphi}$ remain positive. Therefore, the v direction is always timelike and the u direction is always spacelike, in contrast to the normal Schwarzschild coordinates where r is spacelike and t timelike outside the horizon, but r is timelike and t is spacelike within the horizon.

Lines corresponding to $r = 2M$ separate spacetime into four quadrants, labeled I, II, III, and IV.[2] In quadrant I, $r > 2M$ so this is the Schwarzschild spacetime outside the horizon.

---

[2] A manifold with metric geometry is *geodesically complete* if every geodesic beginning from an arbitrary point can be extended to infinite values of the affine parameter in both directions. A manifold is said to be *maximal* if every geodesic can be extended infinitely in both directions, or terminates in an intrinsic singularity. Hence a geodesically complete manifold is also maximal, but the converse need not be true. Minkowski space is geodesically complete and maximal. The Schwarzschild spacetime parameterized with Schwarzschild coordinates or with Eddington–Finkelstein coordinates is not maximal. The Kruskal parameterization is the *unique maximal extension of Schwarzschild spacetime.* However, it is not geodesically complete because it contains intrinsic singularities (see also the discussion of trapped surfaces in Section 14.2.2).

In quadrant II, $r < 2M$ so this is the black hole interior to the horizon and trajectories there must encounter the $r = 0$ singularity in their future since the singularity is spacelike. Thus regions I and II are of direct physical interest. Regions III and IV are of interest mathematically but it is unclear if they have physical implications. Region IV is equivalent to region II but with time inverted and the singularity is in the *past*; this is called a *white hole,* which is the opposite of a black hole, with matter spewing into the Universe from a singularity rather than disappearing into a singularity, and with the horizon forbidding entry rather than exit. There is no astrophysical evidence for white holes (but see Box 12.2).

Region III is an exterior region (outside the horizon) as for region I, but it corresponds to a *different asymptotically flat spacetime.* To understand the connection between regions I and III, imagine moving along the $t = 0$ line (the horizontal axis) in Fig. 11.4(a) from right to left, beginning in region I at $u = +\infty$ and continuing through the origin to $u = -\infty$ in region III. The radial coordinate r corresponds to timelike hyperbolas cutting the axis, from which it is seen that r decreases to a minimum value $r = 2M$ at the origin of the diagram and then begins to increase again after passing through the origin. If this behavior is cast in an embedding diagram as was done in Section 9.1.4, a geometry similar to that in the figure of Box 9.1 is obtained, with a narrow throat connecting two different asymptotically flat spacetimes. This is called an *Einstein–Rosen bridge* (or a *wormhole*).

The Einstein–Rosen bridge is likely not relevant for physics of the Schwarzschild metric. First, it cannot be used to go from region I to region III because objects attempting to move from I to III must stay within their local lightcones, which means they will end up in region II and will not be able to avoid the singularity found there. Second, it cannot be formed in a normal stellar collapse to a black hole because regions III and IV would be inside the star's mass distribution, for which the Schwarzschild solution is not valid [see Fig. 11.6(a)]. Nevertheless, the possibility of forming wormholes under more exotic circumstances has received substantial attention, as discussed further in Box 9.2. Our presentation will henceforth concentrate on regions I and II of the Schwarzschild spacetime.

11.4.3 The Event Horizon

As illustrated in Fig. 11.4(b), interpretation of the event horizon is particularly simple in Kruskal–Szekeres coordinates. Figure 11.5 compares a trip to the center of a spherical black hole in Schwarzschild coordinates and in Kruskal–Szekeres coordinates. The identification of $r = 2M$ as an event horizon is clear in Kruskal–Szekeres coordinates. The lightcones along any worldline make 45-degree angles with the vertical and the horizon also makes a 45-degree angle with the vertical in the Kruskal diagram. Thus, for any event within the horizon its future *must* contain the $r = 0$ singularity and *cannot* contain the $r = 2M$ horizon. Figure 11.6 illustrates spherical collapse to a black hole, in Kruskal–Szekeres and Eddington–Finkelstein coordinates. A distant observer at fixed r observes light signals sent periodically from the surface of the collapsing star. The light pulses, propagating on the dashed lines, arrive at increasingly longer intervals as measured by the outside observer, until at the horizon light signals require infinite time to reach the observer. Once inside the horizon, no light signals can reach the outside observer and the

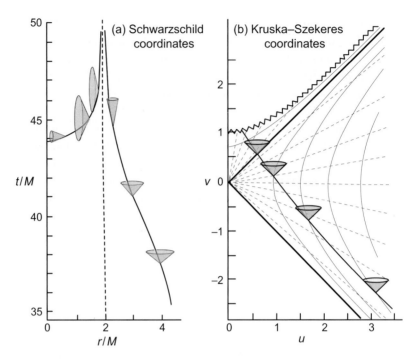

Fig. 11.5 A trip to the center of a spherical black hole in (a) Schwarzschild coordinates and in (b) Kruskal–Szekeres coordinates (adapted from Ref. [156]).

star collapses to a singularity at $r = 0$. Only the unshaded spacetime outside the surface is physically meaningful. Inside the finite matter density the correct solution would not correspond to the Schwarzschild metric.

11.5 Black Hole Theorems and Conjectures

This section summarizes in a non-rigorous way a set of theorems and conjectures concerning black holes (a more technical discussion may be found in Refs. [156, 242] and citations therein). The reader will recognize that some parts of these theorems and conjectures have been used already in earlier discussion.

Singularity theorems: Loosely, these state that any gravitational collapse that proceeds far enough results in a singularity in spacetime [181, 182]. The implications of singularity theorems for the existence of black holes will be discussed further in Section 14.2.

Cosmic censorship conjecture: It is conjectured, though not proved, that the Universe censors all spacetime singularities by hiding them behind event horizons. Thus (it is conjectured), the Universe contains *no naked singularities.*

Area increase theorem: In all processes involving horizons, the area of the horizon (sum of areas if multiple horizons are involved) can never decrease. This theorem is based on

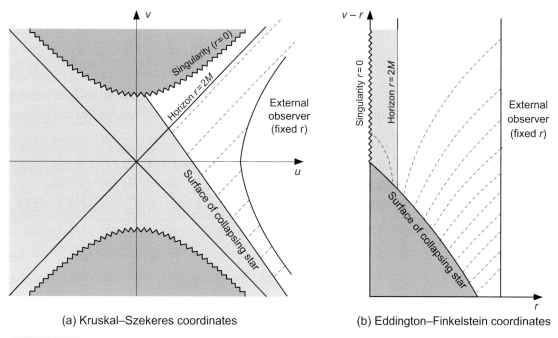

Fig. 11.6 Collapse to a black hole: (a) Kruskal–Szekeres coordinates; (b) Eddington–Finkelstein coordinates.

the assumed positivity of energy (see Box 7.3). Where quantum mechanics is important and this assumption can fail (for example, Hawking radiation, discussed in Chapter 12), the area increase theorem is replaced by the second law of black hole thermodynamics (see Section 12.5): total entropy of the Universe can never decrease, with the entropy of a black hole proportional to the surface area of its horizon.

The no-hair theorem/conjecture: In gravitational collapse to a black hole all non-spherical parts of the mass distribution (quadrupole moments, octupole moments, ...) except angular momentum are radiated away as gravitational waves and eventually the horizon becomes stationary. A stationary black hole is characterized by only three numbers: mass M, angular momentum J, and charge Q. These quantities are determined by fields *outside* the black hole, not by integrals over the interior.[3] That black holes are characterized only by M, J, and Q is often referred to colloquially as the *No Hair Theorem,* meaning that black holes destroy all detailed information (the hair) about the matter that formed them, leaving

---

[3] In principle magnetic charge could be added to the list of quantities characterizing black holes if magnetic monopoles existed. However, there is no experimental evidence for them so this possibility will be ignored here. A spherical black hole with non-zero electrical charge is described by the *Reissner–Nordström* solution of the field equations. The most general black hole solution characterized by mass, angular momentum, and charge is termed a *Kerr–Newman black hole* (the metric is given in Box 13.1). However, the astrophysical processes likely to form a black hole would tend to neutralize any excess charge, so it is usual to assume that all black hole solutions of interest to astronomers are uncharged. Thus astrophysical black holes correspond to Kerr solutions (see Chapter 13), with the Schwarzschild solution being a special case of the Kerr solution for vanishing angular momentum.

behind only global mass, angular momentum, and charge as possible observable external characteristics [126].

Birkhoff's theorem: The Schwarzschild solution is in fact more general than the preceding discussion would suggest. A static and spherical source of the gravitational field was assumed, but even if the source is changing with time *Birkhoff's theorem* asserts that the Schwarzschild solution is the only spherically symmetric solution of the vacuum Einstein equations.

These theorems and conjectures place the mathematics of black holes on reasonably firm ground, though obviously much remains to be understood, particularly about singularities and the conjecture of cosmic censorship, and about the fate of information falling into a black hole that will be discussed in Section 12.7. To place the *physics* of black holes on firm ground these ideas must be tested by observation, which will be taken up beginning in Chapter 14.

Background and Further Reading

Much of the material in this chapter has been adapted from Misner, Thorne, and Wheeler [156]; Hobson, Efstathiou, and Lasenby [122]; Hartle [111]; and Schutz [221]. See also Cheng [65], Raine and Thomas [197], and Taylor and Wheeler [231].

Problems

11.1 Express the Schwarzschild metric in Eddington–Finkelstein coordinates (11.10), starting from the metric in standard coordinates (9.4) and the relation (11.9). ***

11.2 Estimate the proper time to fall from the event horizon to the $r = 0$ singularity in a Schwarzschild black hole of $10\,M_\odot$ (a typical stellar-mass black hole), $62\,M_\odot$ (final black hole in the binary black hole merger that produced the gravitational wave GW150914), 4.3 million $M_\odot$ (mass of the black hole at the center of the Milky Way), and $10^9\,M_\odot$ [typical mass of an active galactic nucleus (AGN) central engine]. ***

11.3 Use the Hawking area theorem for black holes to show that a Schwarzschild black hole cannot split into two smaller Schwarzschild black holes.

11.4 Gravitational waves are emitted when black holes vibrate in a non-spherical way. Suppose an uncharged mass is dropped radially into a Schwarzschild black hole, causing it to vibrate and emit gravitational waves. Is it possible to arrange this experiment such that the amount of energy carried off by the gravitational waves exceeds the energy of the mass dropped in, thereby extracting mass from the black hole?

11.5 Show that $L \equiv g_{\mu\nu}\dot{x}^\mu \dot{x}^\nu$ with $\dot{x}^\mu \equiv dx^\mu/d\tau$ for motion of a test particle having unit mass in a Schwarzschild spacetime can be written as

$$L = -c^2 = -\left(1 - \frac{2M}{r}\right)c^2\dot{t}^2 + \left(1 - \frac{2M}{r}\right)^{-1}\dot{r}^2 + r^2\dot{\varphi}^2,$$

and that this is equivalent to $E^2 = p^2c^2 + m^2c^4$, if the spacetime is flat.

11.6 What mass is required for a spherical black hole to allow an observer in free fall to record one day of elapsed time between crossing the horizon and encountering the singularity (assuming an indestructible clock and observer)?

11.7 Show that the angular velocity with respect to the Schwarzschild coordinates $\omega = d\varphi/dt$ of a particle in a circular orbit outside a Schwarzschild black hole is given by

$$\omega = \frac{1}{r^2}\left(1 - \frac{2M}{r}\right)\left(\frac{\ell}{\epsilon}\right),$$

where ϵ is the energy per unit rest mass and ℓ is the angular momentum per unit rest mass, and that the circular orbit is further constrained by the condition that

$$\epsilon = \sqrt{\left(1 - \frac{2M}{r_0}\right)\left(1 + \frac{\ell^2}{r_0^2}\right)},$$

where the radial coordinate r_0 corresponds to the minimum of the effective potential.

11.8 Assume a black hole of mass $3 \times 10^6\ M_\odot$. What is the corresponding Schwarzschild radius? Assume that a spacecraft executes a circular orbit around this black hole for which the proper distance traveled is 6.283×10^8 km. What is the coordinate radius r corresponding to this circular orbit? The spacecraft then travels inward radially until it reaches a coordinate radius of $r = 10^7$ km. What is the proper distance traveled on this radial path? If the spacecraft then executes a new circular orbit around the black hole, what is the proper distance that it travels around the circle?

11.9 Use the line element corresponding to the 2D event horizon for a Schwarzschild black hole to show that the area of the horizon is $A = 16\pi M^2$, where M is the mass of the black hole. ***

11.10 The latitude–longitude system for the Earth (assumed to have the metric of a 2-sphere) and the Schwarzschild metric expressed in standard Schwarzschild coordinates both have well-known points of singular behavior, the first at the poles (where an infinite number of longitude angles correspond to a single point) and the second at both $r = 2M$ and $r = 0$. Derive formulas for appropriate curvature scalars R and show that they are finite and well-behaved at and near the poles on the Earth, and at $r = 2M$ in the Schwarzschild metric, giving strong support to the assertion that these are coordinate singularities that could be removed by a different choice of coordinate system. Show conversely that the behavior of the curvature scalar near $r = 0$ in the Schwarzschild metric suggests that this is a real (physical) singularity of the metric that cannot be removed by a new choice of coordinates. *Hint*: The Ricci scalar in Eq. (8.17) works for the 2-sphere but not for the Schwarzschild metric, where it vanishes identically because the Schwarzschild metric is a solution of $R_{\mu\nu} = 0$. In that case a more appropriate curvature scalar is $K \equiv R^{\alpha\beta\gamma\delta} R_{\alpha\beta\gamma\delta}$, which is called the *Kretschmann scalar;* see Ref. [117].

11.11 Show that if a 2-sphere of radius R is embedded at the origin of a 3-dimensional cartesian coordinate system, the line element for the 2-sphere may be expressed in the form
$$ds^2 = \frac{R^2 dr^2}{R^2 - r^2} + r^2 d\varphi^2.$$
Is the singularity at $r = R$ physically meaningful, or is it a coordinate singularity? *Hint*: Use the equation of the sphere to eliminate dz and constrain the cartesian line element to the spherical surface; then change to new variables (r, φ) using $x = r \cos\varphi$ and $y = r \sin\varphi$.

11.12 Demonstrate on general (causal) grounds that stationary observers are possible outside the event horizon of a Schwarzschild black hole but not inside. *Hint*: What is the condition that an observer be stationary in space, and can it be maintained both outside and inside the event horizon?

11.13 Show that for a metric defined through the line element
$$ds^2 = -\left(1 - \frac{r^2}{R^2}\right) dt^2 + \left(1 - \frac{r^2}{R^2}\right)^{-1} dr^2 + r^2 d\theta^2 + r^2 \sin^2\theta \, d\varphi^2,$$
$r = R$ defines an event horizon for an observer moving outward from $r = 0$. *Hint*: Examine the causal structure of the spacetime by considering the motion of radial light rays.

11.14 Find an expression for acceleration a of a stationary observer in Schwarzschild spacetime. Show that the length $(a \cdot a)^{1/2}$ of the acceleration 4-vector for a stationary observer tends to infinity at the Schwarzschild radius, implying that infinite acceleration is required to remain stationary at the event horizon of a spherical black hole. *Hint*: In flat spacetime, or a local inertial frame, acceleration may be defined as the derivative of the 4-velocity with respect to proper time.

11.15 In a local inertial frame, the 4-acceleration can be defined by $a^\mu = du^\mu/d\tau$, where u^μ is the 4-velocity and τ is the proper time. Show that in a general curvilinear coordinate system the 4-acceleration generalizes to
$$a^\mu = u^\alpha \left(\frac{\partial u^\mu}{\partial x^\alpha} + \Gamma^\mu_{\alpha\gamma} u^\gamma \right),$$
where $\Gamma^\mu_{\alpha\gamma}$ is the connection coefficient.

11.16 Suppose the Sun were replaced with a Schwarzschild black hole of 1 $M_\odot$. Describe qualitatively what would happen to the Earth.

12 Quantum Black Holes

Classically, the fundamental structure of curved spacetime in the Schwarzschild metric ensures that nothing can escape from within the event horizon. However, this classical picture fails to account for the uncertainty principle and quantum fluctuations of the vacuum that play a central role in quantum mechanics and quantum field theory. General relativity and quantum field theory rule their respective domains but they are fundamentally incompatible. Since we do not yet have a viable theory of quantum gravity, it isn't possible to make definitive statements about quantum effects in general relativity. However, there is reason to believe that we may be able to make at least approximate statements by considering a "semiclassical approximation" in which the curvature of spacetime is assumed to vary slowly so that quantum mechanics can be implemented locally on that slowly varying background. As will now be considered, in such approximations it is found that because of quantum fluctuations of the vacuum state in the vicinity of the event horizon it is possible for a black hole to emit mass. Therefore, as Stephen Hawking discovered [113, 114], if these approximations have any validity black holes are not really black! In this chapter we will outline the basic ideas that underlie this remarkable claim, guided strongly by the excellent presentation in Hartle [111].

12.1 Geodesics and Uncertainty

The discussion so far in this book has been classical in that it assumes free particles to follow geodesics appropriate for the spacetime. But the Heisenberg uncertainty principle implies that microscopic particles cannot be fully localized on classical trajectories because they are subject to a spatial coordinate and 3-momentum uncertainty of the form $\Delta p_i \Delta x_i \geq \hbar$; neither can energy conservation be imposed except with an uncertainty satisfying $\Delta E \Delta t \geq \hbar$, where ΔE is an energy difference and Δt is the corresponding time period during which it exists. This implies an inherent fuzziness in the 4-momenta associated with our description of spacetime at the quantum level. For the Killing vector $K \equiv K_t = (1, 0, 0, 0)$ in the Schwarzschild metric (9.4),

$$K \cdot K = g_{\mu\nu} K^\mu K^\nu = -\left(1 - \frac{2M}{r}\right), \qquad (12.1)$$

from which it may be concluded that

$$K \text{ is } \begin{cases} \text{timelike outside the horizon, since then } K \cdot K < 0, \\ \text{spacelike inside the horizon, since then } K \cdot K > 0. \end{cases} \qquad (12.2)$$

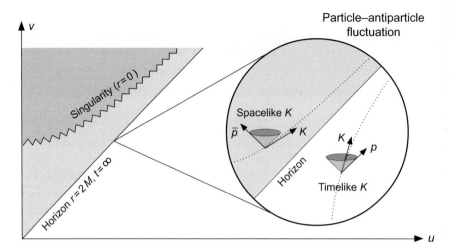

Fig. 12.1 A microscopic fluctuation of the vacuum near the event horizon of a Schwarzschild black hole. The Kruskal–Szekeres coordinates introduced in Section 11.4 are employed since they are particularly useful near the horizon. Adapted from Hartle [111].

It will now be demonstrated that this property of the Killing vector K permits a vacuum fluctuation to be converted into real particles, detectable at infinity as emission of mass from the black hole, while still satisfying the required conservation laws.

12.2 Hawking Radiation

To illustrate the principle we assume that a particle–antiparticle pair is created by a quantum vacuum fluctuation near the horizon of a spherical black hole such that the virtual particle and antiparticle are on opposite sides of the horizon. To be definite we assume that the antiparticle is inside the horizon and the particle outside the horizon, but a similar argument applies if the relative locations are switched. Figure 12.1 illustrates.

12.2.1 4-Momentum Conservation

If the particle–antiparticle pair is created in a small enough region of spacetime there is nothing special implied by this region lying at the event horizon; spacetime there is continuous and in a small enough region it is indistinguishable from Minkowski space because of the equivalence principle. Therefore, we may expect that the normal principles of (special) relativistic quantum field theory are applicable to the microscopic pair creation process.[1] In the Schwarzschild geometry the conserved quantity analogous to the total energy in flat space is the scalar product of the Killing vector $K \equiv K_t = (1, 0, 0, 0)$ with

---

[1] Relativistic quantum field theory is a Lorentz-invariant version of quantum field theory. For an introduction, see [103]. Relativistic in this context means special relativity, so such theories are valid in flat spacetime.

the 4-momentum p. Therefore, if a particle–antiparticle pair is produced near the horizon with 4-momenta p and $\bar p$ for the particle and antiparticle, respectively, the condition

$$K \cdot p + K \cdot \bar p = 0, \qquad (12.3)$$

must be satisfied for any fluctuation of the vacuum, because the quantum numbers carried by the vacuum must be preserved in the virtual pair excitation and the vacuum is assumed to have zero 4-momentum.

Outside the horizon $-K \cdot p$ is required to be positive because it is proportional to an energy that could be measured by an external observer. If the antiparticle also were outside the horizon, by a similar argument $-K \cdot \bar p$ must be positive, in which case the condition (12.3) cannot be satisfied and the particle–antiparticle pair can have only a fleeting existence of average duration $\Delta t \sim \hbar / \Delta E$ (where ΔE is the mass–energy of the pair) as a virtual pair, before being absorbed back into the vacuum. Nothing surprising so far; just garden-variety uncertainty principle arguments. But if instead the virtual antiparticle is *inside* the horizon (see the expanded view in Fig. 12.1), K is *spacelike*, the scalar product $-K \cdot \bar p$ is *not an energy* but rather may be interpreted as a component of spatial momentum, and this means that $-K \cdot \bar p$ could be *either positive or negative*. If it is negative, the conservation law (12.3) can be satisfied and the virtual particle created outside the horizon by a quantum vacuum fluctuation can propagate to infinity as a real, detectable particle, while the antiparticle remains inside the event horizon.

12.2.2 Black Hole Evaporation

By the process described in the preceding paragraph the black hole appears to an observer at infinity to have emitted a portion of its mass $-K \cdot \bar p$. By similar arguments we may imagine the particle to be inside the horizon and the antiparticle to be outside in the quantum vacuum fluctuation, in which case the antiparticle could propagate to infinity and be detected. The outgoing flux of such particles and antiparticles is termed *Hawking radiation* and the corresponding emission of the black hole's mass is termed *black hole evaporation* through Hawking radiation.[2] A fanciful but evocative analogy of this process is described in Box 12.1.

12.2.3 Relative Importance of Quantum Fluctuations

On general grounds we might expect quantum fluctuations for a typical black hole to be negligible. The characteristic length scale upon which quantum effects become important for gravity is the Planck length $\ell_P \sim 10^{-33}$ cm to be discussed in Section 12.6. This is many orders of magnitude smaller than the radius of gravitational curvature near the

---

[2] Production of a virtual particle–antiparticle pair by a fluctuation of the quantum vacuum is more likely for particles with low mass such as photons or neutrinos. Astronomers term massless or nearly massless particles "radiation"; hence the appellation "Hawking radiation" for the flux of mostly low-mass particles and antiparticles emitted from the black hole.

> **Box 12.1** **Shady Transactions of Uncertainty Bank and Trust**
>
> The Hawking mechanism has been described by loose analogy with a rather nefarious financial transaction [231]. Suppose that I am flat out broke (a money vacuum), but I wish to give to you a large sum of money (let's say in Euros, E). Next door is a bank – Uncertainty Bank and Trust – that has lots of money in its secure vault but is afflicted by shoddy lending practices. I could then (1) borrow a large sum of money ΔE from the Uncertainty Bank, which will take a finite time Δt to find that I have no means of repayment (we may hypothesize that $\Delta t \propto 1/\Delta E$, since the bank will be more diligent in checking up on my financial means if the amount borrowed is larger), (2) transfer the money ΔE to your account, and (3) declare bankruptcy within the time Δt, leaving the bank on the hook for the loan. The net effect is that a virtual fluctuation of the money vacuum has caused real money to be emitted (and detected in your distant account!) from behind the seemingly impregnable event horizon of the bank vault.

event horizon of a stellar black hole, implying that the energies of particles created by quantum fluctuations of the gravitational field are tiny compared with the scale set by the spatial curvature energy. However, the age of the Universe is some 60 orders of magnitude larger than the Planck time t_P (see Section 12.6), so we cannot dismiss out of hand the possibility that the effects of quantum fluctuations for black holes could have a non-negligible cumulative effect over cosmological timescales [113]. To evaluate this possibility requires a quantitative measure of the rate of emission caused by quantum processes for a black hole.

12.3 Black Hole Temperatures

Advanced methods in quantum field theory that assume quantum processes to occur on a spacetime background with slowly varying curvature find the remarkable result that the distribution of energies emitted by the black hole as Hawking radiation is equivalent to that of a blackbody with temperature proportional to the *surface gravity* of the black hole. (The surface gravity is a renormalized gravitational acceleration that would be experienced by an observer near the horizon of the black hole.) Specifically, it is found that the blackbody temperature is

$$T = \frac{\kappa}{2\pi} = \frac{\hbar c^3}{8\pi k_B G M}, \qquad (12.4)$$

where the first form uses $\hbar = c = G = k_B = 1$ units and $\kappa = (4M)^{-1}$ is the surface gravity for a Schwarzschild black hole.[3] Notice that Eq. (12.4) combines in a single equation

---

[3] The derivation of Eq. (12.4) is too advanced a task for the present context. However, there are suggestive hints that black holes resemble blackbody radiators from the observations that a black hole acts as a perfect absorber of radiation and that the thermal emission (Hawking radiation) from the black hole originates in random fluctuations; see also Problem 12.9.

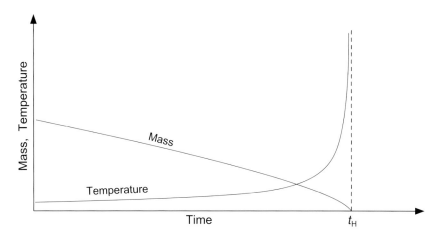

Fig. 12.2 Variation of mass and temperature in the evaporation of a Hawking black hole.

the fundamental constants associated with special relativity (c), gravity (G), quantum mechanics ($\hbar$), and statistical mechanics (k_B). The temperature is proportional to $\hbar$ so it is a quantum effect that vanishes in the $\hbar \to 0$ classical limit. As shown in Problem 12.8, the radiated power is given by

$$P = \frac{\hbar c^6}{15360\pi G^2 M^2}, \tag{12.5}$$

from which the rate of mass emission by Hawking radiation from the black hole is

$$\frac{dM}{dt} = -\lambda \frac{\hbar}{M^2}, \tag{12.6}$$

where λ is a dimensionless parameter. Integrating Eq. (12.6) for a black hole assumed to emit all of its mass by Hawking radiation in a time t_H, the mass of the black hole as a function of time is

$$M(t) = [3\lambda\hbar(t_H - t)]^{1/3}. \tag{12.7}$$

Equations (12.4) and (12.7) indicate that the mass and temperature of the evaporating black hole behave as in Fig. 12.2. The black hole evaporates at an accelerating pace as it loses mass. From (12.6) and (12.4), both the temperature and emission rate of the black hole tend to infinity near the end so a final burst of very high energy (gamma-ray) radiation would be expected to characterize evaporation of a Hawking black hole. From the mass emission rate (12.7) an approximate lifetime for complete evaporation of the Hawking black hole may be estimated:

$$t_H \simeq \frac{M^3}{3\hbar\lambda} = \frac{5120\pi G^2}{\hbar c^4} M^3, \tag{12.8}$$

where in the last step factors of G and c have been restored and a simple blackbody approximation of $\lambda = (15{,}360\,\pi)^{-1}$ has been used. The following example uses these formulas to make some estimates for stellar-size black holes.

Example 12.1 Take 15 $M_\odot$ as a representative mass for a black hole formed from massive star collapse. From Eqs. (12.4)–(12.8), the temperature is then

$$T = \frac{\hbar c^3}{8\pi k_B G M} = 6.2 \times 10^{-8} \left(\frac{M_\odot}{M}\right) \text{ K} = 4.1 \times 10^{-9} \text{ K}, \qquad (12.9)$$

the radiated power is

$$P = \frac{\hbar c^6}{15360\pi G^2 M^2} = 9.0 \times 10^{-29} \left(\frac{M_\odot}{M}\right)^2 \text{ W} = 4 \times 10^{-31} \text{ W}, \qquad (12.10)$$

and the time for the black hole to evaporate all of its mass by means of Hawking radiation is given by

$$t_H \simeq \frac{5120\pi G^2}{\hbar c^4} M^3 = 6.6 \times 10^{74} \left(\frac{M}{M_\odot}\right)^3 \text{ s} = 2.2 \times 10^{78} \text{ s}. \qquad (12.11)$$

This lifetime for Hawking evaporation of a stellar-size black hole is about about *60 orders of magnitude larger than the age of the Universe!* It seems quite safe to assume that the effect of Hawking radiation may be ignored for black holes formed from the collapse of stars or for supermassive black holes. In fact, as you are asked to investigate in Problem 12.7, for normal stellar black holes the temperature is so low that the surrounding cosmic microwave background (Section 20.4) temperature is much higher, so they absorb rather than emit radiation.

Clearly Hawking radiation is negligible for normal black holes. However, we may note that from Eq. (12.11) black holes of initial mass $\sim 10^{14}$ g have evaporation lifetimes comparable to the age of the Universe, and their demise could be observable through a burst of high-energy radiation.

12.4 Miniature Black Holes

The Schwarzschild radius (9.6) for a 10^{14} g black hole is approximately 10^{-14} cm, which is about $\frac{1}{5}$ the size of a proton. To form such a *miniature black hole* requires compressing 10^{14} grams (roughly the mass of a large mountain) into a volume less than that of a proton. No physical process known in the present Universe is capable of doing that! However, in the early moments of the big bang there would have been densities comparable to this and it may be speculated that a population of *miniature black holes* might have formed in the big bang and could be decaying in the present Universe with a detectable signature. From Eq. (12.10) a 10^{14} g black hole would radiate initially with a power of about 3.6×10^{10} W, which is many orders of magnitude less than that of the Sun. Therefore, a Hawking black hole with an initial mass small enough to give it a lifetime less than the age of the Universe would be detectable only if it were relatively nearby when it exploded. No evidence has been found yet for such miniature

> **Box 12.2** **Analog Event Horizons**
>
> In the 1980s William Unruh, inspired by analogies of salmon swimming against a current, recognized that there are many similarities between the equations of motion for long-wavelength sound waves in a flowing medium and those for a scalar field in curved spacetime. In particular, he noted that *event horizons* could exist for motion in a flowing medium when the local flow speed exceeds the wave velocity.
>
> **Analog Black Holes**
>
> Consider a river flowing toward a waterfall, with the flow speed increasing as the waterfall is approached. Now imagine waves propagating against the current with wavespeed c'. At some critical point the speed of the river equals c', meaning that waves upstream from this point can propagate further upstream but waves downstream from this point can't propagate upstream. In the analogy the critical point corresponds to the horizon of a black hole, the flow-velocity gradient at the horizon is the analog of the surface gravity of the black hole, and the waterfall is the singularity. These are sometimes called *sonic black holes*. Various experiments, for example in liquid helium and in Bose–Einstein condensates, or in electromagnetic waveguides, have attempted to study such "black holes" experimentally.
>
> **Analog White Holes**
>
> Now consider a fast river slowing as it flows into the sea. Waves cannot enter the river beyond the point where the flow speed exceeds the wave velocity. This is an analog of a *white hole* (see Section 11.4.2), with a horizon forbidding entry. Although no evidence supports the existence of astronomical white holes, analog white holes have been reported in fiber-optical experiments [186].
>
> **Analog Hawking Radiation**
>
> Unruh showed also that if sound waves are quantized in a fluid-flow "black hole," the faux black hole can emit phonons (quanta of vibrations) that are the analog of Hawking radiation. Thus in principle analog horizons can be used to study Hawking radiation. In practice this is difficult because the Hawking radiation is intrinsically weak. However, it was realized that if analog black hole and white hole horizons were paired, spontaneous amplification of the Hawking radiation (a *black-hole laser*) could be realized between the two horizons [157]. This phenomenon has been reported using an ultracold Bose–Einstein condensate as the fluid and a step potential generated by laser light as an obstacle to alter fluid flow and create the paired analog event horizons [226]. Thus, even if it is difficult to observe astrophysical Hawking radiation, it may be possible to study its fluid analog systematically in the laboratory. Although of great interest in their own right, it is unclear whether such desktop experiments imply any new understanding for astrophysics.

black holes and their associated Hawking radiation. (The gamma-ray bursts discussed in Section 15.7 are much too powerful and have the wrong characteristics to be Hawking radiation; they are associated with a very different mechanism.) Although Hawking radiation has not been observed yet, Box 12.2 discusses analogs that have been studied in fluid flow.

12.5 Black Hole Thermodynamics

The preceding results suggest that classical thermodynamics and the gravitational physics of black holes are closely related. Remarkably, this is the case, with the key unifying principle being the realization by Hawking that a temperature could be associated with a black hole by virtue of the emission of Hawking radiation [114]. It had been noted prior to Hawking's discovery that there were similarities between black holes and blackbody radiators [41, 42]. However, the difficulty with describing a black hole in thermodynamical terms was that a classical black hole permits no equilibrium with the surroundings since it can absorb but cannot emit radiation. Emission of Hawking radiation supplies the necessary equilibrium that ultimately allows a temperature and thermodynamics to be ascribed to a black hole. The rather interesting history of these ideas is summarized in Box 12.3.

12.5.1 Entropy of a Black Hole

If A is the area of the event horizon for a black hole, Hawking has proved a theorem that A cannot decrease in any physical process involving a black hole horizon (see Section 11.5),

$$dA/dt \geq 0. \tag{12.12}$$

From Problem 11.9 the horizon area for a Schwarzschild black hole is $A = 16\pi M^2$. Therefore, $dA/dM = 32\pi M$, which may be expressed in the form

$$dM = \frac{\hbar}{8\pi k_B M} d\left(\frac{k_B A}{4\hbar}\right). \tag{12.13}$$

But $dE = dM$ is the change in total energy of the black hole and from Eq. (12.4) the temperature of the black hole is $T = \hbar/8\pi k_B M$. Therefore, from (12.12) and (12.13),

$$dE = T dS \qquad S \equiv \frac{k_B}{4\hbar} A \qquad dS \geq 0. \tag{12.14}$$

Upon interpreting S as the entropy of the black hole, Eqs. (12.14) are just statements of the first and second laws of thermodynamics: Schwarzschild black holes are blackbody radiators with temperatures given by Eq. (12.4) and entropies given by (12.14)! Although this has been discussed here only for the restricted case of a Schwarzschild black hole, it is in fact expected to be true for all black holes.

12.5.2 The Generalized Second Law

The attentive reader may object that the evaporation of a black hole through Hawking radiation is a blatant violation of the Hawking area theorem: the black hole disappears eventually after radiating away all its mass, which certainly represents a *decrease in the horizon area* for a black hole process. However, the area theorem is predicated on assumptions that are thought to be correct at the classical level but that might break down in quantum processes. Therefore, the correct interpretation of Hawking radiation and the area

> **Box 12.3** **Jacob Bekenstein and Wheeler's Demon**
>
> In 1971 Jacob Bekenstein (1947–2015) was a doctoral student under John Wheeler at Princeton and he was bothered by implications of the no hair theorem (Section 11.5) [42]. As Wheeler summarized it: A wicked creature (Bekenstein called it "Wheeler's demon") could commit the perfect crime by dropping a package of entropy into a black hole, thus decreasing the entropy of the observable Universe and, perforce, violating the second law of thermodynamics. But since by the no hair theorem the black hole exposes only total mass, charge, and angular momentum to an external observer and these are insufficient to reveal the internal entropy of the black hole, no evidence would exist to convict the defendant of violating the second law.[a] This seemed to either neuter the second law (stripping it of predictive power), or raise doubts about whether real black holes existed. Bekenstein liked neither option.
>
> Another Wheeler student, Demetrios Christodoulou, was studying extraction of energy from rotating black holes through the *Penrose process* (see Section 13.3). He found that the most efficient processes were invariably the *most reversible ones*. This, thought Bekenstein, smacks of thermodynamics! Near this time Hawking proved the area theorem (Section 11.5), which suggested that irreversibility is in fact a *general property of processes for real black holes*. It also suggested a formal analogy between entropy and the area of event horizons: *both like to increase*.
>
> Bekenstein's 1972 doctoral thesis assumed that some monotone increasing function of horizon area A might stand in for entropy and reconcile black holes with classical thermodynamics. Initial results suggested that S_{bh} should be proportional to A or $\sqrt{A}$. The latter implies a black hole entropy proportional to its mass [see Eq. (13.11)], as for normal systems. But it is *inconsistent with the area theorem*: for two merging black holes the requirement that the net horizon area cannot decrease could be violated because $\sqrt{A_1} + \sqrt{A_2}$ can exceed $\sqrt{A}$ for the final black hole. Thus Bekenstein chose $S_{bh} \propto A$. For the constant of proportionality he and Wheeler reasoned that it must involve the Boltzmann constant k_B and the Planck length defined in Eq. (12.17), leading to a formula of the form (12.21). Hawking then closed the circle three years later by discovering the Hawking radiation that makes black-hole thermodynamic equilibrium logically possible.
>
> [a] Aha, you might say: what about an eyewitness! But as anyone knows from television drama, with no dead body in evidence eyewitness testimony would show only that *something* was thrown into the black hole and the prosecution's case would collapse.

theorem is that the entropy of the evaporating, isolated black hole decreases with time – because it is proportional to the area of the horizon – but the total entropy of the Universe is increased because of the entropy associated with the Hawking radiation itself. That is, when quantum processes are important, the area theorem of classical general relativity is replaced by a *generalized second law of thermodynamics* requiring that the total entropy of the black hole plus the exterior Universe may never decrease.

12.5.3 The Four Laws of Black Hole Dynamics

The considerations of this section permit the formulation of *four laws of black hole dynamics* that are analogous to the four laws of classical thermodynamics [177]:

1. *Zeroth Law*: The surface gravity κ of a stationary black hole is constant over its event horizon. This is analogous to the zeroth law of thermodynamics that the temperature T is constant for a system in thermal equilibrium.
2. *First Law*: Energy is conserved because of a relation

$$\delta M = \frac{1}{8\pi}\kappa\delta A + \Omega\delta J + \Phi\delta Q, \tag{12.15}$$

where M is the mass, κ is the surface gravity, A is the area of the horizon, Ω is the angular velocity, J is the angular momentum, Q is charge, and Φ is the electrostatic potential for the black hole. (See Box 13.1 for the most general black hole with angular momentum and charge, in addition to mass.) This is the analog of the first law of thermodynamics.
3. *Second Law*: Hawking's area theorem (12.12), or its quantum generalization (12.14). This is the analog of the second law of thermodynamics.
4. *Third Law*: The surface gravity κ of a black hole cannot be reduced to zero by a finite series of operations. This is analogous to the Nernst form of the third law of thermodynamics that the temperature T cannot be reduced to zero in a finite series of operations.

Thus the four laws of black hole dynamics are analogous to the four laws of thermodynamics if an identification is made between temperature T and some multiple of the surface gravity κ, and between the entropy S and some multiple of the event horizon area A. The laws of classical thermodynamics have been tested thoroughly but the four laws of black hole dynamics are more well-grounded theoretical conjecture than established law at this point.

12.6 The Planck Scale and Quantum Gravity

The preceding results for Hawking radiation were derived by assuming that the spacetime in which the quantum calculations are done (typically a Schwarzschild metric) is not influenced by the propagation of the Hawking radiation. This assumption is expected to be valid if $E \ll M$, where E is the average energy of the Hawking radiation and M is the mass of the black hole, but is expected to break down on the *Planck scale* defined in the following expressions:

$$\text{Planck mass:} \quad M_P \equiv (\hbar c/G)^{1/2} = 2.18 \times 10^{-5}\, \text{g}, \tag{12.16}$$

$$\text{Planck length:} \quad \ell_P \equiv (\hbar G/c^3)^{1/2} = \hbar c/E_P = 1.62 \times 10^{-33}\, \text{cm}, \tag{12.17}$$

$$\text{Planck time:} \quad t_P \equiv (\hbar G/c^5)^{1/2} = \ell_P/c = 5.39 \times 10^{-44}\, \text{s}, \tag{12.18}$$

$$\text{Planck energy:} \quad E_P \equiv M_P c^2 = 1.22 \times 10^{19}\, \text{GeV}, \tag{12.19}$$

$$\text{Planck temperature:} \quad T_P \equiv E_P/k_B = 1.44 \times 10^{32}\, \text{K}. \tag{12.20}$$

For a Planck-mass black hole gravity is important even on a quantum ($\hbar$) scale and must be quantized as for other fundamental interactions. If the constants G and c are restored in Eq. (12.14) using $\hbar \to (G/c^3)\hbar$ from Table B.1, the entropy can be written in the form

Box 12.4   The Endpoint of Hawking Black Hole Evaporation

Since the radius of a spherical black hole depends on the mass, near the endpoint of Hawking black hole evaporation the radius approaches the Planck length and the black hole begins to be governed by quantum gravity. Therefore, what actually happens in the final stages of the Hawking evaporation of a black hole is unknown at present. One speculation is that the emission of radiation stops near the Planck scale and a new particle carrying the quantum numbers of the original black hole is left behind. Another is that the evaporation continues to completion (as has been assumed in the preceding discussion). Notice that this latter possibility implies processes that fail to conserve baryon number. A black hole made by collapsing baryonic matter (carrying a finite baryon number) would be expected to evaporate through the Hawking process by emitting equal numbers of baryons and antibaryons, leaving behind baryon number zero since baryons and antibaryons carry opposite baryonic charge. Many theories that extend the Standard Model of elementary particle physics conjecture that the Universe may not conserve baryon number, but thus far there is no experimental evidence supporting that conjecture.

$$S_{\text{bh}} \equiv \frac{k_B}{4\hbar} A = \frac{c^3 k_B}{4\hbar G} A = \frac{k_B A}{4\ell_P^2}, \qquad (12.21)$$

where ℓ_P is the Planck length defined in Eq. (12.17). The presence of $\hbar$, c, G, and k_B in this formula indicates clearly that the entropy of a black hole represents a unique nexus of physics associated with gravity, quantum mechanics, and statistical mechanics.

Example 12.2 Quantum effects become important for a particle of mass m on a length scale set by the Compton wavelength, $\lambda \equiv h/mc = 2\pi\hbar/mc$, and gravity becomes strong on a length scale set by the Schwarzschild radius, $r_s = 2Gm/c^2$. Conjecturing that quantum gravity becomes important when $\lambda = r_s$ and solving for the mass gives $m = \sqrt{hc/2G}$, which is the Planck mass (12.16) within a factor of $\sqrt{\pi}$.

The name of Max Planck (1858–1947) is associated with the Planck scale because he was the first to realize – long before the physics of the present discussion was proposed – that unique combinations of the fundamental constants $\hbar$, c, and G defined length, mass, and time scales. One implication of the Planck scale for the present discussion is addressed in Box 12.4 and a quantum theory of gravity will be considered further in Section 26.3.

12.7 Black Holes and Information

Entropy is related to information content because it is proportional to the logarithm of the number of microscopic configurations that leave the macroscopic description of an object unchanged. Black hole evaporation by the Hawking mechanism leads to apparent paradoxes associated with this relationship. This may be illustrated by noting that Hawking radiation is produced randomly by vacuum fluctuations, so in the simplest picture it contains no information. Thus, the information content of the matter from which the black hole formed appears to be lost to the Universe if the black hole then decays completely away by Hawking radiation. This is a complex issue that is not fully resolved. Some contend that perhaps this means that the Universe does not conserve information. Others have conjectured that a (future) quantum gravity treatment of black hole evaporation may resolve the issue.

12.7.1 The Holographic Principle

An interesting view of the black hole information problem is suggested by Eq. (12.14), which implies that the entropy of a black hole is proportional to the *surface area* of its horizon. But the information content of the black hole is associated with its entropy and the information content of a region of space is usually thought of as being proportional to the *volume* of that region, not to the area of a bounding surface. This has led to a proposed solution of the black hole information paradox originally due to Gerard 't Hooft and Leonard Susskind called the *holographic principle,* which asserts that the description of a volume of space can be thought of as being encoded on a 2-dimensional boundary of that region. For a black hole, this implies that surface fluctuations of the event horizon must in some way contain a complete description of all the objects that have ever fallen into the black hole.

12.7.2 The Holographic Universe

Even more speculatively, the holographic principle has been extended to a cosmological statement that perhaps the entire universe should be thought of as a 2-dimensional information structure painted on the cosmological horizon. (Cosmological horizons are described in Section 18.8.) The extension of the holographic principle suggested by local black hole thermodynamics to a cosmological statement may be motivated by noting that a universe with a cosmological horizon resembles in some ways the interior of a black hole. In this view the entire Universe is a kind of gigantic hologram and our perception that it has three spatial dimensions is an illusion rooted in an effective description of the actual Universe that is valid only at low energies. The *AdS/CFT correspondence* described in Box 26.3 is a specific implementation of such ideas.

Background and Further Reading

The discussion of Hawking black holes in this chapter has been influenced by the presentation in Hartle [111]. Chow [66] gives a nice heuristic discussion of quantum black holes, though there are typographical errors in some equations. Page [177] discusses Hawking radiation and black hole dynamics with particular focus on historical aspects. The treatment of Hawking radiation and quantum black holes has of necessity been rather cursory. Those wishing to pursue more deeply the technical issues associated with black hole thermodynamics and quantum field theory in curved spacetime may start by consulting Chapter 9 of Carrol [63], Chapter 14 of Wald [242], the lectures by Wald [243], Birrell and Davies [47], and the review article by Ford [88]. On the other hand, a popular-level introduction by Hawking himself may be found in Ref. [115]. Bekenstein [42] gives an accessible introduction to the entropy of black holes, including a history of how the idea emerged. Information, black holes, and the holographic universe are introduced in Susskind and Lindesay [229].

Problems

12.1 Estimate the present entropy of the Sun. By how large a factor would the entropy increase if the Sun were compressed into a spherical black hole?

12.2 What is the characteristic density associated with the Planck scale? How does this compare to the density of a neutron star (the most dense object known that is not a black hole)?

12.3 Suppose that spherical Hawking black holes of various masses were created in the big bang. If one of these black holes decayed away by Hawking radiation on a timescale of 500,000 years (near the time for decoupling of radiation and matter; see Section 20.1.2), what was its temperature and the radius of its event horizon at creation?

12.4 What is the gravitational acceleration produced by a Hawking black hole with a lifetime comparable to the age of the Universe at a distance of one meter from the black hole?

12.5 The Sun has a surface temperature of about 5800 K and radiates nearly as a blackbody with a power of 3.828×10^{26} W. If the Sun were replaced by a spherical black hole of the same mass radiating only by the Hawking mechanism, what would the new blackbody temperature of the Sun be and by what factor would the power output of the Sun change?

12.6 What is the initial mass and power of a black hole that decays away completely by Hawking radiation in a time of one second?

12.7 Because a Hawking black hole has a thermodynamic temperature, it can radiate a net mass only if it has a temperature higher than its surroundings. What is the mass of the Hawking black hole that would be in equilibrium with the 2.7 K cosmic microwave background, emitting on average just as much radiation as it received? ***

12.8 Assuming a blackbody temperature given by Eq. (12.4) and Schwarzschild geometry, use the usual Stefan–Boltzmann law to prove that the power radiated by a Hawking black hole is given by Eq. (12.5), and thus that the mass emission rate is given by Eq. (12.6). ***

12.9 Deriving the Hawking temperature

$$T = \frac{\hbar c^3}{8\pi k_B G M}$$

of Eq. (12.4) rigorously from quantum field theory requires more sophisticated theory than has been introduced here. However, its form can be made plausible by the following considerations. Assume a particle–antiparticle pair having a rest mass $\sim m$ to be created near the event horizon of a spherical black hole. Use the uncertainty principle to estimate the separation Δr of the virtual pair after a time Δt permitted by the uncertainty principle, if the particle and antiparticle are assumed to separate at light velocity. Using Newtonian gravity, estimate the differential (tidal) gravitational force ΔF acting over this distance. Then, show that requiring the rest mass energy $E \sim mc^2$ to be comparable to $\Delta F \cdot \Delta r$ near the event horizon implies, up to constant factors, Eq. (12.4). ***

in the absence of a horizon. The existence of such a *naked singularity* is potentially disruptive since one does not know how to discuss physics in the vicinity of such an object. This concern, coupled with many supportive examples, lies at the basis of the *cosmic censorship hypothesis*: Nature conspires to "censor" spacetime singularities in that all such singularities come with event horizons that render them invisible to the outside universe (see Section 11.5). Although there are no clear violations of the cosmic censorship hypothesis either observationally or theoretically, it cannot at this point be derived from any more fundamental concept and must be viewed as only an hypothesis. One kind of thought experiment that supports the cosmic censorship hypothesis is to attempt to add angular momentum to a Kerr black hole in excess of the maximum angular momentum permissible for existence of the horizon. Attempts to do this theoretically in realistic scenarios generally have failed. Such thought experiments give some anecdotal confidence in the censorship hypothesis.

13.1.4 The Kerr Horizon

The area of the horizon for a black hole is significant because of the area theorem discussed in Section 12.5 and 11.5. The Kerr horizon corresponds to a constant value of $r = r_+$ given by Eq. (13.6), and the horizon also is a surface of constant t since the metric is stationary. Setting $dr = dt = 0$ in the Kerr line element (13.1) gives the line element for the 2-dimensional manifold defining the horizon of the Kerr black hole,

$$d\sigma^2 = \rho_+^2 d\theta^2 + \left(r_+^2 + a^2 + \frac{2Mr_+ a^2 \sin^2\theta}{\rho_+^2}\right)\sin^2\theta d\varphi^2$$

$$= \rho_+^2 d\theta^2 + \left(\frac{2Mr_+}{\rho_+}\right)^2 \sin^2\theta d\varphi^2, \quad (13.8)$$

where ρ_+^2 is given by ρ in Eq. (13.2) with r set to $= r_+$. The metric tensor is then

$$g = \begin{pmatrix} \rho_+^2 & 0 \\ 0 & \left(\frac{2Mr_+}{\rho_+}\right)^2 \sin^2\theta \end{pmatrix}. \quad (13.9)$$

This is not the metric of a 2-sphere; even though the horizon has constant Boyer–Lindquist coordinate $r = r_+$, its intrinsic geometry is not spherical (see Problem 13.4).

Example 13.2 From the discussion of invariant integration in Section 3.13.1, the area of the Kerr horizon is

$$A_K = \int_0^{2\pi} d\varphi \int_0^\pi \sqrt{\det g}\, d\theta = 2Mr_+ \int_0^{2\pi} d\varphi \int_0^\pi \sin\theta\, d\theta$$

$$= 8\pi M r_+ = 8\pi M(M + \sqrt{M^2 - a^2}), \quad (13.10)$$

where $\det g$ is evaluated from Eq. (13.9).

The corresponding horizon area for a Schwarzschild black hole is obtained from the result in Example 13.2 by setting $a = 0$ (corresponding to vanishing angular momentum) in Eq. (13.10), in which case $r_+ = 2M$ and

$$A_S = 16\pi M^2, \tag{13.11}$$

as would be expected for the surface area of a sphere having radius $2M$.

13.2 Particle and Photon Motion

A similar approach as for the Schwarzschild metric may be followed to determine the orbits of particles and photons, though the algebra is more complicated in the Kerr geometry because it has less symmetry than the Schwarzschild geometry.

13.2.1 Orbits in the Kerr Metric

In the Schwarzschild geometry the overall spherical symmetry of the metric required that the orbital angular momentum of a test particle be conserved. This in turn meant that particle motion was confined to a plane, chosen conveniently by setting $\theta = \frac{\pi}{2}$. The Kerr metric has only axial symmetry, not full rotational symmetry, and in the general case the orbits in the Kerr metric will not be confined to a plane. Nevertheless, to illustrate basic principles while keeping the discussion simple only motion in the equatorial plane ($\theta = \frac{\pi}{2}$) will be considered here. The Kerr line element subject to that restriction is then

$$ds^2 = -\left(1 - \frac{2M}{r}\right)dt^2 - \frac{4Ma}{r}d\varphi dt + \frac{r^2}{\Delta}dr^2 + \left(r^2 + a^2 + \frac{2Ma^2}{r}\right)d\varphi^2.$$

There are two conserved quantities arising from the symmetries of the metric and corresponding to scalar products of Killing vectors with the 4-velocity u,

$$\epsilon = -K_t \cdot u \qquad \ell = K_\varphi \cdot u, \tag{13.12}$$

where at large distances ϵ may be interpreted as the energy per unit rest mass and ℓ as the angular momentum component per unit rest mass along the symmetry axis (which is the same as the total angular momentum per unit rest mass for the special case of equatorial orbits). The corresponding Killing vectors, K_t and K_φ, were given in Eqs. (13.3)–(13.4). In addition, the norm of the 4-velocity provides the usual constraint $u \cdot u = -1$ for timelike particles. From the Kerr metric (13.1),

$$-\epsilon = g_{00}u^0 + g_{03}u^3 \qquad \ell = g_{30}u^0 + g_{33}u^3, \tag{13.13}$$

and solving for $u^0 = dt/d\tau$ and $u^\varphi = d\varphi/d\tau$ leads to

$$\frac{dt}{d\tau} = \frac{1}{\Delta}\left[\left(r^2 + a^2 + \frac{2Ma^2}{r}\right)\epsilon - \frac{2Ma}{r}\ell\right], \tag{13.14}$$

$$\frac{d\varphi}{d\tau} = \frac{1}{\Delta}\left[\left(1 - \frac{2M}{r}\right)\ell + \frac{2Ma}{r}\epsilon\right]. \tag{13.15}$$

Similar steps as those applied in Section 9.3 to analyze particle motion in the Schwarzschild metric then give for *timelike particles* [111]

$$\frac{\epsilon^2 - 1}{2} = \frac{1}{2}\left(\frac{dr}{d\tau}\right)^2 + V(r, \epsilon, \ell), \tag{13.16}$$

$$V(r, \epsilon, \ell) = -\frac{M}{r} + \frac{\ell^2 - a^2(\epsilon^2 - 1)}{2r^2} - \frac{M(\ell - a\epsilon)^2}{r^3}, \tag{13.17}$$

as the corresponding equations of motion in the Kerr metric. A similar approach can be used to determine the motion of photons in the Kerr metric, with the primary difference being that the massless photons are now required to satisfy $u \cdot u = 0$. Let us skip the details and just quote the result; for *lightlike particles* [111],

$$\frac{1}{\ell^2}\left(\frac{dr}{d\lambda}\right)^2 = \frac{1}{b^2} - V(r, b, \sigma), \tag{13.18}$$

$$V(r, b, \sigma) \equiv \frac{1}{r^2}\left[1 - \frac{a^2}{b^2} - \frac{2M}{r}\left(1 - \frac{\sigma a}{b}\right)^2\right], \tag{13.19}$$

where λ is an affine parameter, $b \equiv |\ell/\epsilon|$ is the impact parameter, and $\sigma = \text{sign}\,\ell$ indicates whether the photon moves with the rotational motion of the black hole ($\sigma = +1$) or against it ($\sigma = -1$).

13.2.2 Frame Dragging

A particularly striking feature of the Kerr solution is *frame dragging*, where the rotating black hole drags space itself with it as it rotates. This concept was introduced earlier in Section 9.12.2, where it was shown that frame dragging by the Earth's weak, slowly rotating gravitational field leads to Lense–Thirring precession of gyroscopes. That was a measurable but tiny effect. As we will now demonstrate, frame dragging can assume dramatic proportions in the Kerr spacetime. This arises ultimately because the metric corresponding to (13.1) contains off-diagonal components $g_{03} = g_{30} \ne 0$ that couple time and angular degrees of freedom. One consequence is that a particle dropped radially into a Kerr black hole will acquire non-radial components of motion as it falls freely in the gravitational field, *even if no forces act on it*.

Let us determine $d\varphi/dt$ for a particle dropped from rest at infinity ($\epsilon = 1$) with zero angular momentum ($\ell = 0$) into a Kerr black hole. For 4-momenta in the Kerr metric

$$p^\varphi \equiv p^3 = g^{3\mu} p_\mu = g^{33} p_3 + g^{30} p_0 \qquad p^0 = mu^0 = mdt/d\tau$$
$$p^t \equiv p^0 = g^{0\mu} p_\mu = g^{00} p_0 + g^{03} p_3 \qquad p^3 = mu^3 = md\varphi/d\tau$$

and combining these relations gives an expression for $d\varphi/dt$,

$$\frac{d\varphi}{dt} = \frac{p^3}{p^0} = \frac{g^{33} p_3 + g^{30} p_0}{g^{00} p_0 + g^{03} p_3}. \tag{13.20}$$

From Eq. (13.13), $p_3 = mu_3 = g_{30} p^0 + g_{33} p^3 = 0$ for an $\ell = 0$ particle and thus

$$\omega(r, \theta) \equiv \frac{d\varphi}{dt} = \frac{g^{30}}{g^{00}} = \frac{g^{\varphi t}}{g^{tt}}. \tag{13.21}$$

The quantity $\omega(r,\theta)$ measures the amount of frame dragging. We may view it as the *angular velocity of a zero angular momentum particle*. Frame dragging has been exhibited here for a Kerr metric but is expected to occur for any metric that has terms depending on angular momentum. In Section 9.12.2 frame dragging effects were derived for a slowly rotating Schwarzschild metric approximated by Eq. (9.72). The corresponding small frame dragging predicted for the Earth's rotating gravitational field was confirmed by Gravity Probe B, as described in Box 9.3.

Example 13.3 The Kerr line element is expressed in terms of the covariant components of the Kerr metric $g_{\mu\nu}$. To evaluate $\omega(r,\theta)$ in Eq. (13.21) requires the contravariant components, which may be obtained by matrix inversion of $g_{\mu\nu}$ [see Eq. (3.41)]. The metric is diagonal in r and θ so

$$g^{rr} = g_{rr}^{-1} = \frac{\Delta}{\rho^2} \qquad g^{\theta\theta} = g_{\theta\theta}^{-1} = \frac{1}{\rho^2}, \qquad (13.22)$$

and it is only necessary to evaluate the matrix inverse

$$g^{-1} = \begin{pmatrix} g_{tt} & g_{t\varphi} \\ g_{\varphi t} & g_{\varphi\varphi} \end{pmatrix}^{-1}$$

to obtain the other non-zero entries for $g^{\mu\nu}$. Letting

$$D = \det g = g_{tt} g_{\varphi\varphi} - (g_{t\varphi})^2, \qquad (13.23)$$

the matrix inverse is

$$g^{-1} = \frac{1}{D} \begin{pmatrix} g_{\varphi\varphi} & -g_{t\varphi} \\ -g_{\varphi t} & g_{tt} \end{pmatrix}. \qquad (13.24)$$

Insertion of explicit expressions for the $g_{\mu\nu}$ and some algebra yields (see Section 11.3 of [221])

$$D = -\Delta \sin^2\theta \qquad g^{tt} = -\frac{(r^2+a^2)^2 - a^2\Delta \sin^2\theta}{\rho^2 \Delta}$$
$$g^{t\varphi} = -a\frac{2Mr}{\rho^2 \Delta} \qquad g^{\varphi\varphi} = \frac{\Delta - a^2 \sin^2\theta}{\rho^2 \Delta \sin^2\theta}. \qquad (13.25)$$

Therefore, the angular velocity (13.21) that measures the amount of frame dragging in the Kerr metric is

$$\omega(r,\theta) = \frac{g^{\varphi t}}{g^{tt}} = \frac{2Mra}{(r^2+a^2)^2 - a^2\Delta \sin^2\theta}, \qquad (13.26)$$

which has the same sign as the Kerr parameter $a = J/M$ and falls off as r^{-3} for large r.

The preceding example indicates that frame dragging effects in the Kerr spacetime are most pronounced at smaller r. We will now demonstrate that a region just outside the event horizon has quite remarkable properties as a consequence of this.

13.2.3 The Ergosphere

With a capable-enough propulsion system an observer could remain stationary with respect to infinity at any point outside the event horizon of a Schwarzschild black hole. However, once inside the event horizon the observer is drawn inexorably to the central singularity and no amount of propulsion can enable an observer to remain stationary inside a Schwarzschild horizon. We shall now show that for a rotating black hole it may be impossible for an observer to remain stationary with respect to infinity, even outside the Kerr horizon. (Practically, infinity may be taken to be an inertial frame defined by the distant stars.) For a stationary observer only the time component of the 4-velocity u^μ is non-zero,

$$u^\mu = (u^0, 0, 0, 0) = (dt/d\tau, 0, 0, 0). \tag{13.27}$$

This 4-velocity is a unit timelike vector with the normalization $u \cdot u = -1$. Writing this expression out for the Kerr line element (13.1) and 4-velocity (13.27) gives

$$u \cdot u = g_{00}(u^0)^2 = -1, \tag{13.28}$$

where g_{00} is explicitly

$$g_{00} = -\left(1 - \frac{2Mr}{\rho^2}\right) = -\left(1 - \frac{2Mr}{r^2 + a^2 \cos^2\theta}\right). \tag{13.29}$$

From Eq. (13.29), g_{00} vanishes if the condition $r^2 - 2Mr + a^2 \cos^2\theta = 0$ is satisfied, implying that g_{00} is equal to zero on the surface given by

$$r_e(\theta) = M + \sqrt{M^2 - a^2 \cos^2\theta}, \tag{13.30}$$

and exceeds zero inside it. Thus, for any observer having a timelike u the condition (13.28) *cannot be satisfied* if $r \leq r_e(\theta)$ and that observer cannot be stationary.

Comparing Eq. (13.6) with Eq. (13.30), if $a \neq 0$ the surface $r_e(\theta)$ lies outside the horizon r_+ except at the poles, where the two surfaces are coincident. Figure 13.1 illustrates. The region lying between $r_e(\theta)$ and the horizon r_+ is termed the *ergosphere,* for reasons that will become more apparent shortly, and the outer boundary (13.30) of the ergosphere is called the *static limit,* since from preceding considerations there can be *no stationary observers within the ergosphere.* The ergosphere for some different values of a/M are illustrated in Fig. 13.2. As illustrated in this sequence, in the limit $a = J/M \to 0$ the angular momentum of the Kerr black hole vanishes, $r_e(\theta) \to 2M$, and it becomes coincident with the Schwarzschild horizon. Rotation has extended the region $r < 2M$ of the spherical black hole where no stationary observers can exist to a larger region $r < r_e(\theta)$ surrounding the rotating black hole where no observer can remain at rest with respect to infinity because of frame dragging effects. But this interesting ergospheric region lies *outside* the horizon, so a particle could enter it and still escape without being captured by the black hole. The implications of this possibility will be explored in Section 13.3.

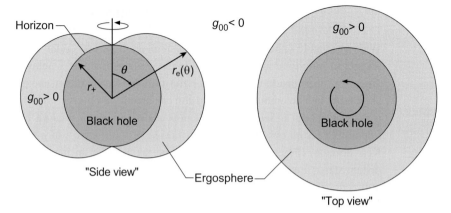

Fig. 13.1 The ergosphere (light gray regions) of a Kerr black hole with $a/M = 0.998$ in Boyer–Lindquist coordinates. The geometry is not correctly illustrated. For example, the horizon has a constant coordinate $r = r_+$ but its geometry is not spherical; see Eq. (13.8).

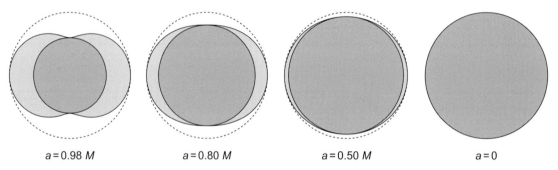

Fig. 13.2 Ergospheres of a Kerr black hole for different values of a/M at constant mass M. The region inside the horizon (the black hole) is indicated by dark gray and the ergosphere is indicated in lighter gray. The dashed circle defines the equivalent Schwarzschild radius $r = 2M$. As $a \to 0$ the outer boundary of the ergosphere and the Kerr horizon approach the event horizon of a Schwarzschild black hole of the same mass.

13.2.4 Motion of Photons in the Ergosphere

We may gain deeper insight by considering the motion of photons within the ergosphere. To simplify the discussion it is assumed that the photons move tangent to a circle at constant r and in the equatorial plane ($\theta = \frac{\pi}{2}$). Since they are photons, $ds^2 = 0$ and from the Kerr line element with $dr = d\theta = 0$,

$$g_{00}dt^2 + 2g_{03}dtd\varphi + g_{33}d\varphi^2 = 0.$$

Dividing both sides by dt^2 gives a quadratic equation in $d\varphi/dt$ that may be solved to give

$$\frac{d\varphi}{dt} = -\frac{g_{03}}{g_{33}} \pm \sqrt{\left(\frac{g_{03}}{g_{33}}\right)^2 - \frac{g_{00}}{g_{33}}}, \qquad (13.31)$$

where the positive sign is associated with motion in the opposite sense and the negative sign with motion in the same sense as the rotation of the black hole. Now g_{00} vanishes at the boundary of the ergosphere and it is positive inside the boundary. Setting $g_{00} = 0$ in Eq. (13.31) gives two solutions,

$$\frac{d\varphi}{dt} = 0 \qquad \frac{d\varphi}{dt} = -\frac{2g_{03}}{g_{33}}, \qquad (13.32)$$

where the first solution corresponds to propagation of the photon opposite to the black hole rotation and the second to propagation in the same sense as the black hole rotation. The photon sent backwards against the black hole rotation at the surface of the ergosphere is stationary in the φ coordinate! Obviously a particle, which must have a velocity less than that of a photon, must revolve with the black hole irrespective of its angular momentum. Inside the ergosphere $g_{00} > 0$, so all photons and particles must rotate with the black hole. In essence, the frame dragging for $r < r_e(\theta)$ is so severe that speeds in excess of lightspeed would be required for an observer to remain at rest with respect to infinity.

13.3 Extracting Rotational Energy from Black Holes

It was argued in Chapter 12 that quantum fluctuations of the vacuum allow mass to be extracted from a Schwarzschild black hole in the form of Hawking radiation, but it is impossible to remove mass from a *classical* Schwarzschild black hole because of the event horizon. However, as will now be explained, the existence of separate surfaces defining the outer boundary of the ergosphere and the horizon for a Kerr black hole implies the possibility of extracting rotational energy from the black hole by an entirely *classical* process.[3]

13.3.1 Penrose Processes

A simple way to demonstrate the feasibility of extracting rotational energy from a black hole is through a *Penrose process*, which is illustrated schematically in Fig. 13.3. In this thought experiment a particle falls into the ergosphere of a Kerr black hole and decays into two particles, one of which falls through the horizon and one of which exits the ergosphere and escapes to infinity [111]. The decay within the ergosphere is a local process that, by equivalence principle arguments, may be analyzed in a freely falling reference frame according to the usual rules of scattering theory. Therefore, energy and momentum are conserved in the decay, implying that in terms of 4-momenta p

$$p_0 = p_1 + p_2 \qquad (13.33)$$

---

[3] Kerr black holes also emit Hawking radiation but the mechanisms to be discussed now are purely classical and operate in addition to the Hawking mechanism. For stellar black holes the Hawking process is negligible but a Penrose process need not be.

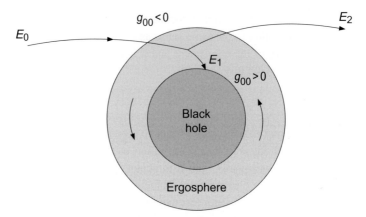

Fig. 13.3 An example of a Penrose process. It can be arranged that $E_2 > E_0$, thereby extracting rotational energy from the black hole.

at the point of decay (the subscripts are particle labels, not coordinate labels). If the particle that scatters to infinity has rest mass m_2, its energy is $E_2 = -p_2 \cdot K_t = m_2 \epsilon$ where Eq. (13.12) has been used. Taking the scalar product of Eq. (13.33) with K_t yields

$$E_2 = E_0 - E_1. \tag{13.34}$$

If particle 1 were to scatter to infinity instead of crossing the event horizon, E_1 would necessarily be positive and therefore $E_2 < E_0$. However, for the Killing vector $K_t = (1, 0, 0, 0)$ in the Kerr spacetime,

$$K_t \cdot K_t = g_{\mu\nu} K_t^\mu K_t^\nu = g_{00} K_t^0 K_t^0 = g_{00}$$

and K_t is a *spacelike vector within the ergosphere*, since then $K_t \cdot K_t = g_{00} > 0$. By arguments similar to those given in Section 12.2.1 concerning Hawking radiation, $-E_1$ is not an energy within the ergosphere but may be interpreted as a component of spatial momentum, which can have either a positive or negative value. For those decays where the trajectories are arranged such that $E_1 < 0$, Eq. (13.34) implies that $E_2 > E_0$ and net energy is extracted in the Penrose process.

It may be shown (for example, by analyzing data taken by a set of observers co-rotating with the ergosphere) that the energy extracted in the Penrose process comes at the expense of the rotational energy of the Kerr black hole. For those trajectories having $E_1 < 0$, the particle captured by the black hole adds a negative angular momentum and energy, thus reducing the angular momentum and total energy of the black hole by amounts that just balance the angular momentum and total energy carried away by the escaping particle. In principle, a series of such Penrose events can extract all the angular momentum of a Kerr black hole, thereby reducing it to a Schwarzschild black hole. No further energy can then be extracted from the resulting spherical black hole (except by Hawking radiation).

13.3.2 Practical Energy Extraction

The Penrose process is important because it establishes proof of principle in a simple model that the rotational energy of Kerr black holes is accessible in classical processes. Practically, it is thought that Penrose processes are not likely to be important as energy sources because the specific conditions required for net energy to be emitted are not easily realized in known astrophysics environments. Instead, it is likely that the primary sources of emitted energy from black hole systems come from complex electromagnetic coupling of rotating black holes to external accretion disks and jets, and from the gravitational energy released by accretion onto black holes. Chapter 15 will take seriously the possibility of extracting energy from black holes and will address them, not so much in terms of the mind-bending alteration of our understanding of space and time that has been emphasized to this point, but rather as very practical "little engines that could," capable of powering some of the most energetic and important phenomena in astrophysics.

Background and Further Reading

Much of the discussion in this chapter has been patterned after that in Schutz [221] and Hartle [111]. At a more advanced level, see Misner, Thorne, and Wheeler [156]. The geometry of Kerr black holes is treated at a more mathematical level in O'Neill [169] and Visser [239].

Problems

13.1 The Kerr line element is sometimes written as

$$ds^2 = -\frac{\Delta}{\rho^2}\left(dt - a\sin^2\theta\, d\varphi\right)^2 + \frac{\sin^2\theta}{\rho^2}\left((r^2+a^2)d\varphi - a\,dt\right)^2 + \frac{\rho^2}{\Delta}dr^2 + \rho^2 d\theta^2$$

(see, for example, Eq. (33.2) of Reference [156]). Show that this form of the line element is equivalent to Eq. (13.1).

13.2 Derive an expression for the upper limit on the energy that could be obtained from a Kerr black hole by extracting all of its angular momentum. Show that for an extremal Kerr black hole up to 29% of its mass could in principle be extracted. *Hint*: The horizon cannot shrink, by entropy conservation. The maximum energy extraction will occur if the horizon size remains constant as angular momentum is removed.

13.3 Supply the missing steps leading from Eqs. (13.14) and (13.15) to Eqs. (13.16) and (13.17) that describe particle motion in the equatorial plane of the Kerr metric.

13.4 Use the horizon surface defined in Eq. (13.8) to find the distance around the equator, the distance around a meridian through the poles, and their ratio for an extremal Kerr black hole. ***

13.5 Classical centripetal accelerations (which can lead to distortions of gravitating bodies) vary as the square of the angular momentum J. The Kerr metric has terms that are linear and higher in J, so for very slow rotations the metric may be approximated by keeping only terms linear in J. To this order in J there is no centripetal distortion but the metric is altered by the rotation. Show that in this slow-rotation limit the Kerr metric reduces to the (spherical) Schwarzschild metric plus a rotational correction term proportional to J. Show, by interpreting the angular momentum of the slowly rotating spherical body in Newtonian terms, that the correction to the spacetime curvature outside the spherical body produced by this term is approximately $(v/c)C_0$, where C_0 measures the curvature produced by the spherical mass distribution alone and v is the velocity associated with the rotation. Thus show that the curvature of spacetime outside a rotating body depends on both its mass and its rotational velocity. Show that this effect, which is representative of *gravitomagnetic effects* (see Section 9.12.1), is one order higher in powers of $(1/c)$ than gravitational redshift effects.

13.6 Find the relationship between the proper radial distance and the Boyer–Lindquist coordinate radial distance dr for the Kerr metric if $\theta = \frac{\pi}{2}$. Evaluate it for an extremal Kerr black hole. Show that it reduces to the corresponding relationship in the Schwarzschild metric if the angular momentum of the black hole vanishes.

13.7 Consider a particle dropped from rest at infinity with zero initial orbital angular momentum into a Kerr black hole. Derive an expression for $d\varphi/dr$. *Hint*: Invoke the conservation laws implied by the Killing vector relations (13.12).

13.8 Show that the Kerr metric (13.1) reduces to the Minkowski metric for flat space

$$ds^2 = -dt^2 + dr^2 + r^2(d\theta^2 + \sin^2\theta\, d\varphi^2),$$

in the limit $r \gg M$ and $r \gg a$. Thus the Kerr spacetime is asymptotically flat. ***

14 Observational Evidence for Black Holes

The Schwarzschild and Kerr spacetimes have been shown to have black hole solutions. In addition there was earlier work such as the numerical simulations by Oppenheimer and collaborators beginning in the late 1930s giving the first glimmers that the collapse of stars having sufficient mass would lead to final states that would now be termed black holes (see Box 11.1). Thus by the early 1960s when the Kerr solution was obtained there was ample theoretical reason to believe that black holes might exist and be observable. However, the observational astronomy community paid little attention and general relativity remained, as it always had been, an esoteric backwater of theoretical physics with a rather small set of practitioners. This chapter describes how that situation was changed dramatically by a series of observational and theoretical discoveries that began in the 1960s.

14.1 Gravitational Collapse and Observations

Three major discoveries pushed gravitational collapse and general relativity to the forefront of observational astronomy: (1) the realization in 1963 that quasars (Section 15.4) were enormous energy sources lying at great distance, (2) the discovery of pulsars (Section 10.4) in 1967, and (3) discoveries beginning in the 1970s of X-ray binaries containing massive unseen companions (Section 14.4). Attention soon focused on the possibility that gravitationally collapsed objects (black holes in the first case, neutron stars in the second, and black holes or neutron stars in the third) could explain these new discoveries. Quasars in particular elevated the question of whether black holes actually formed in the Universe to a top priority because of the proposal that they might be powered by rotating, extremely massive black holes (see Chapter 15).[1]

14.2 Singularity Theorems and Black Holes

But do black holes *really* exist? After all, both Einstein (the most famous physicist of his day) and Eddington (the most famous astronomer of his day) felt strongly that such

---

[1] The Kerr solution was published in 1963, the same year that the interpretation of quasars as distant sources emitting enormous energy was established. The events in the 1960s and 1970s described in these paragraphs led – a decade after Einstein's death – to a resurgence of interest in general relativity that continues today. Indeed, some would say that general relativity has enjoyed a "golden age" since the 1960s, culminating in 2015 with establishment of gravitational wave astronomy as a new observational discipline (see Chapter 24).

solutions of the field equations existed mathematically but would never be realized in nature. Let's play devil's advocate for a moment. The Einstein equations can be solved to yield Schwarzschild or Kerr black holes primarily because they are *vacuum solutions with a high degree of symmetry.* Since the Kerr solution reduces to the Schwarzschild solution in the limit of vanishing angular momentum, it may be taken as representative.

As discussed in Chapter 13, because of its high symmetry the Kerr solution is remarkably simple, being described by only two parameters, one representing the effective mass of the gravitational-field source and one corresponding to angular momentum per unit mass. This led to worries that the gravitational collapse producing a black hole was an anomaly created by unrealistically high symmetry. Hence the theoretical black holes of general relativity might not exist in the wild because real gravitational collapse would likely not proceed with such high symmetries, and more realistic asymmetric solutions might somehow avoid the formation of singularities. For example (it was argued), perhaps in a more realistic treatment of collapse for spinning matter centrifugal effects might partially offset gravity and prevent collapse to a singularity. Help in this regard emerged from highly mathematical work that began in the 1960s and is associated with the names of Roger Penrose, Stephen Hawking, Brandon Carter, and others.

14.2.1 Global Methods in General Relativity

The classical formulation of general relativity is in terms of *local differential equations.* To build up a global picture of a spacetime one traditionally performs local computations around each event in spacetime and then patches these solutions together to give the global structure. This local formulation of general relativity was almost the only approach until the 1960s, when it began to be realized that general relativity was also subject to powerful *global laws* that were often remarkably simple and elegant. These techniques permit the global properties of a spacetime to be investigated *directly,* rather than by building up from local properties.

Example 14.1 The conciseness and elegance of results obtained using global methods may be illustrated by the *second law of black hole dynamics* discussed in Section 12.5.3, which can be stated simply as "In an isolated system, the sum of areas for black hole horizons can never decrease." This result was derived by Hawking using the global methods to be discussed in Section 14.2.3. It is a global and not local statement about the spacetime because establishing whether a true horizon exists is not possible using only local properties of the spacetime, as discussed in Box 14.1.

14.2.2 Singularities and Trapped Surfaces

In 1965, Roger Penrose introduced *global topological techniques* into the study of spacetimes, and used these new methods to prove the first of the singularity theorems mentioned in Section 11.5. The details are far too technical for the present discussion but a brief qualitative overview will prove useful. A key idea proposed by Penrose was

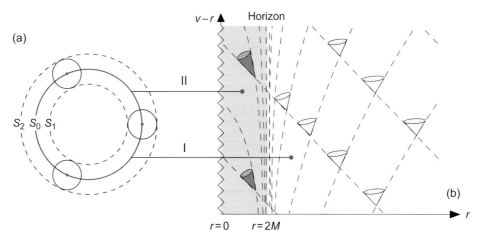

Fig. 14.1 (a) Illustration of light emission by a 2-sphere S_0 in Schwarzschild spacetime. (b) The Schwarzschild spacetime in Eddington–Finkelstein coordinates from Fig. 11.3. Each point corresponds to a 2-sphere of angular coordinates. There are two solutions at each point. Outside the horizon at $r = 2M$ one solution is ingoing and one is outgoing; inside $r = 2M$ both solutions are ingoing. Cases I and II correspond to S_0 located outside and inside the horizon, respectively. For a trapped surface all emitted light moves inward. This is fairly obvious for the highly symmetric Schwarzschild solution, but Penrose showed that such trapped surfaces can form under more general conditions.

that of a *trapped surface,* which is a closed, spacelike 2-surface from which any light emitted always moves inward [182]. The basic idea is illustrated by Fig. 14.1, which is adapted from the discussion in d'Inverno [76]. Suppose a 2-sphere S_0 corresponding to a point in Fig. 11.3 of a Schwarzschild spacetime emits light. Photons will propagate away from S_0 in 2-spheres whose envelopes will form the 2-spheres S_1 and S_2. In case I the surface S_0 is outside the event horizon at $r = 2M$ and the areas of the 2-spheres will be ordered $S_2 > S_0 > S_1$. However, S_0 is inside the horizon for case II and the orientation of lightcones implies that both the emitted wavefronts will implode and in general both S_2 and S_1 will have areas *less than* S_0. In this latter case S_0 is termed a *closed trapped surface,* since all light emitted from it (even "outgoing" light) travels inward toward the singularity. The outermost trapped surface for a spacetime is termed the *apparent horizon.* Box 14.1 discusses the relationship between the event horizon and the apparent horizon for a black hole.

The preceding illustration is based on the highly symmetric Schwarzschild solution but Penrose showed that under very general conditions (like positivity of energy, so that light is always focused by gravity; see Box 7.3), *once a trapped surface forms a singularity must form inside it.* Spacetimes admitting trapped surfaces exhibit *geodesic incompleteness,* meaning that some worldlines for particles or light can be extended only for a finite proper time (finite affine parameter for light) for an observer on the geodesic. Penrose interpreted this to be the end of time (no future) at a singularity where physical law breaks down. Because this derivation invokes minimal assumptions, it indicates that singularities (with attendant horizons if cosmic censorship is valid) are generic features

> **Box 14.1** **Apparent Horizons and True Horizons**
>
> If it exists, the outermost trapped surface defines an *apparent horizon* for a spacetime. How are apparent and true event horizons for a black hole related?
>
> **Local and Global Criteria**
>
> From the discussion in Section 14.2.2, the apparent horizon is a *local* concept because it depends on tests made locally. However, the black hole event horizons discussed in Chapters 11–13 are defined *globally*, since they depend on constructing null geodesics and determining whether they reach infinity or not. Thus determining the true event horizon requires knowing the complete history of the spacetime.
>
> **Apparent and True Horizons Might or Might Not Coincide**
>
> For the simple case of stationary Schwarzschild spacetime the apparent and true horizons are located at the same radial coordinate, $r = 2M$, so we can speak of just "the horizon," but in more complex situations this might not be true. An example is a black hole accreting a spherical shell of matter. Imagine a light ray emitted radially from just outside the horizon before the shell accretes. When emitted it is not trapped, but after the shell accretes the gravitational field acting on the light is increased and may trap the light, even though it was not trapped when emitted. In this case the apparent horizon is inside the true event horizon. Roughly, we may think of the apparent horizon as defining the light-trapping radius for the black hole now, while the true event horizon is that radius in the future.
>
> **Utility of Apparent Horizons**
>
> When computing the dynamical evolution of a spacetime from initial data in numerical relativity (Section 8.8.3), the full global history of the spacetime is not available as the solution is advanced. In such a context the local definition of an apparent horizon is useful because it permits determining the presence of a black hole at any time during the numerical simulation. This is so because it can be shown that if the cosmic censorship conjecture is valid (Section 11.5) and the null energy condition is satisfied (Box 7.3), an apparent horizon signals the existence of a true event horizon that either coincides with or lies outside the apparent horizon [26].

of any collapse where gravity is sufficiently strong, not just for ones with special initial conditions.

The overall picture that emerged from such considerations was that uncharged collapsing matter radiates away its irregularities as gravitational waves, leaving finally a Kerr black hole with a singularity clothed by an event horizon. Thus it was suggested that the final equilibrium configuration of *any* fully collapsed, uncharged object is a Kerr black hole (or its $J \to 0$ limit, a Schwarzschild black hole), irrespective of initial asymmetries. These ideas made it much more plausible that gravitational collapse could form black holes under real-world conditions. This in turn spurred increased interest in observational evidence for the existence of black holes that will be the subject of this chapter and the next.

14.2.3 Generalized Singularity Theorems

Singularity theorems typically assume: (1) a global structure for spacetime, (2) a restriction placed on allowed energies such as those discussed in Box 7.3, and (3) gravity sufficiently strong to trap a region of spacetime in the sense described above. By making different choices for these assumptions, different singularity theorems may be obtained. Such generalizations were used to show that *classically* (this might be modified by quantum gravity at the Planck scale) the expanding Universe must have a singularity in its past, even if initial conditions were far less symmetric than normally assumed. This will be relevant to discussion of the big bang in Chapter 20. They also were used by Hawking to prove the area theorem for black hole horizons which, along with Hawking's introduction of quantum mechanics into black hole physics (Chapter 12), was crucial to the development of black hole thermodynamics (see Box 12.3).

14.3 Observing Black Holes

Armed with some theoretical justification that black holes might form for realistic instances of gravitational collapse, we now turn to whether observations support their existence. An isolated black hole would be difficult to observe directly. However, black holes often should be accreting surrounding matter and interacting gravitationally with their environment, which are potentially observable properties. Thus, although the direct observation of a natural black hole is difficult and the production of a black hole in a laboratory is far beyond any present technology (unless the behavior of gravity at very short distances is fundamentally different from that observed so far), there are strong reasons to believe that black holes of astrophysical origin exist and are being observed indirectly on a regular basis. Most such observations fall into three categories:

1. Binary star systems that are strong X-ray sources where there is evidence for an unseen companion too massive – as inferred from its gravitational influence on the accompanying visible star – to be a white dwarf or neutron star.
2. Observational anomalies in the centers of many galaxies, where very large masses inferred from average star velocities (or even direct measurement of individual star orbits in the center of the Milky Way) exist, often accompanied by evidence of an enormous energy source in the center of the galaxy. These are difficult to explain by any hypothesis simpler than the presence of a supermassive black hole.
3. Observation of gravitational waves with measured properties indicating that they could have originated only in the merger of two black holes.

We now consider some of the reasons why these observations give confidence that black holes exist. This chapter will address evidence from the first two of the above categories, and the third category of gravitational waves as a probe of black hole properties will be

14.4 Stellar-Mass Black Holes

A large amount of indirect evidence suggests the existence of *stellar black holes*: objects having the properties of black holes with masses comparable to that of the more massive stars ($\sim$10–100 $M_\odot$). Much of this evidence comes from observation of high-mass X-ray binary (HMXB) systems, which are binary star systems having large total mass that are strong X-ray sources (see Chapter 18 of Ref. [105] for an introduction). Most such systems have been observed as *spectroscopic binaries*, in which one object is a visible star and the other is a compact object – usually a neutron star or black hole – that is not imaged directly but is inferred from periodic Doppler shifts of spectral lines in the visible companion.

14.4.1 Masses for Compact Objects in X-Ray Binaries

From Kepler's laws, the *mass function* $f(M)$ for a spectroscopic binary star system may be related to the observed radial velocity curve through

$$f(M) = \frac{(M \sin i)^3}{(M + M_c)^2 (1 - e^2)^{3/2}} = \frac{PK^3}{2\pi G}, \tag{14.1}$$

where K is the semiamplitude and P the period of the observed radial velocity curve, e is the eccentricity and i is the tilt angle of the orbit relative to the observer, M_c is the mass of the visible companion star, and M is the mass of the unseen component. The utility of the mass function is that in Eq. (14.1) the value of the right side can be obtained by direct observation of the radial velocity curve and the left side is a function of the masses in the system. Thus, by measuring the velocity curve it is possible to learn something about the masses in the binary. In particular, it will be shown in Section 14.4.3 that the measured value of the mass function places *lower limits* on the mass, even absent further information.

The X-ray emission from HMXB systems results from significant accretion onto the compact companion, which implies a relatively small separation between components of the binary. Many close binaries have eccentricity $e \simeq 0$ because tidal interactions tend to circularize elliptical orbits, so to keep things simple in the examples to be discussed the orbit of the binary will usually be assumed circular, in which case Eq. (14.1) reduces to

$$f(M) = \frac{(M \sin i)^3}{(M + M_c)^2} = \frac{M \sin^3 i}{(1 + q)^2} = \frac{PK^3}{2\pi G} \quad \text{(circular orbits)}, \tag{14.2}$$

where for convenience in later discussion the *mass ratio* $q \equiv M_c/M$ has been introduced. The tilt angle i is illustrated in Fig. 14.2(a) and is defined so that a binary orbit seen edge-on has $i = \frac{\pi}{2}$ and one perpendicular to the line of sight has $i = 0$. An observed velocity curve for a binary system is shown in Fig. 14.2(b). For a spectroscopic binary the angle i is not known unless eclipses are observed implying that $i \sim \frac{\pi}{2}$. (Conversely, absence of

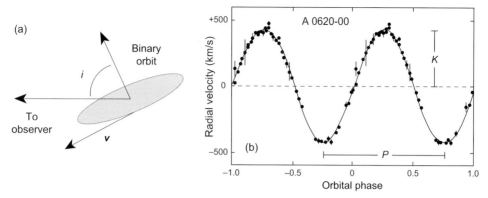

Fig. 14.2 (a) Tilt angle i of a binary orbit. (b) Observed heliocentric (referenced to position of the Sun) radial velocity curve for the black hole binary candidate A 0620–00 [150]. The period corresponds to $P = 0.323$ days and the semiamplitude to $K = 433 \pm 3$ km s^{-1}. The orbital phase corresponds to the fraction of one complete orbit. Reproduced from Blandford and Gehrels, Revisiting the Black Hole, *Physics Today* **52**, 6, 40 (1999), with the permission of the American Institute of Physics.

eclipses excludes angles near $i = \frac{\pi}{2}$.) Hence the mass function alone places only a lower limit on the mass of the unseen component of the binary.

Example 14.2 Let's compute the mass function for a binary with a period $P = 6$ hr and radial-velocity semiamplitude $K = 400$ km s^{-1}. Expressing Eq. (14.2) in convenient units,

$$f(M) = \frac{PK^3}{2\pi G} = 1.036 \times 10^{-7} \left(\frac{P}{\text{day}}\right) \left(\frac{K}{\text{km s}^{-1}}\right)^3 M_\odot. \tag{14.3}$$

Inserting $P = 0.25$ day and $K = 400$ km s^{-1} gives $f(M) = 1.66$. Then from Eq. (14.2) the mass M is constrained by $M^3 (\sin i)^3 = 1.66 (M + M_c)^2$. If the companion mass M_c and the tilt angle i can be inferred from other information, solution of this cubic equation gives an estimate for the mass M of the unseen companion.

Let us now illustrate the use of the mass function supplemented by additional information to deduce the mass of an unseen compact object in a binary system.

14.4.2 Masses from Mass Functions

Assume that from observation and Doppler-shift measurements of the visible component of a spectroscopic binary the quantity

$$F = F(P, K) = \frac{PK^3}{2\pi G} \tag{14.4}$$

has been determined by inferring the period P and velocity semiamplitude K from the velocity curve [see Fig. 14.2(b)]. Then from Eq. (14.2), $M^3 \sin^3 i / (M + M_c)^2 = F$ and the unknown compact-object mass M is defined by the cubic equation

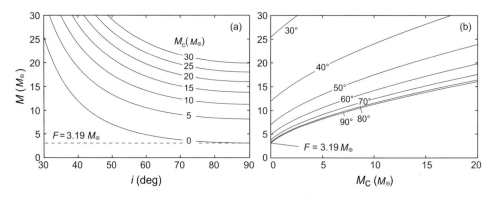

Fig. 14.3 Mass plots for A0620–00 using Eq. (14.6) with a measured $F = 3.19\,M_\odot$. (a) Mass M of the unseen component versus the tilt angle i for different values of the companion mass M_c. (b) Mass M of the unseen component versus M_c for different values of i.

$$M^3 + aM^2 + bM + c = 0$$

$$a = -\frac{F}{\sin^3 i} \qquad b = -\frac{2FM_c}{\sin^3 i} \qquad c = -\frac{FM_c^2}{\sin^3 i}. \tag{14.5}$$

This equation has three roots but the solution of interest for this problem is given by the real root

$$M(F, M_c, i) = \left(R + \sqrt{Q^3 + R^2}\right)^{1/3} + \left(R - \sqrt{Q^3 + R^2}\right)^{1/3} - \frac{a}{3}$$

$$R \equiv \frac{1}{54}(9ab - 27c - 2a^3) \qquad Q \equiv \frac{1}{9}(3b - a^2). \tag{14.6}$$

If F is measured, M is a function of the mass of the visible companion M_c and the tilt angle i for the orbit. If these can be measured or estimated in some way, Eq. (14.6) provides the mass M of the unseen component, or at least a constraint on it.

14.4.3 An Example: A0620–00

The soft X-ray transient A0620–00 is an interacting binary system in which a spectral class K main sequence star is transferring mass to an accretion disk around an unseen compact object [112, 150, 151, 152, 168]. A velocity curve is shown in Fig. 14.2(b). The orbital period is $P = 7.75 \pm 0.0001$ hr and the semiamplitude of the radial velocity curve is taken to be $K = 457 \pm 8\,\text{km s}^{-1}$ [151].[2] Thus from Eq. (14.4) the observed value of the mass function for A0620–00 is $F = 3.19\,M_\odot$. In Fig. 14.3 the solution $M(F, M_c, i)$ from Eq. (14.6) is plotted as a function of i and M_c for $F = 3.19\,M_\odot$. These figures illustrate clearly (1) the degeneracy of the unseen mass M in the parameters i and M_c, and (2) that the measured value of the mass function $F = 3.19\,M_\odot$ is the minimum possible mass for

---

[2] The value of K determined from Fig. 14.2(b) in Ref. [150] is $K = 433\,\text{km s}^{-1}$. Different authors determine values of K ranging from about 430 to 460 km s^{-1}. The example discussed here has used consistently the value $K = 457\,\text{km s}^{-1}$ from Ref. [151].

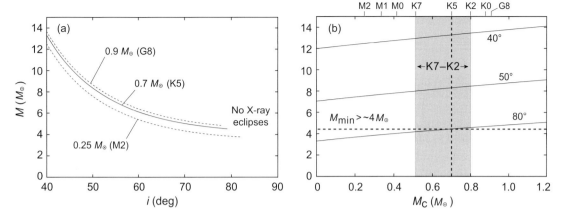

Fig. 14.4 Mass plots for A0620–00 using Eq. (14.6) with a measured $F = 3.19\,M_\odot$ and constraints on M_c and i. (a) Mass M of the compact object versus tilt angle i for various companion masses M_c. From systematics, the companion is of spectral class K5, with $M_c \sim 0.7\,M_\odot$. A region on the right is excluded by the absence of X-ray eclipses. (b) Mass M versus M_c for different values of i. The most likely spectral class K5 is indicated by the dashed vertical line, the uncertainty range K7–K2 is indicated by the shaded region, and the minimum implied mass M_{min} for the compact object is indicated by the horizontal dashed line.

the unseen component [corresponding to $f(M)$ when $M_c = 0$ and $i = \frac{\pi}{2}$], since all other combinations of i and M_c give a larger value for $M(F, M_c, i)$.

This already is a powerful constraint but a more precise statement about M is possible if further information can be obtained about M_c and i. Neither quantity is measured directly but two general considerations can be brought to bear. (1) The distance to the system is known to be ~ 770 pc [168] and no eclipses of the X-ray source are observed, which by geometry excludes values of i for a range too near $\frac{\pi}{2}$. (2) The companion is determined observationally to be a main sequence star with spectral class lying in the range K7–K2; from stellar systematics this requires the mass to lie in the range $0.5\,M_\odot < M_c < 0.8\,M_\odot$. In Fig. 14.4 the plots of Fig. 14.3 are repeated with these constraints displayed.

In Fig. 14.4(a) the mass M is plotted as a function of the tilt angle for companion masses M_c corresponding to a conservative range of spectral classes M2–G8, with the corresponding mass from stellar systematics indicated. It is concluded in Ref. [151] that the spectral class of the companion is K5, with an uncertainty ranging from K7 to K2. From stellar systematics, this implies a mass M_c indicated by the solid curve.[3] The curves are truncated for tilt angles greater than about $80°$ by the constraint that no X-ray eclipses are observed for the system, ruling out nearly edge-on orientations.

In Fig. 14.4(b) the mass M is plotted as a function of the companion mass M_c for several tilt angles $i \leq 80°$. The spectral classes corresponding to M_c are indicated at the top, the range of spectral classes consistent with the data is indicated by the shaded box, and the most likely spectral class K5 is indicated by the vertical dashed line. Since the

---

[3] The mass M is not very sensitive to the companion mass M_c in this example. This is generally true for massive binaries in which the companion is much less massive than the compact object (implying a highly asymmetric mass ratio q).

Table 14.1 Some black hole candidates in X-ray binaries [51]

X-ray source	Period (days)	$f(M)$	$M_c(M_\odot)$	$M_h(M_\odot)$
Cygnus X-1	5.6	0.24	24–42	11–21
V404 Cygni	6.5	6.26	~0.6	10–15
GS 2000+25	0.35	4.97	~0.7	6–14
H 1705–250	0.52	4.86	0.3–0.6	6.4–6.9
GRO J1655–40	2.4	3.24	2.34	7.02
A 0620–00	0.32	3.18	0.2–0.7	5–10
GS 1124–T68	0.43	3.10	0.5–0.8	4.2–6.5
GRO J0422+32	0.21	1.21	~0.3	6–14
4U 1543–47	1.12	0.22	~2.5	2.7–7.5

absence of X-ray eclipses rules out $i > 80°$ for spectral class K5 [see Fig. 14.4(a)], the dashed horizontal line $M \sim 4\,M_\odot$ corresponds to the minimum mass $M_{\min}$ of the unseen companion. This minimum holds for $i \sim 80°$; from Fig. 14.4, M can be considerably larger than $4\,M_\odot$ if the (unknown) tilt angle is greater than $80°$. In a more extensive analysis employing additional observational information and systematics [112], it was concluded that $4.16 \pm 0.01\,M_\odot \leq M \leq 5.15 \pm 0.015\,M_\odot$. Since it is expected that no neutron star can have a mass larger than $2-3\,M_\odot$, we deduce that the unseen companion must be a black hole.

14.4.4 Some Black Hole Candidates

Table 14.1 lists some binary systems where observations suggest a compact unseen companion of mass $M = M_h$ that is too massive to be a white dwarf or neutron star, and thus must be a black hole. The procedure by which it was concluded that one of these candidate binaries, A0620–00, harbors a black hole was described in Section 14.4.3, and a similar analysis for Cygnus X-1 is outlined in Box 14.2. As will be discussed in Chapter 24, it is now possible to detect gravitational waves from binary black hole mergers and to determine parameters such as the masses of the merging and final merged black hole from analysis of the data. A summary of more than 35 black hole masses determined from X-ray binary and gravitational wave data will be given later in Fig. 24.21.

14.5 Supermassive Black Holes

Many observations of star motion near the centers of galaxies indicate the presence of large, unseen mass concentrations. The most direct is for our own galaxy, where individual stars have been tracked for some time in the vicinity of the strong radio source Sgr A*, which is thought to lie very near the center of the galaxy.

Box 14.2 Swans and Black Holes: Cygnus X-1

The first-discovered and most storied stellar black hole candidate is Cygnus X-1 (Cyg X-1), the first X-ray source found in the constellation Cygnus (the Swan). It was discovered using Geiger counters during short rocket flights in the 1960s [54], but the initial systematic study was by the first X-ray satellite Uhuru in the early 1970s. Briefly, the steps used to conclude that Cyg X-1 contains a black hole were:

1. In the early 1970s the X-ray source Cygnus X-1 was investigated using the Uhuru satellite, with no optical counterpart identified [166].
2. In 1972 a radio source was discovered near Cyg X-1 and identified optically with the blue supergiant HDE 226868 (spectrum–luminosity class O9.7 Iab) [55, 119]. Time correlation of radio and X-ray activity suggested that HDE 226868 and Cyg X-1 were members of a binary star system. Doppler radial velocities for HDE 226868 indicated a 5.6-day orbital period, which was supported by periodic brightening and dimming of the two sources with the same periodicity [53, 245].
3. The X-ray intensity fluctuates on millisecond timescales so, by causality (Box 15.3), the source diameter is less than several hundred km. The small size and X-ray emission indicate that Cyg X-1 is a neutron star or black hole accreting from its companion; the figure below left displays an artist's conception [2].

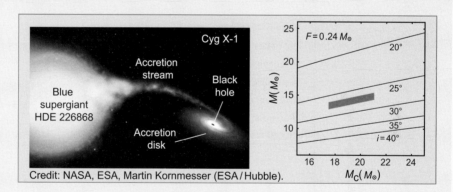

Credit: NASA, ESA, Martin Kornmesser (ESA/Hubble).

4. The mass M_c of HDE 226868 may be estimated from stellar systematics; then the mass function $f(M)$ (see Section 14.4, the right side of the figure above, and Problem 14.6) constrain the mass M of the unseen component. A more extensive analysis has concluded that the tilt angle $i = 27.1 \pm 0.8°$ and $M_c = 19.2 \pm 1\,M_\odot$ (shaded box in above-right figure), and that $M = 14.8 \pm 1.0\,M_\odot$ [172].
5. No plausible equation of state supports a neutron star with $M > 2$–$3\,M_\odot$, so it may be concluded that the $M = 14.8\,M_\odot$ unseen companion is a black hole.

This indirect reasoning builds a strong case that Cyg X-1 contains a black hole. Note finally that a comprehensive analysis has concluded that the black hole has a spin greater than 95% of the maximal value discussed in Section 13.1.2 [101]. Thus, Cyg X-1 may be a *near-extremal Kerr black hole*.

14.5.1 The Black Hole at Sgr A*

The star denoted S0-2 (also called S2) is a 15 solar mass main sequence star in a highly elliptical Keplerian orbit with Sgr A* near a focus. The orbit has been measured precisely using adaptive optics and speckle interferometry at near-IR wavelengths, that are able to penetrate the dust clouds that hide the center of the galaxy at optical wavelengths. The positions observed for S0-2 through 2002 are shown in Fig. 14.5, with dates in fractions of a year beginning in 1992. The orbit drawn in Fig. 14.5 corresponds to the projection of the best-fit ellipse with Sgr A* at a focus [218]. At closest approach (periapsis) S0-2 has an orbital velocity of 5000 km s^{-1} and its separation from Sgr A* is only 17 lighthours.

From the fits to the orbit of S0-2 assuming Keplerian motion, the mass contained inside the orbit of the star is determined to be approximately $4.3 \times 10^6 \, M_\odot$ [97]. The observed size of the orbit indicates that this mass must be concentrated in a region that cannot be much larger than the Solar System, and little luminous mass can be seen in this region. Constraints from orbits of other stars in this region place an upper limit of 45 AU on the size of the $4.3 \times 10^6 \, M_\odot$ source of the gravitational field [57], and long-baseline radio

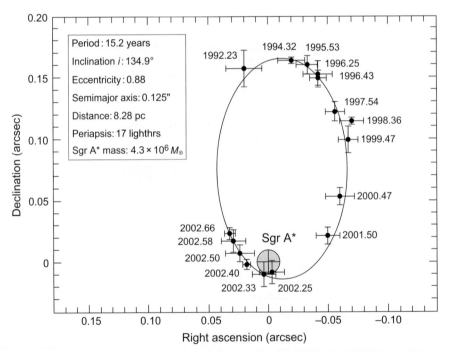

Fig. 14.5 Orbit of the star S0-2 around the radio source Sgr A* through 2002 [218]. The filled circle indicates the position uncertainty for Sgr A* as determined from the elliptical orbit assuming a point mass located at the focus to be responsible for the orbital motion. The star completed this orbit in 2008 and the parameters displayed in the box are those obtained from the completed orbit [97]. *Periapsis* is the general term for closest approach of an orbiting body to the center of mass about which it is orbiting. Reprinted by permission from Macmillan Publishers Ltd: *Nature* **419**, 694–696, copyright (2002).

> **Box 14.3** **Hair and Sgr A***
>
> It has been proposed that stars orbiting the black hole at Sgr A* could be used to test the no-hair theorems (Section 11.5) of black hole physics. The orbital planes for stars in short-period, high-eccentricity orbits near the black hole will precess because of frame dragging and quadrupolar gravity of the (assumed rotating) black hole. By analyzing the orbits of two or more such stars systematically, the angular momentum J and the quadrupole moment Q_2 could be determined, and that could be used to test a requirement of no-hair theorems that $Q_2 = -J^2/M$, where M is the mass of the black hole [252].

interferometry discussed below constrains the region to a size comparable to the event horizon. By far the simplest explanation is that the radio source Sgr A* coincides with an approximately 4.3 million solar mass black hole (which has a Schwarzschild radius only about 18 times larger than the Sun's radius).[4] Note that this periapsis distance is still well outside the tidal distortion radius for the star of about 16 light minutes, and is about 1500 times larger than the event horizon of the black hole.

A number of stars in addition to S0-2 are being tracked near Sgr A* and the accumulating data should allow further tests of general relativity. The star that comes closest to Sgr A* goes by the name S0-102 [154]. It has an orbital period of 11.5 years, which is 30% shorter than the period of S0-2, but S0-102 is more difficult to track because it is 16 times fainter than S0-2. As of 2016 both S0-2 and S0-102 had been tracked for more than one complete orbit, offering a more precise determination of the gravitational potential associated with Sgr A*. As you are asked to show in Problem 14.7, the gravity experienced by a star like S0-2 at closest approach to the black hole is approximately 100 times stronger than the largest gravitational fields found in the Solar System or in binary pulsars. Thus, these stars near Sgr A* offer the possibility of testing general relativity in much stronger gravity than in any context other than the black hole mergers to be discussed in Chapter 24. The general effects that can be measured are associated with curvature (orbital precession) and gravitational redshifts, though advances in instrumentation and analysis will probably be required to attain the required accuracy and precision. In addition, testing more exotic predictions of general relativity has been suggested; one possibility is discussed in Box 14.3.

As we will discuss below, even larger black holes with masses of order billions of solar masses have been inferred from systematic analysis of average stellar motion near the center of other large galaxies. However, the information from precise study of motion for individual stars near the center of our own galaxy is more direct and represents one of the best current proofs that black holes exist. Thus it is amusing that some of the best evidence for the existence of black holes may be for more exotic supermassive black holes than for "garden variety" black holes formed from collapse of massive stars.

---

[4] In principle other forms of mass could fit into the required volume (though it needs to be rather invisible), but simulations indicate that if the mass distribution were not a black hole it would collapse to a black hole on a timescale less than the age of the galaxy. Thus a black hole is the most plausible explanation.

14.5.2 The Water Masers of NGC 4258

The galaxy NGC 4258 (also cataloged as the Messier object M106) is an Sb spiral visible through a small telescope, but its nucleus is moderately active and it is also classified as a Seyfert 2 active galactic nucleus (AGN; see Section 15.5). It lies near the Big Dipper in the constellation Canes Venatici at a distance of around 7 Mpc, which makes it one of the nearest AGNs. This galaxy is remarkable because a set of *masers* has been observed in its central region. (A maser is the analog of a visible-light laser for light in the microwave region of the spectrum.) The maser emission in NGC 4258 originates in clouds of heated water vapor, so these are termed *water masers*. Such naturally occurring masers are important in astronomy because they produce sharp spectral lines, allowing their Doppler shifts to be measured precisely. Because of the sharp spectral lines, and because microwaves are not strongly attenuated by the gas and dust near the nucleus of the galaxy, observation of the water masers has permitted the motion of gas near the center of NGC 4258 to be mapped precisely using the Very Long Baseline Array (VLBA). The observational situation for NGC 4258 is summarized in Box 14.4.

The nucleus of the galaxy produces radio jets that appear to come from the dynamical center of the rotating disk and are approximately perpendicular to it [artist's conception (b) in the figure displayed in Box 14.4]. The radio emission at 1 cm wavelength is superposed on the drawing of the warped disk in image (a) of the figure in Box 14.4. The position of the radio jets, coupled with the precise location of the center of the disk, permits the position of the central black hole engine to be determined within the uncertainty of the black circle shown in image (a) of the figure in Box 14.4. This black circle denotes the uncertainty in location of the black hole, *not its size*. A black hole with $M \sim 10^7 \, M_\odot$ would have an event horizon orders of magnitude smaller than this.

The preceding results taken in concert make NGC 4258 one of the strongest cases for the presence of supermassive black holes in galactic cores. The argument for NGC 4258 and that for the star S0-2 described in Section 14.5.1 are similar conceptually. In both cases the magnitude of an unseen very large mass is being determined by following the motion of a "test particle" in the resulting gravitational field and assuming that the motion is Keplerian. In the first case the test particle is a star; in the second it is a set of water masers.

14.5.3 The Virial Theorem and Gravitating Mass

For distant galaxies we have insufficient telescopic resolution to track individual stars like S0-2, and we may not have the luck of finding powerful masers as in NGC 4258. But it is still possible to learn something about the mass contained within a particular region by observing the *average velocities of stars* in that region. This follows because it may be expected on conceptual grounds that the larger the gravitational field that stars feel, the faster they will move. This intuitive idea may be quantified in a precept known as the *virial theorem*, as will now be described.

According to the virial theorem applied to a group of gravitating objects like stars in galaxies or a globular cluster [62],

Box 14.4 Masers and the Black Hole in NGC 4258

Observation of maser emission from the central region of NGC 4258 using the VLBA leads to the results summarized in the following figure. Image courtesy of NRAO/AUI.

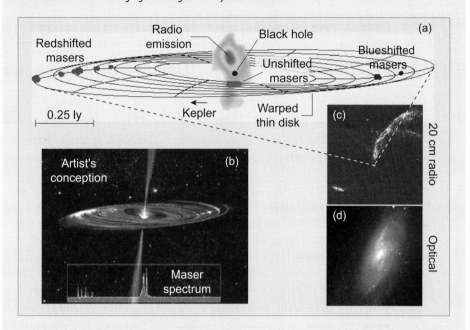

The maser positions are mapped with greater than 1 mas precision and their radial velocities indicate bulk gas motion near the center with $v \sim 1000$ km s^{-1}. From the Doppler shifts of the masers the rotation of the gas disk may be inferred (the label "Kepler"). The line-of-sight velocity inferred from Doppler shifts and the precise locations of the masers allow the orbital motion of gas around the galactic nucleus to be determined with high confidence. These measurements indicate that the masers are embedded in a dusty gas disk that is very thin and is warped. The orbits of the masers are found to obey Kepler's laws to better than 1%. From the Keplerian motion the central mass about which the masers revolve is determined to be $\sim 3.9 \times 10^7 \, M_\odot$.

$$\frac{1}{2}\left\langle \frac{d^2 I}{dt^2} \right\rangle - 2\langle K \rangle = \langle U \rangle \qquad K = \frac{1}{2}\sum_{i=1}^{N} m_i v_i^2 \qquad U = \frac{1}{2}\sum_{i}\sum_{j \neq i} G\frac{m_i m_j}{r_{ij}}, \quad (14.7)$$

where K is the kinetic energy and U is the potential energy, the angle brackets indicate a time average, and the moment of inertia is given by $I = \sum_i m_i r_i^2$. If the system is assumed to be in equilibrium $\langle d^2 I/dt^2 \rangle = 0$, and $-2\langle K \rangle = \langle U \rangle$. For enough stars approximately in equilibrium the central region will look the same at all times and the time average may be dropped. Therefore,

$$-2K = -2\sum_{i=1}^{N} \tfrac{1}{2} m_i v_i^2 = U.$$

If there are N stars of equal mass m and spherical symmetry is assumed with a radius R,

$$-\frac{m}{N}\sum_{i=1}^{N} v_i^2 = \frac{U}{N}.$$

Observationally, radial velocities are typically measured through Doppler shifts. Assuming that a given radial velocity is just as probable as a velocity in the two orthogonal directions gives $\langle v^2 \rangle \simeq 3\langle v_r^2 \rangle$, so that

$$\frac{1}{N}\sum_{i=1}^{N} v_i^2 = \langle v^2 \rangle = 3\langle v_r^2 \rangle \equiv 3\sigma_r^2 \qquad -3m\sigma_r^2 = \frac{U}{N},$$

where σ_r is the *radial velocity dispersion*. But for a spherical mass distribution $U \simeq -\tfrac{3}{5}GM^2/R$ and utilizing $M = Nm$, the velocity dispersion and the mass are related by

$$\sigma_r^2 \simeq \frac{GM}{5R}. \tag{14.8}$$

The mass M appearing in this equation is termed the *virial mass;* it may be viewed as an average mass responsible for the observed motion in a group of stars or other self-gravitating objects.

The virial theorem may be used to determine the masses of clusters of galaxies (see Example 17.3) but most relevant to the present discussion is that these methods may be applied statistically to high-resolution imaging of the central regions of galaxies, even when individual stars cannot be tracked, and thus used to estimate the mass contained in these regions. One example is given in Box 14.5 and others are given in Problem 14.4. The general outcome of many such studies is that large galaxies tend to have central regions containing enormous amounts of non-luminous mass. These masses are typically estimated to be millions to billions of solar masses, and to occupy regions that are comparable in size to the Solar System or smaller. The obvious interpretation is that most large galaxies have at their centers supermassive black holes of millions to billions of solar masses.

14.6 Intermediate-Mass Black Holes

Some evidence exists for black holes of mass intermediate between stellar black holes (~ 10–$100\,M_\odot$) and galactic black holes ($\sim$ millions to billions of $M_\odot$). A pulsar has been discovered orbiting an unseen mass concentration in the globular cluster 47 Tucanae [133], and the precise timing of the pulsar indicates that the magnitude of the unseen mass concentration is $2200^{+1500}_{-800}\,M_\odot$. This may be the first conclusive evidence for an intermediate-mass black hole. There is no electromagnetic signal from this object, so if it is a black hole it must not be accreting significant amounts of matter at present.

Box 14.5 Evidence for a Supermassive Black Hole in M87

As an example of virial theorem methods, let us consider the giant elliptical galaxy M87, located 16.5 Mpc away in the Virgo cluster of galaxies.

The Center of M87

The left portion of the figure below is a Hubble Space Telescope (HST) image showing the center of M87.

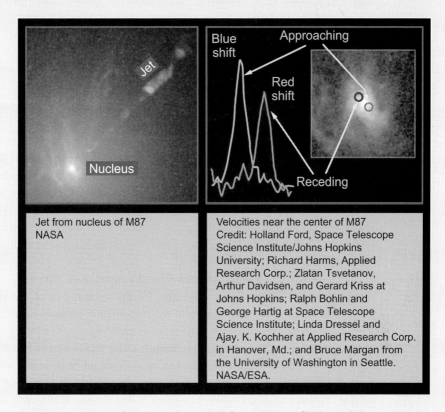

Jet from nucleus of M87
NASA

Velocities near the center of M87
Credit: Holland Ford, Space Telescope Science Institute/Johns Hopkins University; Richard Harms, Applied Research Corp.; Zlatan Tsvetanov, Arthur Davidsen, and Gerard Kriss at Johns Hopkins; Ralph Bohlin and George Hartig at Space Telescope Science Institute; Linda Dressel and Ajay. K. Kochher at Applied Research Corp. in Hanover, Md.; and Bruce Margan from the University of Washington in Seattle. NASA/ESA.

The diagonal line emanating from the nucleus in the left image is a jet of high-speed electrons (synchrotron jet) approximately 2 kpc long. This can be explained if matter is swirling around a supermassive black hole, with part of it falling into the black hole and part being ejected in the high-speed jet seen coming from the nucleus of M87.

Central Velocities

The right side of the above figure illustrates measurements on the central region of M87 suggesting rapid motion of matter. The measurement studied how light from the disk is redshifted and blueshifted by the Doppler effect, using the Faint Object Spectrograph aboard the HST. One side of the swirling disk spins in Earth's direction and the other side spins away from Earth, thus causing opposite Doppler shifts. The gas on one side of the disk is moving away at ~ 550 km s^{-1} (a redshift). The gas on the other side of the disk is approaching at the same speed (a blueshift).

> **A Large Central Virial Mass**
>
> The observed high velocities suggest a gravitational field produced by a mass concentration at the center of M87 far larger than can be ascribed to the effect of the visible matter there. Application of the virial methods described in Section 14.5.3 to these observations indicates approximately three billion solar masses concentrated in a region at the core of M87 that is only about the size of the Solar System. This is far too much mass to be accounted for by the visible matter and the simplest interpretation is that a $3 \times 10^9 \, M_\odot$ black hole lies at the center of M87.

14.7 Black Holes in the Early Universe

A comprehensive survey of black holes at large redshift is displayed in Fig. 14.6. This is a composite image of Chandra X-ray Observatory data for a region of the sky in Fornax having an angular extent somewhat less than the diameter of the full Moon; it is the furthest X-ray image ever obtained.[5] The black holes are of course invisible in this image but X-rays produced by gas being accelerated as it accretes mark the position of black holes. The central part of this image is thought to contain about 5000 black holes, which translates to of order one billion black holes potentially visible in X-rays out to this redshift over the whole sky.

The study combined long-exposure Chandra X-ray data with very deep field optical images from the Hubble Space Telescope for the same region of the sky. This allowed Chandra to study X-ray emission from more than 2000 galaxies identified by Hubble as being 12 to 13 billion light years distant (corresponding to redshifts $3.5 < z < 6.5$) [240]. About 70% of the X-ray objects are thought to be supermassive black holes ranging in mass from 10^5 to $10^{10} \, M_\odot$. The remainder are thought to be stellar-size black holes in X-ray binaries hosted by unresolved galaxies. In addition to providing further evidence for the existence of black holes, this deep X-ray survey of the early Universe reached two conclusions of potential importance for the role of supermassive black holes in the evolution of the Universe.

1. The preferred seeds for the growth of supermassive black holes were black holes of mass 10^4 to $10^5 \, M_\odot$, rather than the 10 to 100 $M_\odot$ seeds expected if supermassive black holes formed from merger of stellar black holes.
2. The growth of supermassive black holes in the first several billion years after the big bang tended to occur in sporadic episodes.

These conclusions are significant if they can be corroborated. The first favors the "heavy seeds" model of supermassive black hole formation (where intermediate-mass black

---

[5] Chandra data from different exposures at the positions of optically selected galaxies were stacked, allowing effective exposure times of $\geq 10^9$ s to be achieved [3, 240].

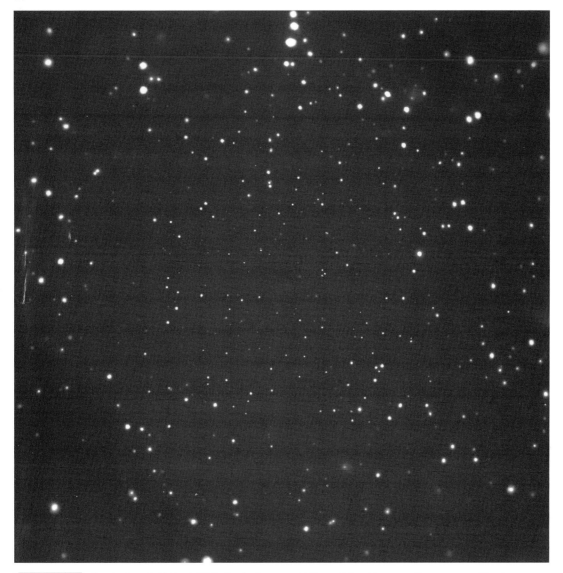

Fig. 14.6 Chandra Deep Field–South [3]. Each point lies at a redshift $z \geq 3.5$ and is a strong X-ray source marking the position of a supermassive black hole, or of a galaxy containing multiple X-ray binaries with accreting black holes. Credit: X-ray: NASA/CXC/Penn state/B. Luo *et al*.

holes form by direct collapse of gas clouds, or possibly by rapid merger of black holes formed from massive first-generation stars), and suggests that perhaps there is no simple connection between solar mass black holes that form in galaxies and supermassive black holes that form in the centers of those same galaxies. The second suggests that the rapid growth of supermassive black holes in the early Universe may have been greatly assisted by galaxy collisions and mergers, during which central black holes grew at very high rates, with much slower growth during intervening periods. However, this is but one set of data and the mechanism by which supermassive black holes form remains an open question.

14.8 Show Me an Event Horizon!

It may be concluded from the preceding discussion that compelling evidence exists for objects exhibiting many of the features expected for a black hole, and that these objects would be difficult to explain by any other hypothesis. However, the essential distinguishing observational characteristic of a black hole is its *event horizon,* which is expected to have properties unlike anything else encountered in the Universe.[6] Therefore, from a rigorous perspective, proof that black holes exist will require "imaging" the event horizon of one. This is a challenging task, since a black hole emits no light directly and is very small, and no nearby black hole candidates have been discovered.

The best prospects are for the approximately 4 million solar mass black hole at the center of the galaxy (at Sgr A*; see Section 14.5.1 above), which has a Schwarzschild radius of about 0.1 AU and a larger angular size as seen from Earth than for any other black hole candidate. The challenge is to resolve an object less than 20 times the diameter of the Sun at a distance of 8 kpc. This is daunting but perhaps not impossible. Very Long Baseline Interferometry (VLBI) using radio telescopes spread across the globe is beginning to achieve resolutions in the right ballpark, so imaging of the event horizon of the black hole at Sgr A* may be possible soon. The angular size undistorted by gravitational lensing would be about 10 microarcseconds (µas), but if the Sgr A* radio emission were collocated with the black hole gravitational lensing would be expected to increase the apparent size of the black hole radio image by at least a factor of 4 or 5. VLBI at 1.3 mm wavelength has reported an upper limit of 37^{+16}_{-10} microarcseconds for the size of the radio source [77]. This is somewhat smaller than the expected lensed event horizon, so it is speculated that the bulk of the radio emission from Sgr A* may have a source offset from the black hole, perhaps originating in an accretion disk around the black hole or in a jet emitted by it.

Suppose that a resolved image of the black hole has been obtained. What would we expect to see in this image that would identify it unequivocally as the event horizon of a black hole? From detailed analysis of the data from the gravitational wave event GW150914 (corresponding to the merger of 29 and 36 $M_\odot$ black holes) to be described in Chapter 24, a computer simulation of how the merger might have appeared to a nearby observer has been constructed. A frame extracted from that simulation is shown in Fig. 14.7. In this image all stars are in the distant background and are unaffected by the black holes, but gravitational lensing severely distorts the path of their light and hence

---

[6] Normal objects have *surfaces,* but a black hole does not. It is useful conceptually to view its event horizon as a one-way membrane but this is nothing more than analogy; the event horizon is not a material object but rather is entirely a consequence of curved spacetime (recall: Schwarzschild and Kerr black holes are solutions of the *vacuum* Einstein equations). The absence of a true surface might lead to observational effects allowing a black hole to be distinguished from a neutron star (in a spectroscopic binary, for example) [197]. Such arguments also have been used to conclude that the luminosity of Sgr A* at submillimeter wavelengths is too large to be explained by accretion onto a surface, implying that Sgr A* has no surface and is a black hole [57].

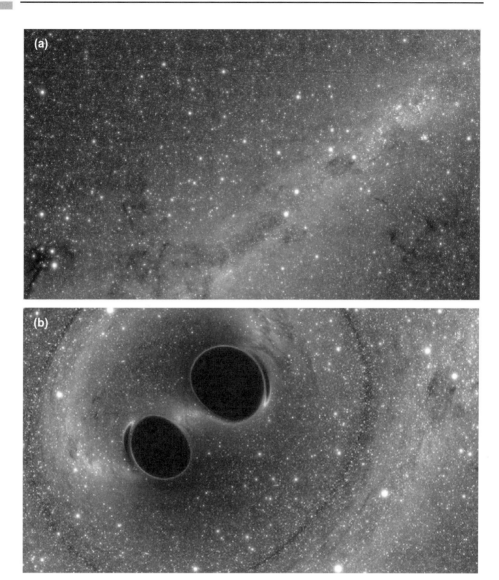

Fig. 14.7 Computer simulation of how the two black holes might have appeared to a nearby observer just prior to merger in the gravitational wave event GW150914, which will be described in Chapter 24. (a) Background stars without the black holes. (b) Image including black holes. The ring around the black holes is an *Einstein ring* resulting from strong gravitational lensing of the light from stars behind the black holes. Image extracted from video in [69]. A more extensive sequence of images for the merger is given later in Fig. 24.8. Image credit: SXS, the Simulating eXtreme Spacetimes (SXS) project (www.black-holes.org).

apparent positions in (b). The dark, well-defined shapes are the shadows of the event horizons blocking all light from behind. Flattened dark features around them and the marked displacement of the apparent background star images are caused by gravitational lensing arising from the strong curvature near the black holes.

For the simulation illustrated in Fig. 14.7 the two black holes are assumed to be isolated, with any accompanying material long since having been accreted by the black holes, so the only other objects are distant stars in the background. Hence the image is dominated by shadowing of light from background stars by the black hole event horizons, and by the extremely strong gravitational lensing effects produced by the black holes. The environment near Sgr A* is not likely to be as pristine, since it is a single black hole that is likely accreting some surrounding matter and producing radiation from this accretion at various wavelengths. It might be expected that the dominant feature would be the complete and sharply defined shadowing of light by the event horizon of the black hole, strongly distorted by gravitational lensing, as suggested by Fig. 14.7. It is unknown how the local environment of Sgr A* will alter this picture, and whether features clearly suggestive of an event horizon would survive.

14.9 A Circumstantial but Strong Case

The evidence cited in this chapter is highly suggestive but is not yet iron-clad proof that black holes exist. The defining characteristic of a black hole is an event horizon and no known black hole candidates are near enough to make imaging the event horizon easy, though angular resolution sufficient to resolve the event horizon of the Milky Way's central black hole at Sgr A* may be eminent. Nevertheless, data on X-ray binary systems and on the motions of stars and other objects in the central region of our galaxy and other nearby ones provide extremely strong circumstantial evidence for the existence of black holes. To put it succinctly, "The existence of black holes in our universe is generally accepted – by now it would be hard for astronomers to run the universe without them" [169]. In Chapter 24 evidence will be presented that gravitational waves from the merger of two black holes into a single Kerr black hole now have been observed. This provides even more direct evidence for the existence of black holes and of their properties, since the strongest gravitational waves in this merger would have been emitted from near the event horizons of the black holes.

Background and Further Reading

An accessible introduction to observational evidence for black holes may be found in Blandford and Gehrels [51]. The methodology of determining the mass of the unseen companion in binary systems is illustrated in Haswell *et al.* [112] and McClintock *et al.* [151].

Problems

14.1 The binary system Cygnus X-1 has a total mass $M \sim 30 - 50\, M_\odot$, an orbital period of 5.6 days, and is a strong X-ray source with fluctuations in the X-ray intensity on a timescale of milliseconds. Assuming the validity of Kepler's laws and the total mass to be 40 $M_\odot$, what is the average separation of the components and what is the maximum size of the X-ray emitting region based on these observations? *Hint*: See Box 15.3.

14.2 Show that the binary star mass function $f(M) = PK^3/2\pi G$ may be expressed in convenient units as

$$f(M) = 1.036 \times 10^{-7} \left(\frac{P}{1 \text{ day}}\right) \left(\frac{K}{\text{km s}^{-1}}\right)^3 M_\odot,$$

where P is the binary orbital period in days, K is the semiamplitude of the radial velocity in units of 100 km s^{-1}, and $f(M)$ is in solar mass units [see Eq. (14.2) and Fig. 14.2]. From the data in Fig. 14.2 and Table 14.1, estimate the mass function and thus the lower limit for the mass of the black hole in the X-ray binary A 0620–00.

14.3 Use the parameters for S0-2 given in Fig. 14.5 to compare the distance of closest approach with the event horizon and tidal distortion radii of the black hole, assuming it to be spherical with a mass $4.3 \times 10^6 M_\odot$. How does the approximately 5000 km s^{-1} orbital speed of S0-2 at closest approach to the black hole compare with the average speed of Earth on its orbit? How does the horizon radius of the Sgr A* black hole compare with the radius of the Sun and the radius of the orbit of Mercury?

14.4 (a) The central 0.8 pc of the spiral galaxy M31 (Andromeda) exhibits an average velocity dispersion of approximately 240 km s^{-1}. Estimate the mass contained in this region using the virial theorem. (b) A typical velocity dispersion in the nucleus of a Seyfert galaxy (see Section 15.5.2) is found to be ~ 1000 km s^{-1} within a region less than 100 pc in diameter. Estimate the mass contained within this region. ***

14.5 Use Kepler's laws to derive the mass function relation in Eq. (14.2), assuming circular orbits for the binary. *Hint*: How are the observed radial velocity curves related to the radii and period of the orbits?

14.6 Look up velocity curve information for Cygnus X-1 from Ref. [172] and repeat the analysis of Section 14.4.3 to place limits on the mass of the unseen compact object. Assume for this analysis that the blue supergiant companion mass lies in the range $M_c = 20$–$30\ M_\odot$, and that the tilt angle lies in the range $i = 25$–$35°$. ***

14.7 Use the quantity $\epsilon \equiv GM/Rc^2$ introduced in Eq. (6.5) to characterize the intrinsic strength of a gravitational field to show that the gravity acting on the star S0-2 at closest approach to the black hole at Sgr A* is about two orders of magnitude stronger than either the strongest gravity found in the Solar System (at the surface of the Sun), or the gravity acting on the pulsar in the Binary Pulsar (Section 10.4.1) at closest approach to the other neutron star. ***

15 Black Holes as Central Engines

As we have explored in preceding chapters, the properties of black holes imply a fundamental modification of our understanding of space and time, and bizarre new twists on traditional concepts of causality. But at a more pedestrian level black holes also are of great practical importance in astronomy because they can be prolific sources of power, with an efficiency for converting mass to energy that is exceeded only by that for direct matter–antimatter annihilation (an impractical energy source for astronomy, since the Universe seems to contain almost no antimatter). As we will now discuss, rotating black holes corresponding to the Kerr solution of the field equations are believed to be the engines powering a diverse set of phenomena associated with quasars, active galaxies, and gamma-ray bursts. Thus once again we encounter the paradox highlighted in Chapter 1: many of the most powerful events observed in the Universe are driven by what is by far the weakest force in the Universe. In addition to the practical importance of these black hole engines, many of these phenomena provide strong indirect evidence for the existence of black holes, since in most cases it would be difficult to explain the enormous and compact source of energy without them.

15.1 Black Hole Energy Sources

Since black holes have event horizons that cut their interiors off from communication with the outside world, they might seem unlikely power sources. However, the discussion of Penrose processes in Section 13.3 demonstrates conceptually that it is possible to extract energy from rotating classical black holes. Penrose processes themselves are not likely to lead to large extraction of energy from black holes in astrophysical environments. However, there are two general classes of phenomena that can lead to large energy release for black holes in regions outside the event horizon, where this energy might be accessible to external processes.

1. Accretion onto black holes can convert large amounts of gravitational energy into other usable forms of energy before the matter falls through the event horizon.
2. A black hole that is rotating may be accompanied by strong external magnetic fields that can tap the rotational power of the black hole to accelerate charged particles to high velocity in relativistic jets along the axes of rotation, outside the event horizon. One specific way to do this is through the *Blandford–Znajek mechanism* [50], where compression of magnetic field lines near the event horizon can lead to acceleration of charged particles to relativistic velocities.

A consideration of plausible formation mechanisms for astrophysical black holes suggests that they form with some amount of angular momentum but no net electrical charge. Thus practically we may focus attention on Kerr black holes, with the Schwarzschild solution viewed as the zero angular momentum limit of the Kerr solution. Let us now turn to an overview of extracting energy from rotating black holes, either through release of gravitational energy by accretion, or through coupling of their rotation to external fields.

15.2 Accretion and Energy Release for Black Holes

In astrophysics, accretion of matter onto a compact object such as a neutron star or black hole can be a large source of power because it is an efficient mechanism for gravitational energy conversion. Since angular momentum must be conserved, particles accreting onto a massive object from a binary companion typically form an *accretion disk* of particles orbiting the object. (A more extensive discussion of accretion and accretion disks is given in Chapter 18 of Ref. [105].) Collisions of the particles in the accretion disk heat it and cause it to radiate energy away, allowing particles to spiral inward and accrete onto the compact object.

For white dwarfs or neutron stars, the inspiraling material accumulates on the surface (which can lead in some cases to violent explosions such as novae or Type Ia supernovae for white dwarfs, and X-ray bursts for neutron stars). For black holes there is no surface but the event horizon sets an obvious boundary for energy extraction. Although no energy can be obtained externally once accreting material falls through the event horizon of a black hole, calculations indicate that large amounts of accretion energy (that is, gravitational energy) can in principle be extracted from inspiraling matter before it falls through the event horizon, through emission of radiation or of relativistic jets of matter (which may involve strong magnetic fields).

Accretion also can serve as a *storage reservoir* that can release energy over a longer timescale than is normally associated with direct gravitational collapse. For example, the long-period gamma-ray bursts to be discussed in Section 15.7 can last ~ 100 seconds and are likely powered by core collapse of a massive, rapidly rotating star. The standard picture is that core collapse leads to a rotating black hole surrounded by an accretion disk that emits ultrarelativistic jets on its rotation axis. The black hole jets, powered by accretion of matter from the disk over a period of tens of seconds, produce the gamma-ray burst over a timescale much longer than that for collapse directly to a black hole.

15.2.1 Maximum Energy Release for Spherical Accretion

For a spherical gravitating object of mass M and radius R, the energy released by accreting a mass m onto the object may be approximated as

$$\Delta E_{\text{acc}} = G \frac{Mm}{R}. \tag{15.1}$$

Table 15.1 Newtonian gravitational energy for hydrogen accretion

Accretion onto	Max energy released (erg g^{-1})	Ratio to fusion
Spherical black hole	1.5×10^{20}	25
Neutron star	1.3×10^{20}	20
White dwarf	1.3×10^{17}	0.02
Normal star	1.9×10^{15}	10^{-4}

In Table 15.1 the amount of energy released per gram of hydrogen accreted onto the surface of various objects is summarized, and in the last column this is compared with the energy released per gram by burning hydrogen into helium. The entries in the table show that accretion onto very compact objects is a much more prolific source of energy than is hydrogen fusion or accretion onto normal stars or white dwarfs. The accretion luminosity is

$$L_{\rm acc} = \frac{GM\dot{M}}{R} \simeq 1.3 \times 10^{21} \left(\frac{M/M_\odot}{R/{\rm km}}\right) \left(\frac{\dot{M}}{{\rm g\,s^{-1}}}\right) \ {\rm erg\,s^{-1}}, \qquad (15.2)$$

for a steady accretion rate $\dot{M}$, assuming that all energy generated by conversion of gravitational energy is radiated from the system (efficiency for realistic accretion will be addressed shortly in Section 15.2.3).

15.2.2 Limits on Accretion Rates

Accretion onto compact objects releases energy that is often emitted from the system as radiation. Since radiation produces a pressure, there is a radiation luminosity where the corresponding radiation pressure is just able to counteract gravity and halt the infall of matter. This critical luminosity is called the *Eddington luminosity*. For a spherical gravitating object the Eddington luminosity may be estimated as

$$L_{\rm edd} = \frac{4\pi GMm_{\rm p}c}{\sigma}, \qquad (15.3)$$

where σ is the effective cross section for photon scattering near the surface and $m_{\rm p}$ is the proton mass. For fully ionized hydrogen, σ may be approximated by the cross section for Thomson scattering of electrons, giving

$$L_{\rm edd} \simeq 1.3 \times 10^{38} \left(\frac{M}{M_\odot}\right) \ {\rm erg\,s^{-1}}. \qquad (15.4)$$

If the Eddington luminosity is exceeded (in which case the luminosity is said to be *super-Eddington*), accretion will be blocked by the radiation pressure, implying that there is a maximum accretion rate on compact objects. Equating $L_{\rm acc}$ and $L_{\rm edd}$ gives for this maximum rate

$$\dot{M}_{\rm max} \simeq 10^{17} \left(\frac{R}{{\rm km}}\right) \ {\rm g\,s^{-1}}. \qquad (15.5)$$

Eddington-limited accretion rates for white dwarfs and neutron stars estimated from this formula are given in Table 15.2. The Eddington limit is significant because some of the

Table 15.2 Some Eddington-limited accretion rates

Compact object	Radius (km)	Max accretion rate (g s^{-1})
White dwarf	$\sim 10^4$	10^{21}
Neutron star	~ 10	10^{18}

most luminous objects in the sky may derive their energy from accretion onto a central black hole at rates approaching the Eddington limit.

15.2.3 Accretion Efficiencies

For the gravitational energy released by accretion to be extracted from the system it must be radiated, or matter must be ejected at high kinetic energy (for example, in relativistic jets). Generally, such processes are inefficient and only a fraction of the energy potentially available from accretion can be extracted to do external work. This issue is particularly critical when black holes are the central accreting object, since they have no "surface" onto which accretion may take place and the event horizon makes energy extraction acutely problematic. An efficiency η for conversion of accreted mass into energy may be introduced by writing Eq. (15.2) for the accretion luminosity as

$$L_{\text{acc}} = \frac{GM\dot{M}}{R} = \eta \dot{M} c^2 \qquad \eta \equiv \frac{GM}{Rc^2}. \tag{15.6}$$

Specializing for a spherical black hole, it is logical to take some multiple of the Schwarzschild radius

$$r_{\text{s}} = \frac{2GM}{c^2} = 2.95 \left(\frac{M}{M_\odot}\right) \text{ km}, \tag{15.7}$$

to define the "accretion radius" R, since any energy to be extracted from accretion must be emitted from outside the horizon. Then for a spherical black hole

$$L_{\text{acc}}^{\text{bh}} = \eta \dot{M} c^2 \qquad \eta = \frac{r_{\text{s}}}{2R}, \tag{15.8}$$

and $r_{\text{s}}/2R$ measures the efficiency of the conversion of rest mass to energy by the accreting black hole power source. A typical choice for the accretion radius R is the radius $3r_{\text{s}}$ of the innermost stable circular orbit for the Schwarzschild spacetime, which was given in Eq. (9.37).

15.2.4 Accretion onto Rotating Black Holes

For thermonuclear burning of hydrogen to helium the mass-to-energy conversion efficiency is $\eta \sim 0.007$. For compact spherical objects like Schwarzschild black holes or neutron stars, reasonable estimates suggest $\eta \sim 0.1$. For rotating, deformed black holes it is possible to be more efficient in energy extraction. The property of particle orbits in the Kerr metric that is of most direct interest in astrophysical environments is the binding energy of the innermost stable circular orbit, since this is related to the maximum energy that can be extracted from accretion of matter onto a rotating black hole. (Circular orbits are of most interest because orbits of particles in accretion disks typically are rapidly circularized by interactions.) If the radius of the innermost stable orbit is denoted by R, for circular orbits we require that $dr/d\tau = 0$ and from Eq. (13.16)

$$\tfrac{1}{2}(\epsilon^2 - 1) = V(R, \epsilon, \ell). \tag{15.9}$$

Furthermore, for orbits to remain circular the radial acceleration must vanish and to be a stable orbit the potential must be a minimum, which requires that

$$\left.\frac{\partial V}{\partial r}\right|_{r=R} = 0 \qquad \left.\frac{\partial^2 V}{\partial r^2}\right|_{r=R} \geq 0, \tag{15.10}$$

with equality holding in the second expression for the innermost stable orbit. This set of equations (15.9)–(15.10) may be solved to determine the innermost stable orbit and its binding energy. For extremal black holes ($a = M$), the result is that [111]

$$\epsilon = \frac{1}{\sqrt{3}} \qquad \ell = \frac{2M}{\sqrt{3}} \qquad R_{\text{ISCO}} = M, \tag{15.11}$$

for co-rotating orbits (accretion orbits revolving in the same sense as the rotation of the black hole, which are more stable than the corresponding counter-rotating orbits). This may be compared with the value $R_{\text{ISCO}} = 6M$ given in Eq. (9.37) for the innermost stable circular orbit in a Schwarzschild spacetime.

Example 15.1 In the Kerr solution the quantity ϵ is the energy per unit rest mass measured at infinity, so the binding energy per unit rest mass, B/M, is given by

$$\frac{B}{M} = 1 - \epsilon. \tag{15.12}$$

Equations (15.11) and (15.12) imply that the fraction f of the rest mass that theoretically could be extracted from the energy released in a transition from a distant unbound orbit to the innermost circular bound orbit of a Kerr black hole is

$$f = 1 - \epsilon = 1 - \frac{1}{\sqrt{3}} \simeq 0.42. \tag{15.13}$$

Therefore, in principle 42% of the rest mass of accreted material could be extracted as usable energy by accretion onto an extremal Kerr black hole.

We may expect less efficiency than in the preceding idealized example for actual situations but it should be possible to achieve as much as 20–30% efficiency for optimal accretion scenarios. This should be compared with the maximum theoretical efficiency of 6% for accretion on a spherical black hole and 0.7% for burning hydrogen to helium in the conversion of rest mass to usable energy. Accretion onto rotating black holes is potentially a very efficient mechanism for converting mass to energy.

15.3 Jets and Magnetic Fields

A magnetic field cannot be anchored to a black hole itself because the event horizon quickly pinches off any magnetic field lines that cross it. However, the whirling plasma of the accretion disk lies outside the event horizon and it could have a strong magnetic field. The accretion of matter by the rotating black hole and the direct tapping of rotational energy by the magnetic field can lead to ejection, at velocities approaching the speed of light along the poles of the black hole rotation, for the portion of the matter that does not cross the event horizon. The rotating magnetic field of the accreting disk can be carried away with the ejected matter, leading to bipolar jets that are perpendicular to the accretion axis.

These jets contain charged particles moving at near the speed of light and twisting and spiraling magnetic fields. The magnetic fields probably are essential to focusing and confining the relativistic jets into the narrow cones that are observed, but the details of how this happens are not very well understood. If the jets are not at right angles to the line-of-sight, the one pointed more toward the observer is termed the jet and the opposite one is termed the counterjet. Because of relativistic beaming effects (see Box 15.1 and Problem 4.8 and 15.5) that are a common occurrence when relativistic jets are oriented near the line of sight, the counterjet may be faint and difficult to see relative to the jet.

The charged particles in the jet, which are primarily electrons and positrons since they are more easily accelerated than more massive particles, spiral around the field lines at relativistic velocities and emit synchrotron radiation because of the accelerated spiral motion (see Box 15.2). The resulting synchrotron radiation has a nonthermal spectrum and is partially polarized; it is strongly focused in the forward direction by relativistic beaming, and fluctuations of the jet in time will also be compressed into shorter apparent periods by relativistic effects. For an observer in the general direction of a jet, these effects will exaggerate both the apparent intensity and the time variation of the nonthermal emission. The nonthermal part of the continuum emission originates largely in the synchrotron radiation produced in the jets. (The thermal part of the continuum is typically produced in the accretion disk and the surrounding matter that it heats.) Let us now turn to a description of various observations in astronomy that may require some combination of Kerr black holes, large accretion energy, strong magnetic fields, and relativistic jets for their explanation.

15.4 Quasars

In the 1950s astronomers began to assemble a catalog of objects in the sky that emitted radio waves, with an effort made to correlate the objects emitting radio waves with sources that were visible in optical telescopes. The resolution of the single-dish radio telescopes in use at the time was much poorer than that of large optical telescopes, so there often were many possible optical sources that might potentially be correlated with the relatively uncertain position of a radio source. Nevertheless, some progress was made in these identifications.

Box 15.1   Relativistic Jets and Apparent Superluminal Velocities

Consider the following diagram in which a distant source at B moves at a velocity $v \sim c$ toward B'.

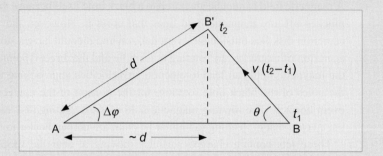

At time t_1 the source at B emits a light signal that is detected at time t'_1 by an observer at A. When the source reaches B' at time t_2, it emits another light signal that is detected by observer A at time t'_2. In Problem 4.8, Lorentz invariance is used to show that the apparent transverse velocity $\beta_T = v_T/c$ is related to the actual velocity $\beta = v/c$ by

$$\beta_T \equiv \frac{v_T}{c} = \frac{\beta \sin\theta}{1 - \beta \cos\theta}.$$

Thus for actual velocity $\beta < 1$, the apparent transverse velocity can exceed c by any amount, since the maximum value $\beta_T^{\max} = \beta/(1 - \beta^2)$ is unbounded as $\beta \to 1$. This illusion of an apparent velocity exceeding that of light is observed frequently for jets in radio astronomy where it is called *superluminal motion*. The figure below

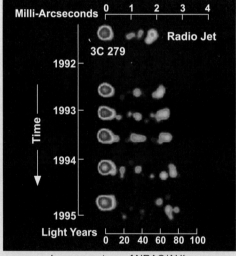

Image courtesy of NRAO/AUI.

shows a radio jet in the blazar 3C 279 (redshift $z = 0.534$) exhibiting apparent superluminal motion, with an inferred transverse velocity $v \sim 7c$ (Problem 15.3).

Box 15.2 — Nonthermal Emission

The Planck law describes *thermal emission* (radiation from a hot gas in thermal equilibrium); the resulting spectrum is a *blackbody spectrum*. The characteristic Planck law curves for thermal emission peak at some wavelength, and fall off rapidly at longer and shorter wavelengths, with the position of the peak moving to shorter wavelength as the temperature of the gas is increased (the Wien law). Light from most stars and from normal galaxies is dominantly thermal in character.

Synchrotron Radiation

In some cases *nonthermal* radiation may be observed, which has a spectrum that increases in intensity at very long wavelengths. The most common nonthermal emission in astronomy is *synchrotron radiation*, where high-velocity electrons (or other charged particles) in a strong magnetic field follow a spiral path around the field lines, radiating energy as highly beamed light. Synchrotron radiation is polarized because it is emitted in a narrow beam in the local plane of the electron's spiral path, and the motion of the electrons as viewed from the side of the spiral is almost side to side in a straight line; see the figure below left. The figure below right contrasts nonthermal emission with a 6000 K thermal (blackbody) spectrum.

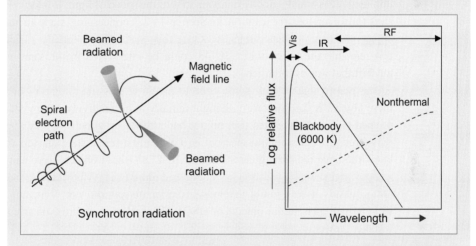

The wavelength of emitted radiation depends on how fast the charged particle spirals in the field. As the particle emits radiation it slows and emits longer wavelengths, explaining the broad distribution in wavelength of synchrotron radiation.

Implications of Nonthermal Emission

A nonthermal spectral component typically signals violent processes and large accelerations of charged particles. High-frequency synchrotron radiation also implies the presence of strong magnetic fields because the frequency increases with tighter electron spirals, which are characteristic of strong fields.

15.4.1 "Radio Stars" and a Spectrum in Disguise

Most radio sources were found to be correlated with certain galaxies but a few of the optical sources identified with radio sources appeared to be points with no obvious spatial extension, just as would be expected for stars. These were often called "radio stars," but they had very strange characteristics for stars. In 1963, the first two of these radio stars were associated with the radio sources 3C 48 and 3C 273, respectively.[1] Although these objects had the appearance of stars in optical telescopes, they had spectra unlike any stars that had ever been observed. There is a very strong continuum across a broad range of wavelengths that is nonthermal in character (see the discussion of nonthermal spectra in Box 15.2). Sitting on top of this continuum are emission lines, but they are very broad with wavelengths that do not correspond to the lines for any known spectra.

Later in 1963, Dutch astronomer Maarten Schmidt found while studying the spectrum of 3C 273 that the strange emission lines were really very familiar spectral lines in disguise. They were lines of the Balmer series of hydrogen and a line in ionized magnesium, but strongly displaced in the spectrum by a redshift $z = 0.158$ (see Section 16.2.2), which implies a recessional velocity of about 15% of the speed of light, if interpreted as a Doppler shift (however, see the caution in Section 16.2.3 concerning such an interpretation for objects at cosmological distances). Once this was realized for 3C 273, it soon became apparent that the spectrum of 3C 48 could be interpreted in the same way, but with a redshift that was even larger.

The reason that it took some time to arrive at this interpretation of the spectra for these objects is that they were thought initially to be relatively nearby stars because they seemed to be pointlike, and thus no one had any reason to believe they should be strongly redshifted. (To set some perspective, in 1963 only a few distant clusters of galaxies were known to have redshifts greater than for 3C 273.) These objects were named *quasistellar radio sources*, which was quickly contracted to *quasars*. After the discovery of the initial radio quasars it was found that many similar objects did not emit radio waves. These were termed quasistellar objects or QSOs. Now, all of these are typically termed quasars, whether they emit radio waves or not.

15.4.2 Quasar Characteristics

Soon other quasars were discovered and it became apparent that they represent a class of distant, highly luminous objects with the following characteristics.

1. Quasars appeared to be star-like in the initial images, but more careful study shows fuzziness and faint jets associated with some quasars.
2. Because quasars emit strongly in the ultraviolet, they are distinctly blue at optical wavelengths.
3. Many of the first quasars discovered were also radio sources but most observed quasars have no detectable radio emission.

---

[1] The radio source names derive from a compilation of radio objects called the *Third Cambridge Catalog*, in which objects were designated by "3C," followed by a number.

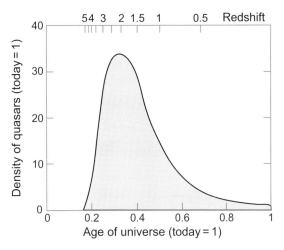

Fig. 15.1 Observed quasar densities as a function of redshift. Quasars were much more abundant earlier in the history of the Universe than they are today. The low quasar abundances at the very earliest times (redshifts larger than about 3) may represent partially the finite amount of time for quasars to begin forming after the formation of the Universe and partially observational bias, since it is more difficult to detect objects at the largest distances and therefore the earliest times. The quasar distribution in this figure indicates that the Universe has changed substantially with time. This will provide support for the validity of the big bang cosmology to be discussed in Chapter 20.

4. Quasars exhibit a nonthermal continuum spectrum that is stronger than most other sources at all wavelengths, varies substantially in time, and exhibits the basic characteristics of synchrotron radiation (nonthermal and polarized; see Box 15.2).
5. Many quasars exhibit large redshifts, implying by the Hubble law (see Chapter 16) that they are at great distances and that the light observed from them was emitted when the Universe was much younger than it is today. Quasar counts as a function of redshift indicate that they were more abundant in the early Universe than in the later Universe, as illustrated in Fig. 15.1.
6. Quasars usually exhibit emission lines that are very broad. If the broadening of the spectral lines is attributed to random motion of sources in the emitting regions, velocities as high as $10,000$ km s^{-1} are indicated by the widths of the emission lines.
7. Quasars also may exhibit absorption lines, usually sharper than the emission lines and with a redshift *less than* the emission lines. These result from absorption of the quasar light in gas clouds between us and the quasar. Thus, they provide a means to examine structure less distant from us than the quasar that often cannot be studied easily by other means, but are not associated directly with the quasar.

The huge energy output, large redshifts, and jets and other structure make it clear that quasars are not stars. The fuzziness and jets suggest that quasars are associated in some way with distant galaxies and that they appeared star-like in the original observations only because they are so far away. Now in favorable cases images have been obtained showing the host galaxy of the quasar. Thus, it is clear that quasars are enormously energetic phenomena that occur in certain galaxies.

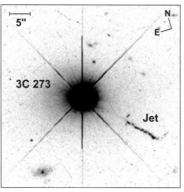 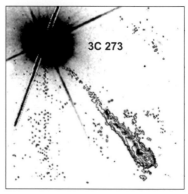

(a) Quasar 3C 273 and jet   (b) Optical jet + radio contours

Fig. 15.2 (a) The quasar 3C 273, which has a redshift of 0.158 [32]. The sharp radial lines from the quasar are optical spike artifacts common in quasar images because of their star-like brightness. (b) Radio contours superposed on the optical jet. Images from Bahcall, J. N. *et al. Astrophysical Journal Letters*, **452**(2). Reprinted with permission from the authors.

Example 15.2 Optical and radio images of the quasar 3C 273 are displayed in Fig. 15.2. Figure 15.2(a) shows the quasar and a jet. Figure 15.2(b), which has been rotated and enlarged relative to the left image, superposes on this optical image contours of radio frequency intensity. The optical jet clearly coincides with a jet structure in the RF map. Despite its distance, 3C 273 has an apparent visual magnitude of +12.9, which implies that it must be exceedingly luminous (see Problem 15.6). At visual wavelengths it is about 3×10^{12} times more luminous than the Sun but 3C 273 emits most of its light at nonvisual wavelengths. When summed over all wavelengths, a quasar like 3C 273 is typically found to be about 1000 times more luminous than bright normal galaxies.

The high luminosity of quasars at all wavelengths implies an enormous energy source. Many quasars also exhibit variability on timescales as small as months, weeks, or even hours. As discussed in Box 15.3, this implies that their energy output originates in a very compact source. For example, if the source exhibits variability over a period of an hour, the table in Box 15.3 implies that the maximum size of the emitting region is about 7 AU (but it could be less, since the timescale for variability sets only an upper limit on size).

15.4.3 Quasar Energy Sources

When it was first realized that quasars possess extremely powerful energy sources and that their variability argues that this energy is produced in a region that can be no larger than the Solar System, it raised serious issues about whether known physical processes could account for the energy characteristics of quasars. To set some perspective, the volume of the visible portion of the Milky Way Galaxy is about 10^{23} times larger than the volume of the Solar System. Taking a typical quasar energy source to be 1000 times more powerful than that for a normal galaxy, the quasar's "source energy density" is some 26 orders of magnitude larger than the source energy density for the normal galaxy. Since the discovery

> **Box 15.3** **Causality and Source Size**
>
> If an energy source exhibits a well-defined period in its luminosity, some signal must travel through the object to tell it to vary. The signal can travel no faster than light velocity, so the maximum size D of an object varying with a period P is the distance that light could have traveled during that time, $D \sim cP$, as illustrated below.
>
>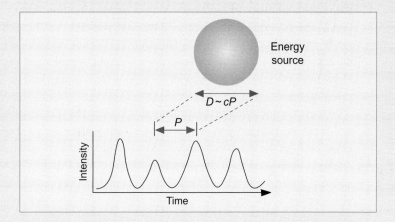
>
> For example, if the source luminosity varies with a a period of a week, the energy-producing region can be no larger than a "lightweek" (1.81×10^{11} km). The distances covered by light for various fixed times are summarized in the following table.
>
> | Time | km | AU | Parsecs |
> | --- | --- | --- | --- |
> | Year | 9.46×10^{12} | 63,240 | 3.07×10^{-1} |
> | Month | 7.88×10^{11} | 5270 | 2.55×10^{-2} |
> | Week | 1.81×10^{11} | 1210 | 5.87×10^{-3} |
> | Day | 2.59×10^{10} | 173 | 8.39×10^{-4} |
> | Hour | 1.08×10^{9} | 7.22 | 3.50×10^{-5} |
> | Minute | 1.80×10^{7} | 0.120 | 5.83×10^{-7} |
> | Second | 3.00×10^{5} | 0.002 | 9.73×10^{-9} |
> | Millisecond | 3.00×10^{2} | 0.000002 | 9.72×10^{-12} |
>
> This places only an *upper limit* on source sizes and the energy-producing region may be smaller than the upper limit imposed by c. But this argument is very powerful because it depends only on general principle (causality).

of quasars a much deeper understanding of black holes has been attained (recall that the Kerr solution was obtained at about the time of the discovery of quasars), and a likely mechanism to produce energy on this scale in such a compact region has emerged. There is relatively uniform agreement that rotating black holes containing of order 10^9 solar masses (which would have event horizons smaller in size than the Solar System) are the most plausible candidates for the central engine of a quasar. However, before turning to a more detailed discussion of this idea, let us examine another class of observational phenomena that appears to have much in common with quasars.

15.5 Active Galactic Nuclei

A large collection of ordinary stars is expected to produce an approximately blackbody spectrum dominated by a continuum that often peaks at visible wavelengths. For example, the radio luminosity of the Milky Way is about a million times smaller than its visible luminosity. In addition, the spectral lines observed for normal galaxies (as for its stars) are mostly absorption lines, with few emission lines. Thus, spectra for normal galaxies, as for the stars that they contain, are typically a continuum peaking at visible-light wavelengths representing thermal emission, with absorption lines superposed on the continuum and few emission lines.

However, some galaxies depart from this norm, exhibiting unusual spectra and evidence that there are extremely violent processes taking place within them, such as nonthermal emission from the radio-frequency to X-ray region of the spectrum and/or jets and unusual structure associated with the visual appearance of the galaxy. These are termed *active galaxies*. Since the source of the activity – though not necessarily all of its consequences – is concentrated in the nucleus of the galaxy, it is also common to refer to them as *active galactic nuclei* or *AGN*. Active galaxies exhibit some combination of the following characteristics:

1. unusual appearance, particularly of the nucleus, often with jets emanating from the nucleus;
2. high luminosities relative to normal galaxies but generally smaller than for quasars, with excess radiation at radio-frequency, infrared, ultraviolet, and X-ray wavelengths;
3. nonthermal continuum emission, often polarized, perhaps with broad and/or narrow emission lines; and
4. rapid variability driven by a very compact energy source located in the galactic nucleus.

Some galaxies with active nuclei are relatively nearby but they are more likely to be found at larger distances, implying that they were more common earlier in the Universe's history. Many of these characteristics also are exhibited by quasars and it will be argued below that quasars may be closely related to the nuclei of active galaxies. There are several general classes of AGN that will now be summarized.

15.5.1 Radio Galaxies

Radio galaxies are AGN associated with a large nonthermal emission of radio waves.[2] Powerful radio galaxies are elliptical and often exhibit jet structure from a compact nucleus. Weaker radio sources are associated with smaller jets in some spiral galaxies. There are two broad classes of radio galaxies. *Core-halo radio galaxies* exhibit radio

---

[2] Strong radio sources are often named by the constellation followed by a capital letter designating the order of discovery in the constellation. For example, the relatively nearby radio galaxy M87 (in the Virgo Cluster of galaxies) is host to the powerful radio source Virgo A, which was the first radio source found in the constellation Virgo.

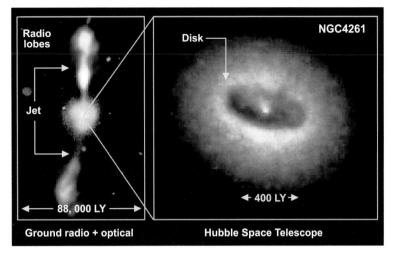

Fig. 15.3 The radio galaxy NGC 4261. On the left the radio map is superposed on the ground-based optical image of the galaxy (bright blob at the center). The image on the right is a high-resolution Hubble Space Telescope image of the core of the galaxy, showing a dusty accretion disk thought to surround a central black hole engine that is driving the jets producing the radio lobes. The putative black hole engine lies deep inside the bright dot at the center of the disk. Credit: HST/NASA/ESA.

emission from a region concentrated around the nucleus of the galaxy that is comparable in size to the optically visible galaxy. *Lobed radio galaxies* display great lobes of radio emission extending in some cases for millions of lightyears beyond the optical part of the galaxy.

Example 15.3 Even powerful radio galaxies may appear to be normal elliptical galaxies at optical wavelengths. The left side of Fig. 15.3 shows enormous and powerful radio emission lobes superposed on the optical image of the elliptical galaxy NGC 4261. Although the optical image of NGC 4261 on the left side suggests a rather normal looking elliptical galaxy, the high-resolution image of the core of the galaxy shown on the right indicates that something very unusual is happening at the center of the galaxy, and that this is the energy source powering the radio jets associated with the galaxy.

As illustrated in the preceding example, abnormalities at optical wavelengths often are obvious only if the core of the galaxy is resolved.

15.5.2 Seyfert Galaxies

Seyfert galaxies are the most commonly observed active galaxies [185]. An example is displayed in Fig. 15.4(a). They are usually spirals with bright, compact nuclei, and exhibit a strong nonthermal continuum from IR through X-ray wavelengths, with emission lines of highly ionized atoms that are sometimes variable. Emission lines suggest a low-density

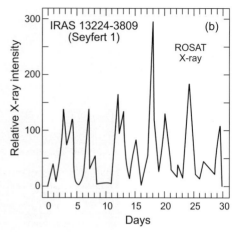

Fig. 15.4 (a) The Seyfert 2 galaxy NGC 7742, which lies about 22 Mpc away in the constellation Pegasus. The nucleus is very compact and bright at optical wavelengths, which is characteristic of Seyfert spirals. Credit: Hubble Heritage Team (AURA/STSU/NASA/ESA). (b) X-ray variability of the Seyfert 1 galaxy IRAS 13224-3809 (data from Ref. [52]).

gas source, while high ionization implies a hot source responsible for the ionization. Some Seyferts have modest jets leading to continuum RF emission. The thousands of Seyfert galaxies that are now known can be classified roughly into two subgroups, Seyfert 1 galaxies and Seyfert 2 galaxies (these are also termed Type 1 and Type 2 Seyferts, respectively).

Seyfert 1 galaxies have both broad and narrow emission lines and are very luminous at UV and X-ray wavelengths. Broad emission lines are associated with allowed transitions in elements like H, He, and Fe. Their width is caused by Doppler broadening, suggesting that the lines are produced in clouds with average velocities as high as $\sim 10,000\,\text{km s}^{-1}$. The narrow lines are associated with forbidden transitions, for example in twice-ionized oxygen (O III), and their widths imply source velocities generally less than $\sim 1000\,\text{km s}^{-1}$. These differences suggest that the broad and narrow spectral lines originate in different physical regions of a Seyfert 1 galaxy.

Seyfert 2 galaxies have relatively narrow emission lines suggesting source velocities less than $\sim 1000\,\text{km s}^{-1}$, and are weak at X-ray and UV wavelengths but very strong in IR. Their continuum emission is weaker than for Seyfert 1 galaxies. In Seyfert 2s both the forbidden and allowed lines are narrow, suggesting that both kinds of lines originate in the same region with relatively low source velocities. Seyfert 2 galaxies appear to be about three times more numerous than Seyfert 1 galaxies.

Seyfert galaxies can exhibit 50% changes in optical brightness over weeks and even larger changes over months. As illustrated in Fig. 15.4(b), variation in X-ray brightness can be even greater. By the arguments given in Box 15.3, the rapid fluctuation on a timescale of approximately a day exhibited in Fig. 15.4(b) implies a very compact energy source for the X-rays, no larger than light days in diameter. This is comparable in size with the Solar System, which has a diameter of about half a light day.

15.5.3 BL Lac Objects

BL Lacertae objects (*BL Lac objects* or just *BL Lacs,* for short) exhibit no emission lines or only very weak ones, but have a strong nonthermal continuum from RF through X-ray frequencies. They are generally radio-loud, and exhibit strongly polarized light. BL Lacs are members of a more general class of AGN called *blazars.* Blazars are relatively uncommon among AGN. For example, there are about 100 times more quasars and 10 times more Seyfert galaxies known than blazars. However, they are extremely powerful, typically being 10,000 times more luminous than the Milky Way and 1000 times more luminous than a bright Seyfert galaxy.

By masking the bright core a spectrum of the faint "fuzz" around many BL Lacs can be acquired. This spectrum and the variation of the light intensity indicate that the fuzz is the outer part of a giant elliptical galaxy in most cases, suggesting that BL Lacs are the active cores of such galaxies. From the redshifts of faint spectral lines it is found that blazars are often more distant than Seyfert or radio galaxies, but they are closer to us on average than quasars. Blazars can exhibit dramatic variability in their light output (and a corresponding strong variability in polarization) on both long and short timescales. Some can vary significantly on timescales as short as days, implying a power source the size of the Solar System or smaller.

15.6 A Unified Model of AGN and Quasars

Some representative spectra of normal galaxies, quasars, and active galaxies are shown in Fig. 15.5. Obviously, these spectra imply that active galaxies and quasars are something very different from normal galaxies. On the other hand, the similarity of the spectra for quasars and Seyfert galaxies implies that there might be a relationship between quasars and active galaxies. That relationship is suggested by the large nonthermal emission from AGN and quasars, which implies a strong and very compact energy source. The only plausible candidate for the engine driving these phenomena is a Kerr black hole of mass millions to billions of solar masses at their center. A large amount of observational data and theoretical understanding now supports this point of view.

Example 15.4 The high efficiency for gravitational energy conversion in black hole accretion provides the most convincing general argument that active galactic nuclei and quasars must be powered by accretion onto rotating supermassive ($M \sim 10^9 M_\odot$) black holes. For example, in Problem 15.1 you are asked to use observed luminosities and temporal luminosity variations to show that a quasar could be powered by accretion of as little as several solar masses per year onto an object of mass $\sim 10^9 M_\odot$, and that this mass must occupy a volume smaller than the Solar System to be consistent with observations.

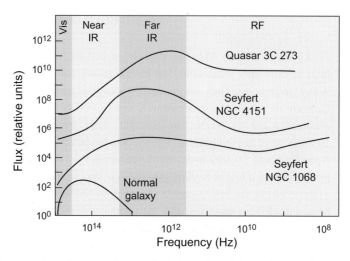

Fig. 15.5 Comparison of some representative spectra for normal galaxies, active galaxies, and quasars. The spectrum of a normal galaxy is approximately thermal but the spectra of the active galaxies and quasars are highly nonthermal. Because electrons must radiate energy continuously as they spiral in magnetic fields, the energy driving the huge sustained synchrotron emission from these nonthermal sources must be replenished constantly. Strong, polarized, nonthermal emission is an indirect sign not only of strong magnetic fields, but also of a very large energy source for the quasars and active galaxies.

The present belief is that – despite significant observational differences – AGN such as Seyfert galaxies, blazars, and radio galaxies are quite closely related, and that quasars in turn are just a particularly energetic form of AGN. All are now thought to be active galaxies with bright nuclei powered by rotating supermassive black holes. In the unified model that we will now discuss, the observational differences among quasars and various AGN mostly reduce to a matter of:

- how rapidly matter is accreting onto the black hole (feeding the monster);
- whether the central engine region is masked by dust (hiding the monster); and
- how far away the AGN or quasar is from us (proximity of the monster).

This suggests an hypothesis. Perhaps all large galaxies have supermassive black holes at their centers. The question of whether the galaxy exhibits an active nucleus is then primarily one of "feeding the monster" at the center. In a quasar, the black hole is accreting matter at a higher rate, leading to very high luminosity. In a normal galaxy like the Milky Way the black hole is rather quiet because it is presently accreting little matter. Seyfert and other active galaxies are somewhere in between: the black hole is active because it is accreting matter, but at a rate lower than that for a quasar. This picture is supported by the finding that Seyfert galaxies are more abundant than average in interacting pairs of galaxies, and that a significant number of Seyferts exhibit evidence for tidal distortion. Both of these observations suggest that interactions with other galaxies may "turn on" an AGN by increasing the fuel flow to the central black hole.

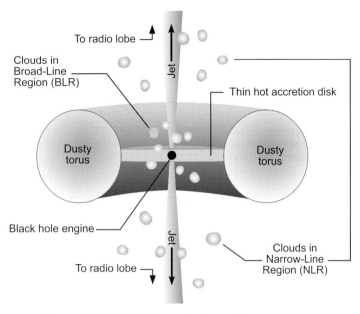

Fig. 15.6   Schematic illustration of the standard AGN black hole central engine paradigm.

15.6.1 The AGN Black Hole Central Engine Model

We now discuss a unified AGN and quasar model where all active galaxies and quasars are powered by central supermassive black holes. In this unified model the differences among various AGN and quasars arise primarily from differences in the orientation angle and local environment for the central engine. The first determines whether the view of the central engine is blocked by dust; the second determines the rate at which fuel flows to the central engine.

Figure 15.6 illustrates the black hole paradigm for AGN and quasars. The black hole itself occupies a tiny region in the center (its size is exaggerated for clarity in this figure). It is surrounded by a dense torus of matter revolving around the hole and a flattened accretion disk inside of that where the matter whirls even faster before disappearing into the black hole. On the polar axes of the spinning black hole jets of matter are ejected, as part of the matter in the accretion disk is sucked into the black hole and part is flung out at high velocity in the jets. Figure 15.7 illustrates qualitatively what the central engine of an AGN or quasar would look like if one could turn the radiation off and clear away the gas and dust. The thin accretion disk surrounding the black hole is very hot because of collisions in the rapidly swirling gas that it contains and radiates strongly in the ultraviolet.

Example 15.5 Radiation from the accretion disk is expected to be dominated by thermal processes. Typical estimates of the temperature for the accretion disk of a $10^8 M_\odot$ black hole imply emitted radiation peaking in the UV region of the spectrum. The accretion disk temperature is larger for less massive black holes. Thus stellar-size black holes have

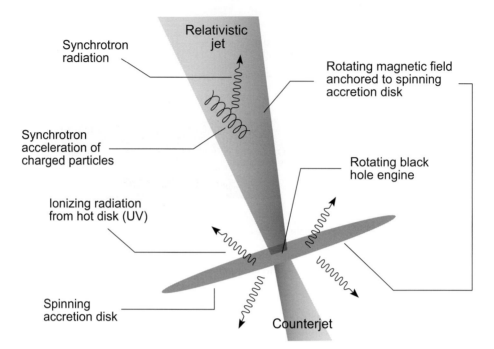

Fig. 15.7 A schematic view of the central engine for an active galactic nucleus or quasar. Beamed synchrotron radiation is emitted by the electrons in the jet and UV photons are emitted by the hot accretion disk.

accretion disks that are even hotter than the ones considered here, and they tend to radiate more in the X-ray region of the spectrum (see Problem 15.7).

Photons from the accretion disk are responsible for producing much of the continuum observed from AGN, either directly or by heating surrounding matter that then re-radiates the energy at longer wavelengths. Photons emitted by the accretion disk also ionize atoms in the nearby clouds of gas where velocities are very high and produce the broad-line emission spectrum of the AGN (the Broad-Line Region or BLR of Fig. 15.6). In addition, they ionize clouds further away from the central engine where velocities are lower and this produces the narrow lines of the emission spectrum (the Narrow-Line Region or NLR of Fig. 15.6). Whether both broad and narrow emission lines are visible to a distant observer will depend on the location of the observer relative to the plane of the accretion disk and the torus that may surround it.

15.6.2 Anisotropic Ionization Cones

If this picture of the central engine for an AGN is correct, there should be observational evidence in active galactic nuclei for ionization in the central region of the galaxy that is concentrated in particular directions from the center. In the simplest case, cone-shaped regions of ionization might be expected corresponding to the directions from the hot accretion disk that are not blocked off by the dusty torus, as displayed schematically in

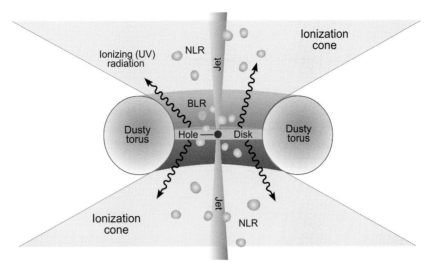

Fig. 15.8 Ionization cones of an AGN. The shaded regions above and below the accretion disk can see directly the UV radiation emitted by the accretion disk. Other regions cannot see the accretion disk directly because it is blocked by the dusty torus. Clouds in the broad-line region are labeled BLR and clouds in the narrow-line region are labeled NLR.

Fig. 15.8. All of the region shaded in light gray (which appears cone-shaped far from the black hole) can see the hot central accretion disk and the ionizing radiation that it is emitting. Therefore this region should be ionized by radiation from the central engine, whereas regions lying outside the ionization cone can be ionized only indirectly or by starlight from the surrounding region and not directly by the continuum emitted from the accretion disk. Various observations suggest the presence of such anisotropic ionization zones near AGN central engines.

15.6.3 A Unified Model

The various threads of this discussion may be pulled together to give the unified model of active galactic nuclei and quasars illustrated in Fig. 15.9.[3] It may be hypothesized that active galaxies are powered by rotating, supermassive black holes in their center that have all or most of the following features:

- an accretion disk, possibly surrounded by a dusty torus;
- high-velocity clouds near the black hole that produce broad emission lines;
- low-density, slower clouds further away that produce narrow emission lines; and
- relativistic jet outflow perpendicular to the plane of the accretion disk.

---

[3] There have been suggestions that the geometrical arguments underlying the simple unified model presented here may be insufficient to account for detailed properties of AGN [84, 153, 202]. For example, a survey of 836 AGN selected for hard X-ray emission concludes that the main differentiator between obscured and nonobscured black holes in AGN is the *accretion rate*, which governs the clearance of dust by radiation pressure [202]. A review of AGN unification attempts is given in Ref. [162].

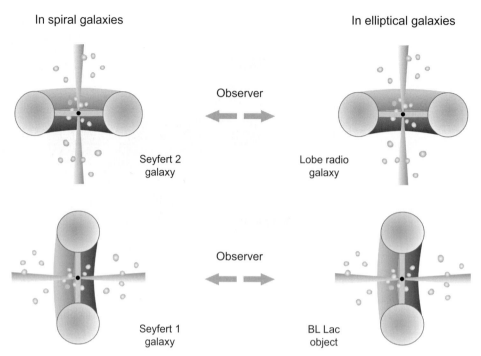

Fig. 15.9 A unified model of active galactic nuclei based on geometry (but see footnote 3 in Section 15.6.3). All cases are powered by supermassive Kerr black holes and observational differences are determined primarily by viewing angles relative to the plane of the black hole accretion disk and the jets. Quasars are then hypothesized to be similar to AGN, but more powerful because the black hole is accreting at a higher rate.

If the orientation of the system relative to the observer is as displayed in the top row of Fig. 15.9, an observer sees a Seyfert 2 galaxy if the active galactic nucleus is in a spiral galaxy and a lobed radio source if the AGN is in an elliptical galaxy and produces strong jets. On the other hand, if the orientation of the active galactic nucleus central engine is as in the bottom row of Fig. 15.9, the observer sees a Seyfert 1 galaxy if the host is a spiral galaxy, a blazar (BL Lac) if the host is an elliptical galaxy and the jet is strong, and perhaps a core-halo radio galaxy if the jets are weaker in an elliptical system.

We are then led to conjecture that quasars are just particularly energetic forms of these active galaxies and are described by the same unified model. The primary difference is that quasars are more luminous because their black hole engines are being fed matter at a higher rate than for typical AGN. This suggests further that quasars are more abundant at larger redshift because large redshift corresponds to earlier times in the expanding Universe's history, when galaxies were more closely spaced and collisions more frequent. Therefore, quasars may be just "better fed" active galactic nuclei dating from an epoch when more fuel was available to power their central engines. As a corollary, we may speculate that many nearby normal galaxies once sported brilliant quasars in their cores, but have since used up the available fuel. Their massive black holes lie dormant, ready to blaze back to life should circumstances like tidal interactions with another galaxy divert matter into the black hole.

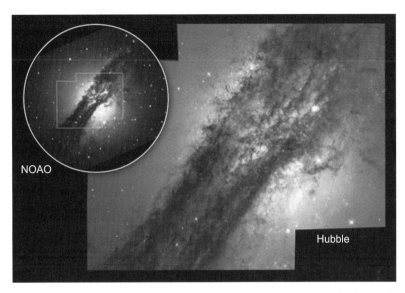

Fig. 15.10 One of the nearest active galactic nuclei, Centaurus A (in NGC 5128). Ground-based image credits (upper left): NOAO; HST image credits (lowe right): E. J. Schreier (STScI) and NASA; Team numbers are E. J. Schreier, A. Marconi, D. Axon, N. Caon, and D. Macchetto (STScI).

This may be true of our own galaxy, which has a ∼4 million solar mass black hole at its center (see Section 14.5.1), but presently has relatively weak nonthermal emission.

Example 15.6 Quasars can derive their power from accretion of a few solar masses per year. This accretion rate implies something important about the characteristic lifetime of a quasar. Even if it is assumed that a significant fraction of the host galaxy mass, let's say 10^9 $M_\odot$, is available to feed the black hole, such a quasar could radiate only for of order 10^8–10^9 years – far less than the age of the Universe.

From the preceding example, it might be expected that quasars should be most abundant in an epoch of duration less than the age of the Universe, as suggested by Fig. 15.1.

15.6.4 Example: Feeding a Nearby Monster

The point of view discussed in the preceding section is exemplified by Fig. 15.10, which displays optical images of one of the nearest AGN: the galaxy NGC 5128, which is only ∼ 4 Mpc away and contains the strong radio source Centaurus A. The area outlined in the ground-based telescopic image upper left in Fig. 15.10 is expanded in the Hubble Space Telescope view in the lower right. The radio lobes of Centaurus A span $10°$ in our sky (20 times the Moon's diameter), and it has a faint optical jet and strong radio jets that are ∼ 10^6 ly in length. At the base of the jets the particle velocities are estimated to be ∼ $\frac{1}{3}c$. This indicates that violent activity at the center has been ejecting gas jets for some time, thus producing the huge radio lobes. Infrared observations suggest that behind the dust

Fig. 15.11 X-ray jet from center of Centaurus A superposed on an optical image of the galaxy. In this Chandra X-ray Observatory map the highest X-ray intensity is indicated by white. Credit: NASA/CXC/SAO.

lanes lies a gas disk some 130 lightyears in diameter, surrounding a black hole of about $5.5 \times 10^7 M_\odot$ that is powering the activity in the core of the galaxy.

A firestorm of starbirth is observed in Fig. 15.10, which is thought to have been precipitated by a collision between a smaller spiral galaxy and a giant elliptical galaxy within the last billion years. (The dark dust lanes may be the remains of the spiral galaxy, seen almost edge-on and encircling the core of the giant elliptical.) The logical conclusion is that the black hole at the center of Centaurus A has been driven into frenetic activity by the collision of the parent galaxy with another. This collision has led to strong radio emission because of relativistic jets powered by enhanced accretion onto the black hole, and to enhanced star formation because of the compression of gas in the collision of the two galaxies. Support for this interpretation is supplied by Fig. 15.11, which shows an X-ray map obtained by the Chandra X-ray Observatory superposed on an optical image of Centaurus A. The Chandra data reveal a bright central region (white ball near the center) that likely surrounds the black hole, and a strong jet oriented to the upper left that is probably being ejected on the polar axis of the black hole. There also is a fainter jet oriented in the opposite direction, which is harder to see because of beaming effects.

15.6.5 High-Energy Photons from AGN

Some AGN emit extremely energetic photons. For example, 10^{12} eV gamma-rays have been detected from Markarian 421, a BL Lac at $z = 0.031$. These are much too high in energy to be produced by thermal processes but the unified AGN–quasar model provides a possible explanation. Box 15.4 illustrates a mechanism called *inverse Compton scattering*, by which lower-energy photons can be boosted to extremely high energy in an AGN. As illustrated there, 10^{12} eV gamma-rays are possible if X-rays produced by irradiation from the accretion disk enter a highly-relativistic jet and are inverse Compton scattered.

Box 15.4   Inverse Compton Scattering and High-Energy Photons

Some AGN emit extremely high-energy photons that are thought to involve a process called *inverse Compton scattering*, illustrated below.

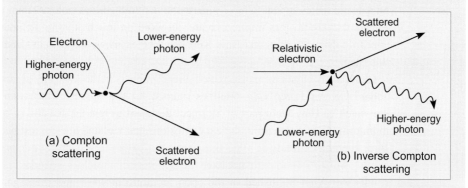

In *Compton scattering* a high-energy photon strikes an electron, giving up energy. Inverse Compton scattering is the reverse: a high-energy electron strikes a photon, imparting energy. An AGN could produce a high-energy photon as follows:

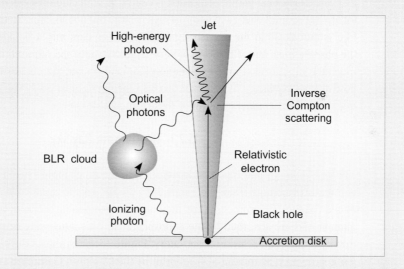

UV from the disk can photoionize clouds in the broad-line region (BLR), which then emit at optical wavelengths. If these photons enter the jet they can be inverse Compton scattered to much higher energies. The frequency boost factor is given by

$$\frac{v}{v_0} = \frac{1}{1 - v^2/c^2}$$

where v is the jet velocity, so in a highly relativistic jet boost factors of 10^6 can be obtained (Problem 15.4). This can convert optical photons into intermediate-energy gamma-rays and X-rays into 10^{12} eV gamma-rays.

15.7 Gamma-Ray Bursts

The preceding discussion introduced a number of examples where impressive phenomena on galactic scales appear to have supermassive rotating black holes as their central engines. Events that are no less impressive are observed on much smaller stellar scales that also appear to derive their power from Kerr black holes, as we will discuss in the remainder of this chapter.

Gamma-ray bursts were discovered serendipitously in the 1960s by gamma-ray detectors aboard US Air Force *Vela* satellites (named for the Spanish verb *velar*, which means "to watch"). They were part of a project designed to test the feasibility of detecting and monitoring from Earth orbit nuclear explosions that violated nuclear test ban treaties. Quite surprisingly, beginning in 1967 these satellites began to see strong bursts of gamma-rays coming from *above* that did not have the signature of a nuclear weapons test. Accumulated results were first published in the open literature in 1973 and these gamma-ray bursts (GRB) became a great scientific puzzle for several decades.

About one burst a day is observable somewhere in the sky but until the 1990s there was little understanding about where they originated or even how far away they were. The distance to a typical gamma-ray burst is central to understanding the mechanism. If they are local, the required energy in gamma-rays is large but not so large. In that case, explanations like "starquakes" in the crusts of galactic neutron stars might be adequate to account for the observations. But if they are at very large distances the required intrinsic luminosity is much too high for a simple explanation like neutron star starquakes to suffice. As we will now discuss, a variety of observations have begun to clear away the mystery and have led to a much deeper understanding of these remarkable events.

15.7.1 The Gamma-Ray Sky

The sky glows in gamma-rays, in addition to the other more familiar wavelengths. Since gamma-rays are high-energy photons, they are not easy to produce and tend to indicate out-of-the-ordinary phenomena; thus they are of considerable astrophysical interest. Because gamma-rays are absorbed by the atmosphere, a systematic study of gamma-ray sources in the sky requires an orbiting observatory. Figure 15.12(b) shows the continuous glow of the gamma-ray sky, as measured from orbit by the Compton Gamma-Ray Observatory.[4]

Superposed on the steady gamma-ray flux illustrated in Fig. 15.12(b) are sudden bursts of gamma-rays. These events can be as short as tens of milliseconds and as long as several minutes, and pour out enormous amounts of gamma-ray energy for that brief

---

[4] It is common to represent the location of sky objects in the galactic coordinates of Fig. 15.12(a) using a 2-dimensional projection as in Fig. 15.12(b), much as the spherical surface of the Earth can be represented in 2D maps. Earth maps often use the Mercator projection but for sky maps it is standard to use projections like the *Mollweide projection*, which is an equal-area pseudocylindrical projection emphasizing accurate representation of area over accuracy of angle and shape. The Mollweide projection is particularly useful for maps depicting global distributions. In contrast, the Mercator projection emphasizes preservation of angles because it was motivated by utility for seventeenth-century ship navigation, but does not preserve area relations.

15.7 Gamma-Ray Bursts

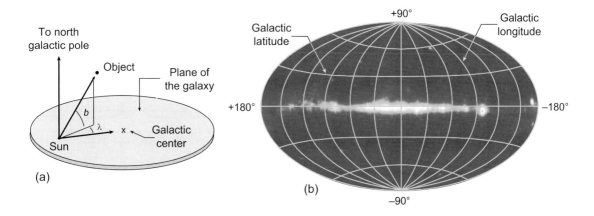

Fig. 15.12 (a) The galactic coordinate system. The angle b is the galactic latitude and the angle λ is the galactic longitude. These may be related to right ascension and declination by standard spherical trigonometry. (b) The sky at gamma-ray wavelengths in galactic coordinates. White denotes the most intense and black the least intense sources. The diffuse horizontal feature at the galactic equator is from gamma-ray sources in the plane of the galaxy. Bright spots to the right of center in the galactic plane are galactic pulsars. Brighter spots above and below the plane of the galaxy are distant quasars. The origins of some of the fainter localized sources are unknown. Credit: NASA EGRET Team.

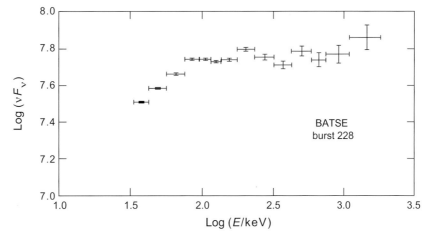

Fig. 15.13 Spectrum of a gamma-ray burst. As is characteristic, the spectrum is nonthermal (see the discussion in Box 15.2). Adapted from *Physics Reports*, **314**(6), Tsvi Piran, Fig. 3 in Gamma-ray bursts and the fireball model, copyright (1999), with permission form Elsevier.

time. The spectrum is nonthermal and dominated by photons in the hundreds of keV to several MeV range, as illustrated in Fig. 15.13. Figure 15.14 shows the position of 2704 gamma-ray bursts observed by the Burst and Transient Source Experiment (BATSE) of the Compton Gamma-Ray Observatory, plotted in galactic coordinates. This plot indicates that the distribution of bursts is isotropic on the sky. If bursts had a more local origin, such as in the disk of the galaxy, they should be concentrated along the galactic equator in this figure,

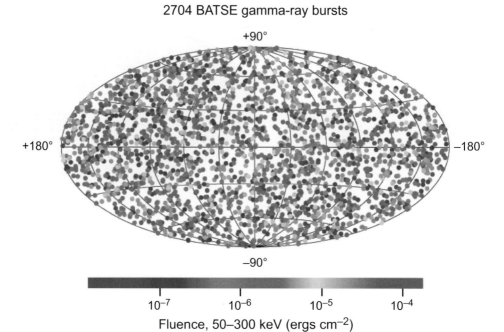

Fig. 15.14 Location on the sky of 2704 gamma-ray bursts plotted in galactic coordinates with the grayscale indicating the fluence (energy received per unit area) of each burst. The data were recorded by the Burst and Transient Source Experiment (BATSE) of the Compton Gamma-Ray Observatory (CGRO). The highly isotropic distribution of GRB events over a broad range of fluences argues strongly that they occur at cosmological distances. NASA.

not randomly scattered over the whole diagram. This indicates that either the gamma-ray bursts come from events at great distances (cosmological distances), or perhaps they come from events in the more spherical halo of our galaxy. It will be shown below that redshifts for spectral lines confirm unequivocally the first interpretation and it is now clear that gamma-ray bursts are cosmological in origin, occurring far outside our galaxy.

15.7.2 Two Classes of Gamma-Ray Bursts

Figure 15.15 illustrates that there appear to be two classes of gamma-ray bursts:

1. *Short-period bursts* last less than two seconds and have harder (higher-energy) spectra.
2. *Long-period bursts* have softer (lower-energy) spectra and last from several seconds to several hundred seconds. The time profile of a long-period burst is shown in Fig. 15.16.

These and other distinctions in the data suggest that GRBs represent a family of events having at least two sources. To determine these sources it was necessary to know where the bursts originated. Initial progress was slow because BATSE could localize a burst position only within several degrees on the sky. Therefore, it was very difficult to know exactly where to point telescopes to find evidence associated with the gamma-ray burst at other wavelengths before that evidence faded from sight. Help in this regard came from a satellite looking, not at gamma-rays, but at X-rays.

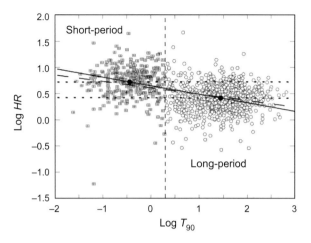

Fig. 15.15 Hardness (propensity to contain higher-energy photons) of the spectrum versus duration of the burst, illustrating the separation of the GRB population into long, soft bursts and short, hard bursts [196]. The parameter HR measures the hardness of the spectrum and T_{90} is defined to be the time from burst trigger for 90% of the energy to be collected. Bursts with T_{90} shorter than 2 seconds are classified as short-period and those with T_{90} greater than 2 seconds are classified as long-period bursts. This plot indicates that shorter bursts generally have a harder spectrum. Reproduced from Qin, Yi-Pins and Xie, Grans-Zhong, The Hardness-Duration Correlation in the Two Classes of Gamma-Ray Bursts, *Publications of the Astronomical Society of Japan*, 2000, **52**(5), by permission of the Astronomical Society of Japan.

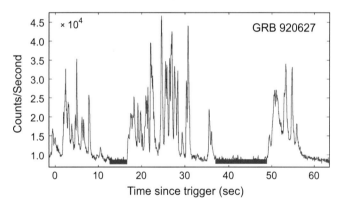

Fig. 15.16 Time profile of a typical gamma-ray burst [189]. This is an example of a long-period GRB. Reprinted figure with permission from Tsvi Piran, *Reviews of Modern Physics*, **76**, 1143, published 28 January 2005. Copyright (2005) by the American Physical Society.

15.7.3 Localization of Gamma-Ray Bursts

A major breakthrough in localizing gamma-ray bursts on the celestial sphere came in 1998 when for the first time it became possible to correlate some gamma-ray bursts with other sources in the visible, RF, IR, UV, and X-ray portions of the spectrum. This was first enabled by a small Dutch–Italian satellite called BeppoSAX that had the ability to pinpoint the location of X-rays following gamma-ray bursts with 2 arc-minute resolution

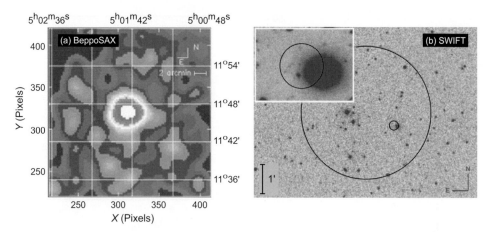

Fig. 15.17 (a) First localization of an X-ray afterglow for a GRB, obtained by the satellite BeppoSAX. Credit: Beppo SAX Team ASI, ESA. (b) Optical association of short-period GRB 050509B with a large elliptical galaxy at a redshift of $z = 0.225$ by SWIFT [96]. The larger circle is the error circle for the Burst Alert Telescope (BAT). The smaller circle is the error circle for the X-Ray Telescope (XRT), which slewed to point at the event when alerted by the BAT. The XRT error circle is shown enlarged in the inset at the upper left, suggesting that the GRB occurred in the outer regions of a large elliptical galaxy (dark blob partially overlapped by the XRT error circle). Reprinted by permisssion from Macmillan Publishers Ltd. *Nature*, **437**, 851–854, Fig. 1 (06 October 2005) Copyright (2005).

in a matter of hours. This permitted other satellite and ground-based instruments to look quickly at the burst site, and this in turn began to allow transient sources ("afterglows") at other wavelengths to be correlated with the burst. Figure 15.17(a) shows an X-ray transient observed by BeppoSAX at the location of a (long-period) gamma-ray burst. These transients are thought to be associated with a rapidly fading *fireball* that is produced by the primary gamma-ray burst. One of the first localizations for a short-period burst by the SWIFT satellite is illustrated in Fig. 15.17(b). Correlation of gamma-ray bursts with sources observed at other wavelengths have allowed distances to be estimated to gamma-ray bursts because spectral lines and their associated redshifts have been observed in the transients after the burst. Assuming these transients to be associated with the gamma-ray bursts and the redshifts to be described by the Hubble law (Chapter 16), these observations show conclusively that gamma-ray bursts are occurring at cosmological distances.

There is rather broad agreement that the afterglow transients observed at various wavelengths following gamma-ray bursts can be accounted for by a relativistic fireball model, as illustrated in Fig. 15.18. In this model some generic central engine that need not be specified explicitly deposits a very large amount of energy in a small volume of space, which produces a fireball expanding initially at relativistic velocities. This fireball is responsible for the observed afterglows.

15.7.4 Necessity of Ultrarelativistic Jets

Gamma-ray bursts *must involve ultrarelativistic jets because observed prompt emission is nonthermal* (see Box 15.2 and Fig. 15.13). If the jet were not ultrarelativistic the ejecta

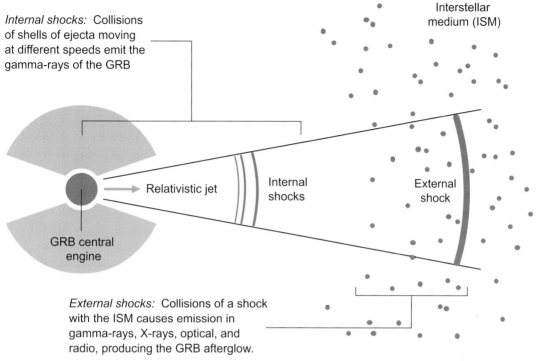

Fig. 15.18 Relativistic fireball model for transient afterglows following gamma-ray bursts. In this model the internal shocks in the relativistic jet produced by clumps of ejected material with different velocities overtaking each other produce the gamma-rays and the external shocks resulting from the jet impacting the interstellar medium (ISM) produce the afterglows.

would be optically thick (see Box 15.5) to pair production, which would thermalize the energy. The requirement that gamma-ray bursts be produced by ultrarelativistic jets (and not thermalized photons) can be understood in terms of the opacity of the medium with respect to formation of electron–positron pairs by $\gamma\gamma \to e^+e^-$. Let us elaborate on this crucial point.

Optical depth for a nonrelativistic burst: We first assume that the burst involves nonrelativistic velocities. The initial spectrum is nonthermal and the number of counts $N(E)$ as a function of gamma-ray energy can be approximated (roughly, but sufficient for this estimate) for particular ranges as a power law,

$$\frac{N(E)}{dE} \propto E^{-\alpha} dE, \qquad (15.14)$$

where the *spectral index* α is approximately equal to 2 for typical cases [188, 189]. Because the observed spectrum is nonthermal the medium must be *optically thin* (small optical depth; see Box 15.5), since scattering in an optically thick (that is, highly opaque) medium would quickly thermalize the photons. Let's consider the optical depth for pair production associated with a typical gamma-ray burst to see whether this condition can be fulfilled.

Box 15.5 Optical Depth

The opacity of a medium can be parameterized in terms of an *optical depth*. Let a flux F be incident on a thin slab of material characterized by a thickness dr, density ρ, and opacity κ, as illustrated in the following figure.

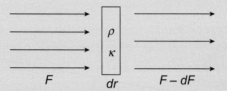

The difference between incident and transmitted flux dF is given by $dF = -\kappa \rho F dr$. Assuming $\kappa \rho$ to be constant over dr, this has a solution

$$F = F_0 e^{-\kappa \rho r} = F_0 e^{-r/\lambda} \qquad \lambda \equiv (\kappa \rho)^{-1},$$

where λ is the mean free path. Introduce a dimensionless quantity τ called the *optical depth* (which generally will depend on wavelength) through the differential equation $d\tau = -\kappa \rho dr$. The optical depth at r is then defined by integrating

$$\tau = \int_r^\infty \kappa \rho dr.$$

Optical depth measures the probability that a photon at r will interact before leaving the star, so it characterizes the transparency of a medium. An opaque medium has a large τ; a transparent medium typically has an optical depth of one or less.

For the reaction $\gamma\gamma \to e^+ e^-$ to occur, energy conservation requires the two photons with energies E_1 and E_2, respectively, to satisfy $(E_i E_2)^{1/2} \gtrsim m_e c^2$, where m_e is the mass of an electron. Let f be the fraction of photon pairs that fulfill this condition. The optical depth with respect to $\gamma\gamma \to e^+ e^-$ is then [188]

$$\tau_0 = \frac{f\sigma_T F D^2}{R^2 m_e c^2} \simeq \frac{f\sigma_T F D^2}{\delta t^2 m_e c^4}, \tag{15.15}$$

where $\sigma_T = 6.652 \times 10^{-25}$ cm^2 is the Thomson scattering cross section for electrons, F is the observed fluence for the burst, D is the distance to the source, and R is its size, which can be related to the observed period δt for time structure in the burst by $R = c\delta t$, using the arguments of Box 15.3. As shown in Problem 15.8, a typical optical depth estimated using this formula is *enormous* ($\tau \sim 10^{14}$), and therefore completely inconsistent with the low optical depth ($\tau \lesssim 1$) required by the nonthermal GRB spectrum. Nonrelativistic jets cannot power a gamma-ray burst, but what about relativistic jets?

Optical depth for an ultrarelativistic burst: The above considerations will be altered in two essential ways if the burst is instead ultrarelativistic with a Lorentz factor $\gamma \gg 1$, so that special relativistic kinematics apply:

1. The *blueshift* of the emitted radiation will modify the fraction f of photon pairs that have sufficient energy to make electron–positron pairs.
2. The *size* R of the emitting region will be altered by relativistic effects.

Specifically, the observed photons of frequency v and energy $E = hv$ have been blueshifted from their energy in the rest frame of the GRB by a factor γ. Thus the source energy E_0 was lower than the observed energy E by a factor of γ^{-1} and $E_0 \sim hv/\gamma$, meaning that fewer photon pairs have sufficient energy in the rest frame of the GRB to initiate $\gamma\gamma \to e^+e^-$ than was inferred from the observed energy assuming nonrelativistic kinematics. From the spectrum (15.14), this means that the factor f in Eq. (15.15) should be multiplied by a factor of $\gamma^{-2\alpha}$. Furthermore, relativistic effects increase the size of the emitting region by a factor of γ^2 over that inferred from the time period δt, so R in Eq. (15.15) should be multiplied by a factor of γ^2. Incorporating these corrections, the ultrarelativistic modification of Eq. (15.15) is

$$\tau \simeq \frac{\tau_0}{\gamma^{4+2\alpha}}, \qquad (15.16)$$

where τ_0 is evaluated from Eq. (15.15) with no ultrarelativistic correction. Thus, even if τ_0 is very large an optically thin medium can be obtained if γ is large enough.

Example 15.7 Taking the estimate from Problem 15.8 that $\tau_0 \sim 10^{14}$ and assuming $\alpha = 2$ for the spectral index gives $\tau \sim 1$ if $\gamma = 56$ and $\tau \sim 0.01$ if $\gamma = 100$.

Thus it may be concluded that consistency with observations requires that the GRB involve an ultrarelativistic jet with a Lorentz γ of order 100 or more.

Direct confirmation of large Lorentz factors: Observational confirmation that gamma-ray bursts are associated with the large values of γ deduced from the preceding theoretical analysis comes from the observed location of *breaks* in the lightcurves for afterglows. These breaks are thought to indicate the time when the initially relativistic afterglow begins to slow rapidly through interactions with the interstellar medium, which can be related to the opening angle of the jet that produced the afterglow. Such analyses typically find jet opening angles in the range $\Delta\theta = 10 - 20°$, and relating these jet opening angles to γ suggests Lorentz factors of order 100 for many gamma-ray bursts. Thus afterglow lightcurve breaks indicate directly that gamma-ray bursts are produced by ultrarelativistic jets, as was surmised from the preceding discussion. Further evidence concerning the beaming of gamma-ray bursts will be discussed in Section 24.7.1.

Implications of ultrarelativistic beaming: The GRB beaming mechanism implies that a fixed observer sees only a fraction of all gamma-ray bursts. Afterglows are not strongly beamed after slowing, so they could be detected even for a GRB not on-axis (not aimed toward Earth). Further complicating the issue is that even if a jet is viewed on-axis, the event may have no prompt high-energy emission if the initial Lorentz factor is not large enough to overcome the pair-production opacity. Such events may create on-axis events without prompt high-energy emission. Ultrarelativistic beaming solves a potential energy-conservation problem. If the energy from detected bursts were assumed to be emitted isotropically, from the energy fluxes detected on Earth total energies exceeding 10^{54} erg would be inferred for some gamma-ray bursts (comparable to the entire rest mass energy of the Sun), which would be difficult to explain by any mechanism that conserves energy. However, if GRBs are assumed to be emitted as collimated jets, then the total energy released would be much smaller than that inferred by an observer viewing it on-axis and assuming it to be isotropic, which places it more in the total-energy range of well-studied events like supernova explosions.

15.7.5 Association of GRBs with Galaxies

The localization provided by afterglows has permitted a number of long-period and short-period GRBs to be associated with distant galaxies.

1. Long-period (soft) bursts appear to be strongly correlated with *star-forming regions* (strong correlation with blue light in host galaxies).
2. Short-period (hard) bursts are generally fainter and sampled at smaller redshift than long-period bursts. They do not appear to be correlated with star-forming regions.
3. There is some evidence that long-period bursts are preferentially found in star-forming regions having low metallicity (low abundance of elements other than hydrogen and helium).

These observations provide further evidence that long-period and short-period bursts are initiated by different mechanisms, and will be interpreted further below.

15.7.6 Long-Period GRBs and Supernovae

One of the most remarkable discoveries in the study of gamma-ray bursts has been that there is a relationship between long-period gamma-ray bursts and particular types of core collapse supernovae. As suggested from the time evolution of the spectra in Fig. 15.19(a), there is compelling evidence that in some cases a long-period GRB afterglow spectrum can evolve into one resembling that of a supernova, hinting strongly that the underlying mechanism for long-period bursts may be a supernova explosion of some kind.

There are various types of supernovae classified observationally according to their spectra and lightcurves (see Chapter 20 of Ref. [105] for a more extensive discussion). The supernovae implicated in the mechanism for long-period gamma-ray bursts are classified as Types Ib and Ic, for which the explosion mechanism is thought to involve core collapse in a rapidly rotating, massive (15–30 $M_\odot$) star called a *Wolf–Rayet star*. These stars exhibit large mass loss and can shed their hydrogen and even helium envelopes before their cores collapse, and they are so massive that they are likely to collapse directly to a rotating (Kerr) black hole instead of to a neutron star. Figure 15.19(b) shows a Wolf–Rayet star surrounded by large shells of gas that it has ejected. It is thought that in a Type Ib supernova the H shell has been removed, and that in a Type Ic supernova both the H and He shells have been removed, before the stellar core collapses and triggers the supernova.

The observation that long-period gamma-ray bursts may be associated with star-forming regions of low metallicity has been interpreted in terms of this model of gamma-ray bursts

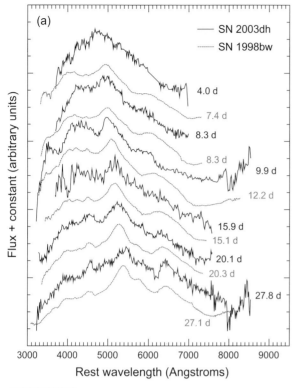

Fig. 15.19 (a) Comparison of time evolution for spectral bumps in the rest-frame optical spectrum of SN2003dh (GRB 030329) in black compared with a reference supernova SN1998bw in gray [120]. The initially rather featureless spectrum of the GRB 030329 afterglow develops bumps similar to those of the supernova SN1998bw over time, suggesting that as the afterglow produced by deposition of the GRB energy fades, an underlying spectrum characteristic of a supernova explosion is revealed. Hence GRB 030329 is also denoted by a supernova label, SN2003dh. Reprinted by permission from Macmillan Publishers Ltd.: *Nature*, **423**, 847–850, Fig. 2, Published 19 June 2003, copyright (2003). (b) A Wolf–Rayet star (black arrow) surrounded by shells of gas that it has emitted. These massive, rapidly spinning stars may be progenitors of Type Ib and Type Ic core collapse supernovae, and hence of long-period gamma-ray bursts. Credit and copyright: P. Berlind and P. Challis (cfA), 1.2-m Telescope, Whipple Obs.

resulting from the core collapse of a Wolf–Rayet star [255]. Low metallicity is expected to suppress mass loss from Wolf–Rayet stars because it decreases the surface photon opacity, making it harder for radiation pressure to eject matter from the star. This has two favorable effects on the *collapsar model* of long-period gamma-ray bursts to be described below:

1. Larger mass favors collapse directly to a black hole rather than to a neutron star.
2. Suppression of mass loss tends to minimize loss of angular momentum, leading to higher spin rates at collapse and facilitating creation of a substantial accretion disk around the black hole. Such an accretion disk may be essential to explain the extended power output of long-period GRBs.

On the other hand, there is little observational evidence that short-period bursts are associated with star-forming regions or supernovae, suggesting that the mechanism responsible for them must be different from the core collapse of Wolf–Rayet stars. As will be discussed further in Section 15.7.8, the favored mechanism for short-period bursts also involves the formation of an accreting Kerr black hole, but one produced by the merger of two neutron stars, or merger of a neutron star and a black hole, rather than by the core collapse of a massive star.

15.7.7 Characteristics of Gamma-Ray Bursts

In preparation for addressing the mechanism for gamma-ray bursts, let's gather the previous discussion into a summary of characteristics that are thought to characterize gamma-ray bursts:

1. *Cosmological origin*: The isotropic distribution on the sky found by BATSE suggests a cosmological origin. This has been confirmed by direct redshift measurements on emission lines in GRB afterglows. The highest spectroscopic redshift observed as of 2018 is $z = 8.2$ for GRB 090423.[5] Their inferred cosmological distances makes gamma-ray bursts quite remarkable, since to even be seen at such large distances they must correspond to events in which energy comparable to that of a supernova is liberated over a short period in the form of gamma-rays.
2. *Nonthermal spectrum*: The spectrum is nonthermal (see Box 15.2). Figure 15.13 illustrates a typical GRB spectrum.
3. *Duration and time structure*: The duration of individual bursts spans at least 5 orders of magnitude: from about 0.01 seconds up to at least several hundred seconds. There is a variety of time structure, from rather smooth to fluctuations on a millisecond timescale (implying a very compact source, by causality; see Box 15.3). Figure 15.16 displays a time profile for a gamma-ray burst.
4. *Beamed bursts*: The gamma-rays of a burst are strongly beamed, suggesting emission from tightly-collimated ultrarelativistic jets. In general, gamma-rays produced in the burst must escape with very little interaction with surrounding matter since even a small interaction would downscatter gamma-rays and thermalize the spectrum.
5. *Two classes of bursts*: Phenomenology suggests that there are two classes of bursts: long-period and short-period bursts, having many common features but with sufficient differences to suggest that the bursts in the two classes are triggered by different events.
6. *Afterglows and the fireball model*: Transient afterglows are observed at various wavelengths after gamma-ray bursts that can be explained phenomenologically by the *relativistic fireball model* illustrated in Fig. 15.18. In this model deposition by some unspecified central engine of a very large amount of energy initiates a relativistic fireball that produces the afterglows.

The next section takes up the issue of what central engines could be capable of producing gamma-ray bursts with these characteristics.

---

[5] The naming convention for gamma-ray bursts is "GRB" followed by a string of three 2-digit numbers indicating the year, month, and day of observation, respectively. For example, GRB 090423 was observed by the SWIFT satellite on April 23, 2009. The redshift of GRB 090423 indicates that it occurred only ∼600 million years after the big bang! One important implication is that the Universe was making massive stars within a few hundred million years of its formation.

15.7.8 Mechanisms for the Central Engine

The central engines responsible for gamma-ray bursts and the associated afterglows are not well understood, but an acceptable model must embody at least the following features:

1. All models require highly relativistic jets to account for observed properties of gamma-ray bursts.

    (a) Lorentz γ-factors of order 100 or larger appear to be required by observations. These are enormous by astrophysical standards; jets observed from most quasars and active galactic nuclei do not exceed $\gamma \sim 10$.
    (b) The jets must have opening angles as small as ~ 0.1 rad and energy as large as $\sim 10^{52}$ erg. Notice that ultrarelativistic jets are naturally highly focused since the kinematics dictate that the opening angle $\Delta\theta \propto \gamma^{-1}$.
    (c) Long-period bursts must (at least sometimes) deliver $\sim 10^{52}$ erg to a much larger angular range (~ 1 rad) to produce the accompanying supernova, and the central engine must operate for 10 seconds or longer in these long-period bursts to account for their duration.

2. The large radiated power and its timescale, particularly for long-period bursts, implies accretion onto a compact object. Thus, acceptable models must produce substantial accretion disks.

Almost the only way known to explain such a rapid release of that amount of energy that is then sustained over periods as long as many tens of seconds is from a collapse involving a compact gravitational source that forms an accretion disk. Two general classes of models are now thought to account for GRBs.

1. Merger of two neutron stars, or a neutron star and a black hole, with jet outflow producing a burst of gamma-rays as the two objects collapse to a Kerr black hole. These are thought to be associated with short-period bursts.
2. A *hypernova,* where a spinning massive star collapses to a Kerr black hole and jet outflow produces a burst of gamma-rays (see the collapsar model discussed below). These are thought to be associated with long-period bursts.

In either case the outcome is a Kerr black hole having large angular momentum and strong magnetic fields, surrounded by an accretion disk of matter that has not yet fallen into the black hole. This scenario likely produces highly focused relativistic jet outflow on the polar axes of the Kerr black hole, powered by some combination of rapid accretion from the disk, neutrino–antineutrino annihilation, and strong coupling to the magnetic field. Thus, the GRB black hole engine may have many similarities with the engine powering AGN and quasars, but on a stellar rather than galactic-core scale.

The collapsar model and long-period bursts: An overview of the collapsar model is given in Fig. 15.20. Simulations of relativistic jets breaking out of a Wolf–Rayet star are shown

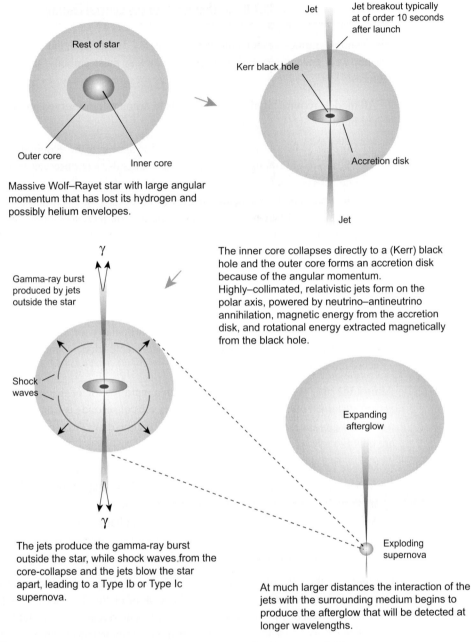

Fig. 15.20 Collapsar model for long-period GRB and accompanying Type Ib or Ic supernova [255].

in Fig. 15.21, in Fig. 15.22(a) a simulation of a Wolf–Rayet star 20 seconds after core collapse is shown, and Fig. 15.22(b) shows a strong nucleon wind blowing off the accretion disk. In the collapsar model the GRB is powered by a relativistic jet, but the accompanying supernova is powered by the disk wind of Fig. 15.22(b), which produces the supernova

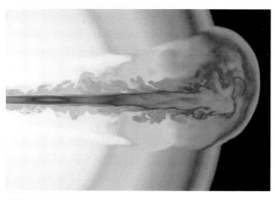
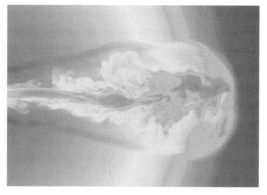

Fig. 15.21 Simulations of relativistic jets breaking out of Wolf–Rayet stars [149, 255]. Breakout of the $\gamma \sim 200$ jet is 8 seconds after launch from the center of a 15 $M_\odot$ Wolf–Rayet star. Reproduced from Woosley and Bloom, the Supernova–Gamma-Ray Burst Connection, *Annual Review of Astronomy and Astrophysics*, **44**, 507–566 (2006), with permission from the Annual Reviews.

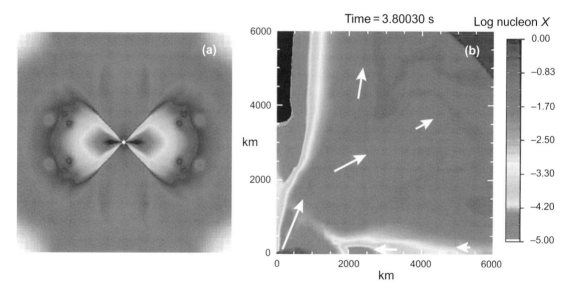

Fig. 15.22 (a) A rapidly rotating 14 $M_\odot$ Wolf–Rayet star, 20 seconds after core collapse. The polar axis is vertical, the density scale is logarithmic, and the 4.4 $M_\odot$ Kerr black hole has been accreting at $\sim 0.1\ M_\odot\ {\rm s}^{-1}$ for 15 seconds at this point [255]. (b) Simulation of the nucleon wind blowing off the accretion disk in a collapsar model [255]. The gray-scale contours represent the log of the nucleon mass fraction X and arrows indicate the general flow. Reproduced from Woosley and Bloom, the Supernova–Gamma-Ray Burst Connection, *Annual Review of Astronomy and Astrophysics*, **44**, 507–566 (2006), with permission from the Annual Reviews.

and synthesizes the ^{56}Ni that powers the lightcurve of the supernova by radioactive decay.

Merging neutron stars and short-period bursts: The collapsar model accounts plausibly for long-period gamma-ray bursts, but is unlikely to be applicable for short-period bursts because they are not observed to be preferentially associated with star-forming regions. The most widely accepted model for production of short-period gamma-ray bursts involves binary neutron stars merging to form a Kerr black hole. Observational evidence in support of this assertion will be given in Section 24.7.

A simulation of a neutron star merger will be shown in Fig. 24.1 and an illustration of how the original magnetic fields of the neutron stars can be magnified in the merger is illustrated in Fig. 15.23. One possible mechanism for powering the jets in GRBs produced by neutron star mergers is neutrino–antineutrino annihilation above and below the plane of the merger disk, as illustrated in Fig. 15.24. Another is accretion onto the rotating black hole from the surrounding disk in the merger, and another is to tap the power of the very strong magnetic fields that are expected (see Fig. 15.23), for example as illustrated in Fig. 15.25. In this model, the frame-dragging effects associated with the Kerr black hole wind the flux lines associated with the magnetic field around the black hole and spiral them off the poles of the black hole rotation axis, thus powering relativistic jets emitted on the polar axes. More detail may be found in Fig. 15.25 and Ref. [222].

A detailed simulation of a neutron-star merger to form a Kerr black hole with strong magnetic fields that lead to magnetically powered polar jets is illustrated in Fig. 15.26 [86, 201]. Successive panels show the evolution in time of the mass density, with magnetic field lines superposed. The first panel shows the state shortly after initial contact and the second displays a high-mass neutron star (HMNS) configuration (one too massive to remain a neutron star for long). In the bottom two panels a Kerr black hole has formed in the center with a disk around it, and the magnetic field is wound up by the disk to a strength of order 10^{15} gauss, with an opening angle for the field lines in the polar direction of about $30°$.

To date, neutron star merger simulations tend to use unphysically large initial neutron star magnetic fields. The motivation is that in the HMNS phase a *magnetorotational instability (MRI)* [36, 37] is expected to develop that amplifies the magnetic field over the lifetime of the HMNS and accelerates the collapse to a black hole [224]. However, there is insufficient resolution available to model the MRI accurately in realistic mergers with present computers, so the MRI effect is approximated by artificially boosting the initial magnetic fields. An additional factor of 100 in resolution is required to deal correctly with the MRI in mergers, which must await the next generation of supercomputers (the *exascale*, with computers 100–1000 faster than current machines) [86].

15.7.9 Gamma-Ray Bursts and Gravitational Waves

Finally we note that neutron star mergers should produce gravitational waves of sufficient strength to be observable in earth-based gravitational wave detectors. This possibility will be discussed further in Chapter 24. This also raises the intriguing possibility of *multi-messenger astronomy* (see Section 24.6) where, for example, a gamma-ray burst might be observed in coincidence with gravitational waves from a binary neutron star merger. Such an observation would presumably have a large impact on the understanding of neutron-star structure, the mechanism for gamma-ray bursts, and gravitational wave sources.

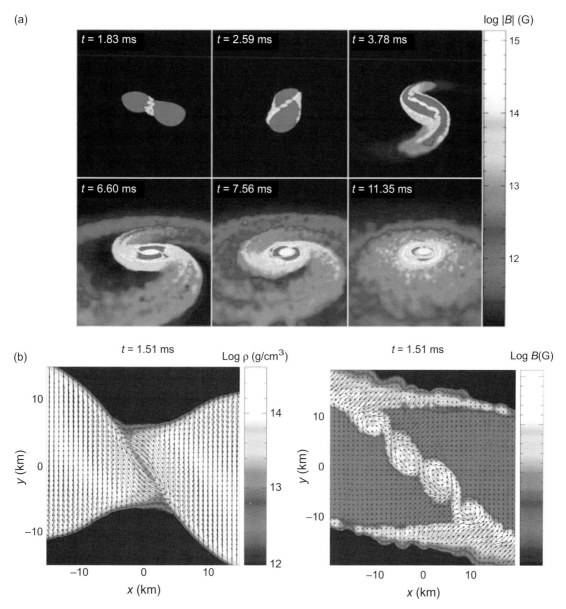

Fig. 15.23 (a) A simulation of merging neutron stars with the magnetic field strength indicated by the grayscale [195]. (b) Amplification of magnetic fields in merging neutron stars for the simulation shown in (a). Arrows indicate the magnitude and direction of the magnetic field. In the simulation the shear produced at the merger boundary is capable of substantially amplifying the already significant magnetic fields that are present. Reproduced from Price and Rosswog, Producing Ultrastrong Magnetic Fields in Neutron Star Mergers, *Science*, **312**(5774), 719–722, 05 May 2006. Reprinted with permission from AAAS.

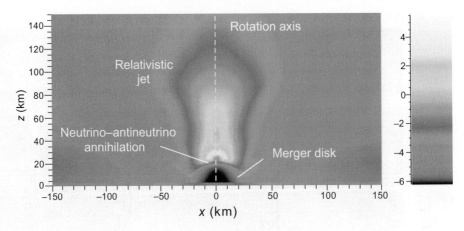

Fig. 15.24 Jet powered by neutrino–antineutrino annihilation in a neutron-star merger [208].

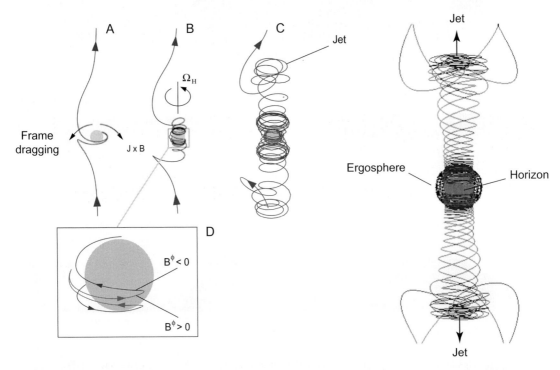

Fig. 15.25 Relativistic jet powered by frame dragging of magnetic fields for a Kerr black hole [222]. Reproduced from Semenov *et al.*, Simulations of Jets Driven by Black Hole Rotation, *Science*, **305**(5686), 978–980, 13 Aug 2004. Copyright © 2004, American Association for the Advancement of Science.

As we shall discuss more extensively in Section 24.7, as this book goes to press a short-period gamma-ray burst and afterglows at various wavelengths have just been observed in coincidence with gravitational waves from a neutron star merger. This represents the first true multimessenger gravitational-wave event observed in astronomy, and provides the

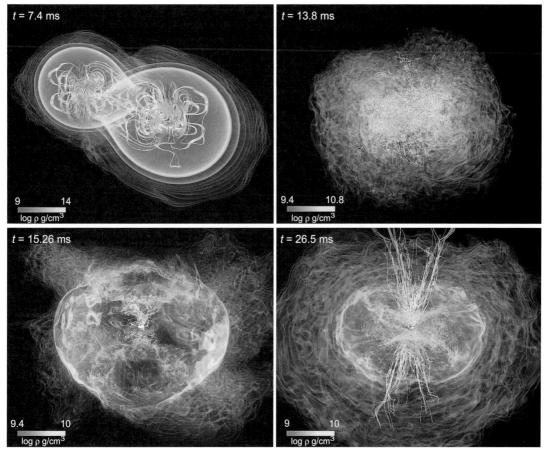

Fig. 15.26 Neutron star merger simulation with strong magnetic fields [86, 201]. Grayscale indicates the log of the density. Reproduced from Rezzolla *et al.*, The Missing Link: Merging Neutron Stars Naturally Produce Jet-Like Structures and Can Power Short Gamma-Ray Bursts, © 2011. The American Astronomical Society. All rights reserved. *Astrophysical Journal Letters*, **732**(1). Reproduced with permission from the authors.

first strong evidence for the conjectured neutron star merger mechanism for short-period gamma-ray bursts. In addition it provides the first direct evidence that many of the heavier elements are synthesized in neutron star mergers, places an extremely strong constraint on any deviation of the speed of gravity from the speed of light, suggests a new way to determine the Hubble constant for expansion of the Universe, and begins to place new constraints on the deformability of neutron-star matter and the neutron star equation of state.

Background and Further Reading

A more general discussion of accretion in stellar environments may be found in the book by Frank, King, and Raine [90]; and in Chapter 18 of Guidry [105]. Hartle [111] gives a basic discussion of accretion on Kerr black holes. An introduction to active galactic nuclei, quasars, and the black hole central engine unified model may be found in Peterson [185] and in Robson [206]. The unified AGN model in simplest form has been called into question by various more-recent observations; the situation as of 2015 has been reviewed in Netzer [162]. An overview of gamma-ray bursts has been given by Piran [188, 189] and an overview of gamma-ray bursts with emphasis on collapsar models may be found in Woosley and Bloom [255]. Neutron star merger models for gamma-ray bursts are discussed in Price and Rosswog [195]; Rosswog [208]; Faber and Rasio [86]; Rezzolla et al. [201]; and the review by Berger [44].

Problems

15.1 Active galactic nuclei (AGN) and quasars can have luminosities of 10^{47} erg s^{-1}, with variations observed in this energy output on timescales of days or less. Use these two observations to show that

(a) If this energy derives from thermonuclear reactions it is necessary that several hundred solar masses be consumed per year, but if it derives from accretion (with accretion efficiencies that are theoretically possible with rotating supermassive black holes), the energy output could be generated by accretion of as little as several solar masses per year.

(b) If the source of the energy powering the AGN is accretion and the source radiates at the Eddington limit (see Section 15.2.2), the mass of the central object onto which accretion occurs must be of order 10^9 $M_\odot$.

(c) The variability timescale and part (b) imply that 10^{47} erg s^{-1} must be produced in a region containing at least 10^9 $M_\odot$, and this region cannot be substantially larger than the Solar System.

Hint: See the general discussion of black hole accretion in Section 15.2. ***

15.2 Verify the entries for energy released per gram by accretion and the ratio of the energy released to that obtained from hydrogen fusion for the entries in Table 15.1.

15.3 From the geometry of the diagram in Box 15.1 for 3C 279, estimate the apparent transverse velocity of the jet by considering how far its leading edge appears to move over a period of several years. ***

15.4 Suppose a highly relativistic jet with $v = 0.9999995c$. What is the maximum frequency to which an optical photon of frequency $v = 5 \times 10^{14}$ Hz could be boosted by inverse Compton scattering in this jet? ***

15.5 The observed flux density for light coming from sources moving at relativistic velocities is altered from that expected in the rest frame by relativistic time dilation. If the spectral energy distribution of the emitted flux is ignored for simplicity, the ratio of observed to emitted flux densities is given by

$$\frac{S_{\rm obs}}{S_{\rm emitted}} = \frac{1}{[\gamma(1-\beta\cos\theta)]^3},$$

where γ is the Lorentz factor, $\beta = v/c$ is the velocity, and θ is the angle between the observer and the direction of motion for the source. Assume light emitted by oppositely directed relativistic jets having the same γ and β. Derive an expression for the ratio of observed flux densities for the two jets as a function of β and the angle θ between the line of sight and the jet most directed toward the observer. Use this to show that if θ is small and β is near one, the oppositely directed jet (the counterjet) will appear to be much fainter than the jet directed more toward the observer, even if the emitted flux density is the same for the jet and counterjet. ***

15.6 Use the Hubble law given in Eq. (16.1), the apparent visual magnitude of $+12.9$, and the observed redshift $z = 0.158$ to estimate the absolute visual magnitude of the quasar 3C 273 (the brightest quasar known). Compare the luminosity of 3C 273 with that of the Andromeda Galaxy (M31), the Sun, and the giant elliptical galaxy M87. Are your results consistent with the statement in this chapter that a typical quasar has 1000 times greater luminosity than a large normal galaxy? ***

15.7 Accretion disks around compact objects are observed to radiate approximately as blackbodies, often at a significant fraction of the Eddington luminosity (15.3). Take as a very crude model of an accretion disk around a compact object a thin disk of radius R radiating as a blackbody at a fraction η of the Eddington luminosity. Derive an expression for the temperature T of such a disk. Use this expression to estimate the temperature and therefore peak emission wavelength for an accretion disk around a neutron star, and for Schwarzschild black holes having masses of order $10 \, M_\odot$ and $10^8 \, M_\odot$, respectively. For the neutron star, approximate R as comparable to the radius of the neutron star, and for the black hole approximate R as comparable to the innermost stable circular radius in a Schwarzschild spacetime. Use these results to explain the observed wavelength ranges associated with the accretion disks for such objects. ***

15.8 Assume for a typical gamma-ray burst that the distance to the source is ~ 3000 Mpc, the fluence is $\sim 10^{-7}$ erg cm^{-2}, the fraction of photon pairs with sufficient energy for $\gamma\gamma \to e^+ e^-$ to occur is of order one, and that the burst intensity exhibits fluctuations on a 10 ms timescale. If the burst is taken (contrary to fact) to not be relativistic, show that the optical depth given by Eq. (15.15) is huge and therefore inconsistent with the observed nonthermal spectrum for gamma-ray bursts. Show that in principle this problem is alleviated if the kinematics are ultrarelativistic and the value of the Lorentz γ-factor is large enough. *Hint*: See Eq. (15.16) and the discussion preceding it. ***

PART III

COSMOLOGY

16 The Hubble Expansion

Observational characteristics, coupled with theoretical interpretation to be discussed further in subsequent chapters, allows the formulation of a *standard picture* of our Universe. In this chapter we introduce this standard picture, concentrating in particular on the expansion of the Universe as codified in the Hubble law.

16.1 The Standard Picture

The standard picture rests on only a few ideas, but they have profound significance for the nature of the Universe. Let's begin by enumerating and briefly discussing these ideas.

16.1.1 Mass Distribution on Large Scales

The Universe appears to be homogeneous (no preferred place) and isotropic (no preferred direction), when considered on sufficiently large scales. Thus, averaged over a large enough volume, no part of the Universe looks any different from any other part. For example, when averaged over distances of order 50 Mpc the fluctuation in mass distribution is of order unity, $\delta M/M \simeq 1$, but when averaged over a distance of 4000 Mpc (which is comparable to the present horizon), $\delta M/M \leq 10^{-4}$ [180]. Even more precise evidence comes from the cosmic microwave background radiation. It is observed to be very nearly isotropic on scales stretching to most of the visible Universe. The idea that the Universe is homogeneous and isotropic on large scales is called the *cosmological principle*.[1]

16.1.2 The Universe is Expanding

Observations indicate that the Universe is expanding; the interpretation of general relativity is that this is because space itself is expanding. As Hubble demonstrated by observing the wavelength of light from distant galaxies, the distance ℓ between conserved particles is changing according to the *Hubble law*

---

[1] The cosmological principle should not be confused with the *perfect cosmological principle,* which was the underlying idea of the *steady state model.* According to the perfect cosmological principle the Universe is homogeneous in both space and time; thus it looks the same not only from any place, but from any time. The steady state theory and perfect cosmological principle fell prey to various observations indicating that the early Universe looked different from the present one, invalidating the idea that the Universe is constant in time.

> **Box 16.1** **What is Expanding?**
>
> The Hubble expansion is a cosmological effect observed in the large-distance behavior of the Universe. Small objects, such as our bodies, are held together by chemical (electrical) forces. They do not expand with the Universe. Even larger objects like planets, solar systems, and galaxies are also held together by forces, in this case gravitational in origin. They do not expand with the Universe either. It is only on much larger scales – beyond superclusters of galaxies – that gravitational forces among local objects are sufficiently weak to cause negligible perturbation on the overall Hubble expansion.

$$v \equiv \frac{d\ell}{dt} = H_0 \ell, \tag{16.1}$$

where H_0 is the *Hubble constant* and v is an apparent recessional velocity obtained by interpreting (incorrectly, as we shall see) the spectral shift as a Doppler shift. As a legacy of some past controversy over the value of H_0, any uncertainty is sometimes absorbed into a dimensionless parameter h by expressing H_0 as

$$H_0 = 100 h \text{ km s}^{-1} \text{ Mpc}^{-1} = 3.24 \times 10^{-18} h \text{ s}^{-1}, \tag{16.2}$$

where h is of order one. There is now some consensus that the correct value of the Hubble constant is in the vicinity of $70 \text{ km s}^{-1} \text{ Mpc}^{-1}$, though different methods give values differing from this by as much as several percent. For this book a value of

$$H_0 = 72 \text{ km s}^{-1} \text{ Mpc}^{-1}, \tag{16.3}$$

corresponding to $h = 0.72$ in Eq. (16.2), often will be adopted when a numerical value is required for illustrations and problems. A *Hubble length* L_H may be defined through

$$L_H = \frac{c}{H_0} \simeq 4000 \text{ Mpc}. \tag{16.4}$$

Thus, for a galaxy lying a Hubble length away from us $v = H_0(c/H_0) = c$, which implies that the recessional velocity of a galaxy further away would exceed the speed of light if the observed redshifts were interpreted literally as Doppler shifts. It will be shown below that in a general relativistic description of the Universe the redshift of the receding galaxies should not be interpreted as a Doppler shift caused by velocities in spacetime. Instead, it is a consequence of the expansion of space itself, which stretches the wavelengths of all light as the expansion proceeds. The lightspeed limit of special relativity applies to velocities *in* space; it does not apply to space itself. In addition, as discussed in Box 16.1, it is important to understand exactly what is expanding and what is not expanding.

16.1.3 The Expansion Is Governed by General Relativity

It is possible to discuss the expanding Universe using only Newtonian physics and insights borrowed from relativity, as will be illustrated in Chapter 17. However, in the final analysis there are serious technical and philosophical difficulties that arise and that require replacement of Newtonian gravitation with general relativity for their resolution (see Problem 16.5). Central to these issues is the understanding of space and time in relativity compared with that in Newtonian gravitation. In relativity, space and time are not separate concepts but enter as a unified spacetime continuum. Even more fundamentally, space and time in relativity are not viewed as a passive background upon which events happen. Relativistic space and time are not "things" but rather are abstractions expressing a relationship between events. In this view, space and time do not have an existence separate from events involving matter and energy.

16.1.4 There is a Big Bang in Our Past

Evidence suggests that our present Universe expanded from an initial condition of very high density and temperature. This emergence of the Universe from a hot, dense initial state dominated by thermal radiation is called the *big bang*. The popular (mis)conception that the big bang was a gigantic explosion is potentially misleading because it conveys the idea that it happened in spacetime and that the resulting expansion of the Universe is a consequence of forces generated by this explosion. This is not the view of modern cosmology. Because the evolution of the Universe is governed by general relativity, the more consistent interpretation of the big bang is that it did not happen *in* spacetime but that space and time themselves are created in the big bang. Thus, questions of "what happened before the big bang?" and "what is the Universe expanding into?" become meaningless because these questions presuppose the existence of a space and time background upon which events happen. The big bang should rather be viewed as an initial condition for the Universe. We may think of it loosely as an "explosion" because of the hot, dense nature of the initial state. But then the explosion should be viewed as happening at all points in space, so that it makes no sense to talk about a "center" for the big bang.

General relativity implies that the initial state was a spacetime singularity. The question of whether this is correct can be settled only by a full theory of quantum gravitation, since general relativity cannot be applied too close to the initial singularity without incorporating the principles of quantum mechanics. However, for most issues in cosmology the question of whether there was an initial spacetime singularity is not relevant. For those issues, all that is important is that the early Universe was very hot and very dense. This hot and dense initial state is what shall be meant by the big bang.

16.1.5 Particle Content Influences the Evolution

The Universe contains a variety of elementary particles and their associated quantum fields. These influence its evolution through processes that require fundamental knowledge from the domains of elementary particle, nuclear, and atomic physics. The "ordinary" matter composed of atoms, ions, and molecules that exists all around us is termed *baryonic matter*.[2] Although baryonic matter is the most obvious matter in everyday experience and

---

[2] Baryons are the strongly interacting particles of half-integer spin such as protons and neutrons that are made from three quarks. The electrons in this matter are leptons (particles of half-integer spin that do not undergo the strong interactions), not baryons. But since nuclei are made of baryons and these dominate the mass of ordinary matter, it is normal in cosmology to refer to ordinary matter as "baryonic."

is what is seen through a telescope, data indicate that only a small fraction of the mass in the Universe is baryonic. The bulk of the matter appears to be in the form of *dark matter*, which is easily detected only through its gravitational influence. It is not clear yet what this dark matter is but one widely held idea is that it is primarily composed of as-yet-undiscovered elementary particles that interact only weakly with other matter and radiation. There is also substantial evidence that the evolution of the Universe is strongly influenced by *dark energy* (not to be confused with dark matter), which is an energy density that permeates even empty space and causes gravity to effectively become repulsive. As for dark matter, the origin of dark energy is shrouded in mystery but a standard guess is that it is associated – in a way not fully understood – with quantum-mechanical fluctuations of the vacuum, or possibly with an undiscovered elementary particle field.

A fundamental distinction for the particles and associated fields that make up the Universe is whether they are massless (or nearly so), or massive. This distinction is important because Lorentz-invariant quantum field theories require that massless particles move at the speed of light and that particles with finite mass move at speeds less than that of light. Therefore, massless particles like photons, gravitons, and gluons, and nearly massless particles like neutrinos, are highly relativistic. In cosmology it is common to refer to such massless or nearly massless particles as *radiation.* Conversely, particles with significant mass have characteristic velocities well below that of light (unless temperatures are extremely high) and are therefore nonrelativistic. It is common in cosmology to refer to nonrelativistic particles as *matter*. Because nonrelativistic matter characteristically moves at low velocity, it exerts little pressure compared with relativistic matter. It is also common to refer to matter that exerts negligible pressure as *dust.* The present energy density of the Universe is dominated by nonrelativistic matter and dark energy. However, this was not always the case: in its earliest stages the Universe was dominated by radiation.

Example 16.1 Particles with rest mass m are nonrelativistic at temperatures T such that $kT \ll mc^2$. For example, electrons have a rest mass of 511 keV and are nonrelativistic below about 6×10^9 K, but protons have a rest mass of 931 MeV and are nonrelativistic up to $T \sim 10^{13}$ K. Conversely, massless photons, gravitons, and gluons, and nearly massless neutrinos are always relativistic.

16.1.6 There is a Cosmic Microwave Background

The most important feature of the standard picture – other than the Hubble expansion itself – is that the Universe is filled with a photon background that lies in the microwave region of the spectrum and that is extremely smooth and isotropic. Any attempt to understand the standard picture must as a starting point account for the expansion of the Universe and for this *cosmic microwave background (CMB).* But the CMB is not *completely* smooth, and precise measurements of tiny fluctuations in the CMB are turning cosmology into a highly quantitative science. Although the CMB currently peaks in the microwave region of the spectrum, its wavelength has been redshifting steadily since the big bang and it was originally radiation of much higher energy. For example, near the time when the temperature dropped low enough for electrons to combine with protons the present microwave background peaked in the near-infrared spectrum. In the current Universe, the CMB accounts for more than 90% of the photon energy density, with less than 10% in starlight.

16.2 The Hubble Law

The Hubble law is given by Eq. (16.1). Because the Hubble expansion will be interpreted below in terms of an expansion of space itself, it is convenient to introduce a *scale factor* $a(t)$ for the Universe that describes how distances scale because of the expansion of the Universe. Hubble's law for the evolution of the scale factor is illustrated in Fig. 16.1. The slopes of the straight lines plotted there define the Hubble constant entering Eq. (16.1).

16.2.1 The Hubble Parameter

The Hubble constant is characteristic of all space but has possible time dependence as the Universe evolves. One often refers to the *Hubble parameter* $H = H(t)$, meaning a coefficient in (16.1) that varies with time, and to the *Hubble constant* H_0, meaning the value of $H(t)$ today. Hubble's original value was $H_0 = 550 \text{ km s}^{-1} \text{ Mpc}^{-1}$, which is almost an order of magnitude larger than the presently accepted value. The large revision – and a corresponding shift in the distance scale of the Universe – resulted from Hubble's original sample being a poorly determined one of what now would be considered nearby galaxies, and because of confusion over the extra-galactic distance scale in the early twentieth century caused by errors in classifying variable stars and failure to account properly for the effect of interstellar dust on the propagation of light.

16.2.2 Redshifts

If photons emitted from a galaxy at wavelength λ_{emit} are found to be shifted to a wavelength λ_{obs} when observed on Earth, the *redshift parameter* z for the galaxy is defined by

$$z \equiv \frac{\lambda_{\text{obs}} - \lambda_{\text{emit}}}{\lambda_{\text{emit}}}. \tag{16.5}$$

A negative value of z is termed a *blueshift* while a positive value is termed a *redshift*. The Hubble law gives rise only to redshifts. Therefore, any blueshifts that are observed correspond to *peculiar motion* of objects with respect to the general Hubble flow.[3]

---

[3] "Peculiar" here doesn't mean "strange" but rather takes its original meaning of "a property specific to an object." Peculiar velocities can either add to or subtract from the overall Hubble expansion velocities.

Example 16.2 From the SIMBAD Astronomical Database [4], the Andromeda Galaxy (M31) in our Local Group of galaxies is 780 kpc away and it is moving toward us at about $-300 \, \text{km s}^{-1}$ (redshift $z = -0.001$) because of gravitational attraction between M31 and our galaxy. Thus its spectral lines are blueshifted by peculiar motion due to local gravitational attraction. The Virgo cluster is the nearest rich cluster of galaxies, at a distance of ~ 16.5 Mpc with an average redshift $z = +0.0038$. Most of its individual galaxies are redshifted but a few have blueshifts corresponding to negative radial velocities. For example, NGC 4254 is redshifted ($z = +0.008036$) with a radial velocity of $+2399 \, \text{km s}^{-1}$ but NGC 4406 is blueshifted ($z = -0.00061$) with a radial velocity of $-183 \, \text{km s}^{-1}$.

16.2.3 Expansion Interpretation of Redshifts

The redshifts associated with the Hubble law are not Doppler shifts in the normal sense. They are most consistently interpreted in terms of the expansion of space, which may be parameterized by the cosmic scale factor $a(t)$, as illustrated in Fig. 16.1. As will be shown in Chapter 18, if all peculiar motion is ignored the time dependence of the expansion

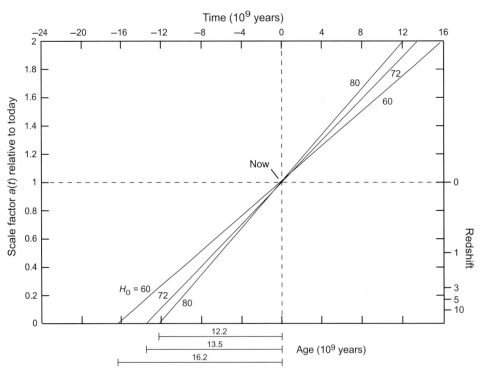

Fig. 16.1 Variation of the scale factor $a(t)$ with time (top axis) for three values of the Hubble constant $H_0 = 60, 72,$ and $80 \, \text{km s}^{-1} \, \text{Mpc}^{-1}$. The corresponding Hubble times (see Section 16.2.4) estimating the age of the Universe are indicated below the lower axis. The redshift (16.5) corresponding to a given value of the scale factor is indicated on the right vertical axis.

16.2 The Hubble Law

is lodged entirely in the time dependence of $a(t)$ and all cosmological distances simply scale with this factor. An analogy on a 2-dimensional surface will be exploited in later discussion: distances between dots placed on the surface of a balloon all scale with the radius of the balloon as it expands (see Figures 16.2 and 18.2).

Let's look at this more precisely. The appropriate metric for the Universe on large scales is given by Eq. (18.14). Light traveling between two galaxies follows a null geodesic ($ds^2 = 0$) and choosing $d\theta = d\varphi = 0$ in Eq. (18.14) gives

$$c^2 dt^2 = a^2(t) dr^2, \tag{16.6}$$

where $a(t)$ sets the overall scale for distances in the Universe at time t, the radial coordinate distance is r, and spacetime curvature has been neglected by choosing $k = 0$ in Eq. (18.14). Therefore,

$$\frac{c\,dt}{a(t)} = dr. \tag{16.7}$$

Consider a wavecrest of monochromatic light with wavelength λ' that is emitted at time t' from one galaxy and detected with wavelength λ_0 at time t_0 in another galaxy [211]. Integrating both sides of Eq. (16.7) gives

$$c \int_{t'}^{t_0} \frac{dt}{a(t)} = \int_0^r dr = r. \tag{16.8}$$

The next wavecrest is emitted from the first galaxy at $t = t' + \lambda'/c$ and is detected in the second galaxy at time $t = t_0 + \lambda_0/c$. For the second wave crest, integrating as above assuming that the interval between wavecrests is negligible compared with the timescale for expansion of the Universe gives

$$c \int_{t'+\lambda'/c}^{t_0+\lambda_0/c} \frac{dt}{a(t)} = \int_0^r dr = r. \tag{16.9}$$

The right sides of (16.8) and (16.9) are equal so their left sides may be equated to obtain

$$\int_{t'}^{t_0} \frac{dt}{a(t)} = \int_{t'+\lambda'/c}^{t_0+\lambda_0/c} \frac{dt}{a(t)}, \tag{16.10}$$

which may be rewritten as

$$\int_{t'}^{t'+\lambda'/c} \frac{dt}{a(t)} + \int_{t'+\lambda'/c}^{t_0} \frac{dt}{a(t)} = \int_{t'+\lambda'/c}^{t_0} \frac{dt}{a(t)} + \int_{t_0}^{t_0+\lambda_0/c} \frac{dt}{a(t)},$$

and therefore as

$$\int_{t'}^{t'+\lambda'/c} \frac{dt}{a(t)} = \int_{t_0}^{t_0+\lambda_0/c} \frac{dt}{a(t)}.$$

Because the interval between wave crests is negligible compared with the characteristic timescale for expansion, the factors $a(t)^{-1}$ may be brought outside the integrals to give

$$\frac{\lambda'}{\lambda_0} = \frac{a(t')}{a(t_0)}, \tag{16.11}$$

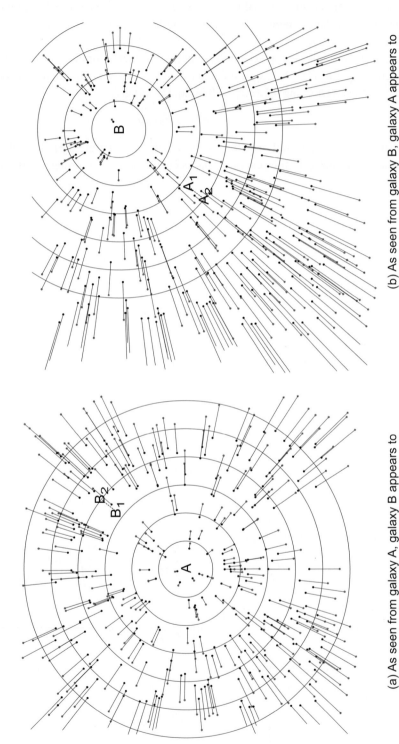

(a) As seen from galaxy A, galaxy B appears to recede from B_1 to B_2 over the time interval shown

(b) As seen from galaxy B, galaxy A appears to recede from A_1 to A_2 over the time interval shown

Fig. 16.2 The *same uniform 2-dimensional Hubble expansion* as seen from two different vantage points. (a) As observed from galaxy A. (b) As observed from galaxy B. Although the patterns in (a) and (b) look very different, it is exactly the same expansion seen from two different perspectives, and the two observers will deduce the same Hubble constant governing the expansion (see Fig. 16.3).

which demonstrates explicitly the stretching of wavelengths caused by the expansion of the Universe, and that the cosmological redshift is *not* a Doppler shift (the result does not depend on a relative velocity of the galaxies). Using the definition (16.5) for the redshift z,

$$1 + z = 1 + \frac{\lambda_0 - \lambda'}{\lambda'} = \frac{\lambda_0}{\lambda'} = \frac{a(t_0)}{a(t')}. \tag{16.12}$$

It is conventional to express the redshift as

$$z = \frac{1}{a(t')} - 1, \tag{16.13}$$

by normalizing the scale parameter so that its value in the present Universe is unity, $a(t_0) \equiv 1$, where t_0 is the present time.

The preceding derivation shows that the redshift entering the Hubble law depends *only* on the ratio of the scale parameters at the time of emission and time of detection for the light; it is independent of the details of how the scale parameter changed between the two times. The ratio of the scale parameters at two different times is determined by the cosmological model in use. Measuring the redshift of a distant object is then equivalent to specifying the scale parameter of the expanding Universe at the time when the light was emitted from the distant object, relative to the scale parameter today.

Example 16.3 Suppose that a redshift $z = \Delta\lambda/\lambda = 5$ is determined from the spectrum of a distant quasar. Then from Eq. (16.13) the scale factor of the Universe at the time that light was emitted from the quasar was $\frac{1}{6}$ of the scale factor for the current Universe.

The preceding discussion indicates that the scale factor $a(t)$ or the redshift z may be used interchangeably as time variables for a Universe in which the scale parameter changes monotonically with time (compare the right and left axes of Fig. 16.1).

16.2.4 The Hubble Time

The Hubble constant has the dimension of inverse time: $[H_0] = [\text{km s}^{-1} \text{ Mpc}^{-1}] = \text{time}^{-1}$ and H_0^{-1} defines a time called the *Hubble time* τ_H,

$$\tau_H \equiv H_0^{-1} = 9.8 h^{-1} \times 10^9 \text{ yr}. \tag{16.14}$$

If the Hubble law were obeyed with a constant value of H_0 for all time, Fig. 16.1 indicates that the intercept of the curve with the time axis defines the time when the scale factor was zero (the big bang). Hence, the value of $\tau_H = 1/H_0$ is sometimes taken as an estimate of the age of the Universe. This is simply a statement that if the expansion rate today is the same as the expansion rate since the big bang, the time for the Universe to evolve from the big bang to today is the inverse of the Hubble constant. In general the Hubble time is not a correct age for the Universe because the Hubble parameter can remain constant only in a Universe devoid of matter, fields, and energy. The realistic Universe contains all of these and the expansion of the Universe is accelerated (positively or negatively, depending on the details) because of gravitational interactions. In later cosmological models it will be seen that the true age of the Universe may be substantially longer or shorter than τ_H, depending on the details of the matter and energy contained in the Universe.

16.2.5 A 2-Dimensional Hubble Expansion Model

Figure 16.2 illustrates a 2-dimensional Hubble expansion, as viewed from two different vantage points. In this simulation 300 galaxies are distributed randomly and then the entire 2-dimensional space has been expanded by a factor of 1.25. Each pair of dots connected by a line corresponds to the position of a galaxy before and after the expansion, as measured relative to an observer at the center of the concentric circles. Two different observing galaxies are indicated, one denoted by A and one by B. In the left diagram the observer is at the center (marked by the A symbol). In the right diagram the same galaxies and the same expanded space are displayed, but the vantage point is now another galaxy (marked by B). Each diagram corresponds to the *same universe,* but viewed from two different vantage points. Each observer concludes that they are the center of the expansion, each observer will find that the expanding space appears to obey a Hubble law, and each observer will extract the same Hubble constant from the data available to that observer (within observational uncertainty) by measuring the recession of other distant galaxies.

Figure 16.3 illustrates the Hubble law extracted from Fig. 16.2. The plot is constructed by choosing representative galaxies and plotting the apparent recessional velocity versus the distance from the observer, with the Hubble constant extracted from the slope of a linear fit to the data. Within a simulated observational uncertainty, the observer at A and the observer at B extract the same Hubble parameter by measuring recessional velocities and distances relative to that observer's position. In this model distances are measured in pixels

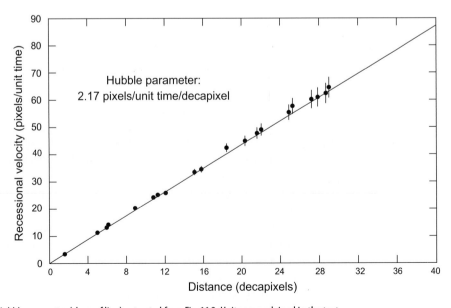

Fig. 16.3 Hubble parameter (slope of line) extracted from Fig. 16.2. Units are explained in the text.

on the computer screen, time is measured in units of the elapsed time for the simulation to expand, and velocities are estimated in units of pixels per unit time by determining the number of pixels the galaxy moves in the time of expansion. Thus, the unit of pixels × time^{-1} × decapixel^{-1} (1 decapixel = 10 pixels) for the Hubble constant in this simulation mimics the standard unit of km s^{-1} Mpc^{-1} used commonly for the actual Hubble constant. In this example the redshifts were generated by expansion of the space but the Hubble parameter was determined by assuming the redshifts to be proportional to a recessional velocity for observed galaxies [literal application of Eq. (16.1)].

16.2.6 Measuring the Hubble Constant

The Hubble constant may be determined by measuring the redshift for spectral lines and comparing that with the distance for objects sufficiently far away that peculiar motion caused by local gravitational attraction is small compared with the motion associated with the Hubble expansion. The redshift is relatively easy to measure but determining the distance to objects that are far enough away to suppress the importance of peculiar motion is much harder. Substantial controversy reigned in astronomy for some time over the exact value of the Hubble constant. Competing camps generally agreed within a factor of less than two, which was a remarkable achievement considering the fundamental nature of the parameter. In recent years a concerted effort to calibrate the distance ladder using Cepheid variables out to larger distances, and application of other methods such as analysis of fluctuations in the cosmic microwave background, have yielded some consensus on the value of the Hubble constant. Most determinations now differ from each other by less than 10%.

Figure 16.4 illustrates determining H_0 by distance ladder methods [91]. The top part shows velocity versus distance for five different secondary distance indicators, as calibrated by Cepheid variable observations out to $\sim$ 20 Mpc. The adopted value is $H_0 = 72 \pm 8$ km s^{-1} Mpc^{-1}. The lower part of the diagram shows the inferred value of H_0 as a function of distance, with 72 km s^{-1} Mpc^{-1} indicated by the horizontal line. The large scatter of points below $\sim$ 100 Mpc is because of peculiar velocities due to local gravitational attraction. Type Ia supernovae provide the data at the largest distances. (In Fig. 19.6 the use of Type Ia supernovae to probe the expansion at even higher redshifts will be demonstrated.) Modern determinations find generally that $H_0 \sim 67$–73 km s^{-1} Mpc^{-1}, with "distance ladder" approaches as in Fig. 16.4 tending to give values nearer the higher end of this range and cosmic microwave background data (see Section 20.4.6) tending to give values closer to the lower end. Gravitational waves can be used to determine H_0 but present data (Section 24.7.1) give $\pm 10\%$ errors and can't yet add meaningful constraints to the value of H_0.

16.3 Limitations of the Standard Picture

The next few chapters will document the successes of the standard world picture. However, we shall find that two aspects of this picture suggest that it is incomplete:

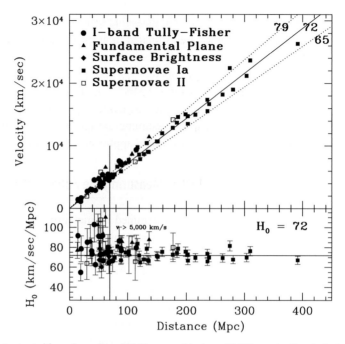

Fig. 16.4 Hubble constant extracted from observations [91]. Three possible slopes (Hubble constant) are indicated in the upper part of the figure. Reproduced from Final Results from the Hubble Space Telescope Key Project to Measure the Hubble Constant. Freedman W.L., Madore, B.F., Gibson, B.K., Ferrarase, L., Kelson, D.D., Sakai, S., Mould, J.R., Kennicutt, R.C., Jr., Ford, H.C., Graham, J.A., Huchra, J.P., Hughes, S.M., Illingworth, G.D., Macri, L.M., and Stetson, P.B. 2001. *Astrophysical* Journal, **553**, 47–72. © 2001. American Astronomical Sociey. All rights reserved.

1. To get the big bang to lead to the present universe, certain assumptions about initial conditions must be taken as given. While not necessarily wrong, some of these assumptions seem unnatural by various standards. It would be much more satisfying if these assumptions had a more fundamental justification.
2. As the expansion is extrapolated back in time, eventually a state of sufficient temperature and density is reached that a fully quantum-mechanical theory of gravitation would be required. This is called the *Planck era*, and the corresponding scales of distance, energy, and time are called the *Planck scale* (see Section 12.6). Since a consistent wedding of general relativity to quantum mechanics doesn't yet exist, the presently understood laws of physics may be expected to break down on the Planck scale; thus the standard picture says nothing about the Universe at those very early times.

In later chapters these issues will be addressed and it will be found that certain modifications of the standard picture can ameliorate some of its problems while retaining its successes, but quantum gravity remains an unsolved problem.

Background and Further Reading

Ryden [211], Roos [207], and Liddle [145] are very good basic introductions to modern cosmology with emphasis on concepts rather than mathematics. Some common misconceptions concerning the expanding Universe are discussed in Davis and Lineweaver [73].

Problems

16.1 Suppose a static, euclidean space in which all galaxies have intrinsic luminosity L and are distributed uniformly with density $n(L_i)$, where $n(L_i)$ is the number density of galaxies with luminosity L_i. The apparent brightness (observed energy flux) of a galaxy at a distance r then is $f = L/4\pi r^2$. Show that the number of galaxies observed per unit solid angle brighter than f is then proportional to $f^{-3/2}$. Express this result in terms of apparent magnitudes.

16.2 Use Lorentz invariance to show that the redshift z obeys

$$1 + z = \frac{1 + v/c}{\sqrt{1 - v^2/c^2}}$$

if interpreted as a Doppler shift. (Such an interpretation is tenable only for small cosmological redshifts.)

16.3 In the steady state model of the Universe it is assumed that the density and the Hubble constant do not change with time (the *perfect cosmological principle*). Show that this implies a universe that is expanding exponentially and that matter must be created continuously from nothing to maintain the postulate of constant density. Estimate the rate at which matter must be created from nothing today for each cubic meter of space if this model were correct. *Hint*: You may assume the present matter density to be $\rho \sim 10^{-30}$ g cm^{-3}. ***

16.4 By *Olber's paradox*, if the Universe were of infinite extent and of infinite age, the night sky should be as bright as the surface of a star because one should see the surface of a star in any direction (just as in a deep forest there is no direction in which the line of sight will not eventually intersect a tree). (a) Explain how an expanding Universe of finite age, even if infinite in extent, can account for Olber's paradox. (b) Two alternative explanations of Olber's paradox have sometimes been proposed:

1. The darkness of the night sky is because intervening dust absorbs light from distant stars.
2. The distribution of distant stars is not uniform but follows a fractal pattern of decreasing density such that the sum of the light from them is finite.

Why are these explanations either wrong or inconsistent with the modern view of cosmology?

16.5 Make a rough estimate of the radius of the visible Universe compared with the radius of gravitational curvature for light propagating in the Universe. From this estimate, would you expect Newtonian gravity to be adequate, or do you expect that general relativity is required to address adequately cosmological questions? *Hint*: Observations suggest that the average density of the Universe is comparable to the critical density given in Eq. (17.6). ***

16.6 The Virgo Cluster of galaxies lies about 16 Mpc from us, which is close enough that the gravitational attraction between our Local Cluster of galaxies and the Virgo Cluster can cause deviations from the Hubble law (peculiar velocity). The observed recessional velocity of the Virgo Cluster is about 985 km s^{-1}. Assuming a Hubble constant of 72 km s^{-1} Mpc^{-1}, what is the peculiar velocity of the Virgo Cluster relative to the Milky Way?

16.7 Hubble's original estimate for the Hubble constant was about 550 km s^{-1} Mpc^{-1}, almost an order of magnitude larger than the presently accepted value. Assuming the validity of the Hubble law, what is the approximate age of the Universe if Hubble's original value for H_0 were correct?

16.8 A distant galaxy has been observed with the Lyman alpha emission line shifted to 968.2 nm. What is the redshift of this galaxy? What was the scale factor of the Universe when the light was emitted? How long after the big bang was this light emitted, assuming the current standard cosmological model? *Hint*: For the current standard cosmological model, see Fig. 19.10.

16.9 Typical peculiar velocities for galaxies are several hundred km s^{-1}. According to the Hubble law, at what distance do galaxies have to be before their peculiar velocity is less than about 10% of their equivalent cosmological recessional velocities?

16.10 What is the approximate distance to SN 2015A (the first supernova observed from Earth in 2015), given that its redshift is $z = 0.0233$?

17 Energy and Matter in the Universe

The history and fate of the Universe ultimately turn on how much matter, energy, and pressure it contains, since these components of the stress–energy tensor couple to gravity and determine how self-gravitation of the Universe influences its evolution. In this chapter we begin to address quantitatively the matter, energy, and pressure contained in the Universe and how that determines its history. As part of this discussion a key concept will be introduced, the critical density of the Universe. We shall show that the Universe is near critical density and thus must be described ultimately by general relativity, but can be understood qualitatively in Newtonian terms.

17.1 Expansion and Newtonian Gravity

The Universe is mostly empty space, which might suggest that Newtonian gravity – which was shown in Section 8.1 to be valid in the weak gravity limit – is adequate for describing its large-scale structure. But in Section 6.4.2 we showed that whether the effects of general relativity were important relative to a Newtonian description could be quantified in terms of the ratio of an actual radius for a massive object compared with its radius of gravitational curvature. Applying such a criterion to the entire Universe, reasonable estimates for the mass–energy contained in the Universe indicate that the actual radius of the known Universe and the corresponding gravitational curvature radius are comparable (Problem 16.5). Thus, a description of the large-scale structure of the Universe must be built on a covariant gravitational theory, rather than on Newtonian gravity. Even so, it is possible to understand a substantial amount concerning the role of energy and matter in the expanding Universe, and introduce many important ideas, simply by using Newtonian concepts. In this chapter a largely Newtonian picture of gravity will be employed to that end. Subsequent chapters will extend the analysis to a proper description of cosmology using general relativity.

Consider the test galaxy illustrated in Fig. 17.1 [207]. It lies at a distance r from us and is assumed to be acted upon gravitationally by a uniformly distributed mass density ρ inside the sphere of radius r centered on the Earth, with the radius r increasing with time according to the Hubble law. The gravitational potential acting on the galaxy is then given by

$$U = \frac{-GMm}{r}, \qquad (17.1)$$

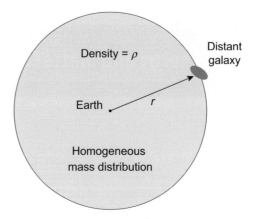

Fig. 17.1 A Newtonian model of the expanding Universe. The product ρr^3 is assumed to remain constant as the Universe expands.

where m is the mass of the galaxy and the total mass contained within the sphere is

$$M = \tfrac{4}{3}\pi r^3 \rho, \tag{17.2}$$

which is constant since r increases with time and ρ decreases, but the product ρr^3 is assumed to be unchanged. Thus

$$U = -\tfrac{4}{3}\pi G r^2 \rho m. \tag{17.3}$$

If the change in distance to the galaxy is caused entirely by the Hubble expansion, its radial velocity relative to the Earth is $H_0 r$ by the Hubble law. This implies a kinetic energy

$$T = \tfrac{1}{2} m v^2 = \tfrac{1}{2} m H_0^2 r^2, \tag{17.4}$$

where m is the mass of the galaxy. The total energy of the galaxy is then the sum of kinetic and potential energies,

$$\begin{aligned} E = T + U &= \tfrac{1}{2} m H_0^2 r^2 - \tfrac{4}{3}\pi G r^2 \rho m \\ &= \tfrac{1}{2} m r^2 \left(H_0^2 - \tfrac{8}{3}\pi G \rho \right). \end{aligned} \tag{17.5}$$

Now this expression may be used to determine the critical density that would just halt the expansion, assuming the Universe to be held together by Newtonian gravity.

17.2 The Critical Density

If the expansion is to halt, $E = 0$ and thus $H_0^2 = \tfrac{8}{3}\pi G \rho$. Solving for ρ, the *critical density* that will just halt the expansion is

$$\rho_c = \frac{3 H_0^2}{8\pi G} \simeq 1.88 \times 10^{-29} h^2 \text{ g cm}^{-3}. \tag{17.6}$$

The corresponding critical energy density is

$$\epsilon_c = \rho_c c^2 = 1.69 \times 10^{-8} h^2 \text{ erg cm}^{-3}$$
$$= 1.05 \times 10^{-2} h^2 \text{ MeV cm}^{-3}. \qquad (17.7)$$

The critical density (17.6) is remarkably small; it corresponds to an average concentration of six hydrogen atoms per cubic meter of space or about $10^{11} M_\odot$ per cubic megaparsec.[1] In this simple model three qualitative regimes may be distinguished for the actual density ρ:

1. If $\rho > \rho_c$ the Universe is said to be *closed* and the expansion will stop in a finite time.
2. If $\rho < \rho_c$ the Universe is said to be *open* and the expansion will never halt.
3. If $\rho = \rho_c$ the Universe is said to be *flat* (or *euclidean*) and the expansion will halt, but only asymptotically as $t \to \infty$.

Thus, in this simple picture the fate of the Universe is determined by its present matter density. (As will be demonstrated in Chapter 18, this conclusion is modified fundamentally by the presence of dark energy in addition to the energy density associated with matter and radiation.) It will prove convenient to introduce the dimensionless *total density parameter* evaluated at the present time

$$\Omega \equiv \frac{\rho}{\rho_c} = \frac{\epsilon}{\epsilon_c} = \frac{8\pi G \rho}{3 H_0^2}, \qquad (17.8)$$

where ρ is the current total density that couples to gravity. Thus the closure condition implies that $\Omega = 1$. Note for future reference that a subscript "0" is often used on Ω and ρ (and other cosmological parameters) to indicate explicitly that they are evaluated at the present time; where possible we suppress these subscripts to avoid notational clutter. Unless otherwise noted, cosmological parameters are assumed evaluated at the present time.

17.3 The Cosmic Scale Factor

As was shown in Section 16.2, the Hubble expansion makes it convenient to introduce a *cosmic scale factor* $a(t)$ that sets the global distance scale for the Universe as a function of time. Then, if all peculiar motion is ignored the time dependence of the expansion is governed entirely by $a(t)$ and all distances simply scale with this factor. For example, if the present time is denoted by t_0 and the present value of the scale factor by a_0, a wavelength of light λ that was emitted at a time $t < t_0$ is scaled to a wavelength λ_0 at $t = t_0$ by the universal expansion according to (see Section 16.2.3)

$$\frac{\lambda_0}{a_0} = \frac{\lambda}{a(t)}. \qquad (17.9)$$

---

[1] A critical density of six hydrogen atoms per cubic meter sounds rather small, and it is; that such a small density can close the Universe gravitationally is testament to the weak but relentless pull of gravity. Bright galaxies have masses in the vicinity of 10^{11}–$10^{12} M_\odot$ and separations of order 1 Mpc, so the average density of a cluster of galaxies is comparable to the critical density of $10^{11} M_\odot$ Mpc^{-3}. This observation is our first clue that the Universe may contain an average density not too different from the critical density.

Likewise, if r_0 and ρ_0 are the present values of r and ρ, these scale as

$$\frac{r(t)}{r_0} = \frac{a(t)}{a_0} \qquad \frac{\rho(t)}{\rho_0} = \left(\frac{a_0}{a(t)}\right)^3, \tag{17.10}$$

and so on. This permits all dynamical equations to be expressed in terms of the scale factor. For example, from (17.1) and (17.2) the gravitational force acting on the galaxy is radial with magnitude

$$F_G = -\frac{\partial U}{\partial r} = -G\frac{Mm}{r^2} = -\frac{4}{3}\pi G \rho r m, \tag{17.11}$$

and the corresponding gravitational acceleration is radial with magnitude

$$\ddot{r} = \frac{F_G}{m} = -\frac{GM}{r^2} = -\frac{4}{3}\pi G \rho r. \tag{17.12}$$

Then from (17.12) and (17.10), the acceleration of the scale factor is given by

$$\ddot{a} = -\frac{4}{3}\pi G \rho_0 a_0^3 \left(\frac{1}{a^2}\right). \tag{17.13}$$

Dropping the subscript on ρ_0 and utilizing (17.8), the acceleration of the scale factor (17.13) also may be expressed in terms of the density parameter Ω,

$$\ddot{a} = -\frac{H_0^2 a_0^3 \Omega}{2a^2}. \tag{17.14}$$

From Eq. (17.14), the time evolution of the scale factor is given by (Problem 17.12)

$$\dot{a}^2 = H_0^2 a_0^2 \, f(\Omega, t), \tag{17.15}$$

with the definition

$$f(\Omega, t) \equiv 1 + \Omega \frac{a_0}{a(t)} - \Omega, \tag{17.16}$$

which must obey the condition $f(\Omega, t) \geq 0$, since $\dot{a}^2$ can never be negative. This condition may be used to enumerate different possibilities for the history of the Universe, as we discuss in the following section.

17.4 Possible Expansion Histories

Consider as an example a dust-filled universe (a universe containing only pressureless, nonrelativistic matter). There are then three qualitatively different histories for this highly idealized universe, depending on the value of Ω.

1. $\Omega < 1$ (*undercritical*): In this case, $f(\Omega, t) \to 1 - \Omega > 0$ as $a(t) \to \infty$. Thus $\dot{a}$ never vanishes, implying an open, ever-expanding (because it is observed to be expanding now) universe.
2. $\Omega = 1$ (*critical*): For this case, as $a(t) \to \infty$ one obtains $f(\Omega, t) \to 0$, but it reaches 0 only as $t \to \infty$. Hence, if $\Omega = 1$ the universe is ever-expanding but the rate approaches zero asymptotically as $t \to \infty$.

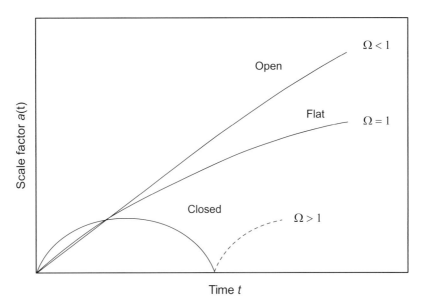

Fig. 17.2 Qualitative behavior of the scale factor $a(t)$ as a function of time for a dust-filled universe. Mathematically the $\Omega > 1$ solution can exhibit multiple oscillations. Whether this is physical is not very meaningful since our Universe does not correspond to this solution.

3. $\Omega > 1$ (*overcritical*): Now as t increases $f(\Omega, t) \longrightarrow 0$, but in a *finite* time $t_{\max}$. Beyond this time the condition $f(\Omega, t) \geq 0$, still must be satisfied. Thus, if $\Omega > 1$ the expansion turns into a *contraction* at time $t_{\max}$ and the universe begins to shrink. It follows that this universe is closed.

Integration of the scale factor equation (17.15) for this dust model gives for these three scenarios [62, 68]:

(1) For a flat universe with $\Omega = 1$,

$$a(t) = \left(\frac{3t}{2t_H}\right)^{2/3}. \tag{17.17}$$

This behavior is sketched as the $\Omega = 1$ curve in Fig. 17.2.

(2) For a closed universe with $\Omega > 1$,

$$a(\psi) = \frac{1}{2}\frac{\Omega}{\Omega - 1}(1 - \cos \psi) \tag{17.18}$$

$$t(\psi) = \frac{1}{2H_0}\frac{\Omega}{(\Omega - 1)^{3/2}}(\psi - \sin \psi), \tag{17.19}$$

where $\psi \geq 0$ parameterizes the solution. The first maximum for $a(\psi)$ occurs for $\psi = \pi$,

$$\frac{a_{\max}}{a_0} = \frac{\Omega}{2(\Omega - 1)}(1 - \cos \pi) = \frac{\Omega}{\Omega - 1}, \tag{17.20}$$

and the time at which this occurs is

$$t_{max} = t(\pi) = \frac{\pi \Omega}{2H_0(\Omega-1)^{3/2}}. \quad (17.21)$$

This case is sketched as the $\Omega > 1$ curve in Fig. 17.2. It is symmetric around t_{max} and has zeroes at $\psi = 0, 2\pi, 4\pi, \ldots$. Thus the time from the beginning of the expansion to the "big crunch" is $t_f = 2t_{max}$.

(3) For an open universe with $\Omega < 1$,

$$a(\psi) = \frac{\Omega}{2(1-\Omega)}(\cosh\psi - 1) \quad (17.22)$$

$$t(\psi) = \frac{1}{2H_0}\frac{\Omega}{(1-\Omega)^{3/2}}(\sinh\psi - \psi), \quad (17.23)$$

where ψ is a parameter. This behavior is sketched as the $\Omega < 1$ curve in Fig. 17.2.

Example 17.1 As shown in Problem 17.3, Equations (17.17)–(17.23) may be used to derive equations for the lifetime of a dust-filled universe. The results are

$$\frac{t_0}{\tau_H} = \frac{2}{3} \quad (17.24)$$

for the dust-filled, flat universe with $\Omega = 1$,

$$\frac{t_0}{\tau_H} = \frac{\Omega}{2(\Omega-1)^{3/2}}\left[\cos^{-1}\left(\frac{2-\Omega}{\Omega}\right) - \frac{2}{\Omega}(\Omega-1)^{1/2}\right] \quad (17.25)$$

for the dust-filled, closed universe with $\Omega > 1$, and

$$\frac{t_0}{\tau_H} = \frac{\Omega}{2(1-\Omega)^{3/2}}\left[\frac{2}{\Omega}(1-\Omega)^{1/2} - \cosh^{-1}\left(\frac{2-\Omega}{\Omega}\right)\right] \quad (17.26)$$

for the dust-filled, open universe with $\Omega < 1$.

The behaviors illustrated in Fig. 17.2 for a dust universe are not realistic for the actual Universe because the discussion must be extended to a covariant description that includes the contribution of radiation and dark energy. However, these simple examples illustrate clearly that the history of the Universe is tied intimately to the mass and energy density that it contains.

17.5 Lookback Times

Because light travels at a finite and constant velocity, telescopes are time machines. The *lookback time* t_L represents how far back in time one is looking when an object having a redshift z is observed,

$$t_L = t(0) - t(z), \quad (17.27)$$

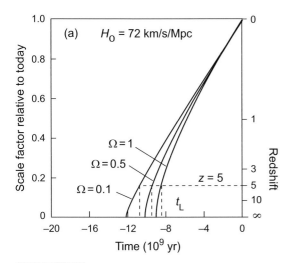
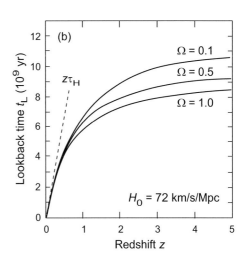

Fig. 17.3 (a) Geometrical interpretation of the lookback time t_L for $z = 5$ in a dust universe with three different values of the density parameter Ω. (b) Lookback time as a function of redshift for different assumed density parameters in a dust model. The dashed line gives the result for Hubble's law.

where $t(z = 0)$ is the present age of the Universe and $t(z)$ is the age when light that is observed today with a redshift z was emitted from its source.

Example 17.2 In a flat dust-filled universe the time $t(z)$ as a function of redshift parameter z is given by (see Problem 17.2)

$$\frac{t(z)}{\tau_H} = \frac{2}{3}(1+z)^{-3/2} \qquad \frac{t(0)}{\tau_H} = \frac{2}{3},$$

where $\tau_H = 1/H_0$ is the Hubble time. Then the lookback time for an object observed to have a redshift z corresponds to the difference

$$\frac{t_L(z)}{\tau_H} = \frac{t(0)}{\tau_H} - \frac{t(z)}{\tau_H} = \frac{2}{3} - \frac{2}{3}(1+z)^{-3/2} = \frac{2}{3}\left(1 - \frac{1}{(1+z)^{3/2}}\right).$$

As a representative example, this result and Eq. (17.24) imply that (for a flat, dust-filled universe) the light from an object observed today to have a redshift $z = 6$ was emitted when $t_L/t_0 \sim 0.946$, when the Universe was only about 5% of its present age t_0, and at the time that the light was emitted the cosmic scale factor of the Universe was $1 + z = 7$ times smaller than it is today.

The lookback time as a function of redshift is interpreted graphically for a dust model in Fig. 17.3(a), and is plotted for various assumed values of the density parameter Ω in Fig. 17.3(b). For small redshifts $t_L \simeq z\tau_H$, as would be expected from the Hubble law, but for larger redshifts the curves in Fig. 17.3(b) are seen to differ substantially from this approximation.

17.6 The Inadequacy of Dust Models

The preceding discussion has applied Newtonian gravity to a universe containing only pressureless matter (dust). Until roughly the last decade of the twentieth century a covariant version of such theories was the favored model for cosmology. For example, one often-discussed model was the *Einstein–de Sitter universe*, which was a covariant version of the $\Omega = 1$ solution with exactly a closure density of dust and zero curvature that we described briefly in noncovariant form in Section 17.4. However, the observational evidence of that period indicated that there was not nearly enough visible matter in the Universe to constitute a closure density, suggesting an open-Universe cosmology with $\Omega < 1$. We now know that the actual Universe contains additional components that influence its gravitational evolution in a highly nontrivial way, and that as a result plots like Fig. 17.2 can give a very misleading picture of the actual history of the Universe. To understand the *new cosmology* that emerged from these discoveries, we begin by taking inventory of these additional components. Our starting point will be the evidence that much of the matter in the Universe is not the visible matter of stars and galaxies.

17.7 Evidence for Dark Matter

There is strong observational evidence for large amounts of *dark matter* in the Universe that reveals its presence through gravity, but is not seen by any other probe. Let's review some observations suggesting that most of the matter of the Universe is, in fact, dark matter.

17.7.1 Rotation Curves for Spiral Galaxies

In spiral galaxies, balancing the centrifugal and gravitational forces at a radius R requires

$$v = \sqrt{\frac{GM}{R}}, \qquad (17.28)$$

where R is the radius and M is the enclosed mass. Thus, well outside the main matter distribution of the galaxy the velocities should fall off as $R^{-1/2}$. The velocities can be measured in spiral galaxies using the Doppler effect, both for visible light from the luminous matter and for the 21 cm hydrogen line from non-luminous hydrogen. The general behavior that is observed for many spirals is not the predicted $R^{-1/2}$ fall-off but almost constant velocity at large distances well outside the bulk of the luminous matter, as illustrated in Fig. 17.4 for the galaxy M33 and in Fig. 17.5 for the Andromeda Galaxy (M31).

Direct measurements indicate nearly constant velocities out to ~ 30 kpc in many spirals, and indirect means suggest near-constant velocities out to ~ 100 kpc in some cases. This indicates the presence of substantial gravitating matter distributed in a halo beyond the visible matter. Detailed analysis typically implies that the mass of the unseen matter exceeds the mass of the visible matter by about a factor of ten. The best models suggest that the galaxies are surrounded by a halo of dark matter contained in an isother-

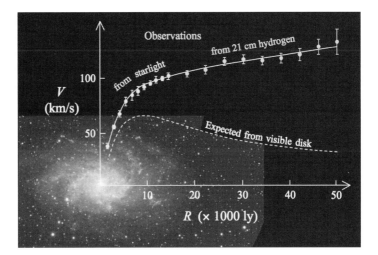

Fig. 17.4 Rotation curve for the galaxy M33 out to a distance of about 15 kpc (50, 000 ly) from the center. Points inside about 15, 000 ly are from visible starlight; points beyond that are from radio frequency (RF) observations. Data are from Ref. [71]. Image by Stefania deluca (own work) [Public domain or CC0], via Wikimedia Commons.

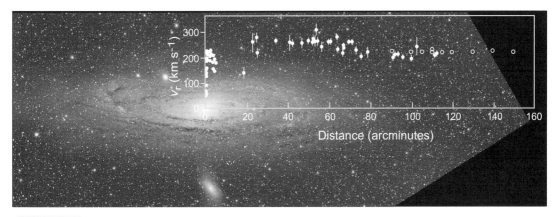

Fig. 17.5 Rotation curve for the Andromeda Galaxy (M31) [209]. Points are measured velocities, with open circles corresponding to RF observations. Converting angular size to distance from the center using a distance of 778 kpc to M31 indicates that the visible matter lies within $60' \sim 14$ kpc of the center but the rotation curve is constant out to at least $150' \simeq 34$ kpc. Reproduced from Seeing Dark Matter in the Andromeda Galaxy, *Physics Today*, **59**(12), 8(2006); https://doi.org/10.1063/1.2435662, with the permission of the American Institute of Physics.

mal (constant-temperature) gas. The full extent of such dark matter halos is unknown at present.

17.7.2 The Mass of Galaxy Clusters

From the virial theorem discussed in Section 14.5.3, the mass within a region can be estimated from the velocity dispersion of objects bound gravitationally in that region. In particular, the equations of Section 14.5.3 may be used to estimate the mass contained within large clusters of galaxies by studying the motion of galaxies within the cluster. Such estimates indicate that clusters of galaxies contain much more mass than the visible matter would indicate.

Example 17.3 The *Coma Cluster* lies at a distance of about 100 Mpc in the constellation Coma Berenices and contains thousands of galaxies. The measured velocity dispersion in a region lying within about 3 Mpc of the center is $\sigma_r \simeq 900 \text{ km s}^{-1}$. Inserting this into Eq. (14.8) leads to the conclusion that the Coma Cluster has a mass of $M \sim 2.8 \times 10^{15} M_\odot$. On the other hand, the visual luminosity of the Coma Cluster is $L = 5 \times 10^{12} L_\odot$, where $L_\odot$ is the luminosity of the Sun. Thus, the ratio of mass to light for the Coma Cluster is

$$\frac{\text{Mass}}{\text{Light}} \equiv \frac{M}{L} = \frac{2.8 \times 10^{15} M_\odot}{5 \times 10^{12} L_\odot} \simeq 560 \frac{M_\odot}{L_\odot},$$

suggesting the presence of large amounts of unseen mass within this cluster of galaxies.

The preceding example is representative and in rich clusters (those containing thousands of galaxies) the measured velocity dispersions typically lie in the range $\sigma_r = 800$–1000 km s^{-1} and the mass to luminosity ratio in solar units is typically found to be several hundred, which should be compared with a value of order one found in the Solar neighborhood. This suggests that clusters of galaxies contain dark matter halos above and beyond the halos of the individual galaxies comprising the cluster.

17.7.3 Hot Gas in Clusters of Galaxies

Observations by X-ray satellites indicate that clusters of galaxies often are immersed in a hot gas filling the space between the galaxies. The X-ray intensity is a measure of the strength of the gravitational field because to produce X-rays requires large velocities of the gas particles. These observations allow the total amount of gravitating matter in the cluster to be estimated and it is found that there must be much more matter than just the luminous matter to account for the hot gas observed systematically in the clusters. Otherwise, the gravitational field would be too weak to trap the gas for extended periods. For example, observations of the Coma Cluster by the X-ray satellite ROSAT indicated that the gas has a temperature of about 10^8 K, which is much too hot for it to be bound by the gravity produced by the visible matter of the Coma Cluster. Thus, the hot and luminous X-ray gas bound in many galaxy clusters also signals the presence of dark matter contributing to the cluster gravitational field.

17.7.4 Gravitational Lensing

As was discussed in Section 9.8, the path of light is altered in curved spacetime. This can cause *gravitational lensing*, where intervening masses act as "lenses" to distort, time-delay,

17.7 Evidence for Dark Matter

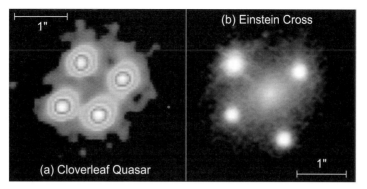

Fig. 17.6 Gravitational lensing of quasars: (a) The Cloverleaf Quasar. The four images are of a single quasar lensed by foreground galaxies too faint to see in this image. NASA/STScI/D. Turnshek. (b) The Einstein Cross. The four outer images are all of a single quasar lensed by a foreground galaxy near the center of the image. Identical spectra confirm that these are images of a single object. ESA/Hubble and NASA.

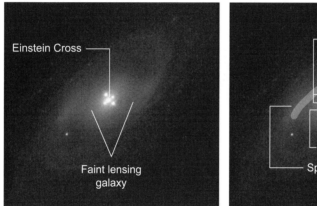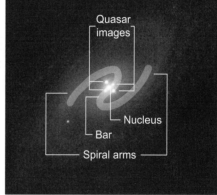

Fig. 17.7 The Einstein Cross and its lensing galaxy. Intensity has been displayed on a logarithmic scale so that the very bright quasar images and the extremely faint bar and arms of the lensing galaxy can be seen at the same time (see annotated image on the right). Adapted from NASA/ESA, J. Westphal, W. Keel.

and magnify the image of distant objects.[2] Two spectacular examples of gravitational lensing are shown in Fig. 17.6, where single objects appears as four images. A distant quasar is positioned behind a massive galaxy (or galaxies) and the intervening mass has created multiple images through gravitational lensing of the light from the quasar. Individual stars in the foreground galaxies may also be acting as gravitational lenses (*microlensing*), which causes the images to change their relative brightness as stars change position in the lensing galaxy. This interpretation of the Einstein Cross is bolstered by Fig. 17.7, which shows in faint outline the foreground lensing galaxy and the four quasar images. The lensing

---

[2] Gravitational lens magnification can allow the study of more distant objects than would otherwise be possible. Magnifications as large as a factor of ~ 100 have been observed. Gravitational lenses converge light like optical lenses but differ in many details. For example, gravitational lenses do not bring light to a unique focus.

Fig. 17.8 Light from a supernova in a distant galaxy at redshift $z = 1.49$ split into four images by gravitational lensing from a cluster of more-nearby galaxies at $z = 0.54$ [129]. The positions of the four supernova images are indicated by arrows in the inset box. The type of the supernova was not determined but the lightcurves suggest that it is not a Type Ia. Most of the galaxies in the image are members of the lensing cluster. Adapted from NASA, ESA, and S. Rodney (JHU) and the Frontier SN team; T. Treu (UCLA), P. Kelly (UC Berkeley) and the GLASS team; J. Lotz (STScI) and the Frontier Fields Team; M. Postman (STScI) and the CLASH team; and Z. Levay (STScI).

galaxy is a relatively nearby barred spiral. Both the spiral arms and the central bar of the foreground galaxy can be seen faintly (see the annotation in the right panel).

Another example of a cross, in this case for lensing of a supernova in a distant galaxy, is displayed in Fig. 17.8. The supernova host galaxy is a spiral at redshift $z = 1.49$ and it is being lensed by a massive elliptical galaxy in a cluster of galaxies named MACS J1149.6+223 at $z = 0.54$ [129]. The gravitational potential of the entire cluster also produces multiple images of the supernova host galaxy. The different magnifications and staggered arrival times[3] of the supernova images carry information about the supernova

---

[3] Light takes multiple paths through the lens so different images have different magnifications (ranging from $\sim 10-30$) and time delays that differ by days to weeks (see Section 9.9). Because the source of the images is a supernova with a lightcurve, the time delays can be measured quantitatively. Computer models suggest that because of the time delays the lensed supernova images could have been seen 50 years ago and 20 years ago, had astronomers been looking, and that a replay of the lensing event should be visible in the cluster field within 1–10 years of the original analysis in 2015.

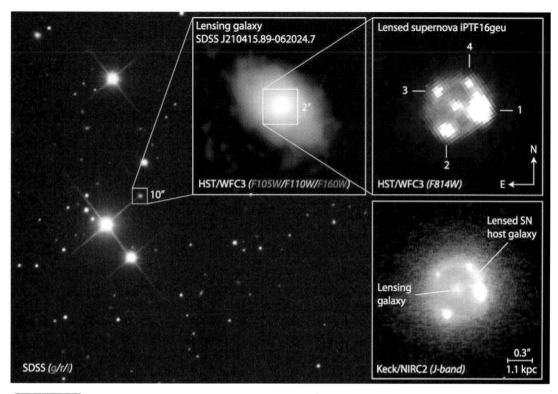

Fig. 17.9 Gravitationally lensed Type Ia supernova iPTF16geu [100]. The supernova and the lensing galaxy are almost collinear, which leads to strong lensing with four images of the supernova in evidence in the upper right panel, and a partial Einstein ring from lensing of the supernova host galaxy in the lower right panel. Data from Goobar et al., iPTFgeu: A Multiply Imaged, Gravitationally Lensed Type Ia Supernova, Science **356**(6335), 291–295, 21 Apr 2017. Copyright © 2017, American Association for the Advancement of Science.

and about the distribution of matter in the lenses, as well as providing a measure of the cosmic expansion rate.

An unusual lensed image of a Type Ia supernova is shown in Fig. 17.9. Normally, when a distant object is lensed gravitationally its true brightness is unknown. However, because of the standardizable candle properties of Type Ia supernovae (see Box 19.2), in this case the actual brightness of the lensed object and its variation with time are well determined. It is found that the supernova has not only been split into four images but its apparent brightness has been amplified by a factor of about 50 relative to its true brightness. Furthermore, by using the known variation of the luminosity with time (the standardized lightcurve of the supernova; see Box 19.2 and Fig. 19.7), the delay times for light in each of the four images can be measured precisely. The difference in arrival times for light in the four lensed images is inversely proportional to the Hubble parameter. Conversely, if a cosmological model is assumed the time delays measure directly the gravitational potential sensed by the four images, which permits reconstruction of the mass distribution in the lens [129].

 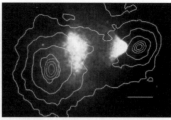

Fig. 17.10 Evidence for dark matter in the Bullet Cluster [67]. The left image shows galaxies in the cluster and total mass contours inferred from gravitational lensing. The right image shows X-ray luminosity superposed on the mass contours. As discussed in the text, the simplest explanation for the displacement of the X-ray luminosity from the mass concentrations is that the majority of the mass is dark matter now found at the two mass centers. Reproduced from Douglas I. Clove, *et al.*, A Direct Empirical Proof of the Existence of Dark Matter, published 30 August 2006 © 2006. The American Astronomical Society. All rights reserved. *Astrophysical Journal Letters*, **648**(2).

The strength of a gravitational lens depends on its total mass – whether that mass is visible or not – so gravitational lenses can serve as excellent indicators of how much unseen matter is present in the lens, and even of the distribution of mass in the lens. Systematic analysis of gravitational lensing leads to conclusions similar to those suggested above by the rotation curves for spiral galaxies and the properties of galaxy clusters: More than 90% of the mass contributing to the strength of large gravitational lenses is dark.

Example 17.4 Strong evidence for dark matter in a cluster of galaxies is displayed in Fig. 17.10. In this image the double cluster of galaxies 1E0657-558 at a redshift $z = 0.296$ (known colloquially as the *Bullet Cluster*) has been studied using gravitational lensing and detection of X-rays. The interpretation of the double cluster is that it represents two galaxy clusters that collided about 100 million years ago. Because of their wide spacings the stars within the galaxies would have been largely oblivious to each other but the gas would have interacted strongly through ram pressure, and if dark matter was present it should have had little interaction with itself or the regular matter as the galaxies collided (it would be collisionless) and so should have continued on unimpeded. The compressed gas, radiating strongly in X-rays, is displaced from the mass centers of the two clusters after the collision: the collision has separated the bulk of the dark matter (at the two local maxima of the mass contours) from the regular matter (concentrated at the sources of X-ray luminosity).

17.7.5 Dark Matter in Ultra-diffuse Galaxies

An array of sensitive telephoto lenses called Dragonfly found evidence for a population of *ultra-diffuse galaxies (UDG)* in the Coma Cluster. These are faint, like dwarf galaxies, but appear to have masses comparable to large galaxies. Now hundreds of UDGs are known. The UDG Dragonfly 44, about 100 Mpc away in the Coma Cluster, is shown in Fig. 17.11. The stellar kinematics for Dragonfly 44 was studied using long observing runs on the Keck II telescope [237]. From the measured velocity dispersion the total mass of

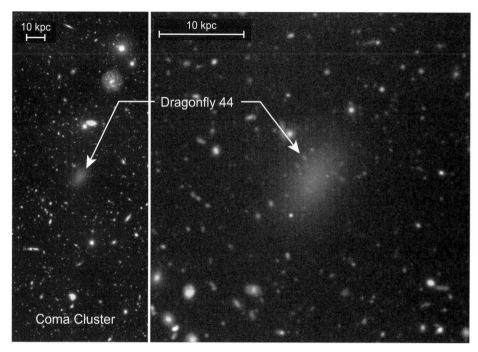

Fig. 17.11 Dragonfly 44, a relatively nearby galaxy that may be almost all dark matter. Although it has a very low surface brightness, it contains of order 100 globular clusters and has a mass comparable to that of the Milky Way galaxy [237]. Gemini/Pieter Dokkum.

the galaxy was estimated to be comparable to that of the Milky Way, and approximately 100 globular clusters were identified. Thus UDGs appear to be a kind of failed galaxy with the size, amount of dark matter, and star clusters of a regular galaxy, but with far less light, suggesting that UDGs may be almost entirely dark matter. For example, from the observations it was estimated that Dragonfly 44 is 98% dark matter.

17.8 The Amount of Baryonic Matter

There seems to be a lot of unseen matter in the Universe. An important question is whether this matter is baryonic (made of neutrons and protons like ordinary matter), or something more exotic. Let us define the ratio η of baryon number density to photon number density,

$$\eta \equiv \frac{n_B}{n_\gamma} \simeq \frac{\Omega_b \rho_c / m_B}{410 \, \text{cm}^{-3}} \simeq 2.76 \times 10^{-8} \Omega_b h^2, \qquad (17.29)$$

where Ω_b is the baryon density parameter (see Section 17.13), ρ_c is the closure density, m_B is the average mass of a baryon, and the density of photons has been approximated by the cosmic microwave background density given in Eq. (20.24). Because of the nucleosynthesis that will be discussed in conjunction with the big bang in Section 20.3, the abundances of light elements such as ^{4}He, ^{3}He, ^{2}H, and ^{7}Li relative to that of ^{1}H are very sensitive to η. From the data discussed in Section 20.3.3, it may be concluded that $\eta \sim 6 \times 10^{-10}$. Therefore, from Eq. (17.29)

$$\Omega_b \simeq (3.6 \times 10^7 h^{-2})\eta \simeq 0.04 \qquad (17.30)$$

and – independent of how much baryonic matter has actually been observed directly – strong nucleosynthesis constraints say that there is far too little of it to close the Universe.

17.9 Baryonic Candidates for Dark Matter

If the Universe is full of dark matter invisible to non-gravitational probes, what could it be? We address first the issue of possible baryonic candidates for the dark matter. The most obvious possibilities can be dismissed rather quickly. If the dark matter were a cold baryonic gas it would not radiate, but a cold gas would eventually be heated by hot gas and radiation, and therefore would become visible. Likewise, dust would be heated and re-radiate light, so it too would become visible. Black holes could hide large amounts of baryonic matter but they would be more likely near the centers of galaxies where densities are higher, not out in the halos where the dark matter seems to be most required.

The most promising candidates for dark baryonic matter are the Massive Compact Halo Objects (MACHOS): Jupiter-like objects, or more massive *brown dwarfs* with too little mass to form stars. If such objects were abundant in the halos of the galaxies, they would be very difficult to detect because of their low luminosities. A typical experiment looks for gravitational lensing effects when the otherwise-invisible MACHO moves in front of a star being imaged in the Large Magellanic Cloud (a nearby satellite galaxy of the Milky Way). The expected signature is brightening over a matter of days or weeks for a star not normally a variable star. Searches indicate the presence of MACHOS in the halo of our galaxy, but too few to account for the required amount of dark matter.

In addition to the absence of direct evidence for sufficient baryonic dark matter, all baryonic solutions to the dark matter problem run afoul of the nucleosynthesis constraint (17.30). Furthermore, there is relatively good observational evidence from the cosmic microwave background radiation favoring $\Omega = 1$. Thus, the dominant matter in the Universe is probably not baryonic and the baryonic matter that we are composed of is but a minor pollutant in the Universe: not only are we not the center of the Universe, we aren't even made of the right stuff!

17.10 Candidates for Nonbaryonic Dark Matter

There are two classes of candidates for nonbaryonic dark matter, each corresponding to either conjectured or known elementary particles: (1) *Cold Dark Matter (CDM),* which consists of particles that decoupled very early, or that were never in thermal equilibrium. (2) *Hot Dark Matter (HDM)*, which consists of low-mass particles that still had relativistic velocities at the time of matter–radiation decoupling.

17.10.1 Cold Dark Matter

Cold dark matter had velocities well below lightspeed when galaxy formation started. Candidates for cold dark matter may be divided into Weakly Interacting Massive Particles (WIMPS), and superlight particles with superweak interactions that were never in equilibrium. Some proposed candidates for WIMPS include several exotically named particles expected for supersymmetric theories (see the discussion in Section 26.1), and a neutrino with mass greater than 45 MeV. No supersymmetric particles have yet been observed and current limits on neutrino masses probably rule out the known neutrinos as a dominant contributor to dark matter. If dark matter consists of superlight particles that were never completely in equilibrium they would have been decoupled from the beginning. A prime candidate for this class of dark matter is the axion, which is a boson required in elementary particle physics theories to prevent unacceptable violation of CP (simultaneous charge conjugation and parity transformation) symmetry in the strong interactions. There is no experimental evidence for axions at present.

17.10.2 Hot Dark Matter

Hot dark matter is relativistic at the time that galaxy formation begins. It could correspond to as-yet undiscovered particles but the neutrinos are obvious candidates, since they do not contribute to luminosity and they – unlike the particles discussed above – are known to exist. The present number density of neutrinos may be estimated by assuming that most neutrinos are in a uniform background of neutrinos analogous to the cosmic microwave background to be discussed in Chapter 20.[4] The number density of neutrinos in this background should be related to the number density of the photons in the microwave background by a factor of $\frac{3}{11}$. Therefore, the neutrino number density for each neutrino family may be estimated as

$$n_\nu \simeq \tfrac{3}{11} n_\gamma \simeq 112 \,\text{neutrinos}\,\text{cm}^{-3}, \qquad (17.31)$$

where Eq. (20.24) was used. The best current data indicate that neutrinos have a mass but it is tiny and the contribution to the closure density of all neutrinos in the Universe can be no more than several percent.

---

[4] Although this low-energy background is not presently detectable, the existence of the cosmic microwave background implies that there should be an analogous neutrino background left over from the big bang.

17.11 Dark Energy

Dark matter is exotic, since it is unclear what it is and why it fails to couple significantly through any force other than gravity. However, in Chapters 18 and 19 evidence will be presented that the evolution of the Universe is being dominated by something even more exotic: *dark energy*. Dark energy (also termed *vacuum energy*) behaves in a fundamentally different way from either normal matter and energy, or dark matter: it has a radically different equation of state that causes the effective gravitational force to become *repulsive*. To understand and to deal adequately with this surprising idea will require a covariant theory of gravity. Therefore, we defer substantial discussion of the role played by dark energy until the following two chapters where a fully covariant cosmology is developed.

17.12 Radiation

Astronomers classify massless and nearly massless particles such as photons, gluons, gravitons, and neutrinos as *radiation*. In addition to visible matter, cold dark matter, and dark energy, the Universe contains radiation that influences its evolution. We shall see in later chapters that the energy density of radiation in the present Universe is very small, but that it dominated the energy density of the very early Universe. Only a small amount of radiation density in the present Universe is found in starlight. The bulk (more than 90%) is in the cosmic microwave background (CMB) radiation, which will be discussed in Section 20.4.

17.13 The Scale Factor and Density Parameters

Anticipating our later treatment of the expansion using general relativity, the density of the Universe may be expected to receive contributions from three major sources: visible and dark matter (with present density denoted by ρ_m), radiation (with present density denoted by ρ_r), and vacuum or dark energy (with present density denoted by ρ_Λ). These component densities may be used to define corresponding component density parameters Ω_i through[5]

$$\rho_\text{r}(a) = \rho_\text{c}\Omega_\text{r} \qquad \rho_\text{m}(a) = \rho_\text{c}\Omega_\text{m} \qquad \rho_\Lambda(a) = \rho_\text{c}\Omega_\Lambda, \qquad (17.32)$$

where ρ_c is the critical density defined in Eq. (17.6). We make no explicit distinction between mass density ρ and the corresponding energy density $\epsilon = \rho c^2$, since they are

---

[5] The matter density may be divided into baryonic and nonbaryonic contributions. Thus the density parameter Ω_b introduced in Section 17.8 is related to Ω_m by $\Omega_\text{m} = \Omega_\text{b} + \Omega_\text{non}$, where Ω_non is the contribution from nonbaryonic matter.

Table 17.1 Cosmological density parameters

Source	Value ($\Omega_i = \rho_i/\rho_c$)
Total matter	$\Omega_m = 0.3$
Baryonic matter	$\Omega_b = 0.04$
Total radiation	$\Omega_r \lesssim 8 \times 10^{-5}$
Total vacuum	$\Omega_\Lambda = 0.7$
Curvature	$\Omega_k \leq 0.01$

numerically the same in $c = 1$ units and which is meant should be clear from the context. In Section 19.2.2 it will be shown that a total density parameter Ω may be approximated as the sum of these three component density parameters, $\Omega = \Omega_r + \Omega_m + \Omega_\Lambda$. Some estimates of the current density parameters for the radiation, matter, baryonic part of the matter, vacuum energy, and curvature are given in Table 17.1.[6]

17.14 The Deceleration Parameter

The preceding discussion indicates that the Universe contains various kinds of gravitating matter and fields. These will cause the Hubble parameter $H(t)$ to change with time and the rate of change will be related directly to the amount of visible matter, dark matter, dark energy, and radiation that the Universe contains. To investigate small changes in the expansion rate we may expand the cosmic scale factor to second order in time,

$$a(t) \simeq a_0 + \dot{a}_0(t - t_0) + \tfrac{1}{2}\ddot{a}_0(t - t_0)^2, \tag{17.33}$$

introduce the *deceleration parameter* $q_0 \equiv q(t_0)$ evaluated at the present time through

$$q_0 \equiv -a_0 \frac{\ddot{a}_0}{\dot{a}_0^2}, \tag{17.34}$$

and utilize the proof in Problem 17.8 that

$$\frac{\dot{a}_0}{a_0} = H_0, \tag{17.35}$$

to obtain the expansion

$$a(t) = a_0 \left[1 + H_0(t - t_0) - \tfrac{1}{2}H_0^2 q_0(t - t_0)^2 + \cdots \right] \tag{17.36}$$

(see Problem 17.10). By expanding a^{-1} to second order in a binomial series the redshift z defined in Eq. (16.13) may be expressed in terms of the expansion (17.36) as

---

[6] The curvature density entry Ω_k arises from spacetime curvature and it will be explained in Chapter 19. Its very small value reflects the observational evidence that the curvature of the Universe on cosmological scales is near zero.

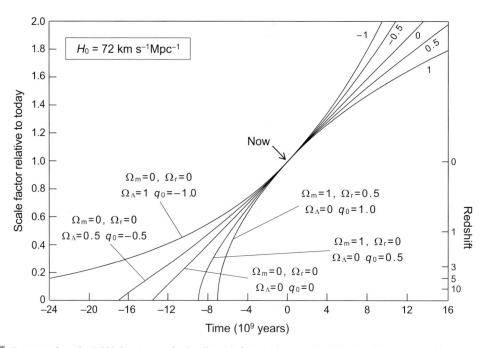

Fig. 17.12 Deviations from the Hubble law to second order. The scale factor is shown on the left axis and the corresponding redshift on the right axis. Time is measured from today. Different curves correspond to different assumed values of the density parameters Ω_i, which are related to the corresponding deceleration parameter q_0 through Eq. (17.42). Each curve has the same linear term but a different quadratic (acceleration) term. Positive values of q_0 correspond to a slowing of the expansion and negative values to an increase in the rate of expansion with time.

$$z = \frac{1}{a} - 1 = H_0(t_0 - t) + H_0^2(1 + \tfrac{1}{2}q_0)(t_0 - t)^2 + \cdots . \qquad (17.37)$$

This is a quadratic equation in $H_0(t_0 - t)$, which may be solved to give

$$H_0(t_0 - t) = z - (1 + \tfrac{1}{2}q_0)z^2. \qquad (17.38)$$

Integrating Eq. (16.7) by using the expansion (17.36) for $a(t)$ to first order, the current proper distance for an object with redshift z is found to be (Problem 17.9)

$$d(t_0) \simeq c(t_0 - t) + \tfrac{1}{2}cH_0(t_0 - t)^2 + \cdots . \qquad (17.39)$$

Inserting Eq. (17.38) in this expression and neglecting higher-order terms gives

$$d(t_0) = \frac{cz}{H_0}\left(1 - \frac{1 + q_0}{2}z\right), \qquad (17.40)$$

where the leading term is the Hubble law linear in z and the second term gives a correction one order higher in z. Thus, measurements at large enough redshift can test the correction terms to the linear Hubble expansion, as illustrated in Fig. 17.12.

17.14.1 Deceleration and Density Parameters

The deceleration parameter q_0 is related to the density parameters Ω_i through

$$q_0 = \frac{\Omega_m}{2} + \Omega_r - \Omega_\Lambda. \tag{17.41}$$

In a dust-only universe $\Omega_r = \Omega_\Lambda = 0$ and

$$q_0 = \frac{\Omega_m}{2} \quad \text{(dust only)}. \tag{17.42}$$

For a flat universe containing radiation, matter, and vacuum energy [Eqs. (19.24)–(19.26)],

$$\Omega = \Omega_r + \Omega_m + \Omega_\Lambda = 1 \quad \text{(flat universe)}, \tag{17.43}$$

which requires that

$$q_0 = \tfrac{3}{2}\Omega_m + 2\Omega_r - 1 \quad \text{(flat universe)}. \tag{17.44}$$

Since Ω_m and Ω_r are non-negative, Eq. (17.41) implies that a negative q_0 (acceleration of the expansion) is possible only if there is a non-zero vacuum energy density. The parameters of Table 17.1 suggest that the present Universe is flat with a negative deceleration parameter,

$$q_0 \simeq \frac{\Omega_m}{2} - \Omega_\Lambda \simeq -0.55, \tag{17.45}$$

implying that the expansion is accelerating. This will be elaborated in Chapters 18 and 19.

17.14.2 Deceleration and Cosmology

Figure 17.13 illustrates that H_0 and q_0 determine the behavior of the Universe only near the present time. Within the shaded box the curves are nearly indistinguishable but at larger redshifts they differ substantially. Until the 1990s the quest of cosmology was to determine reliably the values of H_0 and q_0. However, precision cosmology data acquired through the study of high-redshift Type Ia supernovae and the detailed analysis of anisotropies in the cosmic microwave background radiation now constrain a broader range of parameters than just these two. This will be discussed in more detail in the following chapters.

17.15 Problems with Newtonian Cosmology

As promised, considerable headway has been made in understanding the role of matter and energy in the expanding Universe, largely by using Newtonian gravity. However, the Newtonian approach leads to some problems and inconsistencies. For example:

1. At large distances the Hubble law implies recessional velocities that can exceed the speed of light. How is this to be interpreted?

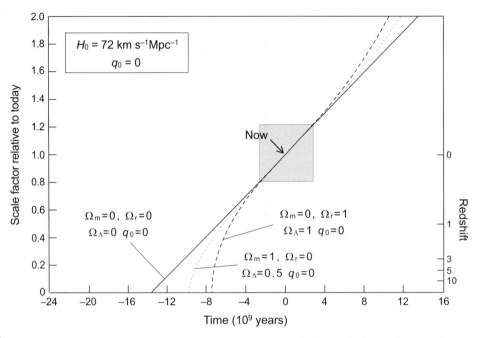

Fig. 17.13 Different matter, radiation, and vacuum energy densities giving the same deceleration for fixed H_0, by virtue of Eq. (17.41). Curves were calculated numerically using the methods described in Section 19.4; they agree near the present time (gray box), but have different long-time behaviors. For example, intercepts of the curves with the lower axis define three ages for the Universe that differ by billions of years.

2. Newtonian gravitation is assumed to act instantaneously but special relativity requires signals to propagate at finite speed, so there should be a delay in the action of gravity.
3. In the Newtonian picture a uniform isotropic sphere expanded into nothing, which causes conceptual problems in interpreting the expansion. Alternatively, if the sphere is of infinite extent there are formal difficulties with even defining a potential.
4. Evidence will be presented that the mass–energy of the Universe receives a large contribution from dark energy. How is the influence of this dark energy to be incorporated into the equations governing the evolution of the Universe?

These difficulties suggest that a better theory of gravity is required for an adequate description of cosmologies built on expanding universes that may contain exotic things like dark energy. In Chapter 18 an understanding of the expanding Universe based on the general theory of relativity will be developed that will be able to deal with these problems.

Background and Further Reading

For further discussion of the material in this chapter see Roos [207]; Ryden [211]; Liddle [145]; and Carrol and Ostlie [62].

Problems

17.1 Calculate the present energy density ϵ_{CMB} of the cosmic microwave background (CMB) and the associated density parameter $\Omega_{\text{CMB}} \equiv \varepsilon_{\text{CMB}}/\varepsilon_c$, where ε_c is the critical density. Estimate the energy density in starlight, ϵ_{stars}, and the ratio $\epsilon_{\text{CMB}}/\epsilon_{\text{stars}}$, assuming that the present luminosity density of galaxies is 2.6×10^{-32} erg s^{-1} cm^{-3} and that this has been true for the entire history of the Universe. *Hint*: The CMB has a temperature of 2.725 K.

17.2 Show that the result (17.17) for the expansion factor in a flat universe follows from Eq. (17.15). Express this in terms of the redshift z, find a general expression for $t(z)/\tau_{\text{H}}$, and use this to show that the age of the Universe for this case is $t_0 = \frac{2}{3}\tau_{\text{H}}$, where $\tau_{\text{H}} = H_0^{-1}$ is the Hubble time. ***

17.3 For the closed and open dust-filled universes defined by Eqs. (17.18)–(17.23), show that for the closed-universe solution ($\Omega > 1$)

$$\frac{t(z)}{t_{\text{H}}} = \frac{\Omega}{2(\Omega-1)^{3/2}}\left[\cos^{-1}\left(\frac{\Omega z - \Omega + 2}{\Omega + \Omega z}\right) - \frac{2(\Omega^2 z + \Omega - \Omega z - 1)^{1/2}}{\Omega + \Omega z}\right],$$

so that the age for a closed, dust-filled universe is

$$\frac{t_0}{t_{\text{H}}} = \frac{\Omega}{2(\Omega-1)^{3/2}}\left[\cos^{-1}\left(\frac{2-\Omega}{\Omega}\right) - \frac{2}{\Omega}(\Omega-1)^{1/2}\right],$$

and that for the open-universe solution ($\Omega < 1$)

$$\frac{t(z)}{t_{\text{H}}} = \frac{\Omega}{2(1-\Omega)^{3/2}}\left[\frac{2(-\Omega^2 z - \Omega + \Omega z + 1)^{1/2}}{\Omega + \Omega z} - \cosh^{-1}\left(\frac{\Omega z - \Omega + 2}{\Omega + \Omega z}\right)\right],$$

so that the age for an open, dust-filled universe is

$$\frac{t_0}{t_{\text{H}}} = \frac{\Omega}{2(1-\Omega)^{3/2}}\left[\frac{2}{\Omega}(1-\Omega)^{1/2} - \cosh^{-1}\left(\frac{2-\Omega}{\Omega}\right)\right].$$

Obtaining these results is not difficult but will require a bit of algebra. ***

17.4 Use the results of Problems 17.2 and 17.3 to verify the lookback times plotted in Fig. 17.3(b).

17.5 Show that a flat rotation curve for a galaxy implies that the enclosed mass increases linearly from the center.

17.6 The Sun moves with a velocity of about 220 km s^{-1} around the center of the galaxy at a distance of about 30,000 light years from the center (assume a circular orbit for the purposes of this problem). (a) Use this information to estimate the mass of the galaxy interior to the Sun's orbit. (b) The velocity curve determined for the Milky Way indicates that the average velocity is still 230 km s^{-1} at a radius of 60,000 light years. Estimate the mass interior to this radius. The mass inferred from these estimates is considerably larger than is suggested by light coming from these regions. This is one aspect of the dark matter problem.

17.7 Assume a set of gravitating objects to have a Gaussian distribution in velocity components,

$$\varphi(v_i) = \frac{1}{\sqrt{2\pi}\sigma_i} e^{v_i^2/2\sigma_i^2} \quad (i = 1, 2, 3),$$

where $v^2 = v_x^2 + v_y^2 + v_z^2$ and $\sigma^2 = \sigma_x^2 + \sigma_y^2 + \sigma_z^2$. Show that $\langle v^2 \rangle = \sigma^2$, where $\langle v_i^2 \rangle$ is the average of v_i^2.

17.8 Prove that if $v \ll c$, the redshift $z \simeq H_0(r/c)$, and that therefore for the scale parameter $a(t)$

$$\frac{\dot{a}_0}{a_0} = H_0,$$

where r is the distance to a distant galaxy and H_0 is the current value of the Hubble parameter. ***

17.9 Derive Eq. (17.39) by integrating Eq. (16.7) using the expansion (17.36). ***

17.10 Demonstrate that Eqs. (17.36), (17.37), and (17.38) follow from Eqs. (17.33) and (17.34). ***

17.11 Show that Eq. (17.40),

$$d(t_0) = \frac{cz}{H_0}\left(1 - \frac{1+q_0}{2}z\right),$$

follows from Eqs. (17.38) and (17.39).

17.12 Show that the differential equation (17.15) for the scale factor $a(t)$ follows from Eq. (17.14). *Hint*: Use the identity $\ddot{a} = \frac{1}{2}d(\dot{a}^2)/da$. ***

18 Friedmann Cosmologies

In the preceding chapter we introduced a cosmological picture based largely on the Hubble law interpreted either explicitly or implicitly within the framework of Newtonian gravity. This permitted an intuitive view of the expanding Universe and the introduction of some important terms and concepts: the critical density of the Universe, dark matter and the role of baryonic and nonbaryonic matter, dark energy, the cosmic scale factor and its dependence on matter and energy densities, lookback times, and the deceleration parameter. But at the same time we highlighted some of the shortcomings of a noncovariant view of spacetime on large scales: the meaning of boundaries for spacetime, the meaning of speeds exceeding that of light in the Hubble law, the unphysical instantaneous action at a distance built into Newtonian gravitation, the formal problem that the largest physical distances in the Universe are comparable to its radius of gravitational curvature, and the question of how dark energy fits into the overall picture. Accordingly, let us now consider solutions of the Einstein equations that may be relevant for describing in covariant fashion the large-scale structure of the Universe and its evolution following the big bang.[1]

18.1 The Cosmological Principle

As for the previous solutions of the Einstein equations that we have obtained, our first step will be to make a choice for the form of the spacetime metric governing the Universe, guided by intuition and observations. The possible forms for the metric of spacetime are thought to be strongly constrained by the *Cosmological Principle*:

> **Cosmological Principle:** When it is observed on sufficiently large scales, the Universe appears to be both homogeneous (no preferred places) and isotropic (no preferred directions).

---

[1] Some authors argue that for cosmology presented at the level of this book the Newtonian picture with suitable reinterpretation of parameters and a bit of finesse yields equations that are formally equivalent to the Friedmann equations that will be derived in this chapter. Thus, the argument goes, it isn't necessary to introduce the heavy machinery of general relativity to understand modern cosmology at this intermediate level. There is some validity to this point of view as a practical matter, but I believe that it has serious drawbacks. Chief among these are that it is built on suspect foundations (similar conceptually to trying to understand event horizons of black holes in terms of Newtonian gravity and escape velocities), it isn't clear how to incorporate dark energy consistently, and it easily leaves the misconception that general relativity is an unimportant luxury instead of the foundation for modern cosmology. But perhaps the most practical argument for doing things correctly is that in modern implementations it isn't really much harder to build cosmology on the Einstein field equations than to build it on a cobbled-together Newtonian framework.

The cosmological principle means that there is a proper time such that at any instant the 3-dimensional spatial line element of the Universe,

$$d\ell^2 = g_{ij} dx^i dx^j \qquad (i, j = 1, 2, 3) \tag{18.1}$$

(where g_{ij} is the spatial part of the metric) is the same in all places and all directions, with a total spacetime metric of the form

$$ds^2 = -dt^2 + a(t)^2 d\ell^2 = -dt^2 + a(t)^2 g_{ij} dx^i dx^j. \tag{18.2}$$

The scale parameter $a(t)$ will describe the uniform expansion or contraction of the spatial metric (see Section 16.2.3). There is no reason for time to pass at different rates for different locations in an isotropic and homogeneous Universe. If it did, this would provide a means to distinguish one region from another, thus contradicting the homogeneity assumption.[2] As discussed in Box 18.1, this permits the definition of a universal *cosmic proper time* and the time term in Eq. (18.2) is simply $-dt^2$ and not a more complicated expression such as in the Schwarzschild metric (9.4).[3] Hence, it is necessary to investigate 3-dimensional curved spaces that are both homogeneous and isotropic (no privileged places or directions). We consider the issue first in two dimensions where visualization is easier, and then generalize to the realistic case of three spatial dimensions.

18.2 Homogeneous and Isotropic 2D Spaces

In two dimensions there are three possibilities for homogeneous, isotropic spaces:

- flat euclidean space,
- a sphere of constant (positive) curvature, and
- an hyperboloid of constant (negative) curvature.

These are illustrated in Fig. 18.1. In each case the corresponding space has neither a special point nor a special direction that can be distinguished from any other. Thus, these are 2-dimensional spaces with underlying metrics that are consistent with the cosmological principle. The 2-sphere may be examined as a representative example. It is common to visualize 2-spheres as a 2-dimensional surface embedded in a 3-dimensional space [see Eq. (7.1)]. But a metric defines internal properties of a space independent of any additional embedding dimensions, as discussed in Section 7.2.2. Therefore, it should be possible to express the metric of the 2-sphere in terms of only two of the three coordinates appearing in Eq. (18.1). From Eq. (7.1), it may be deduced that

$$dz^2 = \frac{(xdx + ydy)^2}{S^2 - x^2 - y^2}.$$

---

[2] Strictly, this statement is true only if the time parameter is suitably averaged over the peculiar motion of galaxies relative to the uniform expansion.

[3] As we discussed in Section 9.1.3, in the Schwarzschild metric neither the coordinate t nor the coordinate r correspond directly to a physical quantity. In the metric Eq. (18.2) the coordinate r is not a physical variable but the cosmic time t is, by virtue of the homogeneity and isotropy assumptions.

Box 18.1 Cosmological Proper Time

By now the reader should have acquired enough skepticism about the definition of time and simultaneity in physical systems to ask whether a global time that applies to the entire Universe is a meaningful concept. All observers in Newtonian mechanics share a universal time, so there is no issue if a Newtonian approximation were valid, and in special relativity time has a well-defined meaning *in an inertial frame*. However, in general relativity there are *no global inertial frames*. Nevertheless, as we now discuss, a global time for the Universe has meaning if a particular set of requirements is met, and a metric embodying the cosmological principle meets those requirements. Then a global time termed the *cosmological proper time* or *cosmic time* can be introduced by a *foliation* of spacetime in terms of a sequence of non-intersecting spacelike 3D surfaces, as illustrated in the following figure for 2+1 spacetime [125].

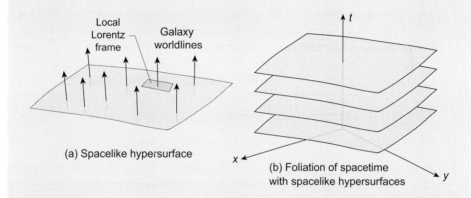

(a) Spacelike hypersurface

(b) Foliation of spacetime with spacelike hypersurfaces

It is assumed that all galaxies lie on such a hypersurface in such a way that the surface of simultaneity for the Lorentz frame of each galaxy coincides locally with the hypersurface. Thus a hypersurface may be viewed as consisting of the smoothly meshed Lorentz frames of all the galaxies, with the 4-velocity of each galaxy orthogonal to the local hypersurface. This series of hypersurfaces may be labeled by a parameter that can be viewed as a global cosmic time for all galaxies on the hypersurface, but *only if the space is homogeneous and isotropic* (which implies further that the spatial curvature is constant). Operationally, the cosmic time is the time measured by any observer who sees the Universe expanding uniformly around her. Such an observer is called a *comoving* or *fundamental* observer.

Fig. 18.1 Homogeneous, isotropic, 2-dimensional spaces. The open and flat surfaces should be imagined to extend infinitely. The open (constant negative curvature) surface cannot be embedded completely in 3-dimensional euclidean space. The saddle-like open surface approximates constant negative curvature only near the center of the saddle.

Therefore, the metric for the 2-sphere may be written in the form

$$d\ell^2 = dx^2 + dy^2 + dz^2 = dx^2 + dy^2 + \frac{(xdx+ydy)^2}{S^2 - x^2 - y^2}. \tag{18.3}$$

This metric depends on only two coordinates (S is a constant). Thus distances on the 2-dimensional surface are specified completely by coordinates intrinsic to that surface, independent of the third embedding dimension.

18.3 Homogeneous and Isotropic 3D Spaces

Spacetime appears to have three rather than two spatial dimensions; thus let us generalize the discussion of the preceding section and consider the embedding of 3-dimensional spaces in four euclidean dimensions.

18.3.1 Constant Positive Curvature

For a 3-sphere the generalization is obvious:

$$x^2 + y^2 + z^2 + w^2 = S^2, \tag{18.4}$$

where w is now a fourth euclidean dimension. The metric of the 3-sphere may be expected to be independent of the embedding and therefore expressible in terms of only three of these coordinates. Proceeding by analogy with the 2-sphere above, the metric may be written as

$$d\ell^2 = dx^2 + dy^2 + dz^2 + \frac{(xdx+ydy+zdz)^2}{1 - x^2 - y^2 - z^2}, \tag{18.5}$$

which has been specialized to a unit 3-sphere ($S = 1$) because the overall spatial scale will be set by the expansion factor $a(t)$ appearing in Eq. (18.2). It it convenient to introduce spherical polar coordinates (r, θ, φ) through the relations

$$x = r\sin\theta\cos\varphi \qquad y = r\sin\theta\sin\varphi \qquad z = r\cos\theta. \tag{18.6}$$

The metric of the unit 3-sphere then takes the form

$$\begin{aligned} d\ell^2 &= \frac{dr^2}{1-r^2} + r^2 d\theta^2 + r^2 \sin^2\theta d\varphi^2 \\ &= \frac{dr^2}{1-r^2} + r^2 d\Omega^2, \end{aligned} \tag{18.7}$$

with the definition

$$d\Omega^2 \equiv d\theta^2 + \sin^2\theta d\varphi^2. \tag{18.8}$$

The 3-sphere, by analogy with the 2-sphere, is a space of constant positive curvature that is homogeneous and isotropic, with great circles as geodesics. An intrepid ant explorer dropped onto the surface of a featureless 3-sphere would find that no point or direction appears any different from any other, that the shortest distance between any two points corresponds to a segment of a great circle, that a sufficiently long journey in a fixed direction returns to the starting point, and that the space has no boundary but the total volume is finite (see Problem 18.3).

18.3.2 Constant Negative Curvature

The generalization of the 2-hyperboloid to three dimensions is given by the equation

$$x^2 + y^2 + z^2 + w^2 = -S^2. \tag{18.9}$$

By a similar argument as above, the spatial metric in this case can be expressed as

$$d\ell^2 = \frac{dr^2}{1+r^2} + r^2 d\Omega^2, \tag{18.10}$$

which describes a space of constant negative curvature that is homogeneous and isotropic, with hyperbolas as geodesics. Our ant explorer would again find no preferred directions or locations, but the shortest paths between points would now be segments of hyperbolas, the ant would never return to the starting point by continuing an infinite distance in a fixed direction, and the ant would find that the volume of the space is infinite.

18.3.3 Zero Curvature

Finally, for a euclidean 3-space the metric may be expressed in the form

$$d\ell^2 = dr^2 + r^2 d\Omega^2. \tag{18.11}$$

The space is homogeneous and isotropic, with straight lines as geodesics, and of infinite extent. Obviously, this space corresponds to the limit of no spatial curvature.

18.4 The Robertson–Walker Metric

The results of the preceding section can be combined and the most general spatial metric in three dimensions that incorporates the isotropy and homogeneity constraints may be written as

$$d\ell^2 = \frac{dr^2}{1 - kr^2} + r^2 d\theta^2 + r^2 \sin^2\theta d\varphi^2, \tag{18.12}$$

where the dimensionless parameter k determines the nature of the curvature:

$$k = \begin{cases} +1 & \text{Hypersphere of positive curvature} \\ 0 & \text{Flat euclidean space} \\ -1 & \text{Hyperboloid of negative curvature.} \end{cases} \tag{18.13}$$

(The *curvature parameter* is k; the actual curvature will be k/a^2.) Finally, combining (18.12) and (18.2) gives the most general metric for 4-dimensional spacetime consistent with the homogeneity and isotropy implicit in the cosmological principle,

$$ds^2 = -dt^2 + a(t)^2 \left(\frac{dr^2}{1 - kr^2} + r^2 d\theta^2 + r^2 \sin^2\theta\, d\varphi^2 \right). \tag{18.14}$$

This is called the *Robertson–Walker (RW) metric*.[4] In matrix form the Robertson–Walker line element is

$$ds^2 = g_{\mu\nu} dx^\mu dx^\nu = (dt\ dr\ d\theta\ d\varphi) \begin{pmatrix} -1 & 0 & 0 & 0 \\ 0 & \dfrac{a^2}{1-kr^2} & 0 & 0 \\ 0 & 0 & a^2 r^2 & 0 \\ 0 & 0 & 0 & a^2 r^2 \sin^2\theta \end{pmatrix} \begin{pmatrix} dt \\ dr \\ d\theta \\ d\varphi \end{pmatrix},$$

with non-vanishing covariant metric tensor components

$$g_{00} = -1 \quad g_{11} = \frac{a^2}{1 - kr^2} \quad g_{22} = a^2 r^2 \quad g_{33} = a^2 r^2 \sin^2\theta, \tag{18.15}$$

and corresponding contravariant components

$$g^{00} = -1 \quad g^{11} = \frac{1 - kr^2}{a^2} \quad g^{22} = \frac{1}{a^2 r^2} \quad g^{33} = \frac{1}{a^2 r^2 \sin^2\theta}. \tag{18.16}$$

As discussed in Box 18.1, the time variable t appearing in the Robertson–Walker metric is the time that would be measured by an observer who sees uniform expansion of the surrounding Universe; it is termed the *cosmological proper time* or the *cosmic time*. The metric is generally a rank-2 symmetric tensor and so has 10 independent components. The symmetries assumed for 3-space in the RW metric have reduced this to specification of only two independent parameters in Eq. (18.14): the scale factor $a(t)$ and the curvature parameter k. Clearly imposition of the cosmological principle has produced a tremendous simplification in our mathematical description of cosmology.

The Robertson–Walker metric may be written in an alternative form by introducing the 4-dimensional generalization of polar angles

$$\begin{aligned} w &= \cos\chi & x &= \sin\chi \sin\theta \cos\varphi \\ y &= \sin\chi \sin\theta \sin\varphi & z &= \sin\chi \cos\theta \end{aligned} \tag{18.17}$$

(with ranges $0 \leq \varphi \leq 2\pi$, $0 \leq \theta \leq \pi$, and $0 \leq \chi \leq \pi$) into Eq. (18.4) with $S = 1$, and

$$\begin{aligned} w &= \cosh\chi & x &= \sinh\chi \sin\theta \cos\varphi \\ y &= \sinh\chi \sin\theta \sin\varphi & z &= \sinh\chi \cos\theta \end{aligned} \tag{18.18}$$

(with ranges $0 \leq \varphi \leq 2\pi$, $0 \leq \theta \leq \pi$, and $0 < \chi \leq \infty$) into Eq. (18.9) with $S = 1$, in which case the RW metric becomes

---

[4] Historically this metric is associated with the names of Alexander Friedmann (1888–1925), Georges Lemaître (1894–1966), Howard Robertson (1903–1961), and Arthur Walker (1909–2001). It is sometimes called the Friedmann–Lemaître–Robertson–Walker (FLRW) metric or the Friedmann–Robertson–Walker (FRW) metric. Since these are rather cumbersome labels, we shall refer to it as the Robertson–Walker or RW metric, while noting that Friedmann and Lemaître were actually the first to use the metric in the context of their solutions of the Einstein equations, while Robertson and Walker later independently put it on a firm mathematical footing using geometrical arguments, independent of general relativity.

$$ds^2 = -dt^2 + a(t)^2 \begin{cases} d\chi^2 + \sin^2\chi(d\theta^2 + \sin^2\theta d\varphi^2) & (k = +1) \\ d\chi^2 + \chi^2(d\theta^2 + \sin^2\theta d\varphi^2) & (k = 0) \\ d\chi^2 + \sinh^2\chi(d\theta^2 + \sin^2\theta d\varphi^2) & (k = -1) \end{cases}, \quad (18.19)$$

where the alternative forms (18.14) and (18.19) are related by the change of variables

$$r = \begin{cases} \sin\chi & (k = +1) \\ \chi & (k = 0) \\ \sinh\chi & (k = -1) \end{cases}. \quad (18.20)$$

It is important to appreciate that the Robertson–Walker metric follows from purely geometrical reasoning, subject to the constraints of isotropy and homogeneity; no dynamical considerations enter explicitly into its formulation. Dynamics will come later from the Einstein field equations with a metric of the Robertson–Walker form, which must be solved for the time dependence of the scale factor $a(t)$.

18.5 Comoving Coordinates

We may exemplify the homogeneous, isotropic expansion of the Universe implied by the cosmological principle and the Hubble law by placing dots on the surface of a balloon and blowing the balloon up, as illustrated in Fig. 18.2 (see also Fig. 16.2 and Box 18.2). As the balloon expands, the spherical coordinates (θ, φ) remain the same but the distance between points changes with the scale factor of the expansion (here parameterized by the radius $R(t)$ of the balloon). For example, imagine two cities on Earth, with their positions defined by latitude and longitude coordinates. If the size of the globe is expanded the actual distance between the two cities will increase but the latitude–longitude coordinates of the two cities remain the same. That is, the *coordinates* are unchanged but the *physical distance* changes because of the increase in the scale factor.

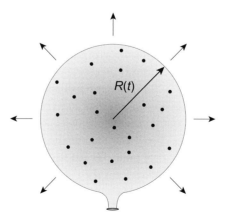

Fig. 18.2 Hubble expansion in two spatial dimensions. Each dot sees other dots receding from it radially and the distance between any two dots grows as $R(t)$ while the balloon expands.

> **Box 18.2** **Some Misconceptions about Balloons**
>
> Although the balloon analogy for an expanding Universe is extremely useful, it is important to guard against some misconceptions that it can generate.
>
> 1. The surface that is expanding is *2-dimensional;* the "center" of the balloon is in the third dimension and is not part of the surface, which has no center.
> 2. The Universe is *not* being expanded by a pressure. For that matter, neither is the balloon. Expansion of the balloon is generated by a *difference in pressure* but in a homogeneous, isotropic universe there can be no pressure differences on global scales. Furthermore, because pressure couples to gravity in the Einstein equation, addition of (positive) pressure to the Universe would *slow,* not increase, the expansion rate.
> 3. If the dots on the balloon represent galaxies, they too will expand as the balloon expands. But real galaxies don't partake of the general Hubble expansion because they are gravitationally bound. A better analogy is to glue solid objects to the surface of the balloon to represent gravitationally bound objects like galaxies, so that they don't expand when the balloon expands.
>
> Particular care should be taken in popular-level discussions, where the audience may not have the expertise to recognize these shortcomings of this useful analogy.

Coordinates exhibiting these properties are said to be *comoving* because they do not change with the expansion. An observer attached to a comoving coordinate is termed a *fundamental observer* or *comoving observer,* and this special coordinate system may be termed a *comoving frame*. Comoving observers will see other points on the surface of the balloon recede over time, but these points will maintain their comoving coordinates as they move away from the observer. An observer at rest in the comoving frame will see the Universe as homogeneous and isotropic (on large scales), but an observer in motion with respect to the comoving frame will *not* observe the Universe to be isotropic. An example is given in Box 18.3. Thus the comoving frame is a preferred frame of particular importance in a universe described by a Robertson–Walker metric.

Generalizing the balloon analogy to three spatial dimensions, the coordinates (r, θ, φ) of the Robertson–Walker metric (18.14), or (χ, θ, φ) of the metric in the form (18.19), are comoving coordinates. As the Universe expands the galaxies (if peculiar motion is ignored) keep the same coordinates (r, θ, φ) or (χ, θ, φ), and only the RW scale factor $a(t)$ changes with time. Just as in the 2-dimensional analogy the galaxies recede from an observer but for fundamental observers attached to a comoving galaxy the receding galaxies maintain their comoving coordinates and the recession is described entirely by the scale factor $a(t)$. Peculiar velocities change the comoving coordinates but they are small on the large scales where the cosmological principle and therefore the RW metric is valid. One consequence of the Robertson–Walker form for the metric is that if the Universe is described by the RW metric at any time t, then absent peculiar motion it will remain so for all time. The proof is simple: $dr = d\theta = d\varphi = 0$ if peculiar motion is negligible and this implies from Eq. (18.14) that $ds^2 = -dt^2$. Therefore, the expansion of the scale depends only on the time and all objects maintain constant comoving coordinates as the Universe evolves.

> **Box 18.3** **The Solar Rest Frame**
>
> In Chapter 20 we will discuss the cosmic microwave background (CMB), which is a blackbody radiation field expected to be isotropic in the comoving frame. However, a small dipole component is observed in the CMB when transformed to the solar rest frame (to eliminate the effect of the periodic motion of the Earth around the Sun), indicating that the CMB is slightly hotter in a particular direction and slightly colder in the opposite direction (by about 3 mK relative to the average CMB temperature of 2.725 K). This implies that the solar rest frame is not comoving but rather has a small velocity of about 600 km s^{-1} with respect to the comoving frame and thus sees a Doppler anisotropy in the observed CMB. This is presumably caused by gravitational interactions of our local group of galaxies with other large mass concentrations relatively nearby. Of course general relativity is frame-independent by construction and physical laws cannot depend on the coordinate system. However, some frames may be preferred because they are more *useful* than others in interpreting physical results. The comoving frame in a homogeneous and isotropic expanding universe is an example of such a preferred frame.

18.6 Proper Distances

We now consider the question of distances between galaxies in the RW metric. Measuring distances, or even defining them, becomes a nontrivial task in a spacetime that is expanding and possibly curved, and we shall discuss several notions of distance before we are through. The first is suggested by the diagram in Box 18.1. We imagine the cosmic time to be held fixed and then use the metric to compute the distance between two galaxies lying on the spacelike hypersurface corresponding to that fixed time. Take one galaxy to be at comoving coordinates $(r, \theta, \varphi) = (0, 0, 0)$ and the other to be at $(r, 0, 0)$, at fixed (cosmic) time t. Thus, $dt = d\theta = d\varphi = 0$ and Eq. (18.14) yields for the *proper distance* ℓ

$$\ell = a(t) \int_0^r \frac{dr}{\sqrt{1 - kr^2}}. \tag{18.21}$$

This is the "rulers end-to-end" distance that would be measured by a set of observers with rulers distributed between the two objects. Although this notion of distance is intuitive and conceptually well-grounded, it is clearly impractical to implement in astronomy where instead essentially all distance information comes from data carried by signals propagating on null geodesics (that is, light). Therefore, it will be necessary later to consider more extensively the meaning of distance and how to specify and measure it practically in observational astronomy.

Example 18.1 Equation (18.21) may be evaluated for the three curvatures given in Eq. (18.13).

1. *Flat euclidean space* ($k = 0$): the integral (18.21) is then trivial and $r = \ell/a$. Thus, r increases without limit as ℓ increases at fixed $a(t)$, implying a universe with no boundary or curvature that is of infinite extent. We say that such a universe is *flat*.

2. *Positive curvature* ($k = +1$): solution of Eq. (18.21) for $k = 1$ gives

$$\ell = a \int_0^r \frac{dr}{\sqrt{1-r^2}} = a \sin^{-1} r, \qquad (18.22)$$

and upon inverting, $r = \sin(\ell/a)$. In a space of constant positive spatial curvature, r returns to the origin whenever $\ell = \pi a$ and this universe has no boundary, but is of finite volume. We say that such a universe is *closed*.

3. *Negative curvature* ($k = -1$): Solution of Eq. (18.21) gives

$$\ell = a \int_0^r \frac{dr}{\sqrt{1+r^2}} = a \sinh^{-1} r, \qquad (18.23)$$

implying that $r = \sinh(\ell/a)$. Therefore, r grows without limit in a Universe of constant negative curvature as ℓ is increased at fixed $a(t)$, implying a universe that has no boundary and is of infinite extent. We say that such a universe is *open*.

Notice from these examples that the observational nature of the Universe depends strongly on whether there is curvature on cosmological scales.

18.7 The Hubble Law and the RW Metric

For a galaxy participating in the Hubble flow, r is a comoving coordinate and so is constant in time. Therefore from Eq. (18.21),

$$\dot{\ell} = \dot{a}(t) \int_0^r \frac{dr}{\sqrt{1-kr^2}}$$
$$= \frac{\dot{a}(t)}{a(t)} a(t) \int_0^r \frac{dr}{\sqrt{1-kr^2}} = \frac{\dot{a}(t)}{a(t)} \ell. \qquad (18.24)$$

We may recognize this immediately as a generalized form of Hubble's law, with

$$v \equiv \dot{\ell} = H\ell \qquad H = \frac{\dot{a}(t)}{a(t)}. \qquad (18.25)$$

From this derivation v clearly is *not a velocity in space* (the comoving coordinates are fixed), but rather a *velocity of expansion for space itself,* parameterized by the scale factor $a(t)$, which contains the only time dependence in the metric (18.14). It follows from this discussion that the Hubble law and the RW metric are linked fundamentally.

> The Robertson–Walker metric leads to the Hubble law. Conversely, the observation of a Hubble law indicates a metric of Robertson–Walker form.

It is of interest to note that when the RW metric was proposed the Hubble law was known but there was not yet strong observational evidence for homogeneity and isotropy of the Universe on large scales.

18.8 Particle and Event Horizons

An important consequence of the metric structure of spacetime and the finite speed of light is the possibility that regions of spacetime may be intrinsically unknowable for a fixed observer, either at the present time, or perhaps for all time. Such limitations are termed *horizons*. Two related concepts may be distinguished: a *particle horizon* and an *event horizon*. The former is of particular importance in cosmology and the phrase "cosmological horizon" or the generic term "horizon" are used often to mean "particle horizon" in the context of cosmology.

18.8.1 Particle Horizons in the RW Metric

A particle horizon is the largest distance from which a light signal could have reached us at time t, if it were emitted at some earliest possible time that we take to be $t = 0$. Imagine a spherical light wave emitted by us at the time of the big bang.[5] Over time it sweeps out over more and more galaxies. By symmetry, at the same instant that those galaxies can see us we can see them, so this spherical light front divides the galaxies into two groups: those inside our particle horizon, for which their light has had time to reach us since the big bang, and those outside our particle horizon, for which their light has not had time to reach us.

Light travels along a geodesic defined by the lightcone equation $ds^2 = 0$. By its definition, a particle horizon depends on time. Choosing the direction $\theta = \varphi = 0$ as before and inserting $ds^2 = d\theta = d\varphi = 0$ in (18.14) gives that

$$dt = a \frac{dr}{\sqrt{1 - kr^2}}.$$

Dividing by a and integrating both sides of this expression,

$$\int_0^t \frac{dt'}{a(t')} = \int_0^{r_p} \frac{dr}{\sqrt{1 - kr^2}}, \qquad (18.26)$$

where r_p is the comoving distance to the particle horizon. Combining Eq. (18.26) and Eq. (18.21) gives the corresponding proper (physical) distance to the particle horizon ℓ_h as,

$$\ell_h = a(t) \int_0^t \frac{dt'}{a(t')}. \qquad (18.27)$$

Whether a particle horizon exists then depends on the behavior of the integral on the right side of Eq. (18.27). Convergence of this integral at its lower limit is not guaranteed because the scale factor in the denominator of the integrand generally tends to zero at the lower limit in many cosmologies that are consistent with the Robertson–Walker metric. For example, it will be found that for matter-dominated or radiation-dominated Friedmann cosmologies the behavior of $a(t)$ near the lower limit gives a convergent integral; thus a particle horizon exists in these cosmologies (see Section 19.3.3). On the other hand, de Sitter cosmologies will be found to have no particle horizon (Section 19.3.1).

---

[5] Anything is possible in a thought experiment!

The lower limit of zero on the integral in Eq. (18.27) assumes that it is possible to see all the way back to the big bang, provided that the light has had time to reach us. Practically, the earliest visible time is later than this because of the opacity of the early Universe to various probes. The Universe was opaque to photons until about 400,000 years after the big bang, so photons from times earlier than that are not observable, even if there has been time for the light to reach us [see Figs. 20.9(b) and 20.10]. Likewise, the Universe was opaque to neutrinos until about one second after the big bang. Only the horizon for observation of gravitational waves would be expected to correspond to signals from near the actual time of the big bang, as will be discussed further in Section 22.1.3.

Example 18.2 As an example of a particle horizon, assume a flat, static Universe. Then the RW metric reduces to the Minkowski metric expressed in spherical coordinates with a equal to a constant and Eq. (18.27) implies that

$$\ell_h = c t_0, \tag{18.28}$$

which is the distance that light would travel in a time t_0 if the universe were flat and not expanding.

The discussion of particle horizons in cosmology is sometimes facilitated by the introduction of an alternative time coordinate η called the *conformal time*[6] through $dt = a(t)d\eta$, which is described in Box 18.4.

18.8.2 Event Horizons in the RW Metric

An event horizon is the most distant present event from which a worldline can *ever* reach our worldline. Proceeding in analogy with the discussion of particle horizons, the comoving distance r_e to an event horizon can be defined through

$$\int_{t_0}^{t_{max}} \frac{dt'}{a(t')} = \int_0^{r_e} \frac{dr}{\sqrt{1-kr^2}}. \tag{18.29}$$

Then the proper distance to the event horizon is

$$\ell_e(t) = a(t) \int_0^{r_e} \frac{dr}{\sqrt{1-kr^2}} = a(t) \int_{t_0}^{t_{max}} \frac{dt'}{a(t')}, \tag{18.30}$$

which differs from Eq. (18.27) for the particle horizon ℓ_h only in the limits of the integral. As for the case of particle horizons, whether an event horizon exists depends on the

---

[6] The transformation to conformal time is a special case of a *conformal transformation*, which is a transformation on the metric of the form $g^{\mu\nu} \to f g^{\mu\nu}$, where f is an arbitrary spacetime function called the *conformal factor*. A conformal transformation may be viewed as a local change of scale that preserves angles locally but not distances. (Example: In cartography the Mercator projection of the Earth's spherical surface onto a flat surface is mathematically a conformal map that distorts distances but preserves angles locally.) A conformal transformation is not a coordinate change but a *change of geometry*. Null geodesics are conformally invariant so conformal transformations have the useful feature that they preserve the lightcone and therefore causal structure of the spacetime.

Box 18.4   Conformal Time and Horizons

We may introduce the *conformal time* η through $dt = a(t)d\eta$. The flat ($k = 0$) Robertson–Walker metric (18.14) then may be expressed as

$$ds^2 = a^2(\eta)(-d\eta^2 + dr^2 + r^2 d\theta^2 + r^2 \sin^2\theta \, d\varphi^2),$$

which is the same form as the metric for a uniformly expanding Minkowski space, since η may be viewed as a time coordinate. Furthermore, this metric implies that in the η–r plane light rays move at 45 degree angles at all times (Problem 18.10), which simplifies discussion of horizon and causality issues. (Recall the discussion of Schwarzschild black holes in Section 11.4, where the transformation of the metric to Kruskal–Szekeres coordinates had a similar effect.) The figure below illustrates the behavior of particle horizons using a conformal time $\eta(t) \equiv \int_0^t dt/a(t)$ [111].

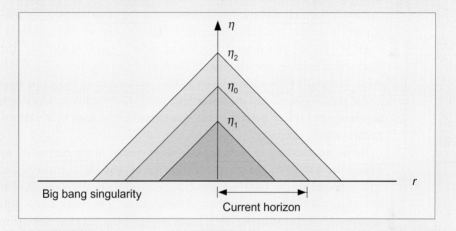

The horizon at the current time η_0 is indicated. Clearly the horizon was smaller at the earlier time η_1 and will be larger at the future time η_2. The $k = 0$ Robertson–Walker universe is related to Minkowski spacetime by a conformal transformation. It is called "flat" for this reason, though its spacetime is not flat (its spacelike slices are flat). Generally, the Robertson–Walker metric is *conformally flat* for all k, meaning that coordinate transformations exist for any k that permit the metric to be cast in Minkowski form.

---

behavior of the integral on the right side of Eq. (18.30). If this integral converges at t_{max} [which depends on the detailed behavior of $a(t)$ in this limit], then there is an event horizon in the cosmology.

Some metrics exhibit event horizons and others do not. The expanding balloon analogy may be used to understand qualitatively how an event horizon could arise in an expanding universe. Suppose that the inhabitants of galaxies on the surface of the balloon can exchange signals that move at constant local velocity. Since the physical distance between galaxies is increasing with time, the exchanged signals must cover a greater distance in going from one galaxy to the next than they would in a static space. But if the space is expanding fast enough, the distance to distant galaxies may be increasing so rapidly that the signal will *never* reach those distant galaxies, even after an infinite amount of time. (The analogy of walking on a treadmill that is moving backwards faster than your walking rate is helpful, if not pushed too far.)

Thus, if the expansion is sufficiently rapid there is an imaginary sphere centered on each galaxy dividing the other galaxies into two sets: those that already have been reached by a signal sent from the galaxy or will be reached by the signal at some point in the future, and those galaxies that will never be reached by that signal. This radius, if it exists, defines an event horizon for the observer, for by symmetry no signals from galaxies beyond this radius will ever reach the observer. Such event horizons bear some resemblance to the event horizons discussed in Chapter 11 for black holes. One important difference is that cosmological event horizons are defined relative to an observer. That is, each observer in a universe containing cosmological event horizons has her own event horizon, defined relative to the worldline of that observer.

Particle horizons and event horizons involve the same integrals but with different limits. A particle horizon represents the largest distance from which light could have reached us *today*, if it had traveled since the beginning of time. An event horizon is the largest distance from which light *emitted today* could reach us at *any time* in the future. Thus cosmological event horizons, as for black hole event horizons, separate regions of spacetime according to the *causal structure* of the spacetime, but particle horizons separate spacetime events according to whether objects in the spacetime can be seen by a particular observer at a particular time and place.

Hence the meaning of particle horizons is similar to our normal meaning of horizons, though the reasons that an observer cannot see beyond his horizon (the finite speed of light and curvature of the Earth's surface, respectively) are different. Horizons on the Earth also clearly depend on the location of the observer. The relationship of particle and event horizons is illustrated in Fig. 18.3. From this diagram it is clear that the space from which signals can be received at the present time is restricted by the particle horizon, and the space at present that could (in principle) be in causal contact with the present spatial position at some point in the future is restricted by the event horizon.

18.9 Einstein Equations for the RW Metric

We now turn attention to specific solutions of the Einstein equations (8.21) that may be relevant for cosmology. A simple class of solutions employs the Robertson–Walker metric with the assumption that the homogeneously and isotropically distributed matter and energy of the Universe constitutes a perfect fluid characterized by an energy density ϵ and a pressure P. The associated cosmologies are known as *Friedmann Cosmologies*.

18.9.1 The Metric and Stress–Energy Tensor

The non-zero covariant components of the Robertson–Walker metric in (t, r, θ, φ) coordinates are given by Eq. (18.15) and the corresponding contravariant components are

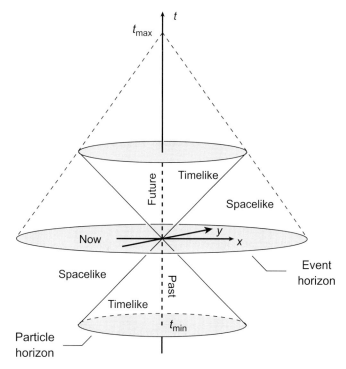

Fig. 18.3 Schematic representation of particle and event horizons (adapted from [207]). Note that in an expanding, possibly curved spacetime null geodesics will correspond to straight lines at ±45° angles like those shown here only if plotted in conformal time (Box 18.4).

given by Eq. (18.16). For a perfect fluid the most general stress–energy tensor for curved spacetime is given by Eq. (7.9) and for a comoving observer the fluid elements are at rest so that $u^\mu = (1, 0, 0, 0)$. Therefore, utilizing the components of the metric (18.15), the stress–energy tensor has only

$$T_{00} = \rho = \epsilon \qquad T_{11} = \frac{Pa^2}{1 - kr^2} \qquad T_{22} = Pr^2 a^2 \qquad T_{33} = Pr^2 a^2 \sin^2\theta \quad (18.31)$$

as non-zero covariant components.

18.9.2 The Connection Coefficients

The required connection coefficients may be obtained by inserting the metric tensor components (18.15) and (18.16) into Eq. (7.30). The following example illustrates the construction of Γ^2_{12}.

Example 18.3 Using Eqs. (18.15) and (18.16), the connection coefficient Γ^2_{12} for the RW metric is evaluated from Eq. (7.30) as

$$\Gamma^2_{12} = \tfrac{1}{2}g^{02}\left(\frac{\partial g_{20}}{\partial x^1} + \frac{\partial g_{10}}{\partial x^2} - \frac{\partial g_{21}}{\partial x^0}\right) \qquad (\nu = 0 \text{ term})$$

$$+ \tfrac{1}{2}g^{12}\left(\frac{\partial g_{21}}{\partial x^1} + \frac{\partial g_{11}}{\partial x^2} - \frac{\partial g_{21}}{\partial x^1}\right) \qquad (\nu = 1 \text{ term})$$

$$+ \tfrac{1}{2}g^{22}\left(\frac{\partial g_{22}}{\partial x^1} + \frac{\partial g_{12}}{\partial x^2} - \frac{\partial g_{21}}{\partial x^2}\right) \qquad (\nu = 2 \text{ term})$$

$$+ \tfrac{1}{2}g^{32}\left(\frac{\partial g_{23}}{\partial x^1} + \frac{\partial g_{13}}{\partial x^2} - \frac{\partial g_{21}}{\partial x^3}\right) \qquad (\nu = 3 \text{ term})$$

$$= \tfrac{1}{2}g^{22}\left(\frac{\partial g_{22}}{\partial x^1} + \frac{\partial g_{12}}{\partial x^2} + \frac{\partial g_{21}}{\partial x^2}\right) = \tfrac{1}{2}g^{22}\left(\frac{\partial g_{22}}{\partial x^1}\right)$$

$$= \tfrac{1}{2}\left(\frac{-1}{r^2 a^2}\right)\frac{\partial}{\partial r}\left(-r^2 a^2\right) = \tfrac{1}{r}.$$

Other connection coefficients may be calculated in a similar way.

All of the non-vanishing $\Gamma^\sigma_{\lambda\mu}$ for the Robertson–Walker metric are summarized in Table 18.1.

18.9.3 The Ricci Tensor and Ricci Scalar

The Ricci tensor may now be constructed from the connection coefficients and Eq. (8.16),

$$R_{\mu\nu} = \Gamma^\lambda_{\mu\nu,\lambda} - \Gamma^\lambda_{\mu\lambda,\nu} + \Gamma^\lambda_{\mu\nu}\Gamma^\sigma_{\lambda\sigma} - \Gamma^\sigma_{\mu\lambda}\Gamma^\lambda_{\nu\sigma}.$$

Utilizing the connection coefficients from Table 18.1, the non-vanishing components of the Ricci tensor are

$$R_{00} = -\frac{3\ddot{a}}{a} \qquad R_{11} = \frac{a\ddot{a} + 2\dot{a}^2 + 2k}{1 - kr^2} \qquad (18.32)$$

$$R_{22} = r^2(a\ddot{a} + 2\dot{a}^2 + 2k) \qquad R_{33} = R_{22}\sin^2\theta,$$

and the Ricci scalar (8.17) is obtained by contracting the Ricci tensor with the metric tensor,

$$R = g^{\mu\nu}R_{\mu\nu} = \frac{6(a\ddot{a} + \dot{a}^2 + k)}{a^2}. \qquad (18.33)$$

This completes the list of ingredients necessary to construct the Einstein equations.

Table 18.1 Friedmann connection coefficients[†]

$\Gamma^0_{11} = a\dot{a}/(1-kr^2)$	$\Gamma^0_{22} = r^2 a\dot{a}$	$\Gamma^0_{33} = r^2\sin^2\theta\, a\dot{a}$
$\Gamma^1_{01} = \dot{a}/a$	$\Gamma^1_{11} = kr/(1-kr^2)$	$\Gamma^1_{22} = -r(1-kr^2)$
$\Gamma^1_{33} = -r(1-kr^2)\sin^2\theta$	$\Gamma^2_{02} = \dot{a}/a$	$\Gamma^2_{12} = 1/r$
$\Gamma^2_{33} = -\sin\theta\cos\theta$	$\Gamma^3_{03} = \dot{a}/a$	$\Gamma^3_{13} = 1/r$	$\Gamma^3_{23} = \cot\theta$

[†]Coefficients are symmetric in their lower indices: $\Gamma^\mu_{\alpha\beta} = \Gamma^\mu_{\beta\alpha}$.

18.9.4 The Friedmann Equations

The Einstein equations (8.21) are $R_{\mu\nu} - \frac{1}{2}g_{\mu\nu}R = 8\pi G T_{\mu\nu}$. Consider the 00 and 11 components. Utilizing the results from Equations (18.31), (18.32), and (18.33) yields two independent equations,

$$H^2 \equiv \left(\frac{\dot{a}}{a}\right)^2 = \left(\frac{8\pi G}{3}\right)\epsilon - \frac{k}{a^2} \tag{18.34}$$

$$\frac{2\ddot{a}}{a} + \frac{\dot{a}^2}{a^2} + \frac{k}{a^2} = -8\pi G P, \tag{18.35}$$

where $H = \dot{a}/a$ from Eq. (18.25) has been used. These are termed the *Friedmann equations*.[7] They represent the most general form of the covariant gravitational equations with the conditions that have been imposed (the 22 and 33 components don't give any new results). In the next section, solutions of these equations will be considered.

18.9.5 Static Solutions and the Cosmological Constant

We first ask if the Friedmann equations have a static solution corresponding to a scale factor that is constant in time. Setting $\ddot{a} = \dot{a} = 0$ in Eqs. (18.34)–(18.35) leads to the requirement that

$$\frac{k}{a^2} = \frac{8\pi G}{3}\epsilon = -8\pi G P. \tag{18.36}$$

From this result it may be concluded that (1) for the energy density ϵ to be positive it is necessary that $k = +1$, and (2) if $\epsilon > 0$, the pressure must be *negative*, $P_0 < 0$! Thus, the Friedmann universe as formulated above has no physically reasonable static solution: it is unstable against either expansion or contraction.

When Einstein found that his field equations did not have a static cosmological solution the expansion of the Universe had not yet been discovered and the natural assumption of the time was that a proper cosmology should give a static solution. This led Einstein to make

---

[7] They are the general Einstein field equations given by Eq. (8.21) with the specific assumptions of a Robertson–Walker form for the metric and a perfect fluid approximation for the stress–energy tensor. These equations were first derived in 1922 and 1924 by the Russian physicist and mathematician Alexander Friedmann (1888–1925). They described a universe that was either expanding or contracting seven years before Edwin Hubble (1889–1953), extending earlier work by Vesto Slipher (1875–1969), published his observations demonstrating expansion of the Universe. Friedmann was in communication with Einstein but Einstein did not take the physics of Friedmann's results very seriously, largely because Einstein thought at the time that the Universe should be static. Friedmann's work was not widely appreciated before his death in 1924 at age 37 from typhoid fever (though it was published in the prestigious journal *Zeitschrift für Physik* with Einstein's help). In 1927 a Belgian priest, Georges Lemaître, derived and published Friedmann's equations independently in a less-read Belgian journal. As Hubble began to present observational evidence that the Universe was expanding in the late 1920s, Lemaître's work caught the eye of Arthur Eddington, who arranged to have the paper translated and republished in the *Monthly Notices of the Royal Astronomical Society*, where it finally began to receive attention. After it was shown that the Universe was expanding, Einstein acknowledged the significance of Friedmann's and Lemaître's results. Today the name "Friedmann equations" is synonymous with the mathematical description of the big bang and expanding Universe, and big bang cosmology is often termed the *Friedmann–Lemaître cosmology*.

> **Box 18.5** **The Cosmological Term and Einstein's Famous Blunder**
>
> From historical accounts it is unclear exactly what Einstein said concerning his dismay over the cosmological constant when it was shown that the Universe was expanding, and whether he might simply have been employing self-deprecating humor [178]. Nevertheless, the "blunder" quote has become an essential piece of cosmological lore. To set the context, when Einstein began to apply general relativity to cosmology in 1917 many astronomers still believed that the Milky Way galaxy was the entire Universe, and the realization that there were other "island universes" (galaxies) with redshifts suggesting that the Universe was expanding awaited the observational discoveries of Vesto Slipher and Edwin Hubble, which lay 5–10 years in the future.
>
> Addition of the cosmological term to the field equations may be justified by recalling from Section 8.3 the properties required of the left side of the Einstein equation (18.37). The cosmological term $\Lambda g_{\mu\nu}$ is a rank-2 tensor since Λ is a scalar, and it has vanishing covariant divergence since Eq. (3.63) requires the covariant derivative of the metric tensor to vanish. Thus it satisfies all the mathematical properties expected for a term in the Einstein equations. Einstein's dislike for this term was primarily aesthetic, since he felt that it spoiled the simplicity of the original field equations. His reasons for including it were invalidated by discovery of the expanding Universe, but modern observations that the expansion is accelerating indicate that something like the cosmological term is in fact required, though there is at present little understanding of its physical origin.

what he reportedly considered to be his greatest blunder (see the discussion in Box 18.5). He modified the field equations (8.21) by adding a term $\Lambda g_{\mu\nu}$ to the left side, where Λ is a scalar called the *cosmological constant*. The modified Einstein equations (8.21) now take the form

$$R_{\mu\nu} - \tfrac{1}{2} g_{\mu\nu} R + \Lambda g_{\mu\nu} = 8\pi G T_{\mu\nu}. \qquad (18.37)$$

Since the new cosmological term $\Lambda g_{\mu\nu}$ depends only on the metric, it is associated with the structure of spacetime itself and, unlike the other terms on the left side, it is finite in the limit of vanishing mass and curvature if $\Lambda \neq 0$. Thus, if this term is moved to the right side of Eq. (18.37) it will act as a gravitational source term, even if $T_{\mu\nu} = 0$. That is, even if the Universe contains no matter, energy, or pressure (a vacuum), $\Lambda g_{\mu\nu}$ can serve as a source for the gravitational field by curving the empty spacetime. For this reason, Λ is also called the *vacuum energy* parameter. If the Friedmann equations are modified to include the vacuum energy a positive value of Λ becomes a repulsive force that counteracts gravity (and a negative value becomes an attractive force that adds to the gravitational force). By proper adjustment of Λ, it is then possible to obtain a static solution (at least for a time; this solution was found later to be unstable on long timescales).

When the expansion of the Universe was discovered, Einstein realized that had he had more confidence in his original field equations he could have *predicted* that the Universe had to be either expanding or contracting. Once Hubble demonstrated that the Universe was expanding Einstein discarded the cosmological term. The data require Λ to be very small if it exists, so it can play a role only over volumes of space that are cosmological in scale; that is why it is called the *cosmological constant*. Since it is equivalent to an energy density associated with the ground state (vacuum state) of the Universe, Λ is also often termed the *vacuum energy density*. Because of this interpretation, in modern applications such as the discussion of dark energy in Section 19.1.2 it is convenient to absorb the effect of the cosmological constant on the left side of Eq. (18.37) into the terms associated with the stress–energy tensor on the right side. Therefore, in the following development the possibility of a finite vacuum energy will be included through a redefinition of the density and pressure variables in the Friedmann equations (see Section 19.1.2 and Section 19.2).

18.10 Resolution of Newtonian Difficulties

We conclude this chapter by remarking that the covariant theory of gravity embodied in Eq. (8.21) and implemented specifically in Friedmann cosmologies implies some essential conceptual differences relative to a Newtonian description of cosmology.

1. The consistent implementation of cosmology in an expanding space using general relativity alleviates inconsistencies associated with apparent recessional velocities that would exceed the speed of light at large distances. Since the recessional velocities are generated by the expansion of space itself, not by motion within space, there is no conceptual difficulty with recessional velocities larger than light velocity.
2. Because of the finite speed of (massless) gravitons implied by Lorentz invariance, gravity is no longer felt instantaneously over large distances. Instead the speed of gravity is equal to the speed of light, a result that has now been confirmed to extremely high precision by the detection of electromagnetic signals in coincidence with gravitational waves from a distant neutron star merger (see Section 24.7).
3. There is no longer the difficulty implicit in Newtonian cosmology of "expansion into nothing." In general relativity spacetime is generated by matter (and energy and pressure), so the idea of a boundary between a homogeneous universe and an empty space "outside" does not arise.
4. As will be seen in later chapters, the cosmological constant term in the Friedmann equations allows for the possibility of incorporating consistently the role of dark energy and cosmic inflation into a gravitational description of the Universe, as required by modern observations.

Therefore, a formulation of cosmology in terms of general relativity resolves several conceptual difficulties with the Newtonian approach to cosmology, in addition to providing a more solid quantitative basis.

Background and Further Reading

For clear introductions to Friedmann cosmologies see Roos [207], Ryden [211], and Hartle [111]. Horizons are discussed in Rindler [204], Islam [125], Weinberg [246], Roos [207], and Hartle [111].

Problems

18.1 Friedmann models in which the pressure and the curvature may be neglected ($P = 0$ and $k = 0$) are called *Einstein–de Sitter universes*. Write the Friedmann equations for this case and use them to show that (a) the time dependence of the scale factor $a(t)$ is

$$a(t) = (6\pi G \rho_0)^{1/3} t^{2/3},$$

where ρ_0 is the present density and it is assumed that $a = 0$ when $t = 0$. (b) Find the Hubble parameter $H(t)$, the deceleration parameter q_0, and the present age of the universe for this model in terms of the present value of the Hubble constant in an Einstein–de Sitter universe.

18.2 Show that for Friedmann cosmologies the constraint $T^{\mu\nu}_{;\nu} = 0$ on the stress–energy tensor implies the conservation of mass–energy equation

$$\dot\epsilon + 3(\epsilon + P)\frac{\dot a}{a} = 0,$$

with ϵ the energy density, P the pressure, and a the scale factor. It will be shown in Chapter 19 that this same relation follows from the Friedmann equations [see Eq. (19.3)]. ***

18.3 Find an expression for the volume of the spatial part of the Robertson–Walker metric in the flat, closed, and open cases. ***

18.4 Demonstrate that for the special case $a(t) = a_0 t^n$ with $\frac{1}{2} \le n \le \frac{2}{3}$, the horizon distance and the Hubble distance differ by a factor of order unity. Argue, however, that in the general case these two quantities could be very different.

18.5 Starting from Eq. (18.4) and its analogs for open and flat geometries, and the polar coordinates (18.17), demonstrate that the Robertson–Walker metric may be written in the form (18.19). Show explicitly the equivalence of the definitions (18.14) and (18.19) for the metric.

18.6 For the Robertson–Walker metric corresponding to the line element

$$ds^2 = -dt^2 + a(t)^2 \left(\frac{dr^2}{1 - kr^2} + r^2 d\theta^2 + r^2 \sin^2\theta d\varphi^2 \right),$$

find explicit expressions for the connection coefficient Γ^2_{02} and the Ricci tensor component R_{11}. Using the results for $R_{\mu\nu}$ quoted in Eq. (18.32), show that the Ricci scalar is given by Eq. (18.33).

18.7 Demonstrate that in the slowly varying, weak-field limit of the Einstein equations with a cosmological term the Poisson equation (8.1) is modified to

$$\nabla^2 \varphi = 4\pi G \rho - \Lambda c^2,$$

which implies a repulsive term increasing linearly with r in addition to the usual attractive $1/r^2$ contribution to the gravitational force. *Hint*: Generalize the result of Problem 8.11.

18.8 Show that evolution of the Universe assuming the Robertson–Walker metric, a stress–energy tensor $T^{\mu\nu}$ describing a perfect fluid, and the requirement $T^{\mu\nu}_{;\nu} = 0$ implies that the Universe obeys a form of the first law of thermodynamics consistent with the cosmological principle. *Hint*: Part of the work required for this problem was already done in Problem 18.2.

18.9 Conformal transformations were introduced in Section 18.8.1. Prove that under a conformal transformation of the metric the angle between vectors is preserved. *Hint*: The angle θ between spacetime vectors u and v is given by a formula analogous to that for euclidean space, $v \cdot u = |v||u|\cos\theta$.

18.10 Show that under a transformation to conformal time η defined through $dt = a(t)d\eta$, the $k = 0$ Robertson–Walker metric of Eq. (18.14) takes the form

$$ds^2 = a^2(\eta)(-d\eta^2 + dr^2 + r^2 d\theta^2 + r^2 \sin^2\theta d\varphi^2)$$

given in Box 18.4, and that the corresponding lightcones always open with 45° angles. ***

19 Evolution of the Universe

In Chapter 18 it was demonstrated that the Friedmann equations correspond to the Einstein equations when the metric takes the Robertson–Walker form and the stress–energy tensor is that of a perfect fluid. Henceforth our discussion of cosmology will be in terms of the Friedmann equations (except for Chapter 21, where we consider extensions of Friedmann cosmology such as inflation) and the resulting cosmologies are commonly called *Friedmann cosmologies*. By construction they represent a covariant description of the Universe on large scales under the physically motivated assumptions of homogeneity, isotropy, and a perfect fluid description of the matter and energy fields contained in the Universe. In this chapter we investigate the history of the Universe by using the Friedmann equations and an assumed *equation of state* representing the behavior of different components of the cosmic fluid. We will first investigate hypothetical universes with a single component of the cosmic fluid, then hypothetical universes with multiple components, and finally a realistic model of the Universe containing matter, radiation, and vacuum (dark) energy in the proportions suggested by observations.

19.1 Friedmann Cosmologies

The Friedmann equations in the form given in the preceding chapter describe the evolution of a universe with a Robertson–Walker metric (when supplemented by an appropriate equation of state, as discussed below), but they are not in the most convenient form for our subsequent discussion. Hence the first step in our investigation will be to reformulate the Friedmann equations in a compact form that will prove more useful for our purposes.

19.1.1 Reformulation of the Friedmann Equations

Consider the Friedmann equations (18.34)–(18.35), where a possible cosmological term will be omitted because its effect will be included in a modification of the energy density, as was discussed in Section 18.9.5. Differentiating (18.34) with respect to time gives

$$2\dot{a}\ddot{a} = \frac{8\pi G}{3}\left(2\epsilon a\dot{a} + \dot{\epsilon}a^2\right), \qquad (19.1)$$

subtracting (18.34) from (18.35) and solving for $\ddot{a}$ gives

$$2\ddot{a} = -\frac{8\pi G}{3}a(\epsilon + 3P), \qquad (19.2)$$

and using this to eliminate $\ddot{a}$ from Eq. (19.1) leads to

$$\dot{\epsilon} + 3(\epsilon + P)\frac{\dot{a}}{a} = 0, \qquad (19.3)$$

which also is often written in the form

$$\frac{\dot{\rho}}{\rho} + 3\left(1 + \frac{P}{\rho}\right)\frac{\dot{a}}{a} = 0, \qquad (19.4)$$

where $\epsilon = \rho c^2 = \rho$ in $c = 1$ units. This is in fact a continuity equation for the conservation of mass–energy in the cosmic fluid, since it can also be derived from the requirement $T^{\mu\nu}_{;\nu} = 0$, as shown in Problem 18.2. Hence Eq. (19.3) [or (19.4)] is termed the *fluid equation*.

Thus, we may study the evolution of the Universe by solving the two Friedmann equations (18.34) and (18.35), or by solving the first Friedmann equation and either (19.3) or (19.4). But these equations have three unknowns (if the integer k is fixed at one of its three possible values): ϵ, P, and the scale factor $a(t)$, and we have only two independent equations. Hence, an additional constraint is required for a unique solution. This is provided by an *equation of state* $P = P(\epsilon)$, which relates the thermodynamical variables P and ϵ. The behavior of the Friedmann universe then can be determined by solving the equations

$$\left(\frac{\dot{a}}{a}\right)^2 = \frac{8\pi G}{3}\epsilon - \frac{k}{a^2} \qquad (19.5)$$

$$\dot{\epsilon} + 3(\epsilon + P)\frac{\dot{a}}{a} = 0, \qquad (19.6)$$

supplemented by an equation of state,

$$P = P(\rho) = P(\epsilon). \qquad (19.7)$$

To proceed it is necessary to examine possible equations of state for the Universe. Typically, such equations of state will use the microphysics of the cosmic fluid to relate thermodynamical quantities to each other.

19.1.2 Equations of State

Cosmology deals with extremely dilute gases so we may expect that the equation of state can be expressed in a linear form

$$P = w\epsilon, \qquad (19.8)$$

where w is a dimensionless constant. The adiabatic sound speed c_s in a dilute gas is given by

$$c_s^2 = \frac{dP}{d\rho} = \frac{dP}{d\epsilon}c^2. \qquad (19.9)$$

Thus, if w is not negative, (19.9) and (19.8) imply that $c_s = c\sqrt{w}$ and the causal requirement that the speed of sound not exceed the speed of light requires that $w \leq 1$.[1] Let

---

[1] If w is negative, as happens if there is a vacuum energy density, the sound speed is imaginary. This means that there are no stable sound waves and pressure disturbances in the medium grow or decay exponentially rather than propagating as waves.

us now consider several possible equations of state that satisfy this condition and could be important in the evolution of the Universe [211].

Low-density gases of matter and radiation: A nonrelativistic gas at low density obeys the ideal gas law $P = (\rho/\mu)kT$, where μ is the average mass of the gas molecules. For a nonrelativistic gas the energy density is contributed almost entirely by rest mass and $\epsilon \simeq \rho c^2$. Therefore, the equation of state for a dilute, nonrelativistic ideal gas is

$$P = \frac{\rho kT}{\mu} = \frac{kT}{\mu c^2}\epsilon = w\epsilon \qquad w \equiv \frac{kT}{\mu c^2} \ll 1. \qquad (19.10)$$

Thus w is the ratio of the thermal energy to the rest mass energy for a gas particle, which is a very small number. Nonrelativistic gases have large energy densities but low pressure and in cosmology it is common to assume $P \sim 0$ for them.

Example 19.1 Consider air at room temperature as an example of a dilute, nonrelativistic gas. Then as shown in Problem 19.3, $P = w\epsilon$ with $w \sim 10^{-12}$.

On the other hand, for a low-density ultrarelativistic gas the equation of state is

$$P = w\epsilon = \tfrac{1}{3}\epsilon. \qquad (19.11)$$

Therefore massless particles (photons, gravitons, ...), or nearly massless particles (neutrinos, ...), have equations of state with $w \simeq \tfrac{1}{3}$, and they exert significant pressure relative to nonrelativistic gases.

Dark energy and the cosmological constant: From Eq. (19.2), the acceleration of the scale factor is

$$\ddot{a} = -\frac{4\pi G}{3}a(\epsilon + 3P). \qquad (19.12)$$

Therefore, if $P = w\epsilon$ and $w < -\tfrac{1}{3}$, the expansion of the Universe *accelerates* (rather than decelerates, as would be expected from a normal gravitational interaction). Any dilute gas with $w < -\tfrac{1}{3}$ is termed *dark energy*. Rewriting Eq. (18.37) as

$$R_{\mu\nu} - \tfrac{1}{2}g_{\mu\nu}R = 8\pi G T_{\mu\nu} - \Lambda g_{\mu\nu} \qquad (19.13)$$
$$= 8\pi G(T_{\mu\nu} - \epsilon_\Lambda g_{\mu\nu}) \qquad (19.14)$$

by utilizing the definition

$$\epsilon_\Lambda \equiv \frac{\Lambda}{8\pi G}, \qquad (19.15)$$

shows explicitly that addition of a cosmological constant Λ to the Friedmann equations is equivalent to adding a component with energy density $\epsilon_\Lambda = \Lambda/8\pi G$ to the cosmic fluid. From the fluid equation (19.6) applied to the dark energy component with energy density ϵ_Λ and pressure P_Λ,

$$\dot{\epsilon}_\Lambda = -3(\epsilon_\Lambda + P_\Lambda)\frac{\dot{a}}{a}. \qquad (19.16)$$

By hypothesis Λ is constant in time so ϵ_Λ is constant too. From Eq. (19.16) this is possible only if $P_\Lambda = -\epsilon_\Lambda$. Therefore, the equation of state associated with the cosmological constant is

$$P = w\epsilon \qquad w = -1, \qquad (19.17)$$

and adding a cosmological constant Λ to the Einstein equation as in Eq. (18.37) is equivalent to adding an energy density corresponding to a component of the cosmic fluid having an equation of state with $w = -1$. This represents a form of dark energy with $P = -\epsilon$, which implies a gas having *negative pressure* if the energy density is positive.

This discussion indicates that dark energy acts as a gas with *positive energy density but negative pressure*. Such equations of state are not unknown. For example, the tension in a stretched rubber band corresponds to a negative pressure since work must be done to stretch the band. What *is* unusual is to find these properties in a dilute gas, as is required for any source of dark energy. All gases ever studied in the laboratory require work to compress, not to expand! Although the cosmological constant is a simple example of dark energy, it is not the only possibility since any gas having an equation of state with $w < -\frac{1}{3}$ will cause the Universe to accelerate and thus qualifies as dark energy.

19.2 Friedmann Equations in Concise Form

The preceding discussion of the Friedmann equations and the cosmological equation of state permits a transparent and compact formulation of the basic equations governing the evolution of the Universe.

19.2.1 Evolution and Scaling of Density Components

Assuming the Universe to be composed of a set of independent components having density parameters Ω_i (see Section 17.13), Equation (19.5) may be expressed as

$$\frac{\dot{a}^2}{a^2} = H_0^2 \sum_i \Omega_i \left(\frac{a_0}{a}\right)^{3(1+w_i)} - \frac{k}{a^2}, \qquad (19.18)$$

where a_0 is the current value of the scale parameter and where the equation of state for each component is given from Eq. (19.8) by $P_i = w_i \epsilon_i$, which implies, by virtue of Eq. (19.6), that

$$\epsilon_i = \epsilon_i(0) \left(\frac{a}{a_0}\right)^{-3(1+w_i)}, \qquad (19.19)$$

where $\epsilon_i(0)$ is the energy density of component i *at the present time* [See Problem 19.1; ultimately (19.19) is a consequence of energy conservation.]

We may give a simple physical interpretation of the scaling of densities for different components of the Universe implied in Eq. (19.19). For radiation (ultrarelativistic particles having negligible mass), $w = \frac{1}{3}$ and the energy density varies as

$$\epsilon_r \simeq \frac{h\nu}{V} = \frac{hc}{V\lambda} \simeq a^{-4}, \qquad (19.20)$$

since the volume V scales as a^3 and the wavelength λ as a. On the other hand, for nonrelativistic matter $w = 0$ and the energy density varies as

$$\epsilon_m = \rho c^2 = \frac{mc^2}{V} \simeq a^{-3}. \qquad (19.21)$$

Thus, the difference in scaling for radiation-dominated and matter-dominated universes is ultimately the additional length-scale factor associated with the redshift of photon wavelengths caused by the expansion.[2] The vacuum energy (corresponding to $w = -1$ in the simplest models) is associated with "empty" space itself. Its density is not changed by the expansion of the Universe, so contribution of a $w = -1$ component is *independent of the scale factor*, unlike the contributions of matter and radiation.

19.2.2 A Standard Model

Observations suggest that (massive) matter, (massless) radiation, and dark energy all influence the evolution of the Universe. Let us take as a minimal idealized model a universe assumed to be composed of a single radiation component with $w = \frac{1}{3}$, a single matter component with $w = 0$, and a single vacuum energy component with $w = -1$; then Eq. (19.18) reduces to

$$\frac{\dot{a}^2}{a^2} = H_0^2 \left(\Omega_r \left(\frac{a_0}{a}\right)^4 + \Omega_m \left(\frac{a_0}{a}\right)^3 + \Omega_\Lambda + \Omega_k \right), \qquad (19.22)$$

where for convenience we have introduced a *curvature density parameter* Ω_k through

$$\Omega_k \equiv \frac{-k}{a^2 H_0^2}. \qquad (19.23)$$

Note though that Ω_k can be *negative* (if $k = 1$), unlike the other Ω_i, which satisfy $\Omega_i \geq 0$. Evaluation of Eq. (19.22) at the present time ($t \to t_0$ and $a \to a_0$) shows that the density parameters are constrained by

$$\Omega_r + \Omega_m + \Omega_\Lambda + \Omega_k = 1. \qquad (19.24)$$

(Recall that the density parameters Ω_i are by definition the ratio of the component density to the critical density evaluated *at the present time*.) Defining a total density parameter Ω through

$$\Omega \equiv \Omega_r + \Omega_m + \Omega_\Lambda \qquad (19.25)$$

permits Eq. (19.24) to be written as

$$\Omega = 1 - \Omega_k. \qquad (19.26)$$

---

[2] This argument is a bit fast and loose since it ignores that photons are not conserved particles and thus can be created continuously, for example by stars. It is still basically correct because the photons in the cosmic microwave background (Section 20.4) vastly outnumber those created by stars. For a more refined argument accounting for this, see Section 5.1 of Ryden [211].

Thus, *for a flat universe* $\Omega = 1$, irrespective of the values of the individual components Ω_r, Ω_m, and Ω_Λ entering Eq. (19.25). This is the fundamental new ingredient brought by dark energy, which invalidates all preceding discussions of curvature and closure for the Universe based only on matter and radiation density:

> With an appropriate density of dark energy the Universe can be flat without a closure density of matter and radiation, and it can be open and expand forever, even if it is flat.

Obviously this permits a much richer set of possibilities for cosmology than those allowed by the earlier considerations of a Universe dominated by matter, so let's now proceed to explore the implications of Eq. (19.22).

19.3 Flat, Single-Component Universes

Equations (19.18) or (19.22) don't have simple analytical solutions if – as in the realistic situation – there are multiple component densities that contribute. In that case, the integrations may be performed numerically (see Section 19.4). However, in the idealized case where only a single component contributes and curvature is neglected, analytical solutions are not difficult to find. These solutions provide insight but they also are of practical importance because evidence suggests that: (1) the Universe is very flat and was even flatter at earlier times, and that (2) at various stages in its evolution the Universe may be dominated by a single component. With that motivation, we now investigate analytical solutions of the Friedmann equations for flat universes having a single component.

Equation (19.18) restricted to a single component described by a density parameter Ω and equation of state parameter w is

$$\dot{a}^2 = H_0^2 \Omega a^{-(1+3w)} - k, \quad (19.27)$$

where we've set $a_0 = 1$. Provided that $1 + 3w$ is positive, the first term on the right side of (19.27) decreases and the curvature term k becomes relatively more important as the Universe expands. As will be seen, for the last several billion years the Universe has been dominated by two components, with the more important of these having $1 + 3w$ negative. However, evidence suggests that the early Universe was dominated by components for which $1 + 3w$ was positive. Furthermore, the curvature term is small today, based on observations, so it was even less important in the early Universe. Therefore, as a first approximation the curvature term in Eq. (19.27) may be neglected, leaving the approximate Friedmann equation

$$\dot{a}^2 = H_0^2 a^{-(1+3w)}, \quad (19.28)$$

where Ω has been set to unity because a flat Universe is assumed. Let us now consider possible solutions to this equation for the single component defined by the equation of state parameter w. Our presentation for single-component universes will parallel that of Ryden [211] and Padmanabhan [174].

> **Box 19.1** **Pure Vacuum Energy with Curvature**
>
> If the curvature term is kept in the pure vacuum energy universe, the general solutions are [179]
>
> $$a(t) = \begin{cases} \sinh(Ht) & (k = -1) \\ \exp(Ht) & (k = 0) \\ \cosh(Ht) & (k = 1) \end{cases}.$$
>
> All solutions evolve toward the $k = 0$ (de Sitter) solution as the universe expands. Generally, H is not the Hubble parameter at arbitrary t unless $k = 0$, but it becomes so rapidly as the hyperbolic functions tend to exponentials in the expanding universe.

19.3.1 Special Solution: Vacuum Energy Domination

For the special choice $w = -1$, Eq. (19.28) reduces to $\dot{a} = H_0 a$, which has the solution

$$a(t) = e^{Ht}. \tag{19.29}$$

Because an exponential curve is self-similar, there is no preferred point at which to normalize the solution (19.29). The integration constant was fixed by requiring that $a(t) = 1$ at $t = 0$. More generally, one could choose a solution $\exp H(t - t_0)$, with the normalization $a(t) = 1$ at $t = t_0$. Also, the variable H_0 has been replaced by H because from Eq. (19.29) $H(t) \equiv \dot{a}/a = H_0$, so for this special solution the Hubble parameter is constant in time.

de Sitter space: The solution for $w = -1$ corresponds to a pure vacuum energy universe (no matter or radiation). Such a universe with no curvature is termed a *de Sitter universe*. (See Box 19.1 for a discussion of a pure vacuum energy universe with finite curvature.) At first glance the de Sitter solution may appear to be rather academic since the Universe clearly does contain matter and radiation, in addition to any vacuum energy. However, it may be argued that if the Universe continues to expand, all matter and radiation will eventually be sufficiently diluted that effectively only vacuum energy will remain [see the dependence on $a(t)$ of the density parameters in Eq. (19.22)]. Therefore, the de Sitter solution may be viewed as a universe in which the density of matter and radiation is negligible compared with vacuum energy, and this is the ultimate fate of any more standard cosmology that expands forever. In addition, evidence will be presented in Section 21.3 that in its first moments our Universe expanded for a short but extremely important period in a de Sitter phase.

Cosmological observables: From the solution (19.29), some quantities of cosmological interest are easily computed. The age of a de Sitter universe is infinite,

$$\text{Age} = \infty, \tag{19.30}$$

since the scale factor is finite all the way to $t \to -\infty$. The redshift for light emitted at time t_e and detected at the present time $t = 0$ in a de Sitter universe is

$$z = \frac{a(t = 0)}{a(t_e)} - 1 = e^{-Ht_e} - 1, \tag{19.31}$$

in terms of which the time of emission t_e is

$$t_e = -\frac{\ln(1+z)}{H} \tag{19.32}$$

and the lookback time is

$$t_L = -t_e = \frac{\ln(1+z)}{H}. \tag{19.33}$$

The proper distance of a light source at the time of observation of a light signal is

$$\ell(t=0) = \int_{t_e}^{0} \frac{dt}{a(t)} = \int_{t_e}^{0} e^{-Ht} dt$$

$$= \frac{1}{H}\left(e^{-Ht_e} - 1\right) = \frac{z}{H}, \tag{19.34}$$

where (19.31) was used. The proper distance at the time of emission of this same light is rescaled by $a(t_e)/a(t_0) = 1/(1+z)$, so

$$\ell(t_e) = \left(\frac{1}{1+z}\right)\frac{z}{H}. \tag{19.35}$$

The particle horizon ℓ_h is the same integral as for $\ell(t=0)$, but with $t_e \to -\infty$ at the lower limit, so

$$\ell_h = \lim_{t_e \to -\infty} \ell(t_0) = \lim_{t_e \to -\infty}\left(\frac{z}{H}\right) = \infty. \tag{19.36}$$

A de Sitter spacetime is infinitely old and has no particle horizon. It is basically the same as the steady state universe described in footnote 1 in Section 16.1.1 and in Problem 16.3, except that the steady state model had in addition a mechanism for creating matter continuously from nothing.

The large redshift limit for the proper distance at the time of detection is

$$\lim_{z \to \infty} \ell(t=0) = \lim_{z \to \infty} \frac{z}{H} = \infty \tag{19.37}$$

but the corresponding large-redshift limit for the proper distance at the time of light emission t_e is

$$\lim_{z \to \infty} \ell(t_e) = \lim_{z \to \infty} \left(\frac{1}{1+z}\right)\frac{z}{H} = \frac{1}{H}. \tag{19.38}$$

In a de Sitter universe, objects with high redshift are at very large proper distance when *observed* but the light that is seen was *emitted* from these objects near a proper distance of c/H (Hubble distance). Once the light source is at greater than the Hubble distance, the expansion in a flat, vacuum energy dominated universe carries it away from the observer at greater than lightspeed and its light can never reach the observer. This constitutes an event horizon for the observer. The energy density in de Sitter space is, by virtue of Eq. (19.19),

$$\epsilon = \epsilon_0 a^{-3(1+w)} = \epsilon_0, \tag{19.39}$$

and is constant. These results for a flat universe containing only vacuum energy are summarized in the second column of Table 19.1. Solution of the Friedmann equations for a pure vacuum-energy universe are shown in Fig. 19.1 as the curve labeled $\Omega = \Omega_\Lambda = 1$.

Table 19.1 Cosmological quantities for flat single-component universes

Property	Vacuum only ($\Omega = \Omega_\Lambda = 1$)	Radiation only ($\Omega = \Omega_r = 1$)	Matter only ($\Omega = \Omega_m = 1$)
Age t_0	∞	$\dfrac{1}{2H_0}$	$\dfrac{2}{3H_0}$
Scale factor $a(t)$	e^{Ht}	$\left(\dfrac{t}{t_0}\right)^{1/2}$	$\left(\dfrac{t}{t_0}\right)^{2/3}$
Time of emission $t_e(z, t_0)$	$-\dfrac{\ln(1+z)}{H}$	$t_0(1+z)^{-2}$	$t_0(1+z)^{-3/2}$
Redshift $z(t_e, t_0)$	$e^{-Ht_e} - 1$	$\left(\dfrac{t_0}{t_e}\right)^{1/2} - 1$	$\left(\dfrac{t_0}{t_e}\right)^{2/3} - 1$
Proper distance at detection $\ell(t_0)$	$\dfrac{z}{H}$	$\dfrac{1}{H_0}\left(\dfrac{z}{1+z}\right)$	$\dfrac{2}{H_0}\left[1 - (1+z)^{-1/2}\right]$
Proper distance at emission $\ell(t_e)$	$\left(\dfrac{1}{1+z}\right)\dfrac{z}{H}$	$\dfrac{z}{H_0(1+z)^2}$	$\dfrac{2}{(1+z)H_0}\left[1 - (1+z)^{-1/2}\right]$
Lookback time t_L	$\dfrac{\ln(1+z)}{H}$	$\dfrac{1}{2H_0}[1 - (1+z)^{-2}]$	$\dfrac{2}{3H_0}\left[1 - (1+z)^{-3/2}\right]$
Particle horizon distance $\ell(t_h)$	∞	$\dfrac{1}{H_0}$	$\dfrac{2}{H_0}$
Energy density $\epsilon(t)$	ϵ_0 (constant)	$\dfrac{3}{32\pi G}t^{-2}$	$\dfrac{1}{6\pi G}t^{-2}$

19.3.2 General Solutions

Now we consider the general solution of Eq. (19.28) for the case where $w \neq -1$ [211]. In that case the form of Eq. (19.28) suggests a power-law solution $a(t) = (t/t_0)^n$. Substitution of this form into (19.28) and comparison of exponents on the two sides indicates that $n = 2/(3 + 3w)$, while comparison of the non-exponent factors indicates that

$$t_0 = \frac{2}{3H_0(1+w)}. \tag{19.40}$$

Thus, the general solution of Eq. (19.28) for $w \neq -1$ is

$$a(t) = \left(\frac{t}{t_0}\right)^{2/(3+3w)}, \tag{19.41}$$

with t_0 given by Eq. (19.40) and the Hubble constant H_0 related to the age of the universe t_0 through

$$H_0 = \frac{2}{3t_0(1+w)}. \tag{19.42}$$

Cosmological observables: Let us now use this solution to calculate some cosmological observables. From Eqs. (19.19) and (19.41)

$$\epsilon = \epsilon_0 \left[\left(\frac{t}{t_0}\right)^{2/(3+3w)}\right]^{-3(1+w)} = \epsilon_0 \left(\frac{t}{t_0}\right)^{-2} = \frac{t^{-2}}{6\pi(1+w)^2 G} \tag{19.43}$$

19.3 Flat, Single-Component Universes

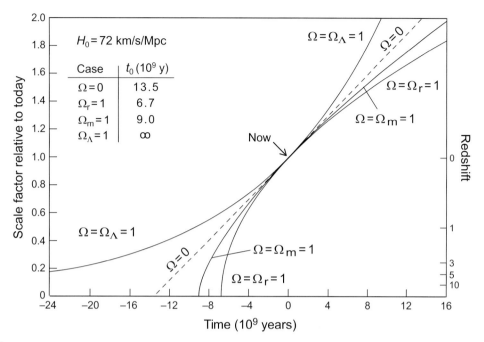

Fig. 19.1 Scale factor $a(t)$ versus time for a universe obeying the Hubble law ($\Omega = 0$), and for single-component critical densities of radiation ($\Omega = \Omega_r = 1$), matter ($\Omega = \Omega_m = 1$), and vacuum energy ($\Omega = \Omega_\Lambda = 1$). The left axis gives the ratio of the scale factor to its value today and the right axis gives the redshift. The corresponding age of the Universe t_0 corresponds to the intercept with the bottom axis and is given in the inset table.

where the present energy density $\epsilon_0 = 3H_0^2/8\pi G$ from Eq. (17.8) with $\Omega = 1$ and Eq. (19.42) were used in the last step. For light emitted at time t_e and detected today at time t_0, the redshift is

$$z = \frac{a_0}{a(t_e)} - 1 = \left(\frac{t_0}{t_e}\right)^{2/(3+3w)} - 1. \tag{19.44}$$

Inverting this, the time of emission is

$$t_e = t_0(1+z)^{-3(1+w)/2}. \tag{19.45}$$

The current proper distance is

$$\ell(t_0) = \int_{t_e}^{t_0} \left(\frac{t}{t_0}\right)^{-2/(3+3w)} dt = t_0 \frac{3+3w}{1+3w}\left[1 - \left(\frac{t_e}{t_0}\right)^{(1+3w)/(3+3w)}\right], \tag{19.46}$$

where $w \neq -\frac{1}{3}$. Equations (19.40) and (19.44) may be used to express the current proper distance in terms of z and H_0,

$$\ell(t_0) = \frac{2}{(1+3w)H_0}\left(1 - (1+z)^{-(1+3w)/2}\right). \tag{19.47}$$

The proper distance at the time of photon emission $\ell(t_e)$ is scaled by a factor $a(t_e)/a(t_0) = 1/(1+z)$ and

$$\ell(t_e) = \frac{1}{1+z}\ell(t_0). \tag{19.48}$$

The lookback time is

$$t_L = t_0 - t_e = t_0\left(1 - (1+z)^{-3(1+w)/2}\right), \tag{19.49}$$

which reduces to the form expected from the Hubble law,

$$t_L \simeq z\tau_H = \frac{z}{H_0} \quad \text{(small } z\text{)}, \tag{19.50}$$

only for small redshift.

Horizons: The particle horizon distance is the same integral as for the proper distance at the current time but with the lower limit $t_e \to 0$, or equivalently, $z \to \infty$, so *if the integral is convergent*

$$\ell_h = \lim_{z \to \infty} \ell(t_0) = \frac{2}{(1+3w)H_0}. \tag{19.51}$$

However, this integral is not convergent for all values of w. Generally one finds

$$\ell_h = \begin{cases} \dfrac{2}{(1+3w)H_0} & w > -\frac{1}{3} \\ \infty & w \leq -\frac{1}{3} \end{cases}. \tag{19.52}$$

Therefore, a flat spacetime with a single component has a particle horizon only if $w > -\frac{1}{3}$. In a flat universe with a single component having $w \leq -\frac{1}{3}$, there is no horizon and an observer can (in principle) see all points in space. The more distant points will be extremely redshifted, however (recall the preceding discussion of the de Sitter solution corresponding to $w = -1$). In a flat universe with a single component having $w > -\frac{1}{3}$, the horizon distance is finite and an observer sees only a portion of an infinite volume (the *visible universe*). Since the horizon distance is proportional to the age of the universe, it grows with time: signals propagating at lightspeed have had time to reach the observer from increasingly distant points as time goes on.

19.3.3 Flat Universes with Radiation or Matter

We now apply the results of the general solutions given in Section 19.3.2 to two specific examples: (1) a radiation-only cosmology with no curvature, corresponding to $w = \frac{1}{3}$, and (2) a matter-only cosmology with no curvature, corresponding to $w = 0$. Finding the observables for these cases requires only substitution of these specific values of w into Eqs. (19.40)–(19.52). For example, the current ages of the Universe for the flat radiation-only and flat matter-only cases are obtained from Eq. (19.40) as

$$\text{Radiation}: \quad t_0 = \frac{1}{2H_0} \quad (\Omega = 1, w = \tfrac{1}{3}). \tag{19.53}$$

$$\text{Matter}: \quad t_0 = \frac{2}{3H_0} \quad (\Omega = 1, w = 0). \tag{19.54}$$

As another example, from Eq. (19.52) the proper horizon distance in a flat pure-radiation universe is

$$\ell_h = \frac{2}{(1+3w)H_0} = \frac{1}{H_0} \quad \text{(radiation; } w = \tfrac{1}{3}\text{)}, \tag{19.55}$$

which happens to be the same as the static result of Eq. (18.28), and the proper horizon distance in a flat pure matter universe is

$$\ell_h = \frac{2}{(1+3w)H_0} = \frac{2}{H_0} \quad \text{(matter; } w = 0\text{)}. \tag{19.56}$$

General results for other cosmological quantities are summarized in the third column of Table 19.1 for the radiation-only case and in the fourth column of Table 19.1 for the matter-only case, and scale factor curves for flat universes containing only matter ($\Omega = \Omega_m = 1$) and only radiation ($\Omega = \Omega_r = 1$) are displayed in Fig. 19.1.

Example 19.2 The preceding results may be used to make a rough estimate of the particle horizon for our current Universe. Assume that: (a) Friedmann cosmologies are applicable, (b) the Universe has been essentially flat for the entire evolution since the big bang, (c) any vacuum energy contribution to the evolution of the Universe can be ignored, and (d) the radiation-dominated era was so short that negligible error is made by taking the entire time since the big bang to have been matter dominated. Equation (19.56) gives for the present particle horizon $\ell_h \sim 2c/H_0 \simeq 8000$ Mpc. The particle horizon is actually at a considerably larger distance ($\sim 14{,}000$ Mpc for the realistic example to be discussed in Section 19.4.7) because this simple estimate fails to account for the effect of vacuum energy, which has dominated the evolution of the Universe for the past four billion years.

The preceding example suggests that single-component cosmological models are useful for qualitative estimates but that a quantitative description requires a more sophisticated treatment. Accordingly, let us consider now more realistic Friedmann models in which the Universe can contain arbitrary combinations of radiation, matter, and vacuum energy.

19.4 Full Solution of the Friedmann Equations

In the general case, corresponding to arbitrary values of the parameters H_0, Ω_r, Ω_m, and Ω_Λ (and therefore to possible non-zero curvature), Eq. (19.22) may be integrated numerically. This is not difficult for standard cosmologies because the differential equations are generally well-behaved except at the initial singularity and standard numerical integration methods such as the Runge–Kutta algorithm will work.

19.4.1 Evolution Equations in Dimensionless Form

For numerical calculations it will be most convenient to rewrite the preceding equations in dimensionless form. If a dimensionless measure $q(t)$ of the scale and a dimensionless measure τ of time are introduced through

$$q \equiv \frac{a}{a_0} \qquad \tau \equiv H_0 t, \tag{19.57}$$

Eq. (19.22) takes the form

$$\frac{1}{2}\left(\frac{dq}{d\tau}\right)^2 + U(q) = E, \tag{19.58}$$

with the definitions

$$U(q) \equiv -\frac{1}{2}\left(\frac{\Omega_r}{q^2} + \frac{\Omega_m}{q} + \Omega_\Lambda q^2\right) \qquad E \equiv \frac{1}{2}(1 - \Omega) = \frac{1}{2}\Omega_k, \tag{19.59}$$

and with the redshift z related to the variable q through

$$z = \frac{1-q}{q}. \tag{19.60}$$

Equations (19.57)–(19.60) represent a solution for the evolution of the Universe in terms of *four independent parameters*, all evaluated at the *present time*:

1. the current value of the Hubble parameter H_0,
2. the current value of the radiation energy density parameter Ω_r,
3. the current value of the matter energy density parameter Ω_m, and
4. the current value of the vacuum energy density parameter Ω_Λ.

These parameters then fix the current curvature density parameter Ω_k through Eq. (19.24).

19.4.2 Algorithm for Numerical Solution

Equation (19.58) may be solved for the evolution of the Universe by the following numerical algorithm:

1. Specify values of the cosmic parameters H_0, Ω_r, Ω_m, and Ω_Λ, which in turn fixes the curvature density parameter Ω_k through Eqs. (19.24)–(19.26).
2. Compute the age of the Universe in units of the dimensionless time τ by evaluating numerically the integral

$$\tau_0 = \int_0^1 \frac{dq}{\sqrt{2(E - U(q))}}. \tag{19.61}$$

3. Integrate the differential equation implied by Eq. (19.58),

$$dq = \sqrt{2(E - U(q))}\, d\tau, \tag{19.62}$$

from the current time (corresponding to $q = 1$) backward to the beginning ($q = 0$ in models with a big bang), and forward to arbitrary times, to determine the past and future time evolution of the dimensionless scale factor $q(\tau)$.
4. Convert back to standard variables t and $a(t)$ as appropriate using (19.57), and use the results to compute quantities of cosmological interest such as redshifts, lookback times, and horizons.

We now consider some examples of using this algorithm to determine the cosmology associated with single-component and multiple-component universes.

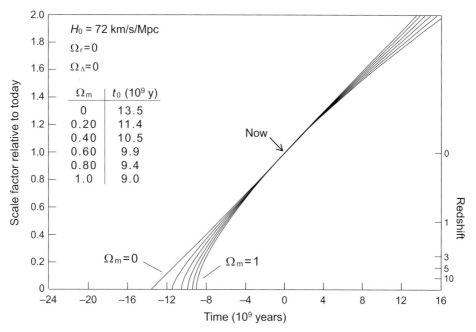

Fig. 19.2 Scale factor versus time for a universe with no vacuum or radiation energy density and matter density parameter Ω_m with values of 0, 0.2, 0.4, 0.6, 0.8, and 1. The left axis is marked in terms of the ratio of the scale factor to its value today and the right axis gives the corresponding redshift. The age of the Universe corresponding to each choice of Ω_m is defined by the intercept of the curve with the lower axis and is given in the inset table.

19.4.3 Examples: Single Component with Curvature

Figures 19.2–19.4 illustrate some numerical calculations that have been carried out according to the prescription given in Section 19.4.2 for single-component universes where the density parameter Ω is not necessarily equal to one. By virtue of Eq. (19.26), these spaces have non-zero curvature if $\Omega \neq 1$. Each of these plots shows the variation of the scale factor as a function of time, with the current value normalized to unity. The time axis has been shifted such that it measures billions of years relative to today. Therefore, the current time corresponds to a scale factor of one and time of zero, and -10 on the horizontal axis means 10×10^9 years earlier than today. In all cases, a Hubble constant of $72 \, \text{km} \, \text{s}^{-1} \, \text{Mpc}^{-1}$ has been assumed. The age of the Universe corresponds to the intercept of the scale factor curve with the bottom axis [that is, the time when $a(t) \to 0$]. In each case, a corresponding redshift is shown on the right axis.

In Fig. 19.2, evolution of the scale factor for a matter-only universe with densities ranging from 0% to 100% of critical is shown. All cases agree in value and slope at the current time ("Now"), but differ substantially in the distant future and past. For example, from the inset table the predicted age of the Universe ranges from 13.5 to 9.0 billion years for this range of densities. By drawing horizontal and vertical intercepts with the scale

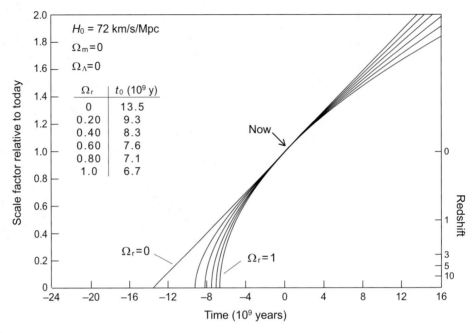

Fig. 19.3 Scale factor versus time for a universe with no vacuum or matter energy density and radiation density Ω_r with values of 0, 0.2, 0.4, 0.6, 0.8, and 1. The left axis is marked in terms of the ratio of the scale factor to its value today and the right axis is marked in terms of redshift. The corresponding age of the Universe is given in the inset table.

factor curve, one sees that for monotonic curves the redshift, time, and scale factor may be used interchangeably as time parameters, and that the relationship among these quantities depends on the cosmological parameters (Ω_m and H_0 for this example). The case $\Omega_m = 1$ is flat with closure density. All other examples correspond to open, curved universes.

In Fig. 19.3, an example similar to that of Fig. 19.2 is displayed, except that now the single component is radiation. The case $\Omega_r = 1$ is flat with closure density and the other curves correspond to open, curved universes. The predicted lifetimes vary over an even larger range than in the matter-only case, from 6.7 billion years for the $\Omega_r = 1$ case, up to 13.5 billion years for the $\Omega_r = 0$ case. Notice that in both Fig. 19.2 and Fig. 19.3, increasing the density causes the age of the Universe to decrease relative to the Hubble time of 13.5 billion years (which corresponds to the intercept of the $\Omega_r = 0$ curve with the bottom axis). Figure 19.4 gives an example similar to Figs. 19.2 and 19.3, but now for varying amounts of pure vacuum energy. Unlike for pure matter or radiation, the curves are concave upward, so the lifetimes for a pure vacuum energy universe are longer than the Hubble time. For the examples shown, the lifetime ranges from 13.5 billion years for $\Omega_\Lambda = 0$ to 40.8 billion years for $\Omega_\Lambda = 0.99$. (As illustrated in Fig. 19.1, for $\Omega_\Lambda = 1$ the universe has no initial singularity and has always existed.)

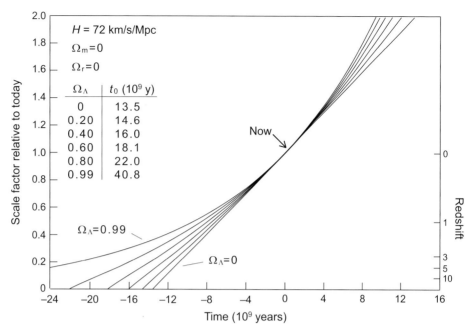

Fig. 19.4 Scale factor versus time for a universe with no radiation or matter but with vacuum energy density Ω_Λ with values of 0, 0.2, 0.4, 0.6, 0.8, and 0.99. The left axis is marked in terms of the ratio of the scale factor to its value today and the right axis is marked in terms of redshift. The corresponding age of the Universe is given in the inset table.

19.4.4 Examples: Multiple Components

Solutions of the Friedmann equations that include more than one component will reflect a superposition of the behaviors illustrated in Figs. 19.1–19.4. In Fig. 19.5 a representative calculation is shown for a hypothetical universe having $\Omega_m = 0.50$, $\Omega_r = 0.10$, and $\Omega_\Lambda = 0.40$. (This is a flat universe, since $\Omega = \Omega_m + \Omega_r + \Omega_\Lambda = 1$.) By comparison with Figs. 19.2–19.4, portions of this curve that are dominated by different components may be identified. From Eqs. (19.19) and (19.22), the different contributions will fall off at different rates with time. Radiation, which has a density scaling as $a(t)^{-4}$, will fall off most rapidly, followed by matter, which has a density scaling as $a(t)^{-3}$. The vacuum energy is constant through all of space, so its density does not change with the expansion and it makes an increasingly larger relative contribution as the Universe evolves.

The curve defining the scale factor in Fig. 19.5 is dominated initially by radiation density, then by matter, and finally by vacuum energy at large times, as indicated in the figure. The relative contribution of each component can be seen more clearly in the inset to the figure, which shows the fraction of the total energy density associated with each component as a function of time (with the total normalized to unity). In this hypothetical universe radiation dominates until about 8 billion years ago (about 1.7 billion years after the big bang). Then matter becomes dominant from about 8 billion years ago until the present time, and finally the vacuum energy will overtake the contribution of the matter

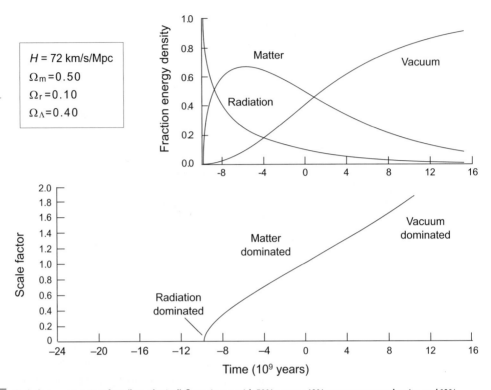

Fig. 19.5 Scale factor versus time for a (hypothetical) flat universe with 50% matter, 40% vacuum energy density, and 10% radiation. The fractions of the total energy densities contributed by matter, radiation, and vacuum energy as a function of time are indicated in the inset plot, with the sum normalized to one. Evolution of this universe is successively dominated by radiation, then matter, then vacuum energy. Its age is only 9.7 billion years. It is flat and contains the same components as the realistic Universe, but in different proportions (more matter and radiation, less vacuum energy than the actual Universe).

within the next billion years or so and increasingly dominate until the universe becomes an exponentially expanding, spatially flat de Sitter space containing negligible matter and radiation density.

19.4.5 Parameters for a Realistic Model

The preceding discussion has introduced the relative contributions of radiation, matter, and vacuum energy to the evolution of the Universe. These have been toy models that have explored ranges of possible parameters. What about the real Universe? What do the data say are the appropriate values to choose for the cosmological parameters and how does the Universe evolve if those values are used in the Friedmann equations? Although the evolution of the actual Universe is undoubtedly more complicated than the 4-parameter model being discussed, there is reason to believe that these four parameters are the most important ones in determining the overall behavior of the Universe. Let's assume that to be the case. Then, observational information is required to fix the current values of (1) the Hubble constant H_0, (2) the radiation density parameter Ω_r, (3) the matter density parameter Ω_m, and (4) the vacuum energy density parameter Ω_Λ.

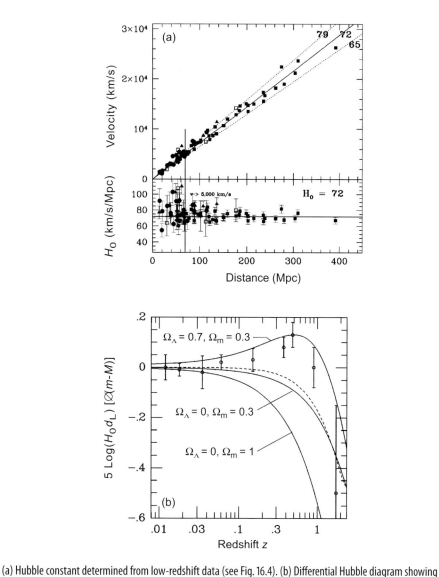

Fig. 19.6 (a) Hubble constant determined from low-redshift data (see Fig. 16.4). (b) Differential Hubble diagram showing distance modulus versus that for an empty universe ($\Omega = 0$) plotted versus redshift. Data points represent the binning of data from more than 200 high-redshift, Type Ia supernovae. The dashed curve corresponds to a non-accelerating flat universe; points above it indicate acceleration of the expansion. The three solid curves correspond to theoretical models with $\Omega_\Lambda = 0.7$ and $\Omega_m = 0.3$ (top curve), $\Omega_\Lambda = 0.0$ and $\Omega_m = 0.3$ (middle curve), and $\Omega_\Lambda = 0.0$ and $\Omega_m = 1.0$ (bottom curve). Figure from Ref. [92], based on data adapted from Ref. [235]. Adapted Figs. 3 and 4 with permission from Freedman W.L. and Turner M.S., *Reviews of Modern Physics*, **75**, 1433, 2003. Copyright (2003) by the American Physical society.

Figure 19.6(a) illustrates one observational determination of the Hubble constant, as was discussed in Chapter 16. Different methods of determining H_0 disagree at the $\sim 5-10\%$ level among themselves but agree that it lies in the range $67-73\,\mathrm{km\,s^{-1}\,Mpc^{-1}}$. We will assume the Hubble constant to have the value $H_0 = 72\,\mathrm{km\,s^{-1}\,Mpc^{-1}}$ for the

> **Box 19.2** **Type Ia Supernovae as Standardizable Candles**
>
> A *standard candle* is a light source that always has the same intrinsic brightness under some specified conditions. A *standardizable candle* may vary in brightness but can be normalized to a common brightness by some reliable method. Thus, once normalized a standardizable candle becomes effectively a standard candle.
>
> **Distance Measurements with Standard Candles**
>
> Standard candles enable distance measurement by comparing observed brightness with the standard brightness using the inverse square intensity law (19.63), $F = L/4\pi r^2$. In a flat, static universe, if the source is a standard candle L is known and measurement of the flux F yields the distance r. As will be shown in Box 19.3, this relationship requires modification in an expanding, curved space.
>
> **Cepheid Standard Candles and Their Limitations**
>
> The preferred method of determining distances beyond those where parallax measurements are feasible is to use the period–luminosity relationship of Cepheid variables to establish them as standard candles. Presently Cepheids are useful out to about 30 Mpc (twice the distance to the Virgo Cluster). On this scale, peculiar motion is substantial and the cosmological principle is violated. For example, gravitational attraction between the Virgo Cluster and our cluster of galaxies causes a net ~ 250 km s^{-1} peculiar velocity relative to the Hubble flow. At a distance of 15 Mpc the Hubble velocity is $v = H_0 d \sim 1100$ km s^{-1}, implying 20% violations of the homogeneity assumption underlying the Hubble law. To reduce such uncertainties to several percent requires measurements at redshifts greater than $z \sim 0.02$ (distances greater than ~ 100 Mpc), which isn't feasible with Cepheid variables. Corrections to the Hubble law require determining higher-order terms in Eq. (17.36), and thus measurements at even larger distances.
>
> **Standard Candles at Distance**
>
> A Type Ia supernova is associated with a thermonuclear runaway under degenerate conditions in white dwarf matter, making them extremely bright and thus visible at great distances. Different Type Ia supernovae have similar but not identical lightcurves, so they are not standard candles. However, empirical methods allow the lightcurves of different Type Ia supernovae to be collapsed to a single curve, as illustrated in Fig. 19.7. Thus, Type Ia supernovae are standardizable candles visible at large distances, which enables precision cosmology with them.

representative calculations done here. The question of a suitable value for the radiation density parameter may be answered quickly. Observations indicate that it is a factor of more than 10,000 less dense than matter or vacuum energy in the current Universe, and therefore Ω_r can be set to zero for the purposes of simple cosmological calculations. This has not always been true. As will be discussed in Chapter 20, radiation dominated the early Universe but its contribution fell off quickly as the Universe expanded because the radiation energy density scales as $a(t)^{-4}$. The values of the remaining parameters Ω_m and Ω_Λ are best determined by: (1) using Type Ia supernovae as standard candles to constrain the evolution of the scale factor at high redshifts, (2) quantitative analysis of the fluctuations in the microwave background to constrain all of the cosmological parameters, and (3) observations of galaxy clusters to constrain the amount of matter in the Universe.

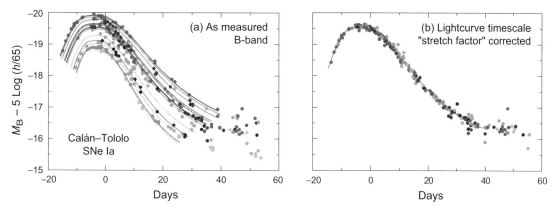

Fig. 19.7 Empirical rescaling of Type Ia supernova lightcurves to make them approximate standard candles. (a) Lightcurves for low-redshift supernovae (Calán–Tololo survey [110]). The intrinsic scatter is 0.3 magnitudes in peak luminosity. (b) After a one-parameter correction the dispersion is reduced to 0.15 magnitudes. Source: Refs. [93, 110, 132]. Reproduced from Frieman *et al.*, Dark Energy and the Accelerating Universe, *Annual Review of Astronomy and Astrophysics*, **46**, 385–432 (2008) with permission from the Annual Reviews.

Type Ia supernovae: The use of Type Ia supernovae as a means to determine very large distances in the Universe is described in Box 19.2. As discussed there, the brightness of Type Ia supernovae can be normalized so that they effectively become standard candles (see Fig. 19.7), which enables distance measurements by comparing observed brightness with the standard brightness using the inverse square intensity law

$$F = \frac{L}{4\pi r^2}. \qquad (19.63)$$

If the source is a standard candle, L is known and measurement of the flux F yields the distance r in a flat, static Universe. As discussed in Box 19.3, the relationship (19.63) must be modified in an expanding, curved space. Type Ia standardizable candles are particularly valuable because they are found in all types of galaxies and their extreme brightness permits distance measurements out to $z \sim 1$ or more. Because of these properties, Type Ia supernovae are at present the most reliable distance indicators at larger redshifts, making them a central tool in modern precision cosmology. Figure 19.6(b) illustrates the use of Type Ia supernova data to constrain Ω_Λ and Ω_m. Data from 200 high-redshift supernovae have been binned into a few points and compared with theoretical calculations. At the higher redshifts there is a clear preference for the solution with a nontrivial vacuum energy density Ω_Λ, which implies an acceleration of the expansion.

CMB fluctuations: In Section 20.4.6 it will be shown that measurements of anisotropies in the cosmic microwave background (CMB) provide a precise and independent determination of cosmological parameters. These analyses of the CMB probe the relative contribution of matter and vacuum energy, but in a different way than for the Type Ia supernova data.

Galaxy clusters: Traditional observational astronomy, augmented by newer techniques such as gravitational lensing, has constrained the amount of matter (visible and dark)

> **Box 19.3** **Luminosity Distance**
>
> In flat, static space, distances can be determined using standard candles (Box 19.2) and Eq. (19.63). However, space is expanding and could be curved, invalidating Eq. (19.63). One way to proceed is to *define* the *luminosity distance* d_L through [211]
>
> $$F = \frac{L}{4\pi d_L^2},$$
>
> where F is the observed flux (energy/area/time) and L is the (assumed known) luminosity (energy/time). For a flat, static universe, the luminosity distance d_L equals the proper distance ℓ. However, three effects alter this relationship: (1) The geometry may be curved. (2) Expansion of the Universe redshifts the energy of photons. (3) Expansion lengthens the time between detection of successive photons. To investigate, let us express the metric as in Eq. (18.19),
>
> $$ds^2 = -dt^2 + a(t)^2[dr^2 + S_k(r)^2 \, d\Omega^2] \qquad S_k(r) = \begin{cases} \sin r & (k = +1) \\ r & (k = 0) \\ \sinh r & (k = -1) \end{cases}$$
>
> and suppose photons to be emitted from coordinates (r, θ, φ) at time t_e and spread over a sphere when detected at $r = 0$ and time t_0. From Problem 19.10, the spherical surface has a proper area $A_p(t_0) = 4\pi S_k(r)^2$. Next, Hubble expansion redshifts photon wavelengths by $\lambda_0 = \lambda_e/a(t_e) = (1+z)\lambda_e$, so energies E_e of emitted and E_0 of detected photons are related by $E(t_0) = E(t_e)/(1+z)$. Finally, time between successive photons is increased by the expansion so that the time interval Δt_0 at detection is related to the interval Δt_e at emission by $\Delta t_0 = (1+z)\Delta t_e$. Collecting these effects, the observed flux is
>
> $$F = \frac{L}{4\pi S_k(r)^2 (1+z)^2},$$
>
> and comparing this with the first equation above gives the luminosity distance as
>
> $$d_L = S_k(r)(1+z) \simeq (1+z)r = (1+z)\ell(t_0),$$
>
> where $\ell(t_0)$ is the current proper distance and, consistent with observation, the actual Universe was assumed flat with $S_k(r) \sim r$. Only for small z does $d_L \sim \ell$.

contained in clusters of galaxies rather tightly. Such observations are relatively insensitive to the amount of dark energy in the Universe.

Remarkably, these different approaches with differing sensitivities to components of the cosmic fluid have reached a consensus on parameters for the amount of radiation, matter, dark energy, and curvature in the Universe. This *concordance model* is described next.

19.4.6 Concordance of Cosmological Parameters

In Fig. 19.8 supernova and CMB data are summarized in terms of confidence level contours. Notice the different dependence of the supernova data on the parameters than

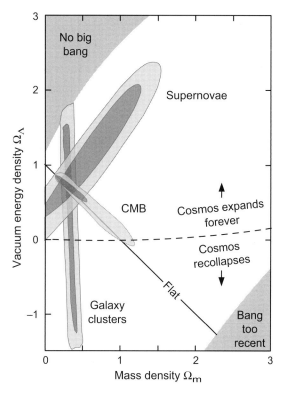

Fig. 19.8 Cosmological parameter space [184]. Supernova data, cosmic microwave background data, and galaxy cluster inventories agree on $\Omega_m \simeq 0.3$, $\Omega_\Lambda \sim 0.7$, and $\Omega_r \sim 0$, implying a cosmos that is spatially flat but will expand forever.

that of the CMB, and that the supernova data and the CMB data taken together imply that the best concordance is for $\Omega_m \simeq 0.3$ and $\Omega_\Lambda \simeq 0.7$, with negligible contribution from radiation. In Fig. 19.8 the aggregate of data from galaxy cluster observations is also displayed. These analyses have almost no sensitivity to the vacuum energy density but high sensitivity to the matter density. Remarkably, the data from galaxy clusters also are consistent with the choice $\Omega_\Lambda \simeq 0.7$ and $\Omega_m \simeq 0.3$. The diagonal line in Fig. 19.8 indicates the flat-space prediction ($\Omega_m + \Omega_\Lambda = 1$) of the inflationary theory to be discussed in Section 21.3, and the dashed line separates an eternally expanding Universe from one that eventually recollapses. Gray regions in the corners indicate ranges of parameters that would either allow no big bang, or would cause it to occur too recently.

Figure 19.9(a) illustrates the parameter concordance for analysis of six Type Ia supernovae lying between redshifts 0.51 and 1.12 [27], CMB fluctuations, and baryon acoustic oscillations (BAO; see Box 19.4) obtained from large-scale surveys of visible matter. The mass density parameter Ω_m and vacuum energy density parameter Ω_Λ corresponding to the overlap of the SNe, CMB, and BAO confidence-interval contours are seen to be comparable to those found in Fig. 19.8. Constraint of the equation of state parameter w from the same comparison is illustrated in Fig. 19.9(b). It was concluded from these data that if the Universe is flat, $w = -0.997^{+0.077}_{-0.082}$, where the error limits combine statistical

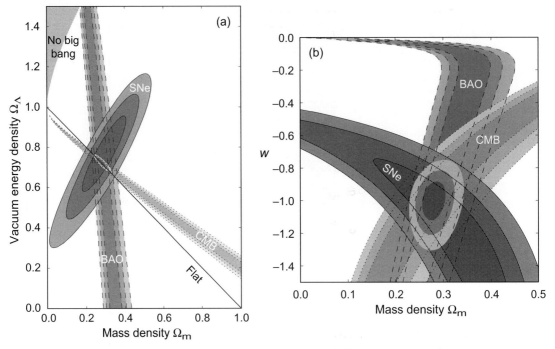

Fig. 19.9 (a) Cosmological parameter space based on comparing data from six Type Ia supernovae lying between redshifts $z = 0.51$ and $z = 1.12$ (SNe), cosmic microwave background fluctuations (CMB; see Section 20.4.6), and baryon acoustic oscillation (BAO; see Box 19.4 and Section 20.4.5). (b) Corresponding equation of state parameter w [see Eq. (19.8)]. Figure adapted from Ref. [27]. R. Amanullah *et al.*, published 2010 May 21 © 2010. The American Astronomical Society. All rights reserved. *Astrophysical Journal*, **716**(1).

Box 19.4   **Baryon Acoustic Oscillations**

Baryon acoustic oscillations (BAO) correspond to periodic fluctuations in the density of visible baryonic matter. Prior to hydrogen recombination in the early universe the pressure of the photons led to acoustic modes in the plasma. When the baryons and electrons combined into atoms, the photons were released from the plasma but both the photons and matter were left in a perturbed state with a preferred length scale corresponding to the distance that a sound wave could have traveled from the beginning of the Universe to the time of decoupling. For the photons the acoustic mode history is manifested in the CMB temperature anisotropies to be discussed in Section 20.4.4. However, the baryons are left in a similar state that mixes with non-oscillating cold dark matter perturbations to leave a small residual imprint on the clustering of matter at late times that will be discussed further in Section 20.4.5.

---

and systematic uncertainties [27]. Summarizing, cosmological data suggest the adoption of a benchmark solution of the Friedmann equations corresponding to the parameters

$$H_0 = 67\text{--}73 \text{ km s}^{-1} \text{ Mpc}^{-1} \qquad \Omega_r = 8 \times 10^{-5} \sim 0$$
$$\Omega_m \sim 0.3 \qquad \Omega_\Lambda \sim 0.7. \tag{19.64}$$

The reliable determination of such parameters changed cosmology from a notoriously qualitative discipline to a precision science in little more than a decade.

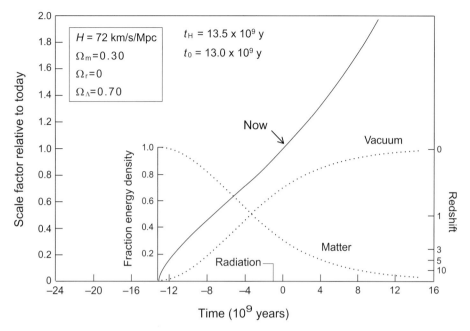

Fig. 19.10 Scale factor versus time for a flat universe with 30% matter, 70% vacuum energy, and negligible radiation. Fractional contributions of components are shown as dotted curves.

19.4.7 Calculations with Benchmark Parameters

In Fig. 19.10 a solution of the Friedmann equations using parameters consistent with Eq. (19.64) is displayed. The left axis gives the scale factor and the right axis gives the corresponding redshift, and the energy density of matter and vacuum energy as a function of time are indicated in the inset by dotted lines. In this model the Universe has been dominated by matter for most of its history since the very early radiation-dominated period (too short to see on this scale). However, about 4 billion years ago the vacuum energy gained ascendancy relative to the nonrelativistic matter and the Universe is presently in an accelerating phase corresponding to steadily increasing dominance of vacuum energy.

The solution displayed in Fig. 19.10 indicates that the Universe first decelerated (dominated first by radiation and then by matter) and then began accelerating as the vacuum energy became dominant. Remarkably, the predicted age of the Universe is almost the same as the simple estimate from the Hubble time of 13.5 billion years (the intersection of the tangent to the scale factor curve today with the bottom axis). This is accidental and results from cancellation between initial deceleration and more recent acceleration in determining the true intercept relative to the prediction of the Hubble law. These results indicate that the geometry of the Universe is flat but it will expand forever because of the influence of vacuum energy, even though it does not have a closure density of matter or radiation.

The calculation displayed in Fig. 19.10 represents a realistic evolution of the actual Universe. However, more recent parameters constrained by cosmic microwave background data indicate a somewhat greater age for the Universe [corresponding to H_0 nearer the lower end of the range in Eq. (19.46)]. For example, the calculation of Fig. 19.10 repeated for the parameters in Table 20.2 gives an age of 13.8 billion years for the Universe, primarily because of choosing $H_0 = 67 \, \text{km s}^{-1} \text{Mpc}^{-1}$. The range of values for H_0 given in Eq. (19.64) is larger than expected from the quoted errors on individual measurements. As of this writing in 2018 it is unclear whether this represents a genuine discrepancy in the determination of H_0 by different means, or an under-estimate of errors.

Background and Further Reading

The approach to solution of the Friedmann equations adopted in this chapter is discussed extensively in Ryden [211], Padmanabhan [174], Hartle [111], and Roos [207]. Frieman, Turner, and Huterer [93] review dark energy and the accelerating Universe.

Problems

19.1 Show that if a component i of the cosmic fluid has a pressure P_i and energy density ϵ_i related by $P_i = w_i \epsilon_i$ in a Friedmann cosmology (with w_i constant), then

$$\epsilon_i = \epsilon_i(0) \left(\frac{a}{a_0} \right)^{-3(1+w_i)},$$

where $\epsilon_i(0)$ is the energy density of component i at the present time and a is the scale factor. *Hint*: This result is a consequence of energy conservation. ***

19.2 Estimate the total energy of matter and radiation contained within the volume of the Solar System and compare with the total vacuum energy contained within that volume. Speculate on the feasibility of a local experiment to measure the vacuum energy density based on these estimates.

19.3 (a) For air at room temperature, show that if the equation of state is approximated as $P = w\epsilon$ then $w \sim 10^{-12}$. (b) Show that Eq. (19.10) also can be written as $w = \langle v^2 \rangle / 3c^2$, where $\langle v^2 \rangle^{1/2}$ is the root mean square velocity for gas particles. ***

19.4 Prove that for a flat, matter-dominated universe ($\Omega = 1$, $w = 0$), the proper distance from the Earth to a photon emitted from the present particle horizon at the time of the big bang first *increases* and then decreases to zero, as time goes from the time of the big bang to the current time, even though the coordinate distance is continually decreasing in that period. Show that the photon reaches a maximum proper distance at about 30% of the current age of this universe and that this distance corresponds to about 15% of the present horizon size for this universe.

19.5 Show that a universe containing only curvature (no matter, radiation, or vacuum energy) is a valid solution of the Friedmann equations, provided that the curvature is negative. Find an expression for the scale factor as a function of time, the age of this universe, and the redshift z. Find expressions for the proper distance to a light source at the time of detection and the proper distance to this source at the time of emission in terms of the redshift. This model is called the *Milne universe*. Although unrealistic compared with observations, it is often used as an example of a very simple solution for the Friedmann equations.

19.6 From the definition of the deceleration parameter given in Eq. (17.34) and the Friedmann equations, show that for a universe with multiple components i having equations of state $P_i = w_i \epsilon_i$, the deceleration parameter is given by

$$q_0 = \frac{1}{2} \sum_i \Omega_i (1 + 3w_i),$$

where Ω_i is the present value of the density parameter for component i. Show that for a 3-component universe with nonrelativistic matter, radiation, and vacuum energy, this reduces to (see Section 17.14.1)

$$q_0 = \frac{1}{2}\Omega_m + \Omega_r - \Omega_\Lambda,$$

that if this universe is euclidean

$$q_0 = \frac{3}{2}\Omega_m + 2\Omega_r - 1,$$

and that acceleration of this universe is possible only if $\Omega_\Lambda > \frac{1}{2}\Omega_m + \Omega_r$. Finally, show explicitly that for the standard cosmology the deceleration parameter indicates a universe that is presently accelerating.

19.7 For most standard cosmologies other than de Sitter the scale factor varies as t^n, with $n < 1$. (a) Show that for such cosmologies the ratio of the particle horizon distance to the scale factor always increases with time. (b) Calculate the present distance to the horizon if the Universe is flat and it is assumed (incorrectly) to contain only matter.

19.8 Use the Friedmann equations to show that the deceleration parameter q_0 and the density parameter Ω are related through $q_0 = \frac{1}{2}\Omega$ for a universe containing only dust.

19.9 From the general solution for a flat, single-component universe discussed in Section 19.3.2, prove that the age of the Universe is greater than the Hubble time if $1 < w < -\frac{1}{3}$.

19.10 Prove that the proper area $A_p(t_0)$ used in Box 19.3 to derive the luminosity distance is given by $A_p(t_0) = 4\pi S_k^2$ if the Universe is described by the metric (18.19). ***

20 The Big Bang

As we have seen in Chapter 19, the general evolution of energy densities with scale factors governed by the Friedmann equations suggests that radiation was much more important in the early Universe than it is today. Furthermore, the observation of a cosmic microwave background is most easily interpreted as evidence of an earlier, much hotter Universe in which radiation was dominant. Such considerations indicate that the Universe began life in a very hot, very dense state called the *big bang*. In this chapter the Friedmann equations are applied to the early Universe in an attempt to understand the most important features of the *big bang model*, which is the cosmologist's "standard model" for the origin of the Universe.

20.1 Radiation- and Matter-Dominated Universes

In Section 19.1.2 the general issue of cosmological equations of state was addressed. Because the influence of vacuum energy grows with expansion of the Universe and because it is only today beginning to dominate, it is safe to assume that vacuum energy was negligible relative to matter and radiation in the early Universe (once the inflationary epoch was over; see Chapter 21). In that case, two extremes for the equation of state provide considerable insight into the history of the Universe:

1. If the energy density resides primarily in light particles having relativistic velocities the Universe is said to be *radiation dominated*; in that case the equation of state relating the pressure P and energy density ϵ is

$$P = \tfrac{1}{3}\epsilon \quad \text{(radiation dominated)}. \tag{20.1}$$

2. If instead the energy density is dominated by massive, slow-moving particles the Universe is said to be *matter dominated*; the corresponding equation of state is

$$P \simeq 0 \quad \text{(matter dominated)}. \tag{20.2}$$

In either extreme, the evolution of the Universe is then calculated easily by solving the Friedmann equations.

20.1.1 Evolution of the Scale Factor

As we have discussed in Section 19.2.1, the density of radiation and the density of matter scale differently in an expanding universe. If the Universe is radiation dominated $P = \tfrac{1}{3}\epsilon$ and Eq. (19.6) implies that $\dot\epsilon/\epsilon + 4\dot a/a = 0$, which has a solution

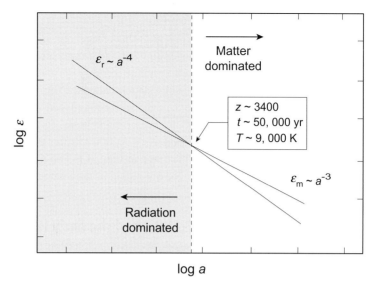

Fig. 20.1 Schematic dependence of matter and radiation energy densities on the scale factor. The time of equivalence corresponds to a redshift of $z_{eq} \sim 3400$ (see Table 20.2).

$$\epsilon(t) \simeq \frac{1}{a^4(t)} \quad \text{(radiation dominated)}. \tag{20.3}$$

On the other hand, if the Universe is matter dominated then $P \simeq 0$ and Eq. (19.6) gives $\dot{\epsilon}/\epsilon + 3\dot{a}/a = 0$, which has a solution

$$\epsilon(t) \simeq \frac{1}{a^3(t)} \quad \text{(matter dominated)}. \tag{20.4}$$

The corresponding behavior of the scale factor with time is

$$a(t) \simeq \begin{cases} t^{1/2} & \text{(radiation dominated)} \\ t^{2/3} & \text{(matter dominated)} \end{cases}, \tag{20.5}$$

as was shown in Section 19.3.3 (see Table 19.1).

20.1.2 Matter and Radiation Density

In the present Universe the ratio of the number density of baryons n_b to number density of photons n_γ is $n_b/n_\gamma \simeq 6 \times 10^{-10}$. However, the rest mass of a typical baryon is approximately 10^9 eV, while most photons are in the ~ 2.7 K cosmic microwave background, with an average energy $E_\gamma \sim kT \sim 2.3 \times 10^{-4}$ eV. Thus the ratio of the energy density of baryons to energy density of photons in the present Universe is $\epsilon_b/\epsilon_\gamma \simeq 10^3$, and the present Universe is strongly dominated by matter (and vacuum energy). But from (20.3) and (20.4), as time is extrapolated backwards relativistic matter becomes increasingly more important until at some earlier time the radiation becomes dominant; Fig. 20.1 illustrates. In this early, radiation-dominated Universe

$$\epsilon \simeq a^{-4} \qquad a \simeq t^{1/2} \qquad P = \tfrac{1}{3}\epsilon,$$

and the density and pressure tend to infinity as $t \to 0$. Furthermore, it will be shown below that for a radiation-dominated Universe $T \simeq a^{-1} \simeq t^{-1/2}$; thus, as time is extrapolated backwards the temperature tends to infinity also. These considerations suggest that the Universe started from a very hot, very dense initial state with $a(t \to 0) \to 0$. The commencement from this initial state is called the *big bang*.[1]

Choosing $t = 0$ when $a = 0$, the transition between the earlier radiation-dominated Universe and one dominated by matter took place (gradually) around a redshift of $z \sim 3400$, corresponding to a time $t \sim 50,000$ yr after the big bang and a temperature of approximately 9000 K, as illustrated in Fig. 20.1. The following section will discuss in more detail the big bang and this early radiation-dominated era of the Universe.

20.2 Evolution of the Early Universe

In the hot and dense early Universe the major portion of the energy density was in the form of photons and other massless or nearly massless particles like neutrinos. We now give a brief description of the most important events in the big bang and the transition from a universe dominated by radiation to one dominated by matter [70, 135].

20.2.1 Thermodynamics of the Big Bang

Neglecting curvature, in the initial radiation-dominated era of the big bang,

$$H^2 \equiv \left(\frac{\dot a}{a}\right)^2 \simeq \frac{8\pi G}{3} \epsilon_r \propto a^{-4} \qquad a \simeq t^{1/2} \qquad H = \frac{\dot a}{a} = \frac{1}{2t}.$$

The evolution may be assumed to correspond approximately to that of an ideal gas in thermal equilibrium, for which the number density n of a particular species is given by

$$dn = \frac{g}{2\pi^2 \hbar^3} \frac{p^2 dp}{e^{E/kT} + \Theta} \qquad \Theta = \begin{cases} +1 & \text{Fermions} \\ -1 & \text{Bosons} \\ 0 & \text{Maxwell–Boltzmann} \end{cases}, \qquad (20.6)$$

where Maxwell–Boltzmann statistics ($\Theta = 0$) apply only if no distinction is made between fermions or bosons in the gas, p is the 3-momentum, g is the number of degrees of freedom per particle (helicity states: two for each photon, quark, and lepton), and the relativistic energy is

$$E = \sqrt{p^2 c^2 + m^2 c^4}. \qquad (20.7)$$

Because of the high temperature the gas may be assumed to be ultrarelativistic ($kT \gg mc^2$ for the particles in the plasma). Then the energy is $E \sim pc$ and the number density is obtained by integrating Eq. (20.6),

---

[1] The name "big bang" was a mildly derogatory term coined by opponents of this cosmology who favored the now-discredited steady state theory. The name stuck, as did the corresponding theory.

20.2 Evolution of the Early Universe

$$n = \int_0^\infty \frac{dn}{dp} dp = \frac{g}{2\pi^2 \hbar^3} \int_0^\infty \frac{p^2 dp}{e^{E/kT} + \Theta}$$
$$= \frac{g}{2\pi^2 \hbar^3} \int_0^\infty \frac{p^2 dp}{e^{pc/kT} + \Theta}. \quad (20.8)$$

Integrals of this form may be evaluated using

$$\int_0^\infty \frac{t^{z-1}}{e^t - 1} dt = (z-1)!\, \zeta(z) \qquad \int_0^\infty \frac{t^{z-1}}{e^t + 1} dt = (1 - 2^{1-z})(z-1)!\, \zeta(z), \quad (20.9)$$

where the Riemann zeta function $\zeta(z)$ has tabulated values

$$\zeta(2) = \frac{\pi^2}{6} = 1.645 \qquad \zeta(3) = 1.202 \qquad \zeta(4) = \frac{\pi^4}{90} = 1.082. \quad (20.10)$$

The results for the number density of species i are

$$n_i = g_i \frac{\zeta(3)}{\pi^2} T^3 \times \begin{cases} 1 & \text{Bose–Einstein} \\ 3/4 & \text{Fermi–Dirac} \\ \zeta(3)^{-1} & \text{Maxwell–Boltzmann} \end{cases}, \quad (20.11)$$

where now $\hbar = c = k = 1$ units are introduced (see Appendix B.3). Likewise, the energy density is given by

$$\epsilon = \int_0^\infty E \frac{dn}{dp} dp = \frac{g}{2\pi^2 \hbar^3} \int_0^\infty \frac{E p^2 dp}{e^{E/kT} + \Theta}$$
$$= \frac{g}{2\pi^2 \hbar^3} \int_0^\infty \frac{E p^2 dp}{e^{pc/kT} + \Theta}, \quad (20.12)$$

which gives for a species i

$$\epsilon_i = g_i \frac{\pi^2}{30} T^4 \times \begin{cases} 1 & \text{Bose–Einstein} \\ 7/8 & \text{Fermi–Dirac} \\ 90/\pi^4 & \text{Maxwell–Boltzmann} \end{cases}. \quad (20.13)$$

The energy density for all relativistic particles is then given by the sum,

$$\epsilon = g_* \frac{\pi^2}{30} T^4 \qquad g_* \equiv \sum_{\text{bosons}} g_b + \tfrac{7}{8} \sum_{\text{fermions}} g_f. \quad (20.14)$$

If all species are in equilibrium and relativistic, the entropy density s is

$$s = \frac{\epsilon + P}{T} = \frac{4\epsilon}{3T} = \frac{2\pi^2}{45} g_* T^3, \quad (20.15)$$

where $s \simeq \sum n_i$, by comparison of Eqs. (20.11) and (20.15). The entropy per comoving volume is constant (adiabatic expansion), $S \simeq sa^3 \simeq$ constant, so

$$\frac{d(sa^3)}{dt} = 0, \quad (20.16)$$

provided that g_* does not change. In fact, as illustrated in Fig. 20.2, g_* is expected to be approximately constant for broad ranges of temperature but to change suddenly at critical temperatures where kT becomes comparable to the rest mass for a species.

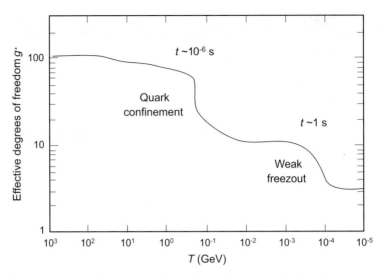

Fig. 20.2 Variation of the effective number of degrees of freedom in the early Universe as a function of temperature.

(The rest masses of particles from the Standard Model that are relevant for the early Universe are summarized in Table 20.1.) From Eqs. (20.15) and (20.16), $T^3 a^3$ is constant and the temperature varies as $T \simeq a^{-1} \simeq t^{-1/2}$, where Eq. (20.5) has been employed. To be precise, the evolution of the ultrarelativistic, hot plasma characterizing the early big bang may be characterized by the equations [135]

$$\frac{\dot{a}}{a} = -\frac{\dot{T}}{T} = \alpha T^2 \qquad t = \frac{1}{2\alpha T^2} = \frac{2.4 \times 10^{-6}}{g_*^{1/2} T^2} \, \text{GeV}^2 \, \text{s}$$

$$\alpha = \left(\frac{4\pi^3 g_*}{45 M_P^2} \right)^{1/2} \qquad M_P \equiv \left(\frac{\hbar c}{G} \right)^{1/2} = 1.2 \times 10^{19} \, \text{GeV},$$

(20.17)

where M_P is the *Planck mass*.

20.2.2 Equilibrium in an Expanding Universe

Strictly, equilibrium cannot hold in an expanding universe. However, we may assume a practical equilibrium to exist as the Universe expands through a series of nearly equilibrated states. It may be expected that both thermal and chemical equilibrium have parts to play in the expansion of the Universe. A system is in thermal equilibrium if its number density is given by Eq. (20.6); a system is in chemical equilibrium if for the reaction $a + b \leftrightarrow c + d$ the chemical potentials satisfy $\mu_a + \mu_b = \mu_c + \mu_d$, and similarly for other reactions. Our primary concern will be with thermal equilibrium and we will consider the equilibrium to be maintained by two-body reactions (the most common situation). The reaction rate is given by $\Gamma \simeq \langle n v \sigma \rangle$, where n is the number density, v is the relative speed, σ is the reaction cross section, and the brackets indicate a thermal average. A species may be expected to remain in thermal equilibrium in the radiation-dominated Universe as long as

Table 20.1 Particles of the Standard Model				
Particle	Symbol	Spin ($\hbar$)	Charge (e)	Mass (GeV/c^2)[†]
– Leptons –				
Electron	e^-	1/2	-1	5.11×10^{-4}
Electron neutrino	ν_e	1/2	0	$< 2.2 \times 10^{-9}$
Muon	μ^-	1/2	-1	0.1057
Muon neutrino	ν_μ	1/2	0	$< 1.7 \times 10^{-4}$
Tau	τ^-	1/2	-1	1.777
Tau neutrino	ν_τ	1/2	0	$< 1.6 \times 10^{-2}$
– Quarks –				
Down	d	1/2	$-1/3$	0.005
Up	u	1/2	$2/3$	0.002
Strange	s	1/2	$-1/3$	0.096
Charm	c	1/2	$2/3$	1.28
Bottom	b	1/2	$-1/3$	4.2
Top	t	1/2	$2/3$	173
– Gauge and Higgs bosons –				
Photon	γ	1	0	0
Charged weak bosons	$W^\pm$	1	± 1	80.4
Neutral weak boson	Z^0	1	0	91.2
Gluons	$G_1, G_2, \ldots, G_8$	1	0	0
Graviton (?)	g	2	0	0
Higgs	H	0	0	125.1

[†]Quark masses are so-called current or Lagrangian masses. Effective quark masses are larger.

$$\Gamma \gg \frac{\dot{a}}{a} \equiv H = \frac{1}{2t}, \qquad (20.18)$$

which means that the characteristic reaction rate is much faster than the characteristic expansion rate of the Universe at that time [see Eq. (20.17)]. In the earliest stages of the big bang the densities, velocities, and many cross sections are large and it is easy to satisfy Eq. (20.18) for most species. However, as the temperature and density drop the number density, velocity, and cross section factors will decrease steadily and at certain reaction thresholds the reaction rate maintaining equilibrium for a particular species will become too small and it can drop out of thermal equilibrium [its number density is no longer given by Eq. (20.6)]. The physical reason is simple: if the reaction rates are slow compared with the rate of expansion, it is unlikely that the particles can find each other often enough to react and maintain equilibrium as the gas expands.

Example 20.1 As an illustration of decoupling from thermal equilibrium, consider weak interactions in the early Universe. At the energies of primary interest the weak interaction strengths scale quadratically with the temperature; thus shortly after the big bang the weak

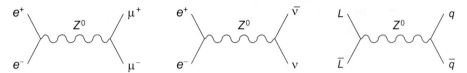

Fig. 20.3 Feynman diagrams for some weak interactions important for maintaining equilibrium in the early Universe. Generic leptons are represented by L and generic quarks by q. Bars denote antiparticles.

interactions are not particularly weak and particles such as neutrinos are kept in thermal equilibrium by reactions like $\nu\bar{\nu} \leftrightarrow e^+e^-$ and others defined by the Feynman diagrams in Fig. 20.3. (Feynman diagrams were described in Box 8.2.) Let's use dimensional analysis to estimate the weak reaction rates. The weak interaction cross sections depend on the square of the weak (Fermi) coupling constant, $\sigma_w \propto G_F^2$, with $G_F \simeq 1.17 \times 10^{-5}\,\text{GeV}^{-2}$. The Fermi coupling carries dimension $[G_F] = [M]^{-2}$ in $\hbar = c = k = 1$ units (k is the Boltzmann constant), where $[x]$ denotes the dimensionality of x in these units and M denotes mass (see Appendix B). In these units mass M and length L have inverse dimensions, $[L] = [M]^{-1}$. Thus, the cross section carries dimension $[\sigma_w] = [L]^2 = [M]^{-2}$, and

$$[\sigma_w] = [G_F]^2 [\text{quantity}] = [M]^{-2},$$

so that $[\text{quantity}] = [M]^2$. The only physical quantity available carrying mass (energy) units is the temperature, T, so up to constants of proportionality,[2]

$$\sigma_w \simeq G_F^2 T^2. \tag{20.19}$$

We may now estimate the reaction rate $\Gamma = \langle n v \sigma \rangle$. From Eq. (20.11) the number density is $n \simeq T^3$, from Eq. (20.19) the cross section is $\sigma_w \simeq G_F^2 T^2$, and the velocity $v \simeq 1$ for the ultrarelativistic plasma, implying that the reaction rate for weak interactions is

$$\Gamma_w \simeq \langle n v \sigma \rangle \simeq (T^3)(G_F^2 T^2)(1) \simeq G_F^2 T^5. \tag{20.20}$$

Utilizing Eq. (20.17) with $H = \dot{a}/a$ and all constants but M_P dropped, the ratio of the reaction rate to the expansion rate is given by

$$\frac{\Gamma}{H} \simeq \frac{G_F^2 T^5}{T^2/M_P} \simeq \left(\frac{T}{1\,\text{MeV}}\right)^3. \tag{20.21}$$

Therefore weak interactions decoupled from thermal equilibrium at $kT \sim 1\,\text{MeV}$, which was about one second after the expansion began (a more careful analysis gives $kT = 0.8\,\text{MeV}$). This is the time marked "Weak freezeout" in Fig. 20.4.

Let's consider now the specific sequence of events that is expected to have occurred in the big bang, along with an approximate timeline.

---

[2] The numerical factors are not important here but may be obtained from a complete evaluation of the Feynman diagrams, which gives $\sigma_w = G_F^2 k^2 T^2/(\pi \hbar^4 c^4)$, with the constants $\hbar$, c, and k restored.

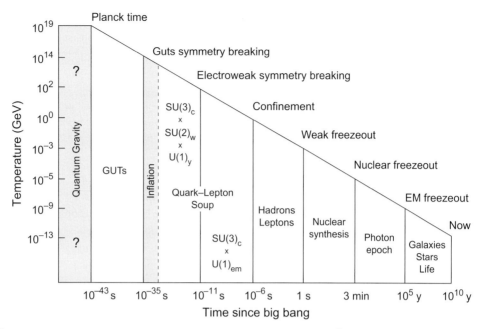

Fig. 20.4 A history of the Universe. The time axis is highly nonlinear and 1 GeV $\simeq 1.2 \times 10^{13}$ K (after D. Schramm).

20.2.3 A Timeline for the Big Bang

Equations (20.17) and known properties of matter allow a reasonably complete reconstruction of the big bang (see Fig. 20.4 for an overview). In a radiation-dominated era particles and their antiparticles are continuously undergoing reactions in which they annihilate each other to produce photons, and photons can collide and create particle and antiparticle pairs. Thus, under these conditions radiation and matter are in equilibrium because they can interconvert freely. Later we shall have more to say about the earlier more speculative Planck and inflationary epochs, but here we pick up the story at about 10^{-6} s after the expansion began, from which point the physics is thought to be quite well understood. The primary characters participating in the big bang drama then include: (1) photons, (2) protons and neutrons, (3) electrons and positrons, and (4) neutrinos and antineutrinos.

Time ∼ 1 microsecond: The temperature is about 10^{13} K and the quarks and gluons, which have been in a deconfined plasma to this point, undergo a confining phase transition to form the hadrons. The Universe becomes filled with protons and neutrons, in about equal numbers.

Time ∼ 0.01 seconds: The temperature is about 10^{11} K. The Universe consists of a hot undifferentiated soup of matter and radiation in thermal equilibrium with an average particle energy of $kT \simeq 8.6$ MeV. The electrons and positrons are in equilibrium with the photons, the neutrinos and antineutrinos are in equilibrium with the photons, antineutrinos are combining with protons to form positrons and neutrons, and neutrinos are combining with neutrons to form electrons and protons through reactions such as

$$e^- + e^+ \leftrightarrow \text{photons} \qquad \nu + \bar{\nu} \leftrightarrow \text{photons}$$
$$\bar{\nu} + p^+ \to e^+ + n \qquad \nu + n \to e^- + p^+.$$

A neutron is more massive than a proton by $Q \equiv (m_n - m_p)c^2 = 1.3$ MeV, so the equilibrium neutron-proton number density ratio is

$$n_n/n_p \sim \exp(-Q/kT).$$

Since $kT \gg Q$ for $T = 10^{11}$ K, at this temperature $n_n \sim n_p$.

Time ~ 0.1 seconds: By now the temperature has dropped to 3×10^{10} K, corresponding to $kT \sim 2.6$ MeV. As a result, the initial balance between neutrons and protons begins to be tipped in favor of protons, with the ratio now down to $n_n/n_p \sim 0.6$. No composite nuclei can form yet because the temperature implies an average energy for particles in the gas of about 2.6 MeV, and deuterium has a binding energy of only 2.2 MeV and so cannot hold together at these temperatures.[3] This barrier to production of composite nuclei allows the free neutrons to be converted steadily to protons until it is surmounted.

Time ~ 1 second: The temperature has dropped to about 10^{10} K as the young Universe continues to expand. At this temperature $kT \simeq 0.9$ MeV, and the neutrinos cease to play a role in the continuing evolution (*weak freezeout* – see Example 20.1 above). The neutron to proton ratio has now decreased to $n_n/n_p \sim 0.2$ and the temperature is still too high for composite nuclei to form.

Time ~ 10 seconds: The temperature now has fallen to several times 10^9 K, corresponding to an average particle energy of about 0.25 MeV. This is too low for photons to produce electron–positron pairs, so they fall out of thermal equilibrium and the free electrons begin to annihilate positrons to form photons, $e^- + e^+ \to$ photons. This reheats all particles in thermal equilibrium with the photons but not neutrinos, which dropped out of thermal equilibrium at $t \sim 1$ second. It is still too hot for composite nuclei to hold together and the neutrons continue to decay to protons.

Time ~ 100–1000 seconds: Finally at around 100 seconds after the big bang the temperature drops to $\sim 10^9$ K, with $n_n/n_p \sim 0.15$. Nuclear reactions begin to combine neutrons and protons to form deuterium and the deuterium that is produced quickly reacts with neutrons and protons to form ^{4}He (alpha particles),

$$n + p^+ \to {}^2_1\text{H} \qquad {}^2_1\text{H} + p^+ \to {}^3_2\text{He} + n \to {}^4_2\text{He} \qquad {}^2_1\text{H} + n \to {}^3_1\text{H} + p^+ \to {}^4_2\text{He},$$

and small amounts of a few other light elements, as illustrated in Fig. 20.5. This production of the light elements occurs only in a narrow window of a few hundred seconds (the *era of nucleosynthesis*) in which the temperature is low enough to prevent composite nuclei from being photodisintegrated by high-energy photons, but high enough for nuclear reactions forming composite nuclei to occur. From Fig. 20.5(a) the density of the Universe during this period of primordial nucleosynthesis is only $\rho \sim 10^{-5}$ g cm^{-3}, which is two orders

---

[3] In fact, because there are so many photons the temperature will have to drop to kT an order of magnitude below the binding energy of deuterium before nuclei can hold together in the sea of big bang photons.

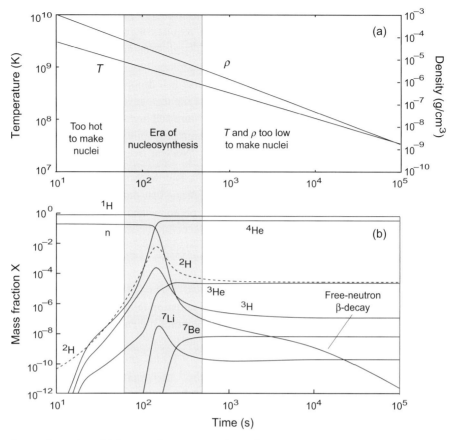

Fig. 20.5 Big bang nucleosynthesis. (a) Variation of temperature and mass density assuming standard big bang evolution and a baryon to photon ratio $\eta = 6 \times 10^{-10}$. (b) Mass fractions as a function of time for the light elements assuming mass fractions of 0.15 for neutrons and 0.85 for protons at the beginning of nucleosynthesis. Methods used to integrate the time evolution of the elemental abundances are described in Ref. [105].

of magnitude lower than the density of air at the Earth's surface. Since a free neutron is unstable against β-decay with a half-life of ~ 10.3 minutes, once nucleosynthesis is complete any neutrons not bound up into composite nuclei quickly β-decay into protons [Fig. 20.5(b) indicates that at the end of nucleosynthesis free neutrons have a mass fraction of $\sim 10^{-7}$ that is decaying steadily with time]. The net effect of this short but intense epoch of nucleosynthesis is to leave the Universe a few hundred seconds after its birth primarily consisting of ^{1}H and ^{4}He, with trace amounts of ^{2}H, ^{3}H (which β^- decays to ^{3}He with a half-life of 12.3 yrs, so it doesn't stick round), ^{3}He, ^{7}Be, and ^{7}Li.

Elements beyond ^{4}He cannot be formed in abundance because of the peculiarity that there are no stable mass-5 or mass-8 isotopes for any element, and because by the time significant ^{4}He has formed the density has dropped too low for reactions that could surmount this isotopic bottleneck to occur with significant probability. The mass-5 and mass-8 isotopic gaps could in principle be bridged by the triple-α reaction

^{4}He + ^{4}He + ^{4}He → ^{12}C, famous as the energy source for red giant stars. But this reaction is effectively 3-body and therefore requires high densities to make it likely that three helium nuclei find themselves mutually close enough together for the reaction to occur (as a consequence, the rate decreases quadratically with decreasing density). Once ^{4}He could form the density of the early Universe had dropped far too low for the triple-α reaction to be likely,[4] halting the synthesis of the elements at ^{4}He. The Universe had to wait until stars formed, with their high-temperature and high-density interiors, for nucleosynthesis to proceed beyond helium with significant probability. The end of the era of nucleosynthesis in Fig. 20.5 is indicated by the label "Nuclear freezeout" in Fig. 20.4.

Time $\sim$ 30 minutes: The temperature is now about 3×10^8 K and the Universe consists primarily of protons, the excess electrons that did not annihilate with the positrons, ^{4}He (about 26% abundance by mass), photons, and neutrinos. The few leftover neutrons continue to β-decay to protons and the neutron mass fraction has decreased now to $\sim 10^{-8}$. There are no atoms yet because the temperature is still too high for protons and electrons to bind.

Time $\sim$ 50,000 years: As indicated in Fig. 20.1, the early Universe was *radiation dominated* but the energy density of radiation falls faster than the energy density of matter as the Universe expands and at some point the energy density of matter began to exceed that of radiation. Observations indicate that this occurred at a redshift of about $z \sim 3400$, corresponding to a temperature of $\sim 9,000$ K and a time roughly 50,000 years after the big bang. From this point onward the Universe begins to be *matter dominated* (vacuum energy will become dominant much later, but it is relatively insignificant at this stage).

Time $\sim$ 400,000 years: The evolution of the Universe is increasingly matter dominated and the temperature has fallen to several thousand K, which is sufficiently low that electrons and protons can hold together to begin forming hydrogen atoms. This is called the *recombination transition* and the time is termed the *recombination era* (both misnomers, since the electrons and protons have never before been combined, but it is enshrined standard terminology). Until this point, matter and radiation have been in thermal equilibrium but now they decouple, so this is also termed the *decoupling transition* or *electromagnetic freezeout* (the time marked "EM freezeout" in Fig. 20.4).[5] As free electrons are bound up in atoms the primary reaction causing photon scattering (Thomson scattering on free electrons) is removed and the Universe, which had been opaque, becomes transparent: light can now travel large distances before being absorbed. The redshift at the decoupling transition is $z \sim 1100$, corresponding to a time $\sim 400,000$ years after the big bang. As will be discussed further below, this redshift represents the furthest back in time that can be seen with photons, since the Universe was completely opaque to light for higher redshifts.

---

[4] In red giant stars, characteristic temperatures and densities for the triple-α reaction are $T \sim 10^8$ K and $\rho \sim 10^5$ g cm^{-3}. From Fig. 20.5, at the end of helium synthesis $T \sim 7 \times 10^8$ K and $\rho \sim 10^{-5} - 10^{-6}$ g cm^{-3}.

[5] To be precise, the recombination transition is defined to have taken place slightly earlier than photon decoupling and electromagnetic freezeout. For our purposes the distinction will not be important and it will be sufficient to assume simply that the decoupling of photons and matter occurred at a redshift $z \sim 1100$. Do not confuse the time when matter gained the ascendancy over radiation in governing the evolution of the Universe, which happened about 50,000 years after the big bang, with the decoupling of matter and radiation, which occurred about 400,000 years after the big bang, in a Universe that had by that time become matter dominated.

Time ∼ 400 million years: After the recombination transition the Universe became largely transparent because it was filled with neutral hydrogen, which absorbs photons only at discrete energies corresponding to bound states. However, observations such as the spectra of high-redshift quasars indicate that at some time after the recombination transition much of the hydrogen once again became ionized. This *reionization transition* is discussed further in Box 20.1, and is thought to have occurred over a period of about 800 million years between redshifts $z \sim 20$ and $z \sim 6$, and to coincide with the appearance of the first stars. The reionized Universe was (and remains until today) still relatively transparent to

---

Box 20.1 — **Reionization of the Universe**

The early Universe went through *two phase transitions* involving its hydrogen. After the decoupling transition at $z \sim 1100$ the Universe was filled with neutral hydrogen, so it became largely transparent. However, there were no stars yet so the only light in the Universe was the background radiation from the big bang. This period between the decoupling transition and formation of the first stars is often termed the *dark ages*. However, observations indicate that the Universe later underwent a *reionization transition* in which neutral hydrogen was ionized again during an *epoch of reionization* lasting from $z \sim 20$ to $z \sim 6$, as illustrated in the following figure.

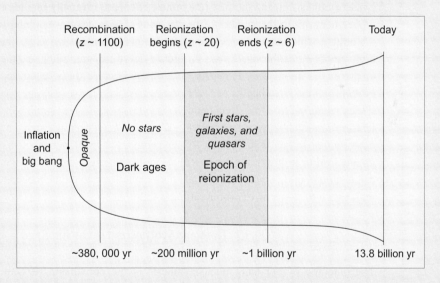

Prime candidates for the agent of this reionization were active galactic nuclei or quasars, and the first generation of stars, often called *Pop III*, which began forming several hundred million years after the big bang. These Pop III stars contained only hydrogen and helium, and simulations indicate that they had masses of hundreds of solar masses or more. Such massive stars would have ionized everything around them with intense UV radiation and would have gone supernova only several million years after their birth, seeding the Universe with heavier elements that could only have been produced in stars. Indeed, high-redshift quasar spectra indicate the presence of heavy elements in the reionization epoch, strongly implicating Pop III stars in the reionization transition.

photons because the expansion diluted the hydrogen sufficiently to reduce the collision rate between photons and free electrons to levels preventing strong scattering.

20.3 Nucleosynthesis and Cosmology

Deuterium (^{2}H) can't form until the temperature drops below a critical value, after which it forms and is converted quickly into ^{4}He (see Fig. 20.5). Therefore, present abundances of helium and deuterium, and other light elements like lithium produced in trace amounts, are a sensitive probe of conditions in the early Universe. The oldest stars contain material that is least altered from that produced originally in the big bang. Analysis of their composition indicates elemental abundances that are in good agreement with the predictions of the hot big bang. This is one of the strongest pieces of evidence in support of the theory.

20.3.1 The Neutron to Proton Ratio

Nucleosynthesis during the first few minutes depends critically on the ratio of the number of free neutrons to the number of free protons. The neutron is slightly more massive than the proton, which favors conversion of neutrons to protons by weak interactions. At very high temperatures the small mass difference doesn't matter much and the ratio of neutrons to protons is about one. However, as the temperature drops neutrons are converted to protons and the ratio begins to favor protons. All neutrons would be converted to protons if they remained free for a long enough period (several hours would be sufficient, once the temperature falls below about 10^{10} K). However, a neutron in a stable nucleus like ^{4}He or deuterium no longer is susceptible to being converted spontaneously to a proton. Therefore, the neutron to proton ratio drops with T until deuterium can hold together and neutrons can be bound up in stable nuclei. Calculations indicate that this happens at $T \sim 10^9$ K, by which time the neutron to proton ratio has fallen to 15–16% [72, 211].

20.3.2 Elements Synthesized in the Big Bang

The outcome of big bang nucleosynthesis is the conversion of neutrons and protons primarily to helium and free protons, with small concentrations of isotopes like deuterium, ^{3}He, and ^{7}Li. How much of each isotope is sensitive to conditions in the short era of nucleosynthesis. For the simulation shown in Fig. 20.5 the mass fractions are 0.253 for ^{4}He and 0.745 for ^{1}H, once nucleosynthesis stops. The simulation is sensitive to parameters having some uncertainty but it basically suggests that the big bang should have left the Universe with about a quarter helium by mass, with almost all the rest hydrogen. That is rather close to the current measured mass fractions for helium and hydrogen. This is an important general test, since processes occurring later in stars cannot change these mass fractions by too much, so a failure of the big bang to leave the Universe consisting of

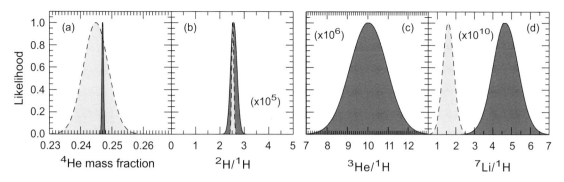

Fig. 20.6 Likelihoods normalized to unity for abundances of light elements. (a) Mass fraction for ^{4}He; (b) Ratio of ^{2}H (deuterium) to ^{1}H abundance; (c) Ratio of ^{3}He to ^{1}H abundance; and (d) Ratio of ^{7}Li to ^{1}H abundance. Dark gray with solid lines are the predictions of Ref. [72] based on big bang nucleosynthesis and CMB analysis. Light gray with dashed lines indicates primordial abundances from astronomical observations, except for ^{3}He where no reliable measurements exist. The agreement between big bang theory (dark gray) and observations (light gray) is very good except for ^{7}Li, where there is a factor of three discrepancy. Adapted with permission from Ref. [72]. Copyrighted by the American Physical Society.

approximately 25% helium and 75% hydrogen by mass would be a serious challenge to the theory.

A review of big bang nucleosynthesis is given in Ref. [72]. Figure 20.6 illustrates a comparison from that review of calculated abundances with observed abundances for the light elements produced mostly in the big bang. The agreement between observation and theory is excellent, except for ^{7}Li where observation is a factor of three smaller than the theoretical prediction. This discrepancy is not understood at present. It occurs for the big bang light element that has by far the tiniest abundance, so it is a small effect that might be explained by observational issues, or by missing physics having a small impact on overall nucleosynthesis.

20.3.3 Constraints on Baryon Density

This agreement between theory and observation for light-element abundances places a limit on the total amount of mass that can be baryonic. That constraint is the basis for the earlier assertion that most of the dark matter dominating the mass of the Universe cannot be ordinary baryonic matter. If enough baryons were present in the Universe to make that true, the distribution of light element abundances would have to differ substantially from what is observed. The implication is that the matter that we are made of (baryonic matter) is but a small impurity compared to the dominant matter in the universe (nonbaryonic matter).

Figure 20.7(a) illustrates calculated abundances for the light elements produced mostly in the big bang as a function of the baryon to photon ratio η, or equivalently the baryon density parameter Ω_b (both for the present Universe). The widths of the curves indicate uncertainty in the theoretical values. By comparing calculated with observed abundances, it is possible both to test big bang nucleosynthesis and to determine the baryon to photon ratio. For example, the analysis indicates that the mass fraction of ^{4}He is 0.247, which is

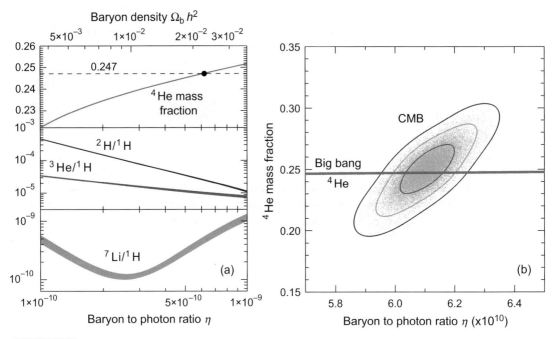

Fig. 20.7 (a) Big bang abundances as a function of the baryon to photon ratio η or baryon density parameter Ω_b. (b) Correlation of helium abundance and η from Planck satellite CMB data. The ^{4}He mass fraction predicted from big bang nucleosynthesis is indicated by the near-horizontal curve. Both plots suggest $\eta \sim 6.1 \times 10^{-10}$, or equivalently $\Omega_b \sim 0.04$. Adapted with permission from Ref. [72]. Copyrighted by the American Physical Society.

indicated as a dashed horizontal line in Fig. 20.7(a). The intersection of this line with the ^{4}He curve implies that $\eta \sim 6 \times 10^{-10}$. As illustrated in Fig. 20.7(b), independent analysis of the cosmic microwave background by the Planck satellite finds a similar value of η (see the discussion in Ref. [72]). From these results it may be concluded that

$$\eta \equiv \frac{n_b}{n_\gamma} \simeq 6.1 \times 10^{-10}, \tag{20.22}$$

implying that there are several billion photons for every baryon in the present Universe. Most of these baryons are neutrons and protons, while most of the photons are in the cosmic microwave background radiation. The total number of each kind of particle is not expected to change in the absence of interactions, so this ratio is also characteristic of that at the time when matter and radiation decoupled.

20.4 The Cosmic Microwave Background

There are three important observables in the present Universe that date back to its early history: the *abundance of light elements* that we have just discussed, the *cosmic microwave background radiation,* and *dark matter.* The microwave background is the faint glow left over from the big bang itself; dark matter appears to represent the major part of the mass in the Universe, but what it is remains a mystery. Both the microwave background and the nature of dark matter provide crucial diagnostics for a fundamental issue in cosmology, the *formation of structure* in the Universe, which will be addressed later in this chapter.

20.4.1 The Microwave Background Spectrum

The cosmic microwave background (CMB) radiation was discovered accidentally by Arno Penzias and Robert Wilson in 1964, while testing a microwave antenna.[6] They initially believed the unusual signal that they detected to be electronic noise associated with an instrumental problem. Once careful tests had ruled out that possibility they were still unaware of the significance of their discovery until it was pointed out that the big bang theory actually predicted that the Universe should be permeated by radiation left over from the big bang itself, but now redshifted by the expansion over some 14 billion years to the microwave region of the spectrum.

Measurements by Penzias and Wilson that are relatively crude by modern standards established that the radiation was coming from all directions in the sky, and had a temperature of approximately 3.5 K (it was initially fashionable to refer to this as the *3-degree background radiation*). Much more precise measurements using the Cosmic Background Explorer (COBE) satellite confirmed an almost perfect blackbody spectrum with an average temperature of $\langle T \rangle = 2.726$ K, where

$$\langle T \rangle \equiv \frac{1}{4\pi} \int T(\theta, \varphi) \sin\theta d\theta d\varphi, \qquad (20.23)$$

with $T(\theta, \varphi)$ the temperature measured at coordinates (θ, φ) on the sky; Fig. 20.8 illustrates. [The temperature was revised to 2.725 ± 0.001 K by the Wilkensen Microwave Anisotropy Probe (WMAP) in 2010.] Figure 20.8 is the *most perfect blackbody spectrum ever measured*. From basic statistical mechanics applied to the spectrum a CMB photon number density

$$n_\gamma \simeq 410 \, \text{photons cm}^{-3}. \qquad (20.24)$$

may be deduced. Theory predicts also a cosmic neutrino background left over from the big bang, but these low-energy neutrinos are not presently observable.

The CMB is the photon remnants of the big bang itself, redshifted into the microwave spectrum by expansion of the Universe, as illustrated in Fig. 20.9(a). The photons detected in the CMB were emitted from the *last scattering surface* illustrated in Fig. 20.9(b), which lies at a redshift $z \sim 1100$ and represents the time when the photons of the present CMB decoupled from the matter (almost 400,000 years after the big bang). At greater redshifts

---

[6] The antenna Penzias and Wilson employed had been used for microwave communication with satellites at a wavelength of 7.35 cm, so that is where the CMB was discovered. As we shall see, the CMB has a blackbody spectrum that is much stronger at shorter wavelengths but the water vapor in the atmosphere absorbs microwaves strongly at wavelengths shorter than about 3 cm. Thus, a quantitative study of the CMB required getting above the water vapor in the atmosphere, first with balloons and high-altitude observatories, and ultimately with satellites operating completely above the atmosphere.

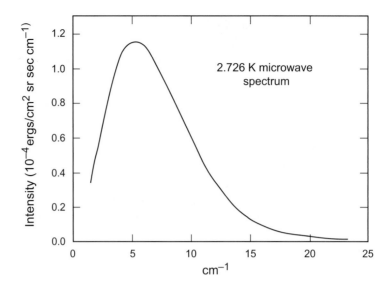

Fig. 20.8 The almost perfect 2.726 K microwave background spectrum recorded by COBE. The data points and the theoretical blackbody curve are indistinguishable on this scale. We acknowledge the use of the Legacy Archive for Microwave Background Data Analysis (LAMBDA), part of the High Energy Astrophysics Science Archive Center (HEASARC).

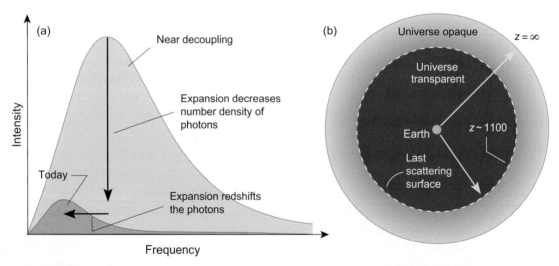

Fig. 20.9 (a) Schematic evolution of the cosmic microwave background. As the Universe expands the spectrum remains blackbody but the photon frequencies are redshifted and the number density of photons is lowered. The 2.7 K CMB is the faint, redshifted remnant of the cosmic fireball in which the Universe was created. Decoupling occurred at $z \sim 1100$. The photon temperature then of about 3000 K is lowered by the redshift factor to the present value of a little less than 3 K. (b) Last scattering surface for the CMB. The Universe is opaque to photons at larger redshift (corresponding to earlier times).

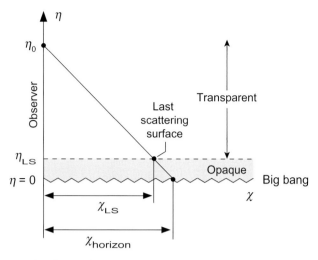

Fig. 20.10 The CMB last scattering surface illustrated in a $\eta-\chi$ spacetime diagram, where χ is the comoving coordinate in Eq. (18.19) and η is a conformal time coordinate (Box 18.4). The present time is η_0, the time of the big bang is $\eta=0$, and the time of last scattering is η_{LS}. The comoving distance to the last scattering surface χ_{LS} (the greatest actual distance from which photons can be detected) is less than the horizon distance $\chi_{horizon}$. The Universe is opaque to photons before η_{LS} and transparent afterwards. Adapted from Ref. [111].

the Universe is opaque to photons because that represents times when matter and radiation were strongly coupled, as elaborated in Fig. 20.10.

20.4.2 Anisotropies in the Microwave Background

The COBE, WMAP, and Planck satellites (as well as high-altitude balloon observations) have measured the angular distribution of the CMB with high precision. It is isotropic down to a dipole anisotropy at the 10^{-3} level corresponding to motion of the Earth relative to the microwave background (blueshifted in the direction of motion and redshifted in the opposite direction). Once the peculiar motion of the Earth relative to the CMB is subtracted, the background temperature is isotropic down to the tens of μK level. The temperature fluctuation at a given point on the sky is defined by

$$\frac{\delta T}{T}(\theta,\varphi) \equiv \frac{T(\theta,\varphi) - \langle T \rangle}{\langle T \rangle}, \tag{20.25}$$

where the temperature averaged over the whole sky $\langle T \rangle$ was defined in Eq. (20.23). Using 7° antennas pointing in directions separated by 60°, COBE measured a root mean square temperature fluctuation of the blackbody spectrum on this angular scale of

$$\sqrt{\langle (\delta T/T)^2 \rangle} = 1.1 \times 10^{-5}. \tag{20.26}$$

Even more precise measurements of the CMB anisotropies have been made, first by WMAP and then by Planck, as summarized in Figs. 20.11 and 20.12.

Fig. 20.11 COBE, WMAP, and Planck maps of the full sky showing fluctuations in the CMB temperature. Notice the increased angular resolution from COBE to WMAP to Planck. WMAP: NASA/WMAP Science Team; Planck: ESA and the Planck Collaboration.

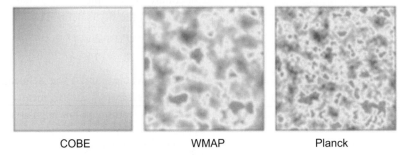

Fig. 20.12 COBE, WMAP, and Planck maps of a 10 degree square portion of the sky showing fluctuations in the CMB temperature at different resolutions. Credit: NASA/JPL-Caltech/ESA.

The primary interest in cosmology is in the statistical correlations of the fluctuations in temperature displayed in Fig. 20.11. Since these are observed on the 2D celestial sphere it is convenient to expand the temperature fluctuations (20.25) as a function of angular position (θ, φ) on the sky using spherical harmonics $Y_{\ell m}(\theta, \varphi)$,

$$\frac{\delta T}{T}(\theta, \varphi) = \sum_{\ell m} a_{\ell m} Y_{\ell m}(\theta, \varphi). \tag{20.27}$$

The *2-point correlation function* $C(\theta)$ is then defined by,

$$C(\theta) \equiv \left\langle \frac{\delta T}{T}(\hat{\boldsymbol{n}}_1) \frac{\delta T}{T}(\hat{\boldsymbol{n}}_2) \right\rangle_{\hat{\boldsymbol{n}}_1 \cdot \hat{\boldsymbol{n}}_2 = \cos\theta}, \tag{20.28}$$

where $\hat{\boldsymbol{n}}_1$ and $\hat{\boldsymbol{n}}_2$ are unit vectors specifying the directions to two points on the celestial sphere separated by an angle θ, and the angular brackets indicate an average over all pairs of points separated by θ. The theory of inflation to be discussed in Section 21.3 predicts as initial conditions for the big bang fluctuations that are maximally random (these are called *gaussian fluctuations*), meaning that the different multipoles in Eq. (20.27) are uncorrelated. With this assumption, and the *spherical harmonic addition theorem*

$$P_\ell(\cos\theta) = \frac{4\pi}{2\ell+1} \sum_{m=-\ell}^{m=+\ell} Y_{\ell m}^*(\hat{\boldsymbol{n}}_1) Y_{\ell m}(\hat{\boldsymbol{n}}_2),$$

where θ is the angle between $\hat{\boldsymbol{n}}_1$ and $\hat{\boldsymbol{n}}_2$, Eq. (20.28) may be evaluated as a multipole expansion in *Legendre polynomials* $P_\ell(\cos\theta)$,

$$C(\theta) = \frac{1}{4\pi} \sum_{\ell=0}^{\infty} c_\ell (2\ell+1) P_\ell(\cos\theta). \tag{20.29}$$

A plot of some function of this correlation versus multipole order is called a *CMB power spectrum*; examples are shown in Figs. 20.13(a) and 20.14. Physically, the $\ell = 0$ component is zero if the average temperature is defined correctly and the $\ell = 1$ (dipole) component corresponds to the overall motion of the Earth with respect to the CMB that is subtracted out. Thus, cosmological interest centers on the multipoles with $\ell \geq 2$. The multipole moments on the horizontal axis of these figures are related to a corresponding angular scale, as illustrated for the top and bottom horizontal axes in Fig. 20.14. Roughly, a multipole moment of order ℓ is sensitive to an angular scale $\sim 180°/\ell$. Thus, the low multipoles in the power spectrum carry information about the CMB on large angular scales and higher multipole components carry information on increasingly smaller angular scales.

Example 20.2 The $\ell = 10$ multipole order in Fig. 20.14 is sensitive to structure on a scale of $180°/10 \sim 18°$ while the $\ell = 1800$ multipole reflects structure on a scale of $180°/1800 \sim 0.1°$. Thus in temperature fluctuation maps such as Figs. 20.11 and 20.12 the lower multipoles in the power spectrum represent (statistically) the larger-scale features and the highest multipoles represent (statistically) the smallest features in the map.

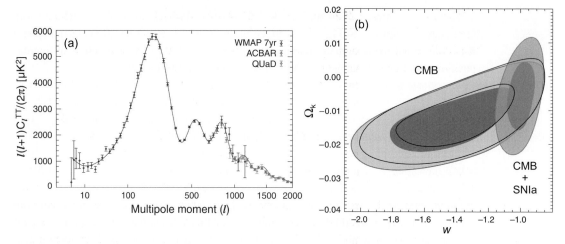

Fig. 20.13 (a) The CMB power spectrum [136]. (b) The curvature density Ω_k versus w for the equation of state $P = w\epsilon$ [136]. The diagonal locus indicates CMB analysis; the more vertical locus combines CMB and Type Ia supernova data. E. Komatsu *et al.*, Seven-Year Wilkinson Microwave Anisotropy Probe (WMAP) Obsevations: Cosmological Interpretation, published 2011 January 11. © 2011. The American Astronomical Society. All rights reserved. *Astrophysical Journal Supplement Series*, **192**(2).

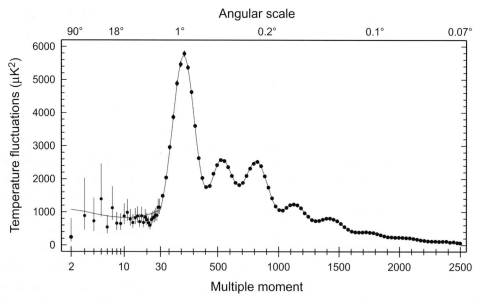

Fig. 20.14 Planck CMB temperature power spectrum [23]. The multipole order ℓ is shown on the bottom axis (with a scale logarithmic below 30 and linear above) and a corresponding angular scale θ on the top axis. The two are related by $\theta \sim 180°/\ell$. Cosmological parameters extracted from these data are shown in Table 20.2. From P. A. R. Abe *et al.*, *Astronomy and Astrophysics*, **594**, 63, 2016. Reproduced with permission © ESO.

20.4.3 The Origin of CMB Fluctuations

The CMB represents photons liberated near the decoupling transition, only about 380,000 years after the big bang, when the Universe was a thousand times smaller in linear dimension and a billion times more dense than it is today. The pattern of CMB temperature fluctuations displayed in Figs. 20.11 and 20.12 reflects the perturbation of the big bang photons at decoupling caused by the density fluctuations present at the time of *last scattering* (the time when statistically photons were unlikely to undergo any further scattering before reaching us). Thus, this pattern contains detailed information about the state of the Universe near the time of decoupling and it is important to understand the cause of the fluctuations. As we shall now discuss, the origin of the CMB fluctuations is different for the low multipoles and high multipoles in Fig. 20.14.

As has been emphasized before, cosmological distances are a matter of *definition*, since what we can measure at high redshift is typically not directly related to a proper distance. In Box 19.3 one such defined distance, the *luminosity distance* d_L, was introduced. In studying the CMB we are often concerned with angular measure and it is useful to introduce another defined distance, the *angular size distance* d_A, which is described in Box 20.2.

Example 20.3 A distance scale relevant at the time of last scattering (LS) is the size of the horizon at that time. A typical estimate of the horizon size at redshift $z = 1080$ is found to be $d_h = 0.25$ Mpc and from a numerical solution of the Friedmann equations using cosmological parameters determined from observations of the Planck satellite that will be summarized below in Table 20.2, the angular size distance is determined to be $d_A = 12.9$ Mpc. A patch on the last scattering surface (LSS) the size of the horizon has an angular size as viewed from Earth of

$$\theta_h = \frac{d_h(t_{ls})}{d_A} = \frac{0.25 \text{ Mpc}}{12.9 \text{ Mpc}} \sim 0.019 \text{ rad} \sim 1°,$$

where d_A is the angular size distance (Box 20.2) and t_{ls} is the time of last scattering. This angular scale corresponds to a multipole order $\ell \sim 180°/1° \sim 180$.

The Sachs–Wolfe effect: From Fig. 20.14 we see that the CMB power spectrum is fundamentally different for multipole order less than $\ell \sim 30$, where it is largely flat without obvious structure, relative to larger multipoles, where various peaks are in evidence. Let us first address the structure of the power spectrum for relatively large angular scales. As you are asked to estimate in Problem 20.11, at decoupling ($z \sim 1080$) vacuum energy was not important and the energy densities for dark matter, radiation, and baryonic matter were approximately in the ratio $\epsilon_{dm} : \epsilon_\gamma : \epsilon_b \sim 5.5 : 1.8 : 1$. Thus gravity was dominated by the dark matter. As we will discuss further in Section 20.5, because the dark matter was not coupled to the photons it could begin to contract under its self-gravity before the baryonic matter could. Thus, small overdensities of dark matter would have grown prior

Box 20.2 Angular Size Distance

In Box 19.3 we defined the *luminosity distance* d_L, which is based on determining distances using a standard candle in an expanding, possibly curved universe. Another distance measure used often in cosmology is the *angular size distance* (also termed the *angular diameter distance*) d_A. The angular size distance is based on observing a *standard ruler,* which is an astronomical object of known transverse size. Then we define the angular diameter distance to be

$$d_A = \frac{D}{\Delta\theta}$$

where $\Delta\theta$ is the observed angular size and D is the physical length of the standard ruler. In static euclidean space d_A is the proper distance but in general relativity we must account for curvature and expansion. Assume the RW metric with an observer at comoving coordinates $(t_0, r, \theta, \varphi) = (0, 0, 0, 0)$, and a standard ruler at redshift z stretching from comoving coordinates $(t_e, r, \theta, \varphi)$ to $(t_e, r, \theta + \Delta\theta, \varphi)$:

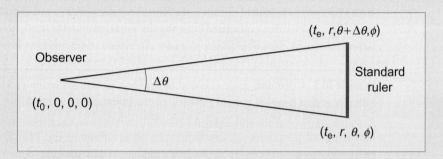

Light is emitted from the standard ruler at time t_e and travels to the observer on a geodesic with $\theta = \varphi =$ constant. The proper length of the standard ruler at time t_e is known to be D, and it also can be calculated by evaluating the line element. Using the metric in the form given in Box 19.3 with $dr = d\varphi = dt = 0$, and that $d\Omega = \Delta\theta$ since φ is fixed, gives

$$D = ds = a(t_e)S_k(r)\Delta\theta.$$

Inserting this in the first equation above, the angular size distance is

$$d_A = \frac{D}{\Delta\theta} = \frac{a(t_e)S_k(r)\Delta\theta}{\Delta\theta} = a(t_e)S_k(r) = \frac{S_k(r)}{1+z}.$$

Comparing with the luminosity distance d_L defined in Box 19.3, we see that

$$d_A = \frac{d_L}{(1+z)^2}.$$

Thus d_A and d_L are very different at large redshift. As shown in Problem 20.14, for a flat Universe d_A is equal to the proper distance *at the time the light was emitted*.

to decoupling. These fluctuations in the dark matter energy density $\epsilon(r) = \epsilon_0 + \delta\epsilon(r)$ would have caused corresponding fluctuations in the gravitational potential. For example, in Newtonian approximation the Poisson equation (8.1) implies a relationship

$$\nabla^2(\delta\varphi) = 4\pi G \left(\frac{\delta\epsilon}{c^2}\right)$$

between density perturbations $\delta\epsilon$ and perturbations of the gravitational potential $\delta\varphi$. A CMB photon climbing out of a local minimum of the potential at decoupling would then be redshifted and one falling off a local potential maximum would be blueshifted, causing fluctuations in the observed temperature of the CMB.

Effective temperature variations caused by a variation in the gravitational potential is called the *Sachs–Wolfe effect*. A general relativistic treatment gives a Sachs–Wolfe temperature fluctuation caused by a variation $\delta\varphi$ of the potential

$$\frac{\Delta T}{T} = \frac{\delta\varphi}{3c^2}. \tag{20.30}$$

This is thought to be the main source of CMB fluctuations for angular scales larger than $\sim 10°$, corresponding to multipole orders below about $\ell = 20$. From Fig. 20.14, the CMB power is nearly constant for $\ell < 20$, indicating that the Sachs–Wolfe fluctuations are nearly constant over a broad range of angular scales and therefore over a broad range of distance scales. This scale invariance of the fluctuations is a central feature of the density perturbation spectrum expected from the theory of inflation to be discussed in Section 21.3. The much more varied fluctuations of the CMB on smaller angular scales (larger multipole order) displayed in Fig. 20.14 are thought to have a different origin than the Sachs–Wolfe fluctuations described above. To understand this, we have to first take a detour to consider the propagation of sound (that is, pressure) waves in the cosmic fluid.

Sound waves: Consider a small region of the young Universe that is slightly overdense relative to the surrounding region [82]. Because it is overdense, it will tend to compress gravitationally. What happens next depends on whether the cosmic fluid is ionized. If it is electrically neutral, as it is after the decoupling transition, there is little radiation pressure and fluctuations are unstable against gravitational collapse as overdense regions attract additional mass more rapidly than the Hubble expansion can carry it away. Over time this causes overdense regions to become more dense and underdense regions to become less dense, and the growing contrast between overdense and underdense regions leads eventually to the formation of superclusters of galaxies surrounding large voids, as observed in the present visible Universe.

The situation is quite different if the cosmic fluid is ionized, as is the case at times earlier than the decoupling transition. If the size of the density perturbation is larger than the mean free path for scattering of photons the baryonic plasma and the photons become coupled by the strong Thomson scattering of the photons into an effective baryon–photon fluid. This fluid has only about half of the density of the dark matter at decoupling, so the baryon–photon fluid moves mostly under the gravitational influence of the dark matter and not its own self-gravity. Gravitational compression of this fluid by a perturbation increases the density, temperature, and pressure, which scatters photons to higher energy and produces a radiation pressure that opposes the gravitational contraction. This drives an expansion of the region that causes the pressure to drop, eventually leading to another gravitational contraction, and so on.

Thus the baryon–photon plasma develops sound waves (pressure fluctuations) called *acoustic oscillations,* for which the driving force is gravity and the restoring force is the radiation pressure of the photons. These sound waves travel at very high speed in the ionized fluid. Near decoupling the temperature is ~ 3000 K, implying a number density of photons $\sim 10^9$ larger than the number density of baryons in the fluid. At constant entropy the speed of sound v_s is expected to be

$$v_s = \left(\frac{dP}{d\rho}\right)^{1/2} \simeq \frac{c}{\sqrt{3}}, \qquad (20.31)$$

since the fluid is dominated by photons and for a photon gas $P = \frac{1}{3}\epsilon = \frac{1}{3}\rho c^2$. Thus the radiation pressure keeps density fluctuations from collapsing gravitationally and creates acoustic oscillations in the plasma before decoupling.

The acoustic scale: The size of these acoustic oscillation regions at decoupling defines a characteristic length called the *acoustic scale* or the *sound horizon* that is set by the distance that sound could have traveled in the cosmic fluid from the big bang until the time of decoupling. As we will see in Sections 20.4.4 and 20.4.5, this preferred length scale is expected to leave its imprint both on the radiation, through temperature fluctuations of the CMB, and on the matter, through baryon acoustic oscillations in the observed distribution of galaxies.

Example 20.4 The preferred length scale corresponding to the sound horizon is set by the proper distance $\ell_s(t_{ls})$ that a sound wave could have traveled from the beginning to the time of decoupling t_{ls}. This may be evaluated from

$$\ell_s(t_{ls}) = a(t_{ls}) \int_0^{t_{ls}} \frac{v_s dt}{a(t)}, \qquad (20.32)$$

where v_s is the speed of sound. As you are asked to prove in Problem 20.13, using the soundspeed from Eq. (20.31) and the cosmological parameters of Table 20.2, the acoustic scale corresponds to a proper distance of $\ell(t_{ls}) \sim 0.144$ Mpc on the last scattering surface, which has expanded to a proper distance of $\ell(t_0) \sim (1 + Z_{ls})\ell(t_{ls}) \sim 156$ Mpc today. Hence a preferred proper distance scale

$$\ell(t_{ls}) \sim 0.144 \text{ Mpc} \quad \longrightarrow \quad \ell(t_0) \sim 156 \text{ Mpc} \qquad (20.33)$$

is established by the physical size of acoustic oscillations at the time of last scattering.

If the fluid within an acoustic oscillation region is at maximum compression at last scattering its density will be higher than average and the photons liberated from it will be slightly hotter than average, while if it is at maximum expansion at last scattering the density will be lower than average and the photons liberated will be slightly cooler than average, since for the photon gas $T \sim \epsilon^{1/4}$. In addition, if at last scattering a region is expanding or contracting there will be a red or blue Doppler shift of the emitted photons corresponding to the acoustic motion.

20.4.4 Acoustic Signature in the CMB

The characteristic acoustic length scale discussed in the previous section is based on well-understood physical principles (the speed of sound in a plasma dominated by photons) and therefore can be taken to define a *standard ruler* in the sense of Box 20.2. This permits an observed angular size to be associated with the acoustic length scale on the surface of last scattering that was estimated in Example 20.4.

Example 20.5 From the solutions of Problems 20.12 and 20.13, fluctuations in the CMB of an angular size $\Delta\theta$ are related to a physical transverse size D on the LSS by

$$D = d_A \Delta\theta, \tag{20.34}$$

where d_A is the angular size distance of Box 20.2. A numerical solution of the Freidmann equations [5] assuming the parameters given in Table 20.2 gives $d_A = 12.9$ Mpc and a proper transverse distance D on the LSS is related to an angle $\Delta\theta$ observed today on the celestial sphere by

$$\Delta\theta = 4.44 \times 10^{-3} \left(\frac{D}{\text{kpc}}\right) \text{ deg}. \tag{20.35}$$

Thus, the *preferred distance scale* established by acoustic oscillations at the time of last scattering translates into a *preferred angular scale* for the CMB observed today. Inserting the acoustic scale of 144 kpc from (20.33) in Eq. (20.35) implies that a preferred angular size of $\Delta\theta \sim 0.6°$ should be observable in the temperature fluctuations of the CMB.

We may conclude that the patterns of hot and cold spots on small angular scales in the CMB displayed in Figs. 20.11 and 20.12 are essentially a snapshot of the sound wave pattern in the cosmic fluid at the time of decoupling that was described in Section 20.4.3. The corresponding CMB power spectrum then contains a wealth of cosmological information. For example, referring to Fig. 20.14 [82, 211]:

1. The first (leftmost) peak is strongest and occurs at a multipole order associated with the preferred angular size of 0.6° (see the horizontal axes of Fig. 20.14). This represents acoustic oscillations that were at maximum compression at the time of last scattering.
2. The harmonic series of peaks represents overtones of the preferred angular scale, much as a harmonic sequence of overtones is produced by a violin string of fixed length [82].
3. Observed angular sizes will depend on curvature of the Universe (see Fig. 20.17). Positive curvature will shift the first peak to larger angle and negative curvature will shift it to smaller angle. Data are consistent at this point with zero curvature.
4. The amplitude of the first peak is increased by lower sound speed v_s and decreased by higher v_s, so its height is a diagnostic of the speed of sound in the cosmic fluid at decoupling. The sound speed is in turn a diagnostic for the baryon density of the fluid.

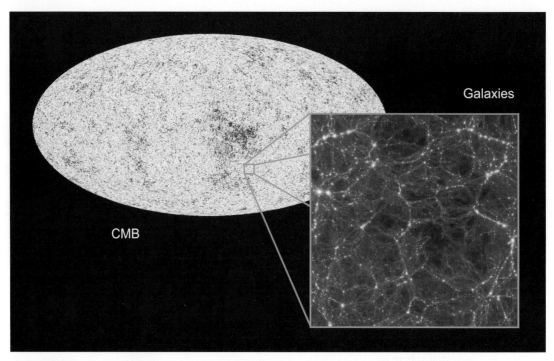

Fig. 20.15 Acoustic oscillations at the decoupling transition leave an imprint on the CMB at last scattering, and on the large-scale distribution of galaxies at later times. NASA, Virgo Consortium.

5. Equal spacings of the peaks in the power spectrum indicates that the initial fluctuations were *adiabatic* (which means in this context that all species varied together in density). Adiabatic fluctuations as initial conditions for the big bang are a key prediction of the theory of inflation that will be described in Section 21.3.
6. The relative heights of the peaks are sensitive to the dark matter and baryon densities: lower dark matter density enhances all peaks but higher baryon density enhances odd-numbered peaks relative to even-numbered peaks.
7. The position of peaks is a measure of the preferred acoustic length scale, which is physical. Thus it can serve as a standard ruler, allowing an independent determination of distance (see Box 20.2).

From the preceding discussion, the characteristic length scale imprinted on the CMB by acoustic oscillations at decoupling also should be visible for the baryonic matter, as illustrated schematically in Fig. 20.15. The possibility of such baryon acoustic oscillations (BAO) was introduced in Box 19.4 and will now be discussed in more detail.

20.4.5 Acoustic Signature in Galaxy Distributions

A remnant signature for the baryon acoustic oscillation should be observable in the clustering of galaxies on large scales. However, as indicated in the solution of Problem 20.13, the characteristic acoustic length on the last scattering surface has been stretched by a factor of

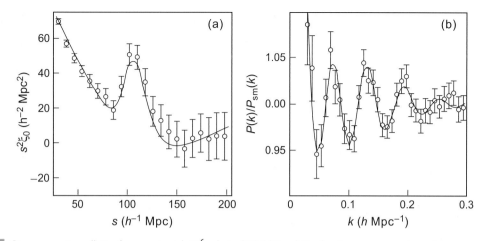

Fig. 20.16 Baryon acoustic oscillation for approximately 10^6 galaxies [28]. (a) Correlation function versus comoving radius. (b) Power versus wavenumber. The data have been reconstructed to remove effects of matter flows and peculiar velocities on intermediate scales and give sharper peaks, as described in Ref. [28]. The curves are calculations for a best-fit BAO model. The galaxy power spectrum (b) is similar in spirit to the CMB power spectrum of Fig. 20.14 except that the 3D data are expanded in a Fourier series rather than in spherical harmonics, and the index k is a wavenumber rather than a multipole moment. Adapted from Anderson et al., The clustering of galaxies in the SDSS-III Baryon Oscillation Spectroscopic Survey: baryon acoustic oscillations in the Data Releases 10 and 11 Galaxy samples. MNRAS (2014) **441**(1), 24–62. By permission of Oxford University Press.

$1 + z_{ls}$ to approximately 150 Mpc today. Using the current average matter density parameter Ω_m as the basis for an estimate suggests that a volume of this diameter contains greater than $10^{17}\ M_\odot$, which is much larger than the mass of a supercluster of galaxies. Furthermore, the original BAO signal has been diluted by mixing with the dominant dark matter and smeared by peculiar local velocities and fluid flows in the evolution of the Universe.

Hence uncovering the BAO signature requires an enormous sample of 3D galaxy locations. Surveys of galaxy angular positions that also determine redshifts are effectively 3D since redshift is a proxy for distance if peculiar motion can be neglected. Determining the baryon acoustic peak from data in such large surveys is described in Refs. [28, 83], and evidence for a BAO signature in a survey of about one million galaxies from the Sloan Digital Sky Survey (SDSS) is shown in Fig. 20.16. The peak in Fig. 20.16(a) indicates a preferred length scale of about 150 Mpc, which is roughly the estimate from Problem 20.13 for the present size of the acoustic length scale established by how far sound waves could have traveled before the decoupling transition.

20.4.6 Precision Cosmology

The WMAP and Planck observations yield precise constraints on important cosmological parameters. For example, Fig. 20.17 illustrates that lensing effects on the CMB distort it in a way that depends quantitatively on the overall curvature of the Universe. As a second example, Fig. 20.13(a) illustrates the power spectrum for observed CMB fluctuations and

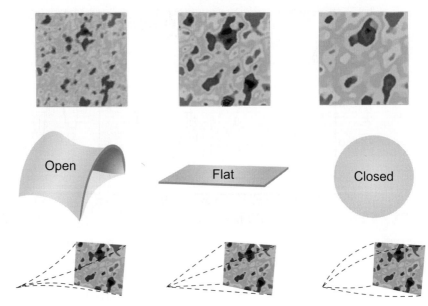

Fig. 20.17 Schematic influence of curvature on observed CMB fluctuations.

Fig. 20.13(b) shows confidence intervals for the curvature density Ω_k and the parameter w from the equation of state $P = w\epsilon$ (see Section 19.1.2) implied by the CMB data. The equation of state parameter extracted from this analysis is $w = -1 \pm 0.02$, which may be compared with $w = -0.997^{+0.077}_{-0.082}$ deduced from Fig. 19.9(b). Within uncertainties, both are consistent with the $w = -1$ expected from a cosmological constant [see Eq. (19.17)]. From Fig. 20.13 it is concluded that the total density parameter is $\Omega = 1 \pm 0.03$, which is consistent with a flat Universe.

Detailed fits to such power spectra using cosmological theories place strong constraints on those theories, and permit cosmological parameters to be determined with high precision. A measurement of the CMB power spectrum by the Planck satellite is shown in Fig. 20.14, some values of cosmological parameters extracted from the Planck data are displayed in Table 20.2, and evolution of the early Universe for a Friedmann cosmology using the parameters of Table 20.2 is illustrated in Fig. 20.18. The precision with which parameters are now being determined from CMB data and from high-redshift supernovae (supplemented by lensing, baryon acoustic oscillation, and other data from more traditional observational astronomy) is unprecedented; over a period of little more than a decade these new observations have transformed cosmology into a quantitative science constrained by precise data.

Example 20.6 As of late 2017 the largest redshift that had been observed was $z = 11.09$ for the irregular galaxy GN-z11 [167]. From Fig. 20.18 this redshift corresponds to light emitted about 400 million years after the big bang, indicating that galaxy formation was already well underway during the reionization of the Universe discussed in Box 20.1.

20.4 The Cosmic Microwave Background

Table 20.2 Cosmological parameters from Planck CMB data [23]

Parameter	Symbol	Value
Hubble parameter	H_0	67.8 ± 0.9 km s^{-1} Mpc^{-1}
Age of the universe	t_0	13.80 ± 0.04 Gyr
Baryon density	Ω_b	0.0484 ± 0.0005
Matter density	Ω_m	0.308 ± 0.012
Dark energy density	Ω_Λ	0.692 ± 0.012
Dark energy equation of state†	w	-1.006 ± 0.045
Sum of neutrino masses	$\sum m_\nu$	< 0.23 eV
Curvature density	$|\Omega_k|$	< 0.005
Redshift for equality of radiation and matter	z_{eq}	3365 ± 44

† The coefficient w in $P = w\epsilon$; see Section 19.1.2.

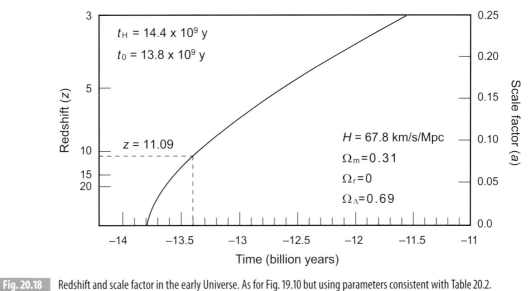

Fig. 20.18 Redshift and scale factor in the early Universe. As for Fig. 19.10 but using parameters consistent with Table 20.2. The redshift $z = 11.09$ corresponds to the most distant galaxy that had been observed as of 2017 and is discussed in Example 20.6.

20.4.7 Seeds for Structure Formation

Fluctuations in the CMB presumably reflect the initial density perturbations responsible for the formation and growth of large-scale structure. As will be seen in Chapter 21, a period of exponential growth in the early Universe called *cosmic inflation* may have been central to producing these density fluctuations and, as we shall now discuss, the gravitational influence of dark matter likely was pivotal for the rapid growth of structure.

20.5 Accelerated Structure Formation

Because dark matter does not couple strongly to photons, it could begin to clump earlier than the normal matter (which was continually perturbed by the high-energy photon bath of the early Universe), and because there is so much more of it, it could clump more effectively. Thus, it is likely that dark matter provided the initial regions of higher than average density that seeded the early formation of structure in the Universe, as illustrated in Fig. 20.19. Condensations of dark matter then served as nucleation centers to hasten the gravitational clumping of normal matter once it decoupled from the photons. As a consequence, the formation of large-scale structure proceeded faster than it could have otherwise. Hot and cold dark matter suggest different paradigms for formation of structure in the early Universe. These are termed *top-down* and *bottom-up*, respectively.

Hot dark matter implies high velocities for the dark particles. They will be stopped only after passing through a volume containing of order $10^{15} M_\odot$ (mass of galaxy superclusters), implying that they can stream freely over 10–100 Mpc distances without substantial interaction. Hot dark matter streaming freely on these scales disfavors structure formation on any smaller scales: in a Universe dominated by hot dark matter the formation of structure occurs first on large scales and this structure must then break apart to form structure on smaller scales. This is called the *top-down* model for structure formation. On the other hand, cold dark matter has relatively low velocity and tends to travel only a small distance before being stopped. This favors the formation of structure on small scales such as galaxies, and then larger structures can form by aggregation of these smaller structures. This is termed *bottom-up* formation of structure.

The top-down scenario has serious problems because observations at the largest distances and thus the furthest back in time see quasars, gamma-ray bursts, and galaxies, not superclusters of galaxies. The best current simulations indicate that the formation of structure was dominated by cold dark matter in a bottom-up scenario, though the details may require some admixture of hot dark matter. The cold dark matter is essential for accelerating the bottom-up scenario because otherwise the Universe would not be old enough to have assembled the large-scale structures seen in the current Universe.

20.6 Dark Matter, Dark Energy, and Structure

We conclude this chapter by summarizing the present understanding of dark matter, dark energy, and the formation of large-scale structure in the Universe. If inflation is correct (see Chapter 21) and the cosmological constant were zero, the matter density of the Universe would be exactly the closure density, which would lead to flat geometry. Current data indicate that the Universe is indeed flat, as predicted by inflation, but that it does not contain a closure density of matter because there is a non-zero cosmological constant. Instead, about 30% of the closure density is supplied by matter and about 70% by dark energy (vacuum energy or a cosmological constant), with the contribution of radiation negligible. Luminous matter is a small fraction of the matter, implying that the present

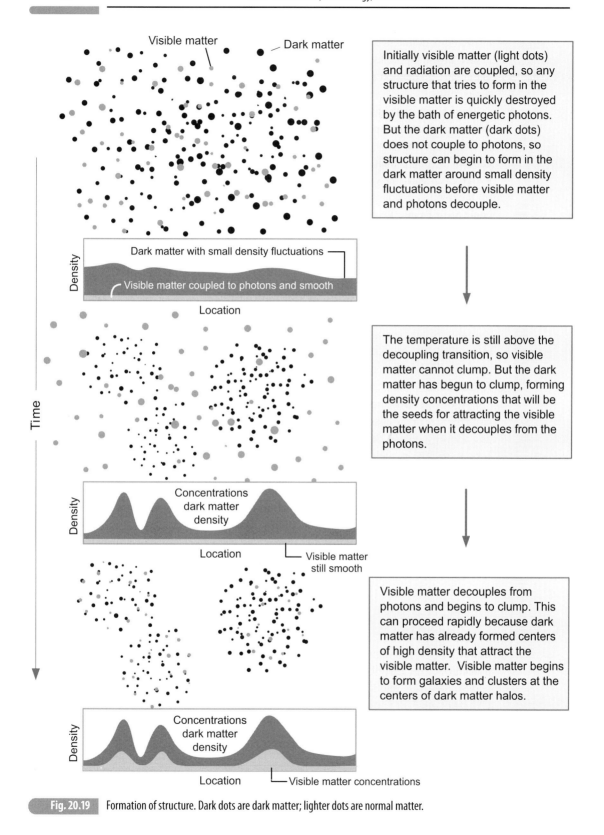

Fig. 20.19 Formation of structure. Dark dots are dark matter; lighter dots are normal matter.

Universe is dominated by dark matter and dark energy. There are no strong clues to the nature of either because neither has been captured in a laboratory. Present knowledge about dark matter and dark energy comes entirely from observations on galactic and larger scales.

The known neutrinos are relativistic (that is, they are hot dark matter) and therefore they erase fluctuations on small scales. They could aid the formation of large structures like superclusters but not smaller structures like galaxies. Thus, they are not likely to account for more than a small fraction of the dark matter. This conclusion is supported by WMAP and Planck data. On the scale of galaxies and clusters of galaxies, 90% of the total mass is not seen. In this case, a significant fraction of the dark matter could in principle be normal (that is, baryonic) and be in the form of small, very low-luminosity objects like white dwarfs, neutron stars, black holes, brown dwarfs, or red dwarfs. However, gravitational lensing observations and searches for subluminous objects generally have not found enough of these "normal" objects to account for the mass of galaxy halos. Further, strong constraints from big bang nucleosynthesis compared with the observed abundances of the light elements indicate that most of the dark matter cannot be baryonic. Thus, a significant fraction of the dark matter is likely to be nonbaryonic and not neutrinos, and to be cold (that is, massive so that it does not normally travel at relativistic velocities). Current speculation centers on not-yet discovered elementary particles as the candidates for this cold dark matter, but none have been found so far.

Large-scale structure and its rapid formation in the early Universe are hard to understand from the size of the cosmic microwave background fluctuations implied by COBE, WMAP, and Planck, unless cold dark matter plays a central role in seeding initial structure formation. The models of structure formation most consistent with current data are the class of ΛCDM models that combine a cosmological constant (denoted by Λ) to model dark energy with cold dark matter (CDM) to give an accelerating but flat universe with cold dark matter to seed structure formation. The origin of the dark energy is a mystery. The most economical explanation would be the energy of vacuum fluctuations, but with our present understanding of how to calculate the vacuum energy that explanation is highly inconsistent with observations (see the discussion in Section 26.2).

Background and Further Reading

See Roos [207], Ryden [211], Perkins [183], and Liddle [145]. A review of the thermodynamics of the early Universe is given in Kolb and Turner [135] and in Collins, Martin, and Squires [70], and a review of the status of big bang nucleosynthesis may be found in Cyburt *et al.* [72]. Clear and pedagogical discussions of acoustic oscillations in the early Universe and how they led to baryon acoustic oscillations in galaxy distributions and to fluctuations in the cosmic microwave background are given by Eisenstein and Bennett [82], and by Ryden [211].

Problems

20.1 Assume that two microwave antennas, pointed in opposite directions, receive radiation from the last scattering surface. Show that the number of horizon distances separating the two sources at the time of emission was of order 100, so the observed homogeneity of the microwave background could not have arisen from a causal connection in the standard big bang. *Hint*: You may assume the Universe to have been matter dominated for most of the history relevant to this calculation.

20.2 Assume that as soon as the deuterium bottleneck is broken (at about $T = 1 \times 10^9$ K) as many free protons and neutrons as possible combine to make ^{4}He, and that nucleosynthesis in the big bang does not proceed beyond the production of ^{4}He. Taking the neutron/proton ratio to be 0.164 when nucleosynthesis begins, estimate the mass fractions of protons, neutrons, and ^{4}He produced by the big bang. *Hint*: Count free neutrons and protons, and those bound in ^{4}He.

20.3 Estimate the redshift at which the radiation density fell below that of baryonic matter in the early Universe. What was the age of the Universe at this time, if for simplicity the role of vacuum energy is ignored in the subsequent evolution of the Universe?

20.4 From observational data, estimate (a) the baryon energy density, (b) the baryon number density, (c) the photon energy density, (d) the photon number density, and (e) the ratio of the photon number density to the baryon number density.

20.5 What were the energy densities (and equivalent mass densities) of nonbaryonic dark matter and baryonic matter, and the energy density of photons, at the time of last scattering for the CMB (corresponding to a redshift $z \simeq 1100$)?

20.6 Assume for estimation that the Universe has been matter dominated for its entire history. Take last scattering (decoupling) of the microwave background to have occurred at a redshift of 1100 and estimate the age of the Universe as the Hubble time. What is the maximum distance that light could have traveled from the big bang to the time of decoupling, and to what size would that distance have been stretched by today because of the expansion of the Universe? Use the maximum distance light could have traveled from the time of decoupling to today to estimate the present distance to the last scattering surface for the CMB. From these two distances, estimate the maximum angular size on the sky of a causally connected region of the CMB.

20.7 Assuming a cosmic neutrino background to exist analogous to the cosmic microwave background, estimate its temperature based on expected conditions in the early Universe and the properties of the CMB. Specifically, assume that

(a) Photons were reheated by electrons annihilating with positrons to produce photons when the temperature dropped below about 0.2 MeV and the reaction $e^+ + e^- \rightleftharpoons 2\gamma$ fell out of equilibrium (because the photons were no longer energetic enough to create electron–positron pairs below that temperature).

(b) The neutrinos were not reheated at this transition, because they had fallen out of equilibrium above that temperature (weak-interaction freezeout).
(c) Entropy was conserved across this electron–positron annihilation transition.

It may be assumed further that the neutrinos and photons had the same temperature up to this transition, and that nothing happened after it to change the ratio of photon to neutrino temperature as the Universe evolved.

20.8 Prove that if the radiation is described by a blackbody spectrum in the early Universe it will continue to be described by a blackbody spectrum as the Universe expands, but the corresponding blackbody temperature will change as the inverse of the scale factor. *Hint*: It may be assumed that the number of photons per comoving volume $na(t)^3$, where n is the photon number density and $a(t)$ the scale factor, is conserved as the Universe expands. Then ask: how do the photon temperature and photon frequencies change as the Universe expands and how does that affect $n(\nu, t) d\nu$?

20.9 Estimate the ratio of contributions for photons and known species of neutrinos and antineutrinos to the energy density of the Universe, taking into account the expected difference in temperatures for the background photons ($T = 2.725$ K) and background neutrinos ($T \sim 1.95$ K) estimated in Problem 20.7.

20.10 The discovery announcement for the Higgs Boson at the Large Hadron Collider in 2012 reported the mass of the Higgs to be 126 GeV. What is the temperature of a gas in equilibrium having this average kinetic energy for its particles? To what time in big bang evolution does this temperature correspond?

20.11 Using the values of the cosmological parameters given in Table 20.2, calculate the energy densities of dark matter ϵ_{dm}, photons ϵ_γ, and baryonic matter ϵ_b at the time of decoupling, and show that they were in the ratio $\epsilon_{dm} : \epsilon_\gamma : \epsilon_b \sim 5.5 : 1.8 : 1$. Thus the Universe was dominated by dark matter at the time of decoupling. ***

20.12 The smallest resolved fluctuations in the Planck data set exemplified by Fig. 20.14 have an angular size $\Delta\theta \sim 5$ arcmin. Using the parameters from Table 20.2 to define the cosmology, estimate the physical size of this fluctuation on the last scattering surface, and the corresponding observed size today. What is the total baryonic mass contained in this volume today, assuming the angular size to define the diameter of a sphere? Compare with the baryonic mass of a typical cluster of galaxies. *Hint*: Use the cosmological calculator available at Ref. [5] with the data from Table 20.2, and the definition of the angular size distance d_A in Box 20.2. ***

20.13 Assume a Friedmann cosmology corresponding to the parameters in Table 20.2. At the time of last scattering for the CMB the distance to the horizon was approximately 0.25 Mpc. Show that the preferred length scale corresponding to the distance sound waves could travel in the cosmic fluid between the big bang and the time of last scattering (the acoustic scale) was about 0.144 Mpc on the last scattering surface, and that this corresponds to a preferred angular scale of $\sim 0.6°$ as observed today. What is the corresponding length scale today and what is the total mass for matter contained within a sphere of this diameter today? This mass scale is important because it sets a lower limit for the number of galaxies that must be surveyed to find the signature of baryon acoustic oscillations in the present distribution of galaxies. ***

20.14 For a flat, expanding Universe, show that the angular size distance equals the proper distance at the time the light was emitted. *Hint*: See Boxes 19.3 and 20.2. ***

21 Extending Classical Big Bang Theory

The big bang is the standard model for the origin of the Universe and has been for half a century. This place is well earned. At a broader conceptual level, all of modern cosmology rests on observations such as the existence of the cosmic microwave background radiation that make sense only if there was a big bang in our past. At a more nuts and bolts level, the standard big bang model accounts quantitatively for a variety of relatively precise observational data – like those associated with specific properties of the cosmic microwave background and with the observed abundances of light elements – that would be difficult to explain with any competing theory. However, there are some aspects of modern cosmology suggesting that the classical big bang is essentially correct as far as it goes, but perhaps is an incomplete picture of the origin and evolution of our Universe. Much of this incompleteness stems from initial conditions that enter the big bang theory as unexplained phenomenology, and from the fundamental puzzle of how the baryons of the present Universe came to be. This chapter addresses some of these issues.

21.1 Successes of the Big Bang Theory

We begin by being more precise about the successes of the standard big bang and its place in the modern picture of cosmology. As has already been discussed in Chapter 16, the standard cosmology rests on a relatively few observations and concepts:

1. The redshifts of distant galaxies imply an expanding Universe described at the simplest level by the Hubble law.
2. On sufficiently large scales (beyond that of superclusters of galaxies) the Universe appears to be both homogeneous and isotropic (cosmological principle).
3. Observations indicate that quasars once were more energetic and more closely spaced than they are today, implying that the properties of the Universe on large scales have evolved with time.
4. The cosmic microwave background (CMB) is all pervading and highly isotropic, but with measurable fluctuations at the one part in 10^5 level, and possesses an almost perfect blackbody spectrum with a temperature of 2.725 K.
5. The elemental composition of the Universe by mass is three parts hydrogen to one part helium, with only small amounts of heavier elements.
6. Because gravitation and not electromagnetism governs the large-scale structure of the Universe, it must be charge-neutral on large scales. Conversely, observations indicate

that the Universe is highly asymmetric with respect to matter and antimatter, with no evidence for significant equilibrium concentrations of antimatter.

7. There are many fewer baryons in the Universe than there are photons.
8. In contrast to the homogeneity of the Universe on very large scales (cosmological principle), matter on scales comparable to superclusters of galaxies and smaller exhibits complex and highly evolved structure, which contrasts sharply with the smoothness of the CMB. Furthermore, observations indicate that this structure began to develop very early in the history of the Universe.
9. Detailed analysis of fluctuations in the CMB and of the observed brightness of distant Type Ia supernovae indicate that:

    (a) The expansion of the Universe is presently accelerating.
    (b) The geometry of the Universe is remarkably flat (euclidean).

10. The bulk of the matter in the Universe is not luminous (dark matter), and is observable only through its gravitational influence. Various observational constraints imply that most of this dark matter is not baryonic.

Observations (1)–(5) provide strong support for the big bang model: they indicate an expanding, isotropic universe that has evolved over time from a beginning in a very hot, very dense initial state. The remaining observations are not necessarily inconsistent with the classical big bang theory but they require either ad hoc imposition of particular initial conditions on the Universe, or assumption of specific microscopic properties for the matter and energy that the Universe contains. Furthermore, we shall see that while observation (4) is one of the greatest triumphs of the theory, it raises a potentially serious problem.

21.2 Problems with the Big Bang

As indicated in the previous section, observational properties (4) and (6)–(10) raise issues that may constitute problems for the classical big bang model. They do not invalidate the basic idea of the big bang but they indicate that the big bang in its minimal form may be incomplete. In most cases it will be found that this incompleteness is likely to originate in an inadequate understanding of how the particle and field content of the Universe couples to its evolution. Specifically, the following problems may be identified.

21.2.1 The Horizon Problem

Observational property (4) represents a potential conflict with causality because the nearly equivalent temperature of the CMB between widely separated parts of the sky is understandable only if those regions were in past causal contact. But in the standard big bang it is easy to show that regions on the sky separated by more than a degree or two in angle could never have been in causal contact (never had sufficient time to exchange light signals) since the big bang (see Problem 21.3). That is, they lie outside each other's horizons and thus cannot have been in causal contact in the past, as illustrated in Fig. 21.1.

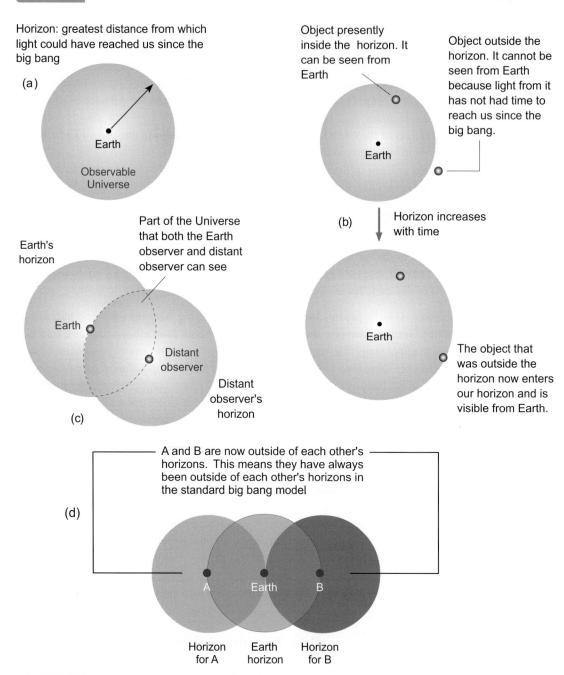

Fig. 21.1 Horizons in an expanding universe. (a) A (cosmological) horizon is the greatest distance from which light could have reached us since the beginning of time. (b) Horizons expand with time, so objects currently outside our horizon may come within our horizon in the future. (c) Cosmological horizons are defined relative to each observer, so each has her own horizon. (d) Horizon problem produced by the CMB having identical temperatures on opposite sides of the sky for an observer: how do A and B know to have the same temperature if they could never have exchanged signals in the past?

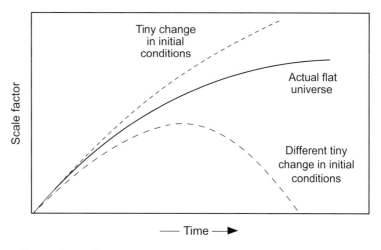

Fig. 21.2 The flatness problem: producing a flat universe today requires remarkable fine-tuning of the initial curvature for the Universe, as implied by Eq. (21.1).

21.2.2 The Flatness Problem

Observational property (9b) implies that the geometry of the Universe is nearly euclidean on large scales, which indicates that the total mass–energy density of the Universe is very near the closure density. While consistent with big bang evolution, this condition can be realized only if parameters are very finely tuned in the early Universe, as illustrated in Fig. 21.2. For example, you are asked to show in Problem 21.2 that the fractional deviation of the density from the critical density at any time in the evolution of the Universe is

$$\frac{\Delta\rho}{\rho} = \frac{\rho - \rho_c}{\rho} = \frac{3kc^2}{8\pi G a^2 \rho}. \tag{21.1}$$

From this result, unless the flatness is tuned to one part in 10^{60} at the Planck time, the Universe ends up in a state that is far from flat today. While possible, this does not seem to be a very natural initial condition.

21.2.3 The Magnetic Monopole Problem

From the theory of elementary particle physics there are reasons to believe that massive particles called magnetic monopoles and others of their ilk could have been produced copiously at phase transitions in the early Universe, as illustrated in Fig. 21.3 and Box 21.1. This has two adverse consequences: (a) Such particles have never been detected in any laboratory experiments. (b) If such particles were produced in high number in the early Universe they would have caused a much earlier transition to a matter-dominated Universe, and this would have modified the evolution of the Universe and negated various successful predictions of the hot big bang model. Of course, it is possible that this is not a real problem since production of massive particles in GUTS phase transitions is a purely theoretical idea at present.

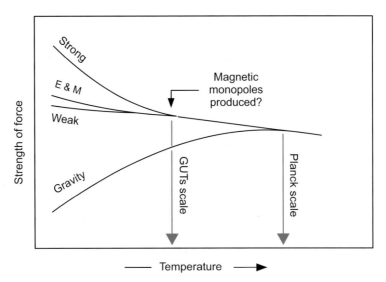

Fig. 21.3 The magnetic monopole problem of the standard big bang: where are the massive relic particles that would be expected to be produced at phase transitions like grand unification (labeled GUTs)?

Box 21.1 — **Topological Defects and Magnetic Monopoles**

Magnetic monopoles are expected to form on the GUT scale because of topological defects characteristic of mismatches between domains undergoing a phase transition (such as the GUT one) that leads to lower symmetry. For example, when water freezes the crystals forming around any one nucleation center have high symmetry but the symmetry axes of adjacent nucleations might not align and the resulting topological defect is 2-dimensional and is called a *domain wall*. It is also possible to have topological defects in other numbers of dimensions: 1-dimensional topological defects are termed *strings* and 0-dimensional topological defects are termed *point defects*. Magnetic monopoles are expected to be point topological defects caused by mismatches between fundamental fields in different regions of space. This is investigated further in Problem 21.4.

21.2.4 The Structure and Smoothness Dichotomy

Observational properties (4) and (8) present a compatibility issue because the remarkably high smoothness of the CMB implies that the early Universe was largely devoid of density perturbations. Where then did the density perturbations responsible for the growth of rich structure in the present Universe on the supercluster and smaller scale originate? The form of the density perturbations required to give the observed structure is known so they could be postulated as initial conditions, but again one would like to know *why*.

21.2.5 The Vacuum Energy Problem

Observation (9a) has a simple explanation if the Universe contains dark energy, which would be most naturally explained if dark energy is a consequence of quantum vacuum fluctuations. However, theoretical estimates of the vacuum energy content of the Universe are spectacularly wrong in comparison with the corresponding observational constraints (see Section 26.2). The accelerated expansion is consistent with the big bang picture if dark energy in the required amount is simply postulated, but it is highly unsatisfying to have no understanding of where this fundamental influence on the evolution of the Universe originates.

21.2.6 The Matter–Antimatter Problem

Observation (6) indicates that the physical Universe contains almost entirely matter with little corresponding antimatter. This might be imposed as an initial condition but that is bothersome, given that matter and antimatter enter modern elementary particle physics on an equal footing and so might be expected to play comparable roles in the structure of the Universe. A closely related problem (because annihilation of baryons with antibaryons produces photons) is Observation 7: how to account for the large excess of photons over baryons in the Universe?

21.2.7 Modifying the Classical Big Bang

In attempting to resolve the problems discussed above it is important to preserve the many successes of the big bang model. A useful observation in that regard is that the successful predictions of the big bang model rest primarily on the evolution of the Universe at times later than about one second after the initiation of the big bang (for example, big bang nucleosynthesis when the universe was several minutes old). Therefore, any modification of our cosmological model that influences the Universe at times earlier than about one second after the big bang will leave the successes of the big bang intact so long as they leave appropriate initial conditions for the subsequent evolution. We now consider a proposed modification of the evolution of the Universe operative only in the first tiny fraction of a second of the big bang that has the potential to resolve the first four problems listed in Section 21.2 in a single stroke.

21.3 Cosmic Inflation

The theory of *cosmic inflation* is based on a simple but dramatic idea that has been discussed already in conjunction with the de Sitter solution found in Section 19.3.1: in the presence of an energy density that is constant over all space, the Einstein equation admits an *exponentially growing solution.* If the Universe underwent a short burst of exponential growth before settling down into more normal big bang evolution, there would be potentially large implications for the subsequent evolution of the Universe. We will take the essential point of inflation to be this general idea of the Universe experiencing an early period of exponential growth. There are many specific versions of inflationary theory that implement this in different ways. For the most part those specifics will be left for the interested reader to pursue in the specialist literature. Our reason is that there is now compelling evidence that the basic idea of inflation is necessary to explain the evolution of the early Universe, but no specific version of inflation currently available gives a completely satisfactory accounting of the cause and detailed effects of the inflationary period.

21.3.1 The Basic Idea and Generic Consequences

From the discussion in Section 19.3.1, a universe with pure vacuum energy expands exponentially, $a(t) = e^{Ht}$, where the Hubble parameter H is constant. The basic idea of inflationary theory is that shortly after its birth the Universe found itself in a situation dominated by a constant (or nearly constant) energy density, which drove an exponential expansion for a very short period of time that cooled the Universe rapidly because of the expansion. Then at the end of this period the Universe exited from the inflationary conditions and reheated (with the mechanism for reheating depending on the version of inflation assumed, but generally involving the rapid conversion of the constant energy density driving inflation into the mass–energy of particles and antiparticles). This then produced a situation dominated by radiation rather than vacuum energy, which caused the Universe to begin evolving according to a standard (hot) big bang scenario with a power-law rather than exponential dependence of the scale factor on time.

In various versions of inflation different reasons are assumed for the initial conditions that triggered the exponential expansion. The original inflationary idea due to Alan Guth and others assumed that inflation was driven by a Lorentz-scalar field associated with a first-order phase transition. This is conceptually simple, but proved to be incompatible with observations (as Guth himself realized) because it was found that the resulting inflation could not halt in a manner that would give something that looks like the real Universe. In subsequent versions of inflation the inflation was often assumed to be driven by a scalar field having a time dependence of a particular form called a *slow rollover transition.* Such theories often give a reasonably good account of data but they suffer from having little connection to scalar fields known already to exist in elementary particle physics, and require extremely fine tuning of parameters to account well for data. In keeping with the philosophy outlined above, discussion of these different forms of inflation will be omitted and we will instead concentrate on the generic consequences of inflationary expansion.

Figure 21.4 illustrates in highly schematic fashion the behavior of the scale factor and temperature during inflation and in the following big bang evolution. During inflation the Universe expanded at a much higher rate than in normal big bang evolution. At the same time, the temperature dropped rapidly in the exponentially expanding Universe. Finally, when the period of inflation halted the Universe first reheated rapidly and then began to decrease in temperature according to the standard scenario already outlined in the radiation-dominated big bang. The question marks represent our substantial lack of knowledge concerning the Universe prior to the inflationary period.

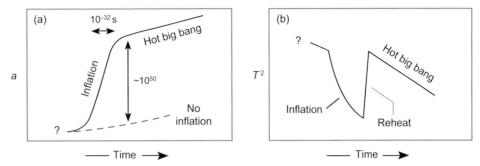

Fig. 21.4 Cosmic inflation. (a) In the inflationary epoch (see Fig. 20.4) the Universe expands exponentially, which can increase the scale factor by 10^{50} or more on a timescale of order 10^{-32} s. (b) The Universe cools as it expands exponentially. At the end of inflation some mechanism – not yet well understood – must reheat the Universe, which then continues a standard hot big bang evolution with an expansion that is a power law rather than exponential in time.

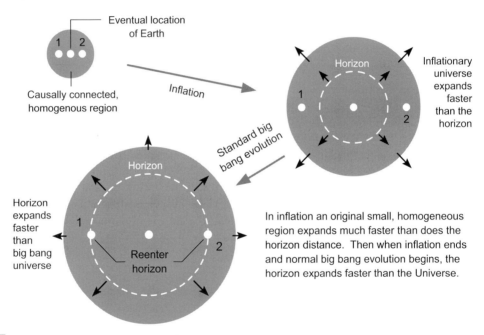

Fig. 21.5 Solution of the horizon problem in the inflationary universe.

21.3.2 Taking the Inflationary Cure

The inflationary hypothesis provides (in principle) a solution to the first four fundamental problems posed above in Section 21.2, as we will now discuss.

Solution of the horizon problem: The solution of the horizon problem is illustrated in Fig. 21.5. The tremendous expansion means that regions observed to be widely separated in the sky now at the horizon were much closer together before inflation, and thus could have been connected to each other by light signals at some time in the past.

Solution of the flatness problem: The tremendous expansion greatly dilutes any initial curvature. Think, for example, of standing on a basketball. It would be obvious that you are standing on a 2-dimensional curved surface. Now imagine expanding the basketball to the size of the Earth. As you stand on it now, it will appear to be flat, even though it is actually curved if you could see it from large enough distance. The same idea extended to 4-dimensional spacetime accounts for the present flatness (lack of curvature) of the Universe. Out to the greatest distances that can be seen the Universe looks flat on large scales, just as the Earth looks approximately flat out to our horizon.

Solution of the monopole problem: The rapid expansion of the Universe tremendously dilutes the concentration of any topological defects that are produced. Simple calculations indicate that they become so rare in any given volume of space that we would be very unlikely to ever encounter one in an experiment designed to search for them. Nor would they have sufficient density to alter the normal radiation-dominated evolution of the Universe following inflation.

Seeds for the formation of structure: One of the most important consequences of inflation is that it suggests an answer to the question that was posed in Section 21.2.4 of how large-scale structure emerged. The inflationary explanation for the origin of structure is in fact rather remarkable. During inflation quantum fluctuations such as that illustrated in Fig. 21.6, which must be present because of the uncertainty principle, end up being stretched from microscopic to macroscopic dimensions by exponential expansion of the space in which they occur. Because this process occurs during the entire time of inflation, density fluctuations of macroscopic size are produced that vary over many length scales. This produces a *scale-invariant spectrum of density fluctuations,* and it is known empirically that this is the spectrum of density perturbations (known in astrophysics as a *Harrison–Zeld'ovich spectrum*) that is most likely to lead to the observed large-scale structure in the Universe. Because of the quantum nature of the fluctuations, they also are *gaussian,* meaning that if they are expanded in a Fourier series or a spherical harmonic expansion (see Section 20.4.2 and Fig. 20.16) the different wavenumbers or multipole

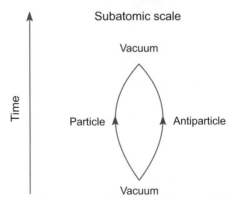

Fig. 21.6 Quantum fluctuations can create a particle–antiparticle pair from the vacuum for a fleeting instant. In inflation such microscopic fluctuations can be stretched to macroscopic scales, producing density perturbations that can later seed the formation of large-scale structure.

orders are not correlated. These density perturbations will generally be expanded beyond the horizon during inflation, but after inflation is over and normal big bang evolution sets in the horizon grows with time (see Fig. 21.1) and eventually these density perturbations come back within the horizon where they can serve as nucleation centers for the formation of structure. Thus inflation ascribes formation of the largest-scale structures in the Universe to quantum fluctuations!

21.3.3 Inflation Doesn't Replace the Big Bang

The theory of inflation does not compete with the big bang. It modifies only the first tiny instants of creation. After the brief period of inflation is complete, hot big bang evolution is assumed to proceed as described earlier. Thus, inflation should be viewed as a modified form of the big bang that accounts for effects due to the properties of elementary particles on the initial conditions that are not included in the standard big bang.

Although there are many specific theories of inflation, each with assumptions of varying plausibility, it is encouraging that simulations of large-scale structure give reasonable results when the effects of inflation are included. Furthermore, the fluctuations in the CMB are described best by theories that include the effect of inflation. For example, Fig. 21.7 compares the angular fluctuations in temperature for the CMB with various models with and without inflation, and with and without dark energy. Models with both dark energy and inflation are clearly favored, while standard models without dark energy and inflation appear to be ruled out by these data.

21.4 The Origin of the Baryons

In the standard big bang model the preponderance of matter over antimatter and of photons over baryons are introduced through initial conditions without fundamental justification. It would be highly desirable to understand the origin of the baryons from a deeper perspective. This has proved to be a highly elusive goal.

21.4.1 Conditions for a Baryon Asymmetry

The Russian physicist Andrei Sakharov (1921–1989) first enumerated the ingredients required to generate baryon asymmetries within the standard big bang model (see Box 21.2 for the symmetries discussed below) [213]:

1. There must exist elementary particle interactions in the Universe that do not conserve baryon number B.
2. There must exist interactions among elementary particles that violate both charge conjugation symmetry (C) and the product of charge conjugation and parity (P) symmetries, which is denoted by CP.
3. There must have been departures from thermal equilibrium during the evolution of the Universe.

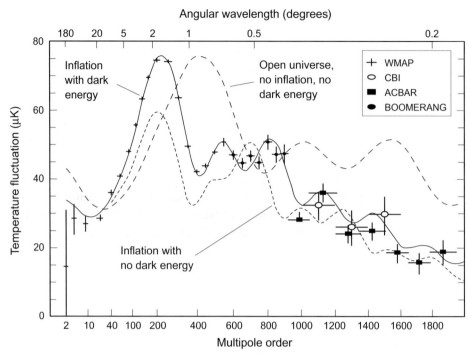

Fig. 21.7 Temperature fluctuations in the cosmic microwave background from satellite and high-altitude and balloon observations compared with various theoretical models [108]. Adapted from Guth and Kaiser, Inflationary Cosmology: Exploring the Universe from the Smallest to the Largest Scales, *Science New Series*, **307**(5711), 884–890. Copyright © 2005, American Association for the Advancement of Science.

Box 21.2   **Baryon Number and CPT Symmetry**

Baryon number B refers to a quantum number that takes the value $+1$ for a baryon and -1 for an antibaryon. Baryon number is then the algebraic sum of these numbers for all the particles in a reaction. Conservation of baryon number (observed in every experiment done so far) means that this number does not change in the interaction. Charge conjugation symmetry (C) is symmetry under exchange of a particle with its antiparticle. Parity symmetry (P) is symmetry under inversion of the spatial coordinate system. CP symmetry is symmetry under inversion of the spatial coordinate system *and* exchange of particle with antiparticle. Most interactions conserve these symmetries to high precision but the weak interactions violate P, C, and CP. It is thought on fundamental grounds that all interactions should conserve the full product CPT of these symmetries and no violation of CPT symmetry has been found.

Departures from thermal equilibrium were likely at times and weak interactions violate both C and CP symmetry, so in principle all ingredients exist to account for baryon asymmetry except for baryon non-conserving reactions. Baryon non-conservation has never been observed experimentally but there are theoretical reasons to believe that baryon conservation might not be exact but is just a symmetry that has not yet been caught in violation. For example baryon non-conservation may occur only at energies not yet reached in laboratory experiments but that could have occurred in the early Universe.

21.4.2 Grand Unified Theories

One class of theories that features baryon-number violating interactions prominently is that of Grand Unified Theories (GUTs), already mentioned in connection with magnetic monopoles. In the Standard Electroweak Theory of elementary particle physics the electromagnetic interactions and the weak interactions have been (partially) unified. This means that at high enough energy (in this case a scale of about 100 GeV) the weak and electromagnetic interactions take on the same properties. A GUT attempts to extend this idea to unify weak, electromagnetic, and strong interactions into a single theory. The characteristic GUT energy scale is very high (10^{14-15} GeV is a common estimate), but on that scale GUTs typically violate baryon number strongly. At one time GUTs were favored as the likely way to account for baryogenesis but there have since been shown to be difficulties with this approach. It is now thought that a viable theory of baryon asymmetry requires a baryon-violating phase transition at a lower energy scale than the GUT scale.

21.4.3 Leptogenesis

A possible alternative to GUTS for generating a baryon asymmetry is *leptogenesis*, which postulates that perhaps the baryon asymmetry was generated by strongly CP-violating processes in the neutrino sector. But it is not clear that this can account for the observed baryon asymmetry of the Universe because the known electroweak interactions do not exhibit interactions with the required characteristics. However, the correct electroweak theory could be more general than the present one (which is not tested exhaustively at energies of 100 GeV or larger, and is now known to be contradicted by the small but finite masses observed for neutrinos), so an improved electroweak theory eventually might be able to account for the baryon asymmetry.

Background and Further Reading

The inception of the theory of cosmic inflation may be followed in Refs. [25, 38, 106, 109, 146]. Very readable introductions to the ideas underlying inflation may be found in books and overview articles by its principal discoverer, Alan Guth [107, 108]. Liddle [145], Roos [207], and Ryden [211] give simple overviews of the basic physics of inflation.

Problems

21.1 For energy density ϵ, pressure P, and cosmic scale factor a, show that $\epsilon + 3P < 0$ is a necessary condition for $\dot{a}$ to increase with time. Hence show that if $\epsilon + 3P > 0$, an object currently outside our horizon has always been outside our horizon. *Hint*: Approximate the horizon as the inverse of the Hubble parameter and compare this with $\dot{a}$.

21.2 One problem of the standard big bang model that the theory of inflation was introduced to cure is the *flatness problem*. Illustrate this by proving that the fractional deviation of the density from the critical density at any time in the evolution of the Universe is given by

$$\frac{\Delta\rho}{\rho} = \frac{\rho - \rho_c}{\rho} = \frac{3kc^2}{8\pi G a^2 \rho},$$

where k is the curvature parameter. Show that this implies (for a Universe dominated by matter or by radiation) that as one goes back in time $\Delta\rho/\rho$ must decrease. For example, show that if $\Delta\rho/\rho$ is of order one today, at the Planck time ($t \sim 10^{-44}$ s) the fractional deviation will be $\Delta\rho/\rho \simeq 10^{-60}$. This is the flatness problem: unless the curvature is tuned to one part in 10^{60} at the Planck time, the Universe evolves to one that is far from flat today, in contradiction to observation. Although this fine tuning does not violate any known laws of physics, it seems incredibly unlikely for an initial condition not set by some principle. ***

21.3 Another problem with the standard big bang that motivated introducing the theory of inflation is the *horizon problem*. Consider the following: at the time of decoupling of the photons from matter (corresponding to $t_d \sim 3 \times 10^5$ yr and $z_d \sim 1100$), how large would the particle horizon have been for an observer? (For purposes of this problem, assume a radiation-dominated Universe.) Now, how large would this horizon size at decoupling have grown to in the present Universe? Finally, estimate the angular size on the sky of this (causally connected) region if it lies at a distance corresponding to the last scattering surface (the imaginary surface on the sky corresponding to the last scattering of a CMB photon from matter in the early Universe). This illustrates the horizon problem, for you should find that the size of causally connected regions is only a degree or two on our sky, but the microwave background is observed to have the same temperature in all parts of the sky to one part in 10^5. It seems incredible that many regions never causally connected in the history of the Universe (in the standard big bang cosmology) would know to have almost exactly the same temperature. ***

21.4 The GUT scale is predicted to be about 10^{15} GeV and to occur at a temperature of $\sim 10^{28}$ K about 10^{-36} s after the big bang. Magnetic monopoles are expected to form on that scale because of topological defects characteristic of mismatches between domains undergoing a phase transition (see Box 21.1). Estimate the number density of such magnetic monopoles at formation, assuming them to have a mass comparable to the GUT scale and for there to be approximately one monopole per causal horizon at formation (since regions outside their respective horizons were not in causal contact, facilitating random mismatches between fields and defect formation at boundaries). Compare to the expected radiation density at the GUT scale. Assuming no inflation, show that at the GUT scale the Universe with magnetic monopoles is still radiation dominated, but that very soon after the GUT transition the massive monopoles would begin to dominate the evolution of the Universe, contrary to evidence that the early Universe was radiation dominated. ***

21.5 In a representative inflationary model the beginning time for inflation is taken to be $t_{GUT} \sim 10^{-36}$ s, the Hubble constant is taken to be $H = t_{GUT}^{-1} \sim 10^{36}$ s^{-1}, and inflation is assumed to continue for ~ 100 Hubble times. What is the factor by which the scale factor expands in inflation for this model? Suppose that the early inflationary period and the present accelerated expansion of the Universe both correspond to de Sitter solutions driven by a vacuum energy. Taking the vacuum energy in each case to be characterized by a different cosmological constant Λ, what is the ratio of vacuum energy densities ϵ_Λ for the present accelerating Universe and for the early-Universe inflationary model assumed above?

PART IV

GRAVITATIONAL WAVE ASTRONOMY

22 Gravitational Waves

Just as Maxwell's theory admits solutions corresponding to propagating waves in the electromagnetic field, general relativity predicts that fluctuations in the metric of spacetime can propagate with the speed of light as *gravitational waves*. Until recently this was a largely theoretical discussion since gravitational waves had been inferred indirectly (see Sections 10.4.2 and 23.2.3), but never observed directly. The absence of direct observation was not because gravitational waves are rare; it was because they interact so weakly with matter, making their detection an enormous technical challenge. This situation changed dramatically when the Laser Interferometry Gravitational-Wave Observatory (LIGO) detected a gravitational wave of duration $\sim$ 200 ms on September 14, 2015 [10]. Detailed analysis of the waveform indicated that the gravitational wave (labeled GW150914) was produced by the merger of two $\sim$ 30 $M_\odot$ black holes.[1] Thus, in a single measurement of unprecedented precision LIGO confirmed the existence of gravitational waves, arguably provided the most direct evidence yet that black holes exist, and opened a fundamental new observational window on the Universe. Not a bad 200 milliseconds of work! In this chapter the Einstein equations will be shown to have propagating solutions corresponding to gravitational waves. In Chapter 23 possible astronomical sources of gravitational waves in the weak-gravity limit will be discussed, and in Chapter 24 gravitational wave sources in the strong-gravity limit and the gravitational waves observed to date will be described.

22.1 Significance of Gravitational Waves

The schematic spectrum of gravitational waves, their possible origin in various astronomical events, and potential methods of detecting them are illustrated in Fig. 22.1. There we see that gravitational waves are expected to be produced by various processes in the Universe ranging in period from milliseconds to timescales comparable to the age of the Universe. Before delving into the theory of gravitational waves and a consideration of their detection in specific events, let's motivate the discussion by summarizing in rather general terms the basic reasons why gravitational waves are thought to be of large scientific importance.

---

[1] GW150914 is a cryptic reference to the date when the gravitational wave (GW) was detected: September 14, 2015.

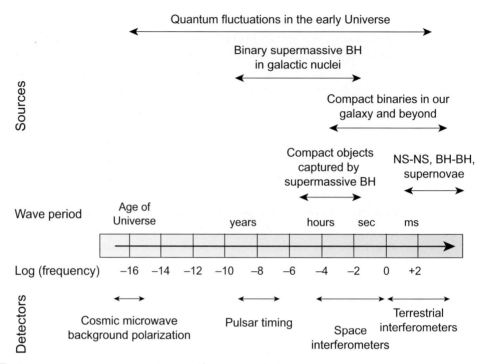

Fig. 22.1 The gravitational wave spectrum with potential sources and methods of detection. Black holes are denoted by BH and neutron stars by NS. Terrestrial interferometers, space-based interferometers, and pulsar timing arrays for detecting gravitational waves will be discussed in Section 22.6. Cosmic microwave background polarization refers to expected signatures imprinted on the CMB by gravitational waves.

22.1.1 Unprecedented Tests of General Relativity

Before the observation of GW150914, gravitational waves represented the last classical prediction of general relativity that had not been tested directly. Although the Binary Pulsar provides strong *indirect evidence* for the emission of gravitational waves from that system, no gravitational wave had been observed directly. General relativity is not only tested by the mere (now confirmed) existence of gravitational waves, but also by their detailed properties. For example, general relativity predicts that gravitational waves can have only two states of polarization and must travel at lightspeed. Some alternative theories of gravity predict gravitational waves with more polarization states and with speeds that can differ from c.

In addition, the detection of GW150914 and its interpretation as a binary black hole merger (see Chapter 24) provides the first test of general relativity in the true *strong gravity, high-velocity, nonlinear limit*.[2] All preceding direct tests: deflection and redshift of light

---

[2] Strong gravity will lead to large gravitational accelerations, so in our discussion high velocities (comparable to c) will be implied in the strong-gravity limit, even if not mentioned explicitly. For example, in the merger of two black holes to produce the gravitational wave GW150914 that is discussed in Section 24.2, the black holes of the binary pair achieved velocities that were a large fraction of c during the inspiral to the merger. Unlike

in gravitational fields, precession of planetary and binary orbits, and so on, correspond to the relatively *weak gravity limit* of general relativity. Furthermore, although black holes themselves are a consequence of strong gravity, most of our compelling but circumstantial evidence for their existence (see Chapter 14) comes from electromagnetic radiation likely emitted from regions of weak gravity far from the event horizons. The strong gravity test is crucial because it is always possible that general relativity is valid for weak gravity (where it has been tested extensively) but breaks down in the untested realm of strong gravity.

The interpretation of GW150914 as the merger of two black holes makes it certain that the detected gravitational waves carried the imprint of very strong gravity near the event horizons of the black holes. Detailed tests of strong gravity will require systematic analysis of gravitational wave events from compact mergers and gravitational collapse, but preliminary analysis of the first detected gravitational wave events suggests no obvious failure of general relativity near the event horizons of black holes.

22.1.2 A Probe of Dark Events

Gravitational waves can signal events having large gravitational energy release but little or no emission of electromagnetic radiation. Indeed, the initial evidence is that GW150914 was an enormously violent event (converting $\sim 200\,M_\odot$ per second into gravitational waves at peak luminosity) for which *no obvious electromagnetic counterpart was observed.* Thus, gravitational waves may provide an alternative – or perhaps the only – probe of some events releasing large gravitational energy.

22.1.3 The Deepest Probe

The weakness of gravitational coupling to matter renders gravitational waves difficult to detect. However, that same weakness means that gravitational waves can be seen that were produced in a much earlier epoch than can be probed directly by observing other forms of radiation, as illustrated in Fig. 22.2. For example, photons decoupled (fell out of equilibrium) several hundred thousand years after the big bang, which implies that direct photons cannot be seen from earlier periods. Neutrinos decoupled at an earlier time because of the weakness of the weak interactions but the weak interactions fell out of equilibrium at about one second after the big bang, so neutrinos can give direct information only back to that time. Gravity becomes strong only on the Planck scale, so gravitational waves emitted after the Planck time could in principle be detectable in the current Universe. For example, it has been proposed that there may exist in the present Universe gravitational waves carrying information about cosmological parameters during the inflationary epoch described in Section 21.3. As indicated in Fig. 22.2, only gravitational wave probes can see that far back in time. However, separating that early-Universe signal from the background produced by other more recent gravitational wave sources represents a large technical challenge.

electromagnetism, general relativity is highly nonlinear, but becomes approximately linear in weak gravity where all previous tests have been carried out. Thus gravitational waves from black hole mergers provide the first good tests of general relativity in the highly nonlinear regime.

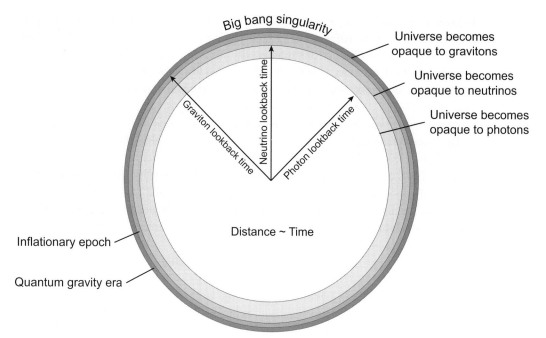

Fig. 22.2 Maximum lookback times (not to scale). We are at the center. Objects observed at greater distance are at earlier times in the evolution of the Universe. The maximum lookback times represent the earliest time since the big bang from which a particular probe could be seen. Generally, more weakly interacting particles can be seen from earlier epochs. Since gravity is the weakest force, gravitational waves can be seen from earlier times than any other signal.

22.1.4 Technology and the Quest for Gravitational Waves

Our primary interest is in the physics of gravitational waves and the astronomy information that they may carry. However, we would be negligent not to at least mention the technology that has been developed to detect gravitational waves. This technology, which is capable of resolving shifts in the distance traveled by light much smaller than the width of the atoms making up the apparatus accomplishing this measurement, is a stunning technical achievement, representing the most sensitive precision measurements that have ever been made. In this context, it is relevant to note that long after gravitational waves were proposed theoretically there was doubt about whether they represented a measurable effect, as discussed in Box 22.1.

22.2 Linearized Gravity

The Einstein equation (8.21) may be written in the equivalent form (see Problem 22.1)

$$R_{\mu\nu} = 8\pi G(T_{\mu\nu} - \tfrac{1}{2}g_{\mu\nu}T^\lambda_\lambda). \tag{22.1}$$

> **Box 22.1** **Existence and Detectability of Gravitational Waves**
>
> The original proposal that the weak-field approximation of general relativity could have gravitational wave solutions was published by Einstein in 1916 and 1918 [79, 80]. However, for many decades there was skepticism that such waves were detectable, at least as a technical matter, and perhaps even as a matter of principle. Einstein himself seems to have had doubts – at least at times – about whether gravitational waves were detectable, or even existed [130, 178].[a]
>
> **Arguments against Gravitational Waves**
>
> Typical (now agreed to be specious) arguments were that a gravitational wave would affect the instrument in precisely the same way as the surrounding space, so no net effect of its passage could be measured, or that the principles of general relativity forbade a gravitational wave to deposit energy in a detector, so a gravitational wave was fundamentally unobservable. This misplaced skepticism led some critics to remark that gravitational waves appeared to travel 'at the speed of thought'.[b]
>
> **The Emergent Observability Consensus**
>
> The tide in the debate began to turn at a crucial conference held in Chapel Hill, North Carolina, in 1957 [45, 215]. There, influential reasoning and simple models presented by several speakers and attendees undermined or refuted the arguments of principle against detection of gravitational waves (they *can* deposit energy in a detector, and the passage through the detector changes proper distances and thus *is measurable*), and began to shift the argument instead to how technically their (challenging) detection could be carried out. This led initially to unsuccessful attempts to measure displacements of large metallic bars caused by passage of a gravitational wave through shifts in the bar's resonant frequencies (positive reports from those measurements were not reproducible and in light of current understanding appear to have been spurious), and eventually to modern laser interferometer systems, which now have been successful in detecting the passage of gravitational waves. The history of this debate, with the emerging agreement by the 1960s that in principle gravitational waves could be detected though the experiment would be very difficult, is summarized in Saulson [215] and commented on by Berti [45].
>
> [a] In 1936 Einstein and Nathan Rosen sent a paper to *Physical Review* concluding that there were no gravitational wave solutions in the full GR field theory. It was returned with a negative referee report. Einstein was incensed, since he typically published without peer review in German journals. He withdrew the paper and submitted it unchanged to the *Journal of the Franklin Institute*. Meanwhile H. P. Robertson (of Robertson–Walker metric fame) gently persuaded Einstein that he and Rosen had made a serious error (poor coordinates, leading to spurious singularities), and that the paper proved the *opposite* of what they claimed. The paper was corrected and published with the conclusion that gravitational waves existed. Einstein never again submitted a paper to *Physical Review*.
> [b] This was an Arthur Eddington quote taken out of context [87, 130]. Eddington had not meant that gravitational waves were spurious but rather that (as Einstein had already discovered and we will investigate in Section 22.2.3) only two states of polarization are physical and the others are coordinate (gauge) effects that can be changed at will by mere thought. Far from being a skeptic, Eddington showed that gravitational waves carry energy, and even corrected a factor of two error in Einstein's 1918 paper.

Because this equation is nonlinear, gravitational waves are themselves a source of spacetime curvature and wave solutions in the general case are difficult to obtain. However, in many instances gravitational waves may be assumed to represent weak perturbations of the spacetime geometry, which permits the metric to be expressed in the form

$$g_{\mu\nu}(x) = \eta_{\mu\nu} + h_{\mu\nu}(x), \tag{22.2}$$

where $\eta_{\mu\nu}$ is the metric of (flat) Minkowski space and $h_{\mu\nu}$ is small. The *linearized vacuum Einstein equation* then results from inserting Eq. (22.2) into the vacuum Einstein equation,

$$R_{\mu\nu} = 0 \qquad (22.3)$$

obtained from Eq. (22.1) by setting the stress–energy tensor $T_{\mu\nu}$ to zero, and expansion of the resulting equations to first order in $h_{\mu\nu}$.

22.2.1 Linearized Curvature Tensor

The Ricci curvature tensor $R_{\mu\nu}$ appearing in Eq. (22.3) is given by Eq. (8.16),

$$R_{\mu\nu} = \Gamma^{\gamma}_{\mu\nu,\gamma} - \Gamma^{\gamma}_{\mu\gamma,\nu} + \Gamma^{\gamma}_{\mu\nu}\Gamma^{\sigma}_{\gamma\sigma} - \Gamma^{\sigma}_{\mu\gamma}\Gamma^{\gamma}_{\nu\sigma}, \qquad (22.4)$$

where the Christoffel symbols $\Gamma^{\gamma}_{\mu\nu}$ are related to the metric tensor $g_{\mu\nu}$ by Eq. (7.30),

$$\Gamma^{\gamma}_{\mu\nu} = \frac{1}{2}g^{\gamma\delta}\left(\frac{\partial g_{\nu\delta}}{\partial x^{\mu}} + \frac{\partial g_{\mu\delta}}{\partial x^{\nu}} - \frac{\partial g_{\mu\nu}}{\partial x^{\delta}}\right). \qquad (22.5)$$

To zeroth order in $h_{\mu\nu}$ the Christoffel coefficients vanish and so does $R_{\mu\nu}$, since $\partial g/\partial x = 0$ for $g \to \eta_{\mu\nu}$. To first order in $h_{\mu\nu}$,

$$\delta\Gamma^{\gamma}_{\mu\nu} = \frac{1}{2}\eta^{\gamma\delta}\left(\frac{\partial h_{\nu\delta}}{\partial x^{\mu}} + \frac{\partial h_{\mu\delta}}{\partial x^{\nu}} - \frac{\partial h_{\mu\nu}}{\partial x^{\delta}}\right). \qquad (22.6)$$

The last two terms in Eq. (22.4) are quadratic in h and may be discarded, giving

$$\delta R_{\mu\nu} = \frac{\partial(\delta\Gamma^{\lambda}_{\mu\nu})}{\partial x^{\lambda}} - \frac{\partial(\delta\Gamma^{\lambda}_{\mu\lambda})}{\partial x^{\nu}} + \mathcal{O}\left(h^2\right), \qquad (22.7)$$

to first order in h.

22.2.2 Wave Equation

Substitution of (22.6) in (22.7) and some algebra yields (Problem 22.3)

$$\delta R_{\mu\nu} = \tfrac{1}{2}(-\Box h_{\mu\nu} + \partial_{\mu}V_{\nu} + \partial_{\nu}V_{\mu}), \qquad (22.8)$$

where the 4-dimensional Laplacian (d'Alembertian operator) is defined by

$$\Box \equiv \eta^{\mu\nu}\partial_{\mu}\partial_{\nu} = -\frac{\partial^2}{\partial t^2} + \nabla^2, \qquad (22.9)$$

with $\partial_{\mu} \equiv \partial/\partial x^{\mu}$, and the V_{ν} are defined through

$$\begin{aligned}V_{\nu} &\equiv \partial_{\lambda}h^{\lambda}_{\nu} - \tfrac{1}{2}\partial_{\nu}h^{\lambda}_{\lambda} \\ &= \partial_{\lambda}\eta^{\lambda\delta}h_{\delta\nu} - \tfrac{1}{2}\partial_{\nu}\eta^{\lambda\delta}h_{\delta\lambda}.\end{aligned} \qquad (22.10)$$

Note the general rule that raising and lowering of indices in linearized gravity is accomplished through contraction with the flat-space metric tensor $\eta_{\mu\nu}$ (see Problem 22.2),

$$h^\lambda_\nu \equiv \eta^{\lambda\delta} h_{\delta\nu}. \tag{22.11}$$

Thus the vacuum Einstein equation to this order yields the wave equation (Problem 22.3)

$$\Box h_{\mu\nu} - \partial_\mu V_\nu - \partial_\nu V_\mu = 0. \tag{22.12}$$

Since $h_{\mu\nu}$ is symmetric under exchange of indices, Eq. (22.12) constitutes a set of 10 *linear* partial differential equations for the metric perturbation $h_{\mu\nu}$. These are termed the *linearized vacuum Einstein equations* and the resulting theory is termed *linearized gravity*. As is clear from the derivation, this is expected to be a valid approximation of the full gravitational theory when the metric departs only slightly from that of flat spacetime.

22.2.3 Coordinates and Gauge Transformations

Equation (22.12) cannot yield unique solutions in its present form because of the freedom to make coordinate transformations: given one solution (metric), another equivalent solution may be generated by a coordinate transformation, since physics cannot depend on choice of coordinate system. As will now be discussed, this ambiguity is related to a similar ambiguity in electromagnetism that is associated with freedom to make gauge transformations without altering the electric and magnetic fields.[3]

If $h_{\mu\nu}$ solves the linearized Einstein equations, so does $h'_{\mu\nu}$ where $h'_{\mu\nu}$ is related to $h_{\mu\nu}$ through a general coordinate transformation $x \to x'$. Such a transformation involves four arbitrary functions $x'^\mu(x)$. This is analogous to the gauge ambiguity in electromagnetism, which is removed by fixing (choosing) a gauge such as the Lorentz gauge or the Coulomb gauge discussed in Section 4.7.3. Here the ambiguity can be removed by fixing the *coordinate system*. The symmetric tensor $h_{\mu\nu}$ has 10 components, while gravitational waves have only two independent degrees of freedom. However, a judicious choice of coordinate system yields a total of eight constraints that allow unique solutions of Eq. (22.12) with two physical degrees of freedom. The freedom to make gauge transformations is of crucial importance in understanding gravitational waves, so let's examine this in a little more detail.

22.2.4 Choice of Gauge

In electromagnetism it is possible to make different choices of the (vector and scalar) potentials that give the same electric and magnetic fields and thus the same classical observables (see the discussion of the Maxwell equations in Section 4.7). The freedom to make these choices is associated with gauge invariance and, by clever choices of gauge, problems in classical electromagnetism often can be simplified. Something similar is possible in linearized gravity. Small changes can be made in the coordinates that leave $\eta_{\mu\nu}$ unchanged in Eq. (22.2), but that alter the functional form of $h_{\mu\nu}$. Under small changes in the coordinates

$$x^\mu \to x'^\mu = x^\mu + \epsilon^\mu(x), \tag{22.13}$$

---

[3] Readers not familiar with electromagnetic gauge transformations may find it helpful to consult the discussion of the Maxwell equations in Section 4.7 before proceeding.

Table 22.1 Gauge invariance in linearized gravity and electromagnetism

	Linearized gravity	Electromagnetism[†]
Potentials	$h_{\mu\nu}$	Vector potential: $\mathbf{A}(t, \mathbf{x})$ Scalar potential: $\Phi(t, \mathbf{x})$
Fields	Linearized Riemann curvature: $\delta R_{\alpha\beta\delta\gamma}(x)$	Electric field: $\mathbf{E}(t, \mathbf{x})$ Magnetic field: $\mathbf{B}(t, \mathbf{x})$
Gauge transformation	$h_{\mu\nu} \to h_{\mu\nu} - \partial_\mu \epsilon_\nu - \partial_\nu \epsilon_\mu$	$A^\mu \to A^\mu - \partial^\mu \chi$
Example of a gauge condition ("Lorentz")	$\partial_\nu \bar{h}^{\mu\nu} = 0$	$\partial^\mu A_\mu = 0$
Field equations in Lorentz gauge	$\Box \bar{h}_{\mu\nu} = 0$	$\Box A^\mu = 0$

[†]The 4-vector potential is $A^\mu \equiv (\Phi, \mathbf{A})$ and χ is an arbitrary scalar function.

where $\epsilon^\mu(x)$ is similar in size to $h_{\mu\nu}$, the metric is changed

$$g_{\mu\nu}(x) = \eta_{\mu\nu} + h_{\mu\nu}(x) \to \eta_{\mu\nu} + h'_{\mu\nu}(x)$$
$$= \eta_{\mu\nu} + h_{\mu\nu}(x) - \partial_\mu \epsilon_\nu - \partial_\nu \epsilon_\mu. \qquad (22.14)$$

The transformation $h_{\mu\nu} \to h'_{\mu\nu}$, with (Problem 22.4)

$$h'_{\mu\nu} = h_{\mu\nu} - \partial_\mu \epsilon_\nu - \partial_\nu \epsilon_\mu, \qquad (22.15)$$

is termed a *gauge transformation,* by analogy with electromagnetism. If $h_{\mu\nu}$ is a solution of Eq. (22.12), so is $h'_{\mu\nu}$. Just as gauge transformations in the Maxwell theory lead to new potentials but the same electric and magnetic fields, the gauge transformations (22.15) lead to new potentials $h_{\mu\nu}$ but to the same fields [the linearized form $\delta R_{\mu\nu\beta\gamma}(x)$ of the Riemann curvature tensor defined in Eq. (8.14)]. The parallels between coordinate ("gauge") transformations in linearized gravity and gauge transformations in classical electromagnetism are summarized in Table 22.1.

A standard gauge choice analogous to choosing Lorentz gauge in electromagnetism[4] permits the linearized vacuum gravitational equations to be replaced by the two equations

$$\Box \bar{h}_{\mu\nu}(x) = \left(-\frac{\partial^2}{\partial t^2} + \nabla^2\right) \bar{h}_{\mu\nu} = 0, \qquad (22.16)$$

$$\partial_\nu \bar{h}^{\mu\nu}(x) = 0, \qquad (22.17)$$

where (22.16) is a wave equation corresponding to the linearized Einstein equation, (22.17) is a (Lorentz) gauge constraint, and the *trace-reversed amplitude* is defined by

$$\bar{h}_{\mu\nu} \equiv h_{\mu\nu} - \tfrac{1}{2}\eta_{\mu\nu}h, \qquad (22.18)$$

---

[4] In electromagnetism Lorentz gauge is chosen by requiring that $\partial_\mu A^\mu = 0$, where A^μ is the 4-vector potential. The constraint (22.17) is of similar form. In linearized gravity this gauge is also called *de Donder gauge* or *harmonic gauge.*

where $h \equiv h_\alpha^\alpha$ is the trace (sum of diagonal elements). As in electromagnetism, the "Lorentz" gauge is really a family of gauges. This will be used to simplify things further below. It may be noted in passing that the preceding formalism treats weak gravity effectively as a field $h_{\mu\nu}$ defined in cartesian inertial coordinates on a fixed *flat Minkowski spacetime* background that is a rank-2 tensor with respect to Lorentz transformations (see Problem 22.2), but not with respect to more general spacetime transformations. This allows weak gravity to be described in mathematical terms similar to those for modern quantum field theories of the electromagnetic, weak, and strong interactions, though none of the latter has a geometrical interpretation in spacetime.

22.3 Weak Gravitational Waves

Solutions of Eqs. (22.16) and (22.17) are expected to be a superposition of plane-wave components in the form[5]

$$\bar{h}_{\mu\nu}(x) = \alpha_{\mu\nu} e^{ik \cdot x} = \alpha_{\mu\nu} e^{ik_\alpha x^\alpha}, \qquad (22.19)$$

where the *polarization tensor* $\alpha_{\mu\nu}$ may be represented by a constant, symmetric, 4×4 matrix and k is the 4-wavevector, $k^\mu = (\omega/c, \mathbf{k})$, where ω is the wave angular frequency. Since it is symmetric, the most general $\alpha_{\mu\nu}$ would have 10 independent components. However, the requirement that Eq. (22.19) satisfy the Lorentz gauge condition (22.17) implies that $ik^\mu \alpha_{\mu\nu} \exp(ikx) = 0$, which is true only if

$$k^\mu \alpha_{\mu\nu} = 0. \qquad (22.20)$$

The four equations (22.20) reduce the number of independent components of $\alpha_{\mu\nu}$ to six. But the gauge (coordinate) degrees of freedom have not yet been exhausted because any coordinate transformation that leaves Eq. (22.17) valid does not alter the physical content of linearized gravity. This may be used to set four linear combinations of the $\bar{h}_{\mu\nu}$ to zero.

22.3.1 Polarization Tensor in TT Gauge

The remaining gauge freedom alluded to at the end of the preceding section may be used to transform to *transverse traceless (TT) gauge* by choosing the four conditions[6]

$$\bar{h}_{0i} = 0 \quad (i = 1, 2, 3) \qquad \mathrm{Tr}\,\bar{h} \equiv \bar{h}_\mu^\mu = 0. \qquad (22.21)$$

In terms of the polarization tensor $\alpha_{\mu\nu}$, the conditions (22.21) correspond to

$$\alpha_{0i} = 0 \qquad \mathrm{Tr}\,\alpha = \alpha_\mu^\mu = 0. \qquad (22.22)$$

---

[5] The right side of Eq. (22.19) is complex and $\bar{h}_{\mu\nu}$ is real, so at the end of the day one should take the real part of the right side.
[6] This is still Lorentz gauge, but now restricted to a particular subset of Lorentz gauges specified by the conditions (22.21). Because of the tracelessness condition, in TT gauge $\bar{h}_{\mu\nu} = h_{\mu\nu}$ and the bar may be dropped on $h_{\mu\nu}$ as long as TT gauge is assumed.

The gauge conditions (22.20) with $\nu = 0$ and $\alpha_{0i} = 0$ then require that $\alpha_{00} = 0$, so four of the $\alpha_{\mu\nu}$ vanish:

$$\alpha_{0\mu} = 0. \tag{22.23}$$

Furthermore, for the spatial components the gauge constraint (22.20) requires the *transversality condition*,

$$k^j \alpha_{ij} = 0. \tag{22.24}$$

Let us take stock of the polarization tensor in the transverse–traceless gauge. Starting with 10 independent components of the symmetric polarization tensor $\alpha_{\mu\nu}$, it was found that:

1. The condition (22.21) that $h_{0i} = 0$ requires that $\alpha_{01} = \alpha_{02} = \alpha_{03} = 0$, and the requirement that Eq. (22.17) be satisfied yields $\alpha_{00} = 0$. Therefore, the four components $\alpha_{0\mu}$ of the symmetric polarization tensor vanish in TT gauge.
2. The trace condition (22.22) implies one constraint and the transversality condition (22.24) with $i = 1, 2, 3$ supplies three additional ones, which gives a total of four constraints.

Therefore, ten $\alpha_{\mu\nu}$, minus four $\alpha_{\mu\nu}$ that are identically zero, minus four constraints on $\alpha_{\mu\nu}$ leaves *two independent physical polarizations* for gravitational waves.

There are many possibilities for choice of gauge so one might well ask "why TT gauge"? The primary answer is that it is convenient to work in TT gauge because the TT gauge conditions fix completely all gauge degrees of freedom. Therefore the metric perturbation $h_{\mu\nu}$ in TT gauge contains only physical (non-gauge) information about the gravitational radiation. Transverse–traceless gauge also emphasizes concisely that gravitational waves (like electromagnetic waves) are transverse with two physical polarizations, as will be elaborated below.

22.3.2 Helicity Components

Further insight into the states of polarization allowed for gravitational waves comes from asking how the $\alpha_{\mu\nu}$ change under rotations of the coordinate system. Helicity is the projection of the spin on the direction of motion. As discussed in more depth by Weinberg [246], a gravitational plane wave can be decomposed into helicity components ± 2, ± 1, and 0, but the components with helicity 0 and ± 1 vanish with a suitable choice of coordinates and only the two helicity components ± 2 are physically relevant. It is in this sense that gravity may be associated with a spin-2 *massless* field. Analogies between states of polarization for gravitational waves and in electromagnetism are discussed further in Box 22.2.

22.3.3 General Solution in TT Gauge

To display the polarizations explicitly, assume the gravitational wave to propagate along the z axis with frequency ω. The requirement that Eq. (22.19) satisfy the wave equation (22.16) implies that (Problem 22.8)

22.3 Weak Gravitational Waves

> **Box 22.2** **Polarization States for Electromagnetic Radiation**
>
> It is instructive to compare polarization states in electromagnetism with those of gravitational waves. As discussed in Section 4.7, electromagnetism is described by a 4-vector field A^μ, suggesting that there should be four independent states of polarization α_μ. However, the requirement that $k_\mu \alpha^\mu = 0$ reduces this to three and the freedom to make gauge transformations that leave the electric field E and the magnetic field $\boldsymbol{B}$ unchanged demonstrates explicitly that the number of independent polarizations is in reality only two. Furthermore, a decomposition under rotations about the z axis yields helicities 0 and ± 1, but only the helicities ± 1 are physically relevant for electromagnetic radiation. These correspond to the two independent states of polarization for a massless vector (spin-1) field.
>
> That there are only two physical states of polarization for the photon is a direct consequence of its masslessness and associated local gauge invariance. A *massive* vector field has three states of polarization and is not locally gauge invariant. Likewise, the gravitational field exhibits only two physical states of polarization because the spin-2 *graviton* thought to mediate the gravitational interaction is massless. A massive spin-2 field would have additional physical polarization states.

$$k_\mu k^\mu = 0, \tag{22.25}$$

so the wavevector for the gravitational wave is a *null vector* with explicit components

$$k^\mu = (\omega, 0, 0, \omega) \tag{22.26}$$

in $c = 1$ units. Since the wavevector is a null vector, the wave propagates at the speed of light.

Example 22.1 Substituting Eq. (22.26) with c restored into Eq. (22.25) gives

$$k^\mu k_\mu = -\omega^2/c^2 + \boldsymbol{k}^2 = 0,$$

from which $\omega/|\boldsymbol{k}| = c$. Thus the group and phase velocities are equal to c, exactly as for electromagnetic waves. The speed of gravity is the speed of light.

As you are asked to show in Problem 22.6, the transversality condition (22.24), the null wavevector Eq. (22.26), Eq. (22.23), and the trace condition from Eq. (22.22) imply that the only non-vanishing components of $\alpha_{\mu\nu}$ are α_{11}, $\alpha_{12} = \alpha_{21}$, and $\alpha_{22} = -\alpha_{11}$. Finally, from (22.26), $ik \cdot x = -i(\omega t - \omega z)$ and the general solution of the linearized Einstein equations for z-axis propagation with fixed frequency ω in transverse–traceless gauge is

$$h_{\mu\nu}(t,z) = \begin{pmatrix} 0 & 0 & 0 & 0 \\ 0 & \alpha_{11} & \alpha_{12} & 0 \\ 0 & \alpha_{12} & -\alpha_{11} & 0 \\ 0 & 0 & 0 & 0 \end{pmatrix} e^{i\omega(z-t)}, \tag{22.27}$$

which exhibits explicitly the transverse and traceless properties, with two independent polarization states. (The bar on $h_{\mu\nu}$ has been dropped since $\bar{h}_{\mu\nu} = h_{\mu\nu}$ in TT gauge.) The part of the wave that is proportional to $\alpha_{xx} = \alpha_{11}$ is called the *plus polarization* (denoted by $+$) and the part proportional to $\alpha_{xy} = \alpha_{12} = \alpha_{21}$ is called the *cross polarization* (denoted by ×). For example, a purely cross-polarized plane wave propagating in the z direction may be represented as

$$h^{\times}_{\mu\nu}(t,z) = \begin{pmatrix} 0 & 0 & 0 & 0 \\ 0 & 0 & \alpha_{12} & 0 \\ 0 & \alpha_{12} & 0 & 0 \\ 0 & 0 & 0 & 0 \end{pmatrix} e^{i\omega(z-t)}. \tag{22.28}$$

The most general gravitational wave is a superposition of waves having the form (22.27) with different ω, directions of propagation, and amplitudes for the two polarizations. In linearized approximation and TT gauge, a gravitational wave propagating in the z direction may be expressed generally as

$$h_{\mu\nu}(t,z) = \begin{pmatrix} 0 & 0 & 0 & 0 \\ 0 & f_{+}(t-z) & f_{\times}(t-z) & 0 \\ 0 & f_{\times}(t-z) & -f_{+}(t-z) & 0 \\ 0 & 0 & 0 & 0 \end{pmatrix}, \tag{22.29}$$

where $f_{+}(t-z)$ and $f_{\times}(t-z)$ are dimensionless functions that characterize the shape and amplitude of the wave.

22.4 Gravitational versus Electromagnetic Waves

Much of the discussion in this chapter has emphasized similarities between gravitational waves and electromagnetic waves. However, it is important to point out also that there are some fundamental differences that have observational consequences [87].

22.4.1 Interaction with Matter

Electromagnetic waves interact strongly with matter but, as has already been emphasized, gravitational waves interact extremely weakly with matter. This has two consequences: (1) Gravitational waves are much harder to detect than electromagnetic waves. (2) Gravitational waves are not significantly absorbed by intervening matter.

22.4.2 Wavelength Relative to Source Size

Electromagnetic waves can be used to form an image because their wavelength is typically smaller than the size of the emitting system. In contrast, the wavelength of gravitational waves is typically greater than or equal to the source size and they generally cannot be used to form an image. Hence detecting gravitational waves has been likened to hearing sound. Ultimately this is because electromagnetic waves are generated by local moving charges within a larger source, while gravitational waves of detectable strength are generated by bulk motion of the entire source.

22.4.3 Phase Coherence

Photons are generated by independent events involving ions, electrons, or atoms, so they are emitted incoherently from a typical astronomical source. In contrast, gravitons are generated by bulk motion of matter or spacetime curvature, so they are typically phase-coherent when emitted. In this sense, gravitational waves are similar to laser light. This coherence has two important observational consequences [87]. (1) If the waveform is well modeled, matched filtering techniques to be discussed in Section 24.3.2 can extend the distance at which sources can be detected by a factor of roughly the square root of the number of cycles observed for the wave. (2) As a result of coherence, the direct observable for a gravitational wave is the *strain* (see Section 22.5.1), which falls off as $1/r$ for a source at distance r. This may be contrasted with observables for incoherent electromagnetic radiation, which are typically energy fluxes falling off as $1/r^2$. Therefore, every improvement in sensitivity for a gravitational wave observatory by a factor of two increases the volume of space visible to the detectors and hence the number of detectable sources by $\sim 2^3$.

22.4.4 Field of View

Electromagnetic astronomy is typically based on deep images with small fields of view, allowing observers to mine large amounts of information from a small region of the sky. In contrast, gravitational wave observatories see essentially the entire sky but with relatively low resolution compared with that typical of electromagnetic astronomy. The difference has been likened to the angular resolution contrast between hearing and seeing [87].

22.5 The Response of Test Particles

Let us now turn to the issue of how to detect gravitational waves. A gravitational wave cannot be detected locally because in a local region the effects of gravity may be transformed away (equivalence principle). Thus the effect of a gravitational wave on a single point test particle has no measurable consequences. Gravitational waves may be detected only by their influence on two or more test particles at different locations.

22.5.1 Response of Two Test Masses

Assume a linearized gravitational wave of pure plus polarization propagating on the z axis in TT gauge,

$$h_{\mu\nu}(t,z) = \begin{pmatrix} 0 & 0 & 0 & 0 \\ 0 & 1 & 0 & 0 \\ 0 & 0 & -1 & 0 \\ 0 & 0 & 0 & 0 \end{pmatrix} f(t-z), \tag{22.30}$$

which implies a corresponding time-dependent line element

$$ds^2 = -dt^2 + [1 + f(t-z)]dx^2 + [1 - f(t-z)]dy^2 + dz^2. \tag{22.31}$$

Thus the geometry of the gravitational wave spacetime fluctuates with this time dependence, which corresponds to a wave of curvature propagating at lightspeed. Consider two test masses, with mass A initially at rest at the origin, $x_A^i = (0,0,0)$ and mass B initially at rest at a point $x_B^i = (x_B, y_B, z_B)$, with a gravitational wave propagating on the z axis. The particles are at rest before the gravitational wave arrives and the initial 4-velocities of the test masses are

$$u_A = u_B = (1,0,0,0). \tag{22.32}$$

The motion in spacetime for a test mass is given by the geodesic equation (7.21),

$$\frac{d^2 x^i}{d\tau^2} = -\Gamma^i_{\mu\nu} u^\mu u^\nu = -\Gamma^i_{\mu\nu} \frac{dx^\mu}{d\tau} \frac{dx^\nu}{d\tau}.$$

The undisturbed spacetime is assumed to be flat, so that the $\Gamma^i_{\mu\nu}$ vanish. From Eq. (22.6), to first order

$$\frac{d^2(\delta x^i)}{d\tau^2} = -\delta\Gamma^i_{\mu\nu} u^\mu u^\nu = -\delta\Gamma^i_{00}$$

$$= -\frac{1}{2}\eta^{i\delta}\left(\frac{\partial h_{\delta 0}}{\partial x^0} + \frac{\partial h_{\delta 0}}{\partial x^0} - \frac{\partial h_{00}}{\partial x^\delta}\right) = 0, \tag{22.33}$$

where Eq. (22.32) and that $h_{\delta 0} = 0$ in TT gauge were used. Thus to this order in TT gauge the *coordinate distance* between A and B is unchanged by the gravitational wave. In essence, the coordinates move with the particle in this gauge (see Problem 22.9). However, coordinate distances have no direct physical meaning and the gravitational wave does affect the *proper (physical) distance* between the test masses. For example, assume A and B to lie on the x axis and to be separated by a distance L_0 before passage of the gravitational wave. From Eq. (22.31), the relevant line element is then

$$ds^2 = -dt^2 + (1 + f(t-z))\, dx^2, \tag{22.34}$$

corresponding to the metric

$$g_{\mu\nu} = \begin{pmatrix} -1 & 0 \\ 0 & 1 + h_{11}(t,0) \end{pmatrix}, \tag{22.35}$$

where $z = 0$ was chosen at time t. Then in the spacetime perturbed by the linearized gravitational wave the proper distance between A and B is

$$L(t) = \int_0^{L_0} (-\det g)^{1/2}\, dx = \int_0^{L_0} (1 + h_{11}(t,0))^{1/2}\, dx$$

$$\simeq \int_0^{L_0} \left(1 + \tfrac{1}{2} h_{11}(t,0)\right) dx = \left(1 + \tfrac{1}{2} h_{11}(t,0)\right) L_0 \tag{22.36}$$

and the fractional change in the distance that light travels between A and B is

22.5 The Response of Test Particles

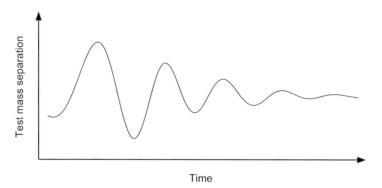

Fig. 22.3 Physical separation of two test masses perturbed by a gravitational wave.

$$\frac{\delta L(t)}{L_0} \simeq \tfrac{1}{2} h_{11}(t,0), \tag{22.37}$$

which oscillates with the time dependence of the gravitational wave given by $f(t-z)$, as indicated schematically in Fig. 22.3.[7] Although this result was derived in TT gauge for convenience, the gauge invariance (that is, coordinate independence) of general relativity implies that it is a general result valid in all gauges.

If spacetime is viewed as an elastic medium, then from the theory of elastic solids the dimensionless quantity $\delta L/L_0$ in Eq. (22.37) may be termed the (fractional) *strain* associated with the deforming wave. However, this strain is very small! It is expected that $\delta L/L_0 \sim 10^{-21}$ for a typical gravitational wave detectable on Earth from a distant astronomical event. This implies that spacetime is an *extremely stiff medium*, since a large energy is required to create a perceptible distortion of the medium. The interferometers to be described below are designed to measure the strain produced by gravitational waves. This has two important implications for such detectors: (1) The gravitational wave strain produced by astronomical events will be extremely small, which will require measurements of exacting precision. (2) Because the strain is proportional to the amplitude of the wave, the strain measured on Earth for events at distance r will fall off as r^{-1} and not as r^{-2}, as would an energy flux characteristic of most kinds of astronomical observations (see Section 22.4.3).

---

[7] To avoid being overly pedantic it is convenient to speak of the test masses moving in describing the detection of a gravitational wave, but it is actually *space that is rippling*, not that the test masses are moving through fixed space. The reason that the interference pattern is shifted in a valid event (see Section 22.6) is that the gravitational wave causes a time-dependent anisotropic distortion of the spacetime metric along the two interferometer arms, which alters the proper distance that light (always moving at constant local speed c) travels through one arm relative to the other. This distinction is reminiscent of the assertion in Section 16.2.3 that cosmological redshifts are caused by the *expansion of space* rather than by Doppler shifts associated with relative motion of light sources *in space* (which astronomers call *peculiar velocities*). In that spirit, spurious events in gravitational wave physics caused by environmental factors such as a car driving into a parking lot are like peculiar velocities in cosmology. As in cosmology, the "peculiar" aspects corresponding to motion through space must be suppressed in order to see clearly the effects due to the time dependence of the spacetime metric itself (periodic in the case of gravitational waves and a uniform rescaling with time in cosmology).

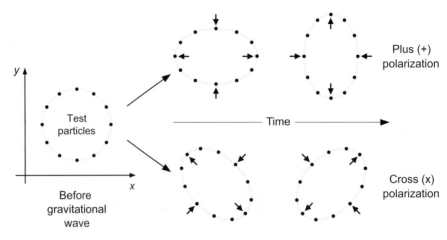

Fig. 22.4 Effect of a gravitational wave incident along the z axis on an array of test masses in the x–y plane for $+$ and $\times$ polarization.

22.5.2 The Effect of Polarization

Gravitational wave polarization may be illustrated by considering the effect of a gravitational wave on an array of test masses arranged in a plane. Like electromagnetic waves, gravitational waves are transverse so only test mass separations in the transverse directions (x and y, if the wave is incident along the z axis) are changed by the gravitational wave. The effect of purely plus-polarized and purely cross-polarized gravitational waves on an initially circular array of test masses is illustrated in Fig. 22.4, with the magnitude of the effect greatly exaggerated. Notice that the gravitational wave acts tidally on the mass distribution, stretching in one direction and squeezing in the other. This is to be expected, since the true gravitational field in general relativity is the tidal field (see Section 6.7.3). Each polarization leads to an elliptical oscillation in the distribution of the test masses, with the cross-polarization ellipse rotated by 45° relative to the corresponding plus-polarization ellipse. The 45° relative orientation is a consequence of gravity being described by a rank-2 tensor field (spin-2 field). In electromagnetism the fields correspond to a rank-1 tensor A^μ (spin-1 or vector field) and the orientation angle between the two independent states of polarization is instead 90°.

22.6 Gravitational Wave Detectors

Modern gravitational wave detectors use laser interferometers with multi-kilometer arm lengths [203]. A typical implementation is illustrated in Fig. 22.5. Light from a laser is split and directed down two arms. Suspended, mirrored test masses reflect the light at the ends of the arms and the reflected light is recombined and interference fringes are analyzed

22.6 Gravitational Wave Detectors

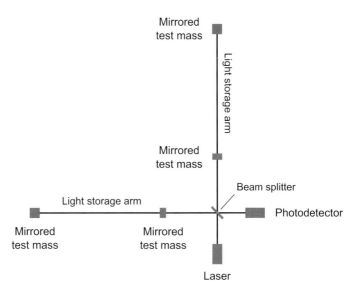

Fig. 22.5 A laser interferometer gravitational wave detector. Light may be multiply reflected in the storage arms, increasing the effective length by as much as a factor of several hundred.

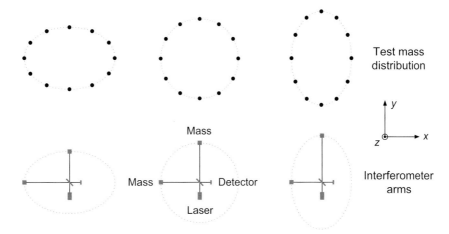

Fig. 22.6 Analogy between interaction of a gravitational wave with a test mass distribution and with an interferometer. The gravitational wave is incident on the z axis (perpendicular to the page).

for evidence indicating changes in the distances to the test masses. Because light traveling over long path lengths is interfering, it is possible to detect extremely small changes. As noted above, this is necessary, since to have a chance to detect gravitational waves from expected astronomical sources such as merging neutron stars or core collapse supernovae, fractional changes in distance of order 10^{-21} or smaller must be measured. (To set things in perspective, 10^{-21} is approximately the ratio of the width of a human hair to the distance to Alpha Centauri!) As illustrated in Fig. 22.6, laser interferometers may be viewed as extremely precise ways to measure the distortions that were illustrated in Fig. 22.4 for a small number of test masses.

Present gravitational wave detectors are more properly termed antennae rather than telescopes, because their sizes are small compared with the wavelengths they are built to detect.[8] Nevertheless, they are commonly referred to as gravitational wave detectors or observatories. During the long technical ramp-up to the sensitivity required to detect gravitational waves, skeptics sometimes objected to the term "observatory" applied to gravitational wave detectors, since they had not observed any valid events over a number of years. Now that six gravitational wave events have been published (at the time of this writing in 2017) by the Laser Interferometer Gravitational-Wave Observatory, it is clear that LIGO has earned the right to be called an observatory!

22.6.1 Operating and Proposed Detectors

Laser interferometers such as LIGO (one detector in Livingston, Louisiana and a second one in Hanford, Washington), Virgo (near Pisa, Italy), and GEO600 (near Hannover, Germany) are now operational and/or being upgraded to greater sensitivity. A largely Japanese consortium is building KAGRA in the Kamiokande Mountain famous for housing the super-Kamiokande neutrino detector, and a third LIGO observatory has been proposed for India.

A proposed space-based array, LISA (Laser Interferometer Space Antenna), would have 2.5 million kilometer interferometer arms. The schematic arrangement for LISA, and its proposed orbit, are illustrated in Fig. 22.7.[9] Each of the three spacecraft constituting LISA (the corners of the triangle in Fig. 22.7) would have optical assemblies composed of two telescopes, two lasers, and two platinum–gold test masses. The optical assemblies of each spacecraft will point at the other two spacecraft, thus forming Michelson laser interferometers. Siting of a gravitational wave detector in space allows for much longer arms and high suppression of ambient noise relative to Earth-based systems. This system could detect low-frequency gravitational waves originating in the merger of supermassive black holes at the centers of colliding galaxies, as well as gravitational waves from ordinary binary stars within the galaxy and the inspiral of compact objects into supermassive black holes (see Fig. 22.1).

The original LISA proposal was a joint effort of the US National Aeronautics and Space Administration (NASA) and the European Space Agency (ESA). It fell to the budget axe when NASA withdrew from the project in 2011. However, the LISA concept has been

---

[8] Gravitational wave interferometers see the entire sky (the Earth is transparent to gravitational waves), so there is no need to "point" a gravitational wave detector as for a normal telescope. However, because interferometers behave as antennae, they are not equally sensitive in all directions.

[9] The LISA array would be 2.5×10^6 km on each side, making the spatial extent of the entire configuration much larger than the size of the orbit of the Moon around the Earth. The array would be situated in an Earth-like heliocentric orbit at a point trailing the Earth in its orbit around the Sun by $20°$, inclined at $60°$ to the ecliptic with an average separation from the Earth of 50×10^6 km. The test masses will be in almost perfect free fall, with the spacecraft kept centered around the test masses using sensors and small thrusters (drag-free technology). This will prevent the test masses from being affected by non-gravitational perturbations such as the solar wind or solar radiation pressure.

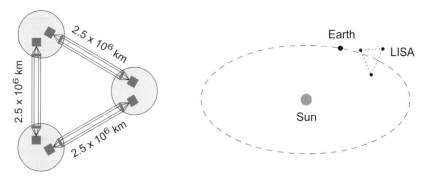

Fig. 22.7 The proposed LISA space-based gravitational wave interferometer. This system consists of three drag-free satellites that implement Michelson laser interferometers having 2.5 million kilometer arms. (Unlike for LIGO, the laser arms will not be multiple-reflection cavities.) Space-based interferometers like LISA would be sensitive to much lower frequencies than those on Earth because: (1) they can have very long interferometer arms, and (2) because there is little interference from environmental noise (in particular seismic noise), which limits the low-frequency performance of Earth-based systems.

resurrected under ESA leadership and a modified LISA proposal is under consideration. The ESA flew the proof of concept LISA Pathfinder mission (a single LISA interferometer arm shortened to fit into a spacecraft), which by 2016 had demonstrated the viability of the technology proposed for the three spacecraft stations illustrated in Fig. 22.7.

22.6.2 Strain and Frequency Windows

The amplitude and frequency ranges for LIGO, Virgo, and LISA, along with ranges expected for some important astrophysical sources of gravitational waves, are illustrated in Fig. 22.8. In this figure "compact binary inspirals" corresponds to the merger of two black holes, the merger of two neutron stars, or the merger of a black hole and a neutron star. "Massive black hole binaries" corresponds to the merger of two of the massive black holes found in the centers of galaxies, which can be initiated by the collision of two galaxies. "Resolved galactic binaries" refers to gravitational waves from individually resolved non-compact binary star systems within the galaxy (see an example in Problem 23.7); there also will be an unresolved continuum in the LISA frequency window generated collectively by galactic noncompact binaries. "Extreme mass ratio inspirals" refers to a stellar compact object on a slowly inspiraling orbit around a much more massive ($\sim 10^5 \, M_\odot$ or greater) black hole. The slow inspiral allows many orbits within the LISA sensitive band, affording the possibility of determining the parameters of the massive black hole with very high precision. Core collapse supernovae fall within the LIGO–VIRGO sensitive bands but Type Ia supernovae lie between the LIGO–Virgo and LISA frequency bands. (Neither type of supernova is very common within detectable ranges, with one expected in the galaxy only once every 50–100 years.) Such potential astronomical gravitational wave sources will be discussed more extensively in Chapters 23 and 24.

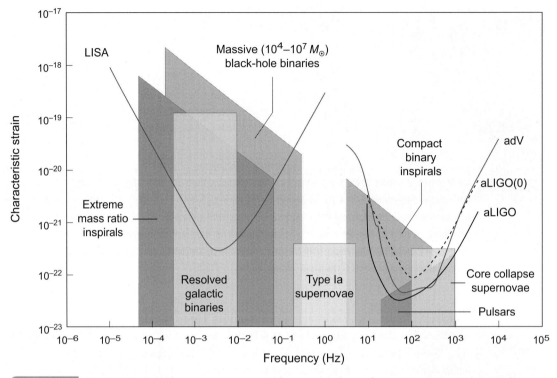

Fig. 22.8 Strain amplitude and frequency ranges expected for gravitational waves from various astronomical sources. (See Ref. [158] for assumptions used in creating the plot.) The minimum strain detection bounds for advanced LIGO (aLIGO) at full design capacity (∼2020), advanced Virgo (adV) at full design capability (∼2020), advanced LIGO in the first observing run after the upgrade [aLIGO(0), indicated by the dashed curve], during which the gravitational wave GW150914 was observed in 2015, and the proposed space-based array LISA are indicated. The figure was generated using the interactive plotter available at http://rhcole.com/apps/GWplotter/, which is described in Ref. [158].

For compact binary mergers that occur in the LIGO–Virgo window of Fig. 22.8, the increasing gravitational wave frequency as the binary spirals toward merger means that at some earlier time the frequency of emission from the inspiral will overlap the LISA window. Therefore, if the strain is large enough it is possible that LISA might be able to provide a forecast alerting other gravitational and non-gravitational observatories to the time and approximate location of the upcoming event. This could maximize the probability of being able to implement the multimessenger astronomy discussed in Section 24.6 for such an event.

22.6.3 Detecting Very Long Wavelengths

Figure 22.8 omits long-wavelength sources with frequencies below 10^{-6} Hz. Sources of considerable astronomical interest in this frequency range include

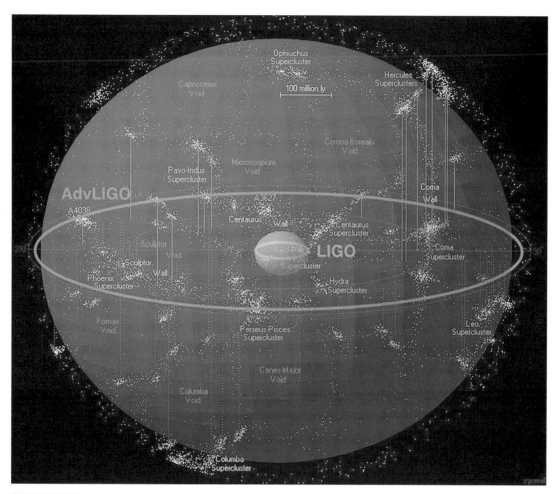

Fig. 22.9 The expected reach of Advanced LIGO for detection of a neutron star merger event when at full design capability in 2020. Courtesy Caltech/MIT/LIGO Laboratory.

1. mergers of the most massive black holes found in the centers of galaxies ($\sim 10^9\ M_\odot$), which are expected to occur in the 10^{-7} to 10^{-9} Hz range with strains as large as 10^{-16};
2. a stochastic background of supermassive black hole mergers relevant to the history of large-scale structure formation in the Universe;
3. a stochastic background associated with cosmic inflation and first-order phase transitions relevant to the early Universe, generally of low frequency because of cosmological redshift.

Detection of some events with very long wavelengths may be feasible with a *pulsar timing array (PTA)*, where millisecond pulsars are viewed as precise clocks and a set of them is monitored for timing changes indicating the passage of a gravitational wave. (Millisecond pulsars are favored because they are faster and more stable than average pulsars.) These PTAs may be thought of as natural interferometers with arms of galactic scale (kiloparsecs) that are sensitive to gravitational waves in the frequency range

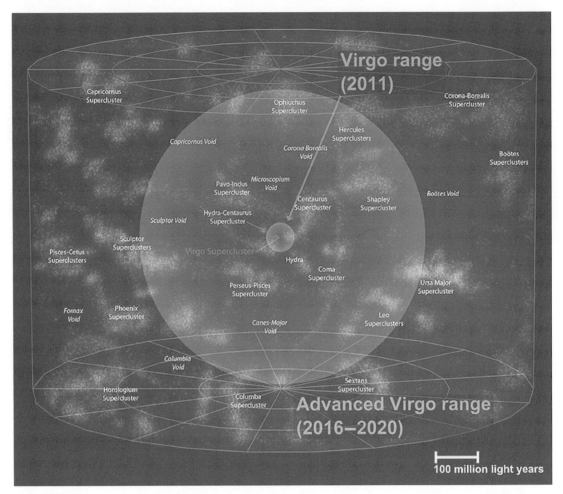

Fig. 22.10 The expected reach of Advanced Virgo (outer sphere) for a neutron star merger event when at full design capacity in 2020. Andrew Z. Colvin.

10^{-9} to 10^{-6} Hz (see Fig. 22.1). An overview of how pulsar timing might be used to detect long-wavelength gravitational waves may be found in Ref. [121].

22.6.4 Reach of Advanced LIGO and Advanced VIRGO

Figure 22.9 illustrates the gravitational wave reach expected for Advanced LIGO when at full design capability in 2020 for binary neutron star mergers. The outer sphere has an approximate radius of 150 Mpc. The inner sphere indicates the corresponding reach of LIGO before the upgrade to Advanced LIGO. Figure 22.10 illustrates the projected reach of Advanced Virgo.

The Advanced LIGO upgrade targeted a 10-fold increase in sensitivity over initial LIGO. In the first observing run with Advanced LIGO beginning in late 2015 the sensitivity had reached a factor of 3–4 improvement over initial LIGO, which translates into a factor of

$\sim (3.5)^3 \sim 40$ increase in sensitive volume. This enabled the first detection of gravitational waves that will be described in Chapter 24. It is anticipated that successive Advanced LIGO observing runs will increase sensitivity incrementally until full design specifications are reached in 2020. When Advanced LIGO is at full sensitivity it should be capable of detecting gravitational waves from a typical neutron star merger anywhere within the outer spherical volume shown in Fig. 22.9. The Virgo gravitational wave detector in Italy was upgraded to similar capability and came online in 2017. When fully operational it will have capabilities similar to those of advanced LIGO.

Background and Further Reading

For general discussions of gravitational waves in linearized gravity at a level similar to that presented here, see in particular Shapiro and Teukolsky [223]; Hartle [111]; Hobson, Efstathiou, and Lasenby [122]; Flanagan and Hughes [87]; and Cheng [65]. See also Schutz [221]; Weinberg [246]; and Misner, Thorne, and Wheeler [156] for more advanced discussions. The original gravitational wave solution was published by Einstein in 1916 and 1918 [79, 80]. The history of the debate over whether gravitational waves were detectable as a matter of principle is summarized in Saulson [215] and Berti [45]. A review of sources, detectors, and searches for gravitational waves is given by Riles [203]. A free interactive plotter that can generate plots similar to Fig. 22.8 for a variety of gravitational wave sources is available from http://rhcole.com/apps/GWplotter/, and is described in Ref. [158].

Problems

22.1 Demonstrate that the Einstein equation (8.21),

$$G_{\mu\nu} = R_{\mu\nu} - \tfrac{1}{2} g_{\mu\nu} R = 8\pi G T_{\mu\nu},$$

may also be written in the form [see Eq. (22.1)]

$$R_{\mu\nu} = 8\pi G(T_{\mu\nu} - \tfrac{1}{2} g_{\mu\nu} T^\alpha{}_\alpha),$$

so that the vacuum Einstein equation is $R_{\mu\nu} = 0$. *Hint*: Contract both sides of the first equation with the metric tensor and solve the resulting equation for the Ricci scalar R. ***

22.2 (a) Prove that in linearized gravity with $g_{\mu\nu} = \eta_{\mu\nu} + h_{\mu\nu}$, under a Lorentz transformation of the metric $h_{\mu\nu}$ behaves as would a rank-2 tensor in special relativity. Hence, the linearized field may be treated as a rank-2 tensor in flat spacetime and indices may be raised and lowered by contraction with $\eta_{\mu\nu}$. (b) The lone exception to the conclusion from part (a) that in linearized gravity indices may be raised or lowered by contraction with $\eta_{\mu\nu}$ is for $g_{\mu\nu}$ itself: show that $g^{\mu\nu} = \eta^{\mu\nu} - h^{\mu\nu}$ to first order in h. ***

22.3 Beginning from Eq. (22.7), supply the missing steps leading to Eq. (22.12) for the wave equation evaluated to first order in a small metric perturbation. ***

22.4 Show that the gauge transformation $h_{\mu\nu} \to h'_{\mu\nu}$, with $h'_{\mu\nu}$ given by Eq. (22.15), results from the coordinate transformation (22.13). ***

22.5 Beginning with Eq. (22.21), supply the missing steps leading to the constraints $\alpha_{0\mu} = 0$ and $k^j \alpha_{ij} = 0$ on the polarization tensor that are provided by Eqs. (22.23) and (22.24).

22.6 Show that in TT gauge for gravitational waves propagating on the z axis, the only non-vanishing components of the polarization tensor are α_{11}, $\alpha_{12} = \alpha_{21}$, and $\alpha_{22} = -\alpha_{11}$. ***

22.7 The gravitational wavelength for optimal response of a laser interferometer may be estimated as the distance traveled by light in the laser arms. This wavelength corresponds approximately to the frequency $f_* = c/2\pi L$, where L is the distance traveled by the laser light. Use this to estimate the optimal frequency for response to gravitational waves for LIGO and for LISA.

22.8 Prove that $\bar{h}_{\mu\nu}$ given by Eq. (22.19) satisfies the wave equation (22.16) only if k is a null vector satisfying Eq. (22.25). Thus, gravitational waves travel at the speed of light. ***

22.9 Consider a gravitational wave incident upon a single test particle that is initially at rest. Demonstrate that in TT gauge the acceleration of the particle by the gravitational wave is identically zero. Therefore, in TT gauge the particle is stationary with respect to the coordinate system and the test particle may be viewed as remaining attached to the coordinates as the wave passes. ***

23 Weak Sources of Gravitational Waves

In Chapter 22 we introduced gravitational waves through a linearized approximation for the vacuum Einstein equations without sources. These equations are valid far from a gravitational wave source but of course we also need to be able to describe the emission of gravitational waves from a source. In this chapter and Chapter 24 possible sources of detectable gravitational waves are considered. To study the production of gravitational waves in the general case we must solve the full nonlinear Einstein equations (8.21) with source terms depending on $T_{\mu\nu}$. This is a formidable problem, generally tractable only for large-scale computation (*numerical relativity* [26, 40]). However, considerable insight can be gained by studying a less complex situation, the linearized Einstein equation with sources. This is an analytically accessible problem that has many parallels with the study of sources for electromagnetic waves. It has been shown by numerical simulation that many key features for the production of weak gravitational waves carry over in recognizable form for the production of gravitational waves in strong-gravity environments. Therefore, our approach will be to concentrate on a more quantitative treatment of sources for weak gravitational waves in this chapter, and then to consider qualitative order-of-magnitude arguments and numerically computed examples for gravitational waves produced by strong-gravity sources in Chapter 24.

23.1 Production of Weak Gravitational Waves

Let's now consider a linearized approximation of the Einstein equations as in Chapter 22, but with a source term included to describe the production of weak gravitational waves.

23.1.1 Energy Densities

In an electromagnetic or Newtonian gravitational field it makes sense to talk about local energy densities. For example, in a local Newtonian field the energy density at a point x is given by

$$\epsilon(x) = -\frac{1}{8\pi G}(\nabla \varphi(x))^2, \qquad (23.1)$$

where $\varphi(x)$ is the Newtonian gravitational potential. There is *no corresponding local energy density in general relativity.* If such a density existed, it would contradict the equivalence principle, which requires gravity to *vanish* in a sufficiently local region (local inertial frame). However, it *does* make sense to speak of an approximate energy density

> **Box 23.1** **Multipolarity of Gravitational Waves**
>
> For gravitational waves monopole radiation is forbidden by conservation of mass and dipole radiation is forbidden by conservation of momentum and angular momentum for the source. For electromagnetic radiation monopole radiation is forbidden by conservation of charge but dipole radiation is permitted because there are electrical charges with two signs ($\pm$), so a dipole can be created by separating a positive and negative charge while still preserving the center of mass. For gravity mass is the charge; it has only one sign and dipole radiation is forbidden. In more detail, for electromagnetism the amplitude of electric dipole radiation depends on to the second time derivative of the electric dipole moment. The gravitational analog of the electric dipole moment is the mass dipole moment, which has the momentum as the first time derivative. Since momentum is conserved the second time derivative is zero and there is no gravitational dipole radiation.

associated with a weak gravitational wave of wavelength λ, provided that λ is much shorter than the characteristic curvature radius R of the background spacetime through which the wave propagates. This approximation becomes very good at large distances from the source of a gravitational wave, where curvature associated with the source becomes negligible and $\lambda/R \to 0$. Therefore, one may formulate a description of energy loss from gravitational wave sources at large distances from the source where an approximate energy density may be associated with a wave by averaging over several wavelengths.

23.1.2 Multipolarities

The lowest-order contribution from a source to electromagnetic radiation corresponds to dipole motion of the source. Like the electromagnetic field, the production of gravitational waves requires non-spherical motion of the charge (which is electrical charge for the electromagnetic field and inertial mass for the gravitational field). However, electromagnetism corresponds to a vector field, while gravity is a rank-2 tensor field. As a consequence, no monopole or dipole component contributes to the generation of gravitational waves and the lowest-order contribution corresponds to time-dependent quadrupole distortions of the source mass, as discussed further in Box 23.1. Hence, many of the formulas for sources of electromagnetic waves and for weak gravitational waves will be similar but not identical.

23.1.3 Linearized Einstein Equation with Sources

Adding a source to the linearized Einstein equation (22.16) in Lorentz gauge gives the wave equation

$$\Box \bar{h}_{\mu\nu} = -16\pi G T_{\mu\nu}, \qquad (23.2)$$

where the stress–energy tensor $T_{\mu\nu}$ describing the source is assumed small, consistent with the linear approximation for the metric, $\Box$ is defined in Eq. (22.9), and $\bar{h}_{\mu\nu}(t, \boldsymbol{x})$ is the trace-reversed amplitude given by Eq. (22.18),

$$\bar{h}_{\mu\nu} \equiv h_{\mu\nu} - \tfrac{1}{2} \eta_{\mu\nu} h^{\lambda}_{\lambda}. \qquad (23.3)$$

Solutions of Eq. (23.2) can be found using Green's function methods similar to those used to solve wave equations in electromagnetism. We skip the details (an outline of the solution may be found in Carrol [63], Section 7.5) and jump to the results, guided by the presentations in Shapiro and Teukolsky [223], and Hartle [111]. The metric perturbation for long-wavelength gravitational waves (wavelengths much larger than the characteristic source size imply $v \ll c$ for mass in the source), far from a nonrelativistic source is found to be

$$\bar{h}^{ij}(t, \boldsymbol{x}) \simeq \frac{2}{r} \ddot{I}^{ij}(t - r), \qquad (23.4)$$

where double dots signify the second time derivative and the *second mass moment* $I^{ij}(t)$ is given by

$$I^{ij}(t) \equiv \int \rho(t, \boldsymbol{x}) x^i x^j d^3 x, \qquad (23.5)$$

where $\rho(t, \boldsymbol{x})$ is the mass density of the source.

23.1.4 Gravitational Wave Amplitudes

The gravitational wave amplitude $\bar{h}^{ij}$ is given by Eq. (23.4) in linear approximation. Let us make some estimates based on this formula using as a simple model for gravitational wave emission a binary star system in an orbit such that the surfaces of the two stars touch (a contact binary). For simplicity, the two stars will be assumed to be of the same radius and mass, and to revolve in circular orbits about their common center of mass, as illustrated in Fig. 23.1. The second mass moment (23.5) becomes a sum of two discrete contributions,

$$I^{ij}(t) = \int \rho(t, \boldsymbol{x}) x^i x^j d^3 x = MR^2 + MR^2 = 2MR^2. \qquad (23.6)$$

The system revolves with a period P and taking the derivative of $I^{ij}(t)$ twice with respect to time gives a factor $\omega^2 \sim 1/P^2$ (see Section 23.2.1)

$$\ddot{I}^{ij} \simeq 2 \frac{MR^2}{P^2}. \qquad (23.7)$$

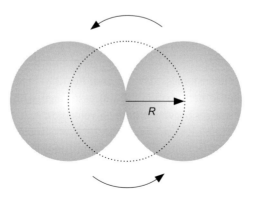

Fig. 23.1 A contact binary as a source of gravitational waves. In this example the two stars are assumed to have the same mass M and same radius R.

Insertion of this approximation in Eq. (23.4) leads to

$$\bar{h}^{ij} \simeq \frac{2}{r} \ddot{I}^{ij} = \frac{4MR^2}{rP^2}, \qquad (23.8)$$

which we can express in terms of source velocities by noting that in one period the center of mass for each star travels a distance $C = 2\pi R$. Therefore, the velocity of the center of mass for a star is

$$v = \frac{2\pi R}{P}, \qquad (23.9)$$

which can be solved for the ratio R/P and inserted into Eq. (23.8) to give

$$\bar{h}^{ij} \simeq \frac{Mv^2}{\pi^2 r}. \qquad (23.10)$$

This is a specialized result obtained assuming some rather crude approximations but more careful analysis indicates that the form of Eq. (23.10) has broader validity than its derivation would suggest. Noting that the Schwarzschild radius and mass are related by $r_S = 2M$, reinserting factors of c and G, and dropping the numerical factors,

$$\bar{h} \simeq \frac{r_S}{r} \frac{v^2}{c^2}. \qquad (23.11)$$

Equation (23.11) may be taken as a qualitative guide to the amplitude of gravitational waves far from a weak source.

Typically, weak gravitational waves are generated by systems that are gravitationally bound or nearly so. Therefore the virial theorem of Section 14.5.3 is applicable and the kinetic and potential energies should be comparable, which implies $\frac{1}{2}Mv^2 \sim GM^2/R$ and

$$\frac{v^2}{c^2} \simeq \frac{2GM^2}{MRc^2} = \frac{r_S}{R} = \epsilon^{2/7}, \qquad (23.12)$$

where an efficiency factor for gravitational wave emission,

$$\epsilon \equiv \left(\frac{r_S}{R}\right)^{7/2} \qquad (23.13)$$

has been defined [the justification for terming this the gravitational wave efficiency will be given below; see Eq. (23.20)]. Therefore, Eq. (23.11) may be written in the form

$$\bar{h} \simeq \frac{r_S}{r} \frac{v^2}{c^2} = \frac{r_S^2}{rR} = \epsilon^{2/7} \frac{r_S}{r}, \qquad (23.14)$$

which is dimensionless. The forms of Eq. (23.11) or Eq. (23.14) indicate that the amplitude of the metric perturbation (strain) associated with the gravitational wave is largest for compact sources moving with high velocities.

23.1.5 Amplitudes and Event Rates

We may use the preceding results to estimate amplitudes and corresponding event rates for candidate gravitational wave events that might be detectable by the current generation of gravitational wave interferometers.

Example 23.1 Inserting $r_S = (2G/c^2)M$ in Eq. (23.14) gives

$$\bar{h} \simeq 9.6 \times 10^{-17} \, \epsilon^{2/7} \left(\frac{M}{M_\odot}\right)\left(\frac{\text{kpc}}{r}\right). \qquad (23.15)$$

Reasonable guesses for a neutron star merger, which is expected to be one type of event with high probability for observation by gravitational wave detectors, are $\epsilon \sim 0.1$ and $M \sim 1 \, M_\odot$ for the mass participating in the generation of gravitational waves. For events within the galaxy a characteristic distance estimate is $r \sim 10$ kpc. Inserting these values in Eq. (23.15) gives $\bar{h} \simeq 5 \times 10^{-18}$. This is a quite detectable strain for modern gravitational wave interferometers but such events occur within the galaxy only infrequently (one estimate is an average of more than 10^5 years between galactic neutron star mergers). Therefore, to obtain a reasonable event rate it is necessary to be able to detect gravitational waves from sources at greater distances, giving a larger sample volume ("detection horizon"). The nearest rich cluster containing thousands of galaxies is the Virgo Cluster, at a distance of 16.5 Mpc. If the range of detection is extended out to hundreds of Mpc, event rates for merging neutron stars should go up substantially because one is now surveying many galaxies (see Figs. 22.9 and 22.10), but the average strain at the detector falls to $\bar{h} \sim 10^{-21}$ to 10^{-22}. Thus, to observe systematic gravitational wave events from neutron star mergers the detectors must be able to sample strains reliably at this level.

23.1.6 Power in Gravitational Waves

The power radiated in gravitational waves for a system that has velocities well below c and weak internal gravity is given by the *quadrupole formula* [111, 223],[1]

$$L = \frac{dE}{dt} = \frac{1}{5} \left\langle \dddot{I}_{ij} \dddot{I}^{ij} \right\rangle = \frac{1}{5} \frac{G}{c^5} \left\langle \dddot{I}_{ij} \dddot{I}^{ij} \right\rangle, \qquad (23.16)$$

where $\langle \, \rangle$ denotes a time average over a wave period, the triple dot signifies a third time derivative, and the *reduced quadrupole tensor* $\textit{I}$ is defined by

$$\textit{I}^{ij} \equiv I^{ij} - \tfrac{1}{3}\delta^{ij}\,\text{Tr}\, I, \qquad (23.17)$$

where $\text{Tr}\, I \equiv I_k^k$. This formula is the gravitational analog of the formula for radiated power in electromagnetism but has a different coefficient ($\tfrac{1}{5}$ instead of $\tfrac{1}{20}$), and corresponds to quadrupole rather than dipole radiation. As shown in Problem 23.3, Eq. (23.16) may be reduced to

$$L = \frac{dE}{dt} \simeq L_0 \frac{r_S^2}{R^2}\left(\frac{v}{c}\right)^6, \qquad (23.18)$$

---

[1] The quadrupole formula was derived originally by Einstein in his 1918 gravitational wave paper [80], by exploiting analogies with the power radiated by electromagnetic multipoles. The formula was questioned at various times over the ensuing decades, but now is firmly established as a reliable first approximation for emitted gravitational wave power.

where the characteristic scale for radiated gravitational wave power is set by

$$L_0 \equiv \frac{c^5}{G} = 3.6 \times 10^{59} \text{ erg s}^{-1}, \tag{23.19}$$

and the total energy ΔE emitted in one period P can then be estimated as

$$\Delta E \simeq L P \simeq Mc^2 \left(\frac{r_S}{R}\right)^{7/2} = \epsilon Mc^2. \tag{23.20}$$

Therefore, as promised earlier, $\epsilon = (r_S/R)^{7/2}$ defined in Eq. (23.13) parameterizes the efficiency for conversion of mass to energy in the form of gravitational waves [223].

23.2 Gravitational Radiation from Binary Systems

In preceding sections some qualitative estimates for gravitational wave emission from binary stars have been made. Let us now derive in a more rigorous fashion a formalism applicable for such systems.

23.2.1 Gravitational Wave Luminosity

A binary system with circular orbits is illustrated in Fig. 23.2, where a_1 and a_2 are the distances of the masses m_1 and m_2, respectively, from the center of mass, and $\varphi = \omega t$ is the azimuthal angle between the line joining the stars and the x axis. The components of the second mass moment are given by Eq. (23.5), which reduces to a sum,

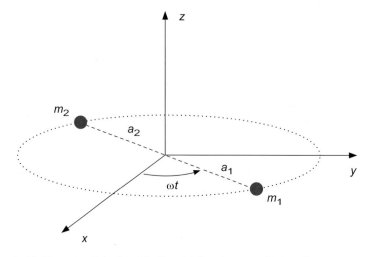

Fig. 23.2 Coordinate system for binary stars in circular orbits. The origin is at the center of mass and $m_1 a_1 = m_2 a_2$.

23.2 Gravitational Radiation from Binary Systems

$$I^{ij} = m_1 x^i x^j + m_2 x^i x^j \qquad (23.21)$$

for two discrete masses (where superscripts label coordinates and subscripts label objects). Introduce polar coordinates for each star i,

$$x_i(t) = a_i \cos \omega t \qquad y_i(t) = a_i \sin \omega t \qquad z_i(t) = 0, \qquad (23.22)$$

and substitute into Eq. (23.21) to evaluate the mass moments. For example,

$$\begin{aligned} I^{xx} = I^{11} &= m_1 a_1^2 \cos^2 \omega t + m_2 a_2^2 \cos^2 \omega t \\ &= (m_1 a_1^2 + m_2 a_2^2) \cos^2 \omega t \\ &= \mu a^2 \cos^2 \omega t \\ &= \tfrac{1}{2} \mu a^2 (1 + \cos 2\omega t), \end{aligned}$$

where $a \equiv a_1 + a_2$, the reduced mass is $\mu = m_1 m_2 / (m_1 + m_2)$, and the identity $m_1 a_1 = m_2 a_2 = \mu a$ was used. Carrying out similar steps for all components gives (Problem 23.4)

$$\begin{aligned} I^{xx} = I^{11} &= \tfrac{1}{2}\mu a^2 (1 + \cos 2\omega t), \\ I^{xy} = I^{yx} = I^{12} &= \tfrac{1}{2}\mu a^2 \sin 2\omega t, \\ I^{yy} = I^{22} &= \tfrac{1}{2}\mu a^2 (1 - \cos 2\omega t). \end{aligned} \qquad (23.23)$$

The trace-reversed amplitude (23.4) requires the second time derivatives of Eqs. (23.23). Computing them (Problem 23.5) and inserting in Eq. (23.4) gives

$$\bar{h}^{ij} = \frac{4\omega^2 \mu a^2}{r} \begin{pmatrix} -\cos 2\omega(t-r) & -\sin 2\omega(t-r) & 0 \\ -\sin 2\omega(t-r) & \cos 2\omega(t-r) & 0 \\ 0 & 0 & 0 \end{pmatrix}. \qquad (23.24)$$

The appearance of 2ω in the arguments of the above equation means that the frequency f of the emitted gravitational wave radiation will be twice the orbital frequency ω because the mass distribution varies in time with a period equal to half the rotational period.[2] The reduced moment $\mathcal{I}^{ij}$ is given by Eq. (23.17) but from (23.23) written as a matrix the trace, $\mathrm{Tr}\, I = I^{xx} + I^{yy} = \mu a^2$, is independent of time. Therefore, $\mathcal{I}^{ij} = I^{ij}$ and from Eq. (23.16) the radiated luminosity is given by

$$L = \frac{1}{5} \left\langle \dddot{I}_{ij} \dddot{I}^{ij} \right\rangle. \qquad (23.25)$$

From the second time derivatives found above, the triple time derivatives are

$$\dddot{I}^{xx}(t) = 4\omega^3 \mu a^2 \sin 2\omega t \qquad \dddot{I}^{yy}(t) = -4\omega^3 \mu a^2 \sin 2\omega t$$
$$\dddot{I}^{xy}(t) = \dddot{I}^{yx}(t) = -4\omega^3 \mu a^2 \cos 2\omega t \qquad (23.26)$$

so that the radiated gravitational wave power is

---

[2] Ultimately this is because to leading order gravitational radiation is generated by quadrupole distortions and the mass quadrupole moment is symmetric under rotations by π.

$$L = \frac{1}{5}\left\langle (\dddot{I}^{xx})^2 + 2(\dddot{I}^{xy})^2 + (\dddot{I}^{yy})^2 \right\rangle$$

$$= \frac{1}{5}\left[\frac{1}{P}\int_0^P \left((\dddot{I}^{xx})^2 + 2(\dddot{I}^{xy})^2 + (\dddot{I}^{yy})^2\right) dt\right]$$

$$= \frac{16\omega^6 \mu^2 a^4}{5P}\int_0^P \left(\sin^2 2\omega t + 2\cos^2 2\omega t + \sin^2 2\omega t\right) dt$$

$$= \frac{32}{5}\omega^6 \mu^2 a^4$$

$$= \frac{32}{5}\frac{G^4}{c^5}\frac{M^3 \mu^2}{a^5}, \tag{23.27}$$

where the second step averages over one period, the integral in line three evaluates to $2P$, in the last step Kepler's third law (23.33) has been used to eliminate ω, factors of G and c have been restored, and $M \equiv m_1 + m_2$. It is often convenient to express Eq. (23.27) in terms of the period P instead of the separation a. Using Kepler's third law again gives

$$L = \frac{128}{5}\, 4^{2/3}\, \frac{G^{7/3}}{c^5}\, M^{4/3}\mu^2 \left(\frac{\pi}{P}\right)^{10/3}, \tag{23.28}$$

where circular orbits have been assumed for the components of the binary. The corresponding luminosity for binary orbits with eccentricity e can be approximated as

$$L = f(e)\bar{L}, \tag{23.29}$$

where $\bar{L}$ is the luminosity calculated from Eq. (23.27) or Eq. (23.28) assuming circular orbits and

$$f(e) = \frac{1 + \frac{73}{24}e^2 + \frac{37}{96}e^4}{(1-e^2)^{7/2}} \tag{23.30}$$

corrects for the eccentricity of the orbit (see Section 16.4 of Shapiro and Teukolsky [223]).

Example 23.2 Assume in the binary system of Fig. 23.2 that $m_1 = m_2 = 1\, M_\odot$, and that the period is 6 hours. From Eq. (23.28) with c and G factors evaluated (Problem 5.8),

$$L = 2.3 \times 10^{45} \left(\frac{M}{M_\odot}\right)^{4/3} \left(\frac{\mu}{M_\odot}\right)^2 \left(\frac{1\,\text{s}}{P}\right)^{10/3} \text{erg s}^{-1}. \tag{23.31}$$

Inserting the masses and period gives a luminosity of 5.2×10^{30} erg s^{-1} for the binary.

23.2.2 Gravitational Radiation and Binary Orbits

Consider gravitational wave emission from the binary illustrated in Fig. 23.2, assuming the orbital motion to be described by Newtonian mechanics. The total energy is negative (it is a bound system), so if the energy is reduced by gravitational wave emission the period P (and size) of a binary orbit must decrease. Although it will not be proved here (see Section 16.4 of Ref. [223]), the emission of gravitational wave radiation also tends to circularize elliptical orbits. From Problem 24.6 the total orbital energy of the system is given by

$$E = -\frac{Gm_1m_2}{2a} = -\frac{G\mu M}{2a}, \qquad (23.32)$$

where $M = m_1 + m_2$ is the total mass, $\mu = m_1m_2/M$ is the reduced mass, and $a = a_1 + a_2$ is the average separation of the binary pair. Utilizing Eq. (23.32) and Kepler's third law for the period P

$$P^2 = \frac{4\pi^2}{GM} a^3, \qquad (23.33)$$

and assuming that $L = -dE/dt$, where L is the gravitational wave luminosity and dE/dt is the rate of energy loss from the orbital motion, the rate of change for the period with time is given by (Problem 23.6)

$$\frac{dP}{dt} = -\frac{96}{5} \frac{G^3}{c^5} \frac{M^2\mu}{a^4} P. \qquad (23.34)$$

Utilizing Eq. (23.33) to eliminate a in favor of P, this can be written as

$$\frac{dP}{dt} = -\frac{192\pi}{5} \frac{G^{5/3}}{c^5} \frac{m_1m_2}{M^{1/3}} \left(\frac{2\pi}{P}\right)^{5/3}, \qquad (23.35)$$

which is a dimensionless measure of how the orbital period is altered over time by the emission of gravitational waves. The preceding formulas have assumed circular orbits. For binary orbits with eccentricity e the variation of the period with time may be approximated by [223]

$$\frac{dP}{dt} = f(e) \left(\frac{dP}{dt}\right)_0, \qquad (23.36)$$

where $f(e)$ is given by Eq. (23.30) and $(dP/dt)_0$ is the result for circular orbits given by Eq. (23.35).

23.2.3 Gravitational Waves from the Binary Pulsar

The Binary Pulsar consists of a pulsar and a compact companion – almost certainly a neutron star but it is not observed as a pulsar – in orbit around their common center of mass. As noted previously in Section 10.4.1, the precise timing afforded by the pulsar clock has allowed unprecedented tests of the general theory of relativity [232, 248, 249].

Example 23.3 If the orbit for the Binary Pulsar were circular, then from Eq. (23.35)

$$\left(\frac{dP}{dt}\right)_0 = -3.66 \times 10^{-6} \frac{(m_1/M_\odot)(m_2/M_\odot)}{(M/M_\odot)^{1/3}} \left(\frac{1\,\text{s}}{P}\right)^{5/3} = -2 \times 10^{-13},$$

where $M = m_1 + m_2$, and the observed masses $m_1 \sim m_2 \sim 1.4 M_\odot$ and orbital period of 7.75 hours were used. However, the orbit of the Binary Pulsar is rather elliptical with an eccentricity $e = 0.617$, so from Eq. (23.36)

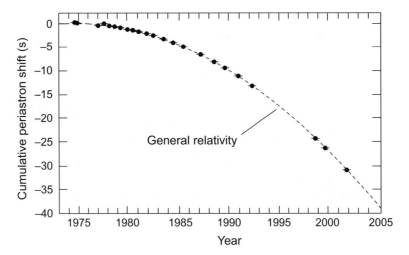

Fig. 23.3 Observed periastron shift of the Binary Pulsar orbit caused by gravitational wave emission. The dashed curve is the prediction of general relativity.

$$\frac{dP}{dt} = f(e)\left(\frac{dP}{dt}\right)_0 = (11.84) \times (-2 \times 10^{-13}) = -2.37 \times 10^{-12},$$

where an elliptical-orbit correction factor $f(e) = 11.84$ was computed using Eq. (23.30). Assuming this rate to be constant, the total change in period over one year is $\Delta P \simeq -7.5 \times 10^{-5}$ s. This decrease is tiny but easy to measure because of the pulsar clock. From this rate of orbital decay the inspiral time to merger is about 3×10^8 years.

Figure 23.3 illustrates the cumulative shift of the periastron time (time for closest approach of the pulsar to its companion) for the Binary Pulsar as a function of elapsed time, compared with a calculation assuming emission of gravitational waves from the system to be responsible for this shift. The quality of the data (note the error bars), and the agreement with the prediction of general relativity are remarkable. This analysis of the Binary Pulsar provided the first strong evidence that gravitational waves exist and that they have properties in quantitative agreement with the predictions of general relativity but it is indirect, since gravitational waves from the Binary Pulsar are too weak to detect directly (see Problem 23.9). In Chapter 24 the first *direct* observation of gravitational waves emitted from a much stronger source will be discussed.

Background and Further Reading

The presentation in this chapter is patterned largely after the discussion of gravitational waves in Shapiro and Teukolsky [223], and Hartle [111]. See also Schutz [221]; Weinberg [246]; Carrol [63]; and Misner, Thorne, and Wheeler [156].

Problems

23.1 Estimate the gravitational wave power emitted by a 20 meter long steel beam of mass 10^5 kg, rotating at near its breakup speed of 30 radians per second. Do you think it feasible to detect gravitational waves produced in the laboratory?

23.2 Estimate the gravitational wave luminosity of the Binary Pulsar (Hulse–Taylor binary), taking into account the orbital eccentricity $e = 0.617$.

23.3 Starting from Eq. (23.16), supply all intermediate steps in deriving Equations (23.18)–(23.20) for gravitational wave luminosity. ***

23.4 Derive the second mass moments (23.23) for a binary system using the coordinate system (23.22) and the definition (23.21). *Hint*: First prove that $m_1 a_1 = m_2 a_2 = \mu a$, where $a = a_1 + a_2$ and use this to formulate the problem in terms of the reduced mass $\mu = m_1 m_2/(m_1 + m_2)$.

23.5 Starting from Eq. (23.23) for the second mass moments, compute explicitly the xx, xy, yx, and yy components of the trace-reversed amplitude (23.24), and the triple time derivatives (23.26). ***

23.6 The modification of a binary orbital period by gravitational wave emission is given by Eq. (23.35) for circular orbits in Newtonian approximation. Supply all steps in deriving Eqs. (23.34) and (23.35) from the orbital energy (23.32) and Kepler's law (23.33). ***

23.7 For binary mergers the strongest gravitational wave radiation will occur at the end of the inspiral, where the two masses are very close and revolving with a short period. This raises the issue of whether gravitational waves could be detected from binaries of noncompact stars that are *contact binaries* (binaries with small enough separation that the stars are touching). Estimate the amplitude and frequency of gravitational wave radiation observed from Earth for the W Uma-type contact binary 44 Boo. For this system the masses of the stars are 0.5 $M_\odot$ and 1.0 $M_\odot$, the period is 6.4 hours, and the location is 12.8 pc away in the constellation Boötes [214]. You may assume that the effective mass contributing to the gravitational wave radiation is $M \sim 0.5\, M_\odot$ and that the effective separation is $R \simeq 2R_\odot$. ***

23.8 Assuming the validity of Newtonian gravity for the orbit, show that Eq. (23.10) also can be expressed as

$$\bar{h}^{ij} = 1.47 \times 10^{-4} \left(\frac{M}{M_\odot}\right)^{5/3} \left(\frac{\text{s}}{P}\right)^{2/3} \left(\frac{\text{km}}{r}\right).$$

Hint: Start from Eq. (23.8)

23.9 Make a rough estimate of the strain and frequency for gravitational waves emitted by the Binary Pulsar. Could gravitational waves from this system be detected by advanced LIGO, advanced Virgo, or LISA? ***

24 Strong Sources of Gravitational Waves

The preceding chapters introduced the basic idea of gravitational waves in terms of a linearized approximation to gravity, and discussed sources of gravitational waves in weak gravity. This chapter addresses the more difficult but more interesting problem of describing gravitational waves originating in events involving strong gravity. The sources that are expected to provide the most likely events that could be observed cannot be described in the near-source region by a linear approximation to gravity. In mergers involving some combination of neutron stars and black holes, or the core collapse of a massive star, the curvature of spacetime and the mass velocities can become very large in the region where the strongest gravitational waves are produced. In this strong-gravity domain, reliable calculations are possible only through complex numerical simulations. Nevertheless, those simulations show that many of the features that have been inferred about gravitational waves in the linear regime survive in some form in the nonlinear strong-gravity regime. In particular, available calculations suggest that many basic qualitative features of gravitational wave production in strong gravity can be obtained by dimensional analysis.

24.1 A Survey of Candidate Sources

In this section we summarize some likely strong sources of gravitational waves and give some results of numerical simulations. These fall into two categories: (1) merger of binary compact objects when their orbits decay because of gravitational wave emission, and (2) the core collapse of massive stars. Events in both categories may exhibit rapid asymmetric motion of large compact masses, which is the essential condition for production of strong gravitational waves. Because the asymmetric motion of the mass in these events is expected to differ in characteristic ways, the detailed gravitational wave pattern should carry information about the type of event that produced it.

24.1.1 Merger of a Neutron Star Binary

Gravitational waves from merging neutron stars should be detectable in Earth-based gravitational wave detectors, if the events occur in our galaxy or in relatively nearby galaxies. The possibility of two neutron stars merging might seem a remote one but once a neutron star binary is formed its orbital motion radiates energy as gravitational waves, the orbits must shrink, and inevitably the two neutron stars must merge. The most relevant question is the timescale for the merger. One typically speaks of merger within a Hubble time [$\sim$ 14 billion years from Eq. (16.14)], meaning that the orbital separation decays fast enough for the merger to occur on a timescale that is less than the age of the Universe and thus possibly observable. Presently several observed binary systems are predicted to have a merger timescale less than the Hubble time. For example, the precisely measured orbit of the Binary Pulsar has shrunk to a separation at nearest approach of about 750, 000 km, with a decay rate that should lead to merger in about 3×10^8 years (Section 23.2.3).

Formation of the neutron star binary (more generally any binary involving neutron stars and/or black holes) is not easy, however. Either a binary, or multiple-star system with more than two stars, must form with two stars massive enough to become supernovae and produce two compact objects, and the compact objects thus formed must remain bound to each other through the two supernova explosions that may eject large amounts of mass from the system, or the binary must result from gravitational capture in a dense cluster.[1] Although these are improbable scenarios, calculations and the observation of compact-object binaries indicate that they are not impossible. Some theoretical estimates indicate that formation of a neutron star binary can happen often enough to produce about one neutron star merger each day in the observable Universe. The probability of forming a neutron star binary in any one region of space is very small, but the Universe is a very big place.

Figure 24.1 shows a numerical simulation of a merger in a binary neutron star system. The orbit of the binary has decayed steadily as a result of gravitational wave emission, causing the stars to spiral together at a rapidly increasing rate near the end. The sequence of images in Fig. 24.1 illustrates the merger over a period of milliseconds near when the surfaces of the neutron stars first touch. The very large asymmetric mass distortion, the high velocities generated by the revolution on millisecond timescales, and the highly compact nature of the mass distribution, imply that mergers of neutron stars will be a very strong source of gravitational waves. As Fig. 22.8 illustrates, the frequencies for gravitational waves emitted from neutron star mergers (the region labeled compact binary inspirals) are expected to lie in the range accessible to Earth-based interferometers like LIGO or Virgo.

The outcome of such neutron star mergers will depend on the total mass of the two neutron stars. Presently, our (relatively poor) understanding of neutron star equations of state suggests an upper limit $\sim 2-3\ M_\odot$ for the mass of a rapidly spinning neutron star. If the total merger mass is less than this the merger will likely produce a rapidly spinning neutron star. If the total merger mass is more than $\sim 2-3\ M_\odot$, the likely outcome will be a Kerr black hole with a large spin. Merger of neutron stars also may produce a short-period gamma ray burst (see Section 15.7.8), raising the possibility of detecting a gamma-ray burst and gravitational waves in coincidence. The first observation of such an event will be discussed in section 24.7.

---

[1] Capture can occur either through tidal dissipation of kinetic energy in a binary encounter, or in a 3-body encounter in which one object carries off enough energy to leave the other two bound gravitationally.

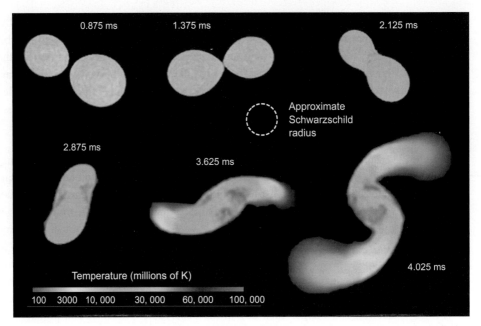

Fig. 24.1 Simulation of the merger of two neutron stars. The elapsed time is about 3 ms and the approximate Schwarzschild radius for the combined system is indicated. The rapid motion of several solar masses of material with large quadrupole distortion and sufficient density to be compressed near the Schwarzschild radius indicates that this merger should be a strong source of gravitational waves. Simulation by Stephan Rosswog; visualization by Richard West.

24.1.2 Stellar Black Hole Mergers

Merging stellar black holes (formed from the collapse of stars and having mass less than $\sim 100 \, M_\odot$) should be strong sources of gravitational waves. Such mergers are expected to have many qualitative similarities to the merging neutron stars described in the preceding section and are expected to leave behind a rapidly spinning Kerr black hole. There is one essential difference, however. The collision of two neutron stars involves dense matter and large gravitational energy, but both Schwarzschild and Kerr black holes correspond to *vacuum solutions* of the Einstein equations. Thus, there is only gravity – that is, curved spacetime – with no matter involved in the black hole collisions (neglecting the possibility of mass accreting from other nearby objects), since all mass is concentrated at the black hole singularities, which are hidden by the event horizons.

It follows that the merger of two black holes should be a simpler event than the merger of two neutron stars, for which the dense matter and the corresponding equation of state would enter into the description of the merger. Indeed, in the simplest picture the initial configuration for the merger of two black holes is characterized only by the spins and masses of the two black holes, and the binary orbital parameters (the no-hair theorem for black holes). The final configuration is even simpler, since the single Kerr black hole will be characterized by its mass and spin, once it has damped into its equilibrium

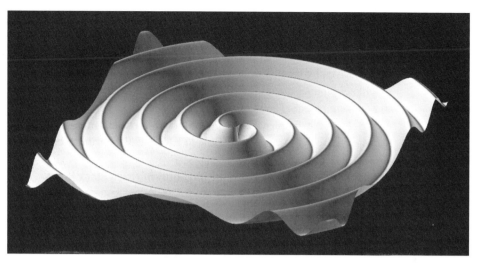

Fig. 24.2 Snapshot of gravitational wave emission from inspiral of a binary black hole system. Image from https://commons.wikimedia.org/wiki/File:Wavy2.gif – GNU Free documentation licence.

configuration by emission of gravitational radiation. Figure 24.2 illustrates gravitational wave emission from an inspiraling black hole pair based on numerical simulations of general relativity. The black holes revolving around their center of mass are at the center and the spirals trailing from them are the outwardly propagating gravitational waves that are being generated by the revolving binary system. In Section 24.2 observational evidence for gravitational waves from the merger of two stellar black holes will be discussed.

24.1.3 Merger of a Black Hole and a Neutron Star

Just as there is evidence for the existence of black hole binaries and neutron star binaries, it is possible to have a binary system in which one object is a neutron star and one a stellar black hole. As for other compact binaries, the orbit will decay by gravitational wave emission, leading eventually to a merger. Equations of state and observations suggest an upper limit of around 2 $M_\odot$ for neutron stars, but there is in principle no limit on the mass of the black hole. Since the black hole component of the binary already exceeds the upper mass limit for a neutron star, it is likely that the outcome of a merger between a black hole and a neutron star is a rapidly spinning Kerr black hole. As for the neutron star merger, the dense matter and associated equation of state for the neutron star likely makes this merger more complex than that of two black holes.

24.1.4 Core Collapse in Massive Stars

In the core collapse of a massive star that leads to a supernova the collapsing core has densities comparable to or greater than that of a neutron star and in the collapse a solar mass or more of material may be set in motion with velocities of order 10% of light velocity.

If such a collapse were to proceed with spherical symmetry, no gravitational waves would be produced. However, numerical simulations of core collapse generally find large asymmetries generated by phenomena such as large-scale supersonic convection (which is in turn driven by large entropy or concentration gradients in the core). These asymmetries will likely produce quadrupole distortions in the mass distribution that vary rapidly in time and thus are potential sources of strong gravitational waves [137, 138, 173]. As indicated in Fig. 22.8, the frequencies of gravitational waves from core collapse supernovae are expected to lie in the range accessible to LIGO and other Earth-based interferometers.

24.1.5 Merging Supermassive Black Holes

As we discussed in Section 14.5, the cores of many – perhaps all – massive galaxies host black holes containing 10^6–10^9 solar masses. For example, the resolved orbits of individual stars at the center of the Milky Way indicate the presence of a $4 \times 10^6 \, M_\odot$ black hole and analysis of average velocity fields near the centers of giant elliptical galaxies often indicate that billions of solar masses of non-luminous material is packed into a region comparable in size to the Solar System. There also is very strong observational evidence that galaxy mergers have been common in the past history of the Universe. Therefore, it may be expected that the merger of supermassive black holes can occur as a consequence of galaxy collisions. The expected frequency of the gravitational waves is too low for observatories like LIGO or Virgo, but the proposed space-based LISA array would be sensitive to such events with black hole masses less than about $10^7 \, M_\odot$ (see Fig. 22.8), while the pulsar arrays discussed in Section 22.6.3 might be able to detect gravitational waves from even more massive mergers.

Quantitative investigation of supermassive black hole merger events requires large-scale computer simulations but dimensional analysis arguments based on the quadrupole power formula (23.16) and supported by numerical simulation suggest that the peak luminosity is set by the fundamental scale c^5/G defined in Eq. (23.19). For supermassive black holes this peak luminosity is expected to be

$$L \propto L_0 \simeq \eta \frac{c^5}{G} \simeq \eta \times 10^{59} \text{ erg s}^{-1}, \tag{24.1}$$

with the factor η accounting for details of the merger and being of order 1% in typical cases. If such events occur, their luminosities would likely be greater than for any other event in the Universe for a period of days. To set Eq. (24.1) in perspective, if $\eta \sim 0.01$ this luminosity is about a million times larger than the photon luminosities associated with supernovae and gamma-ray bursts, ten billion times larger than quasar luminosities, and about 23 orders of magnitude larger than the luminosity of the Sun.

24.1.6 Sample Gravitational Waveforms

Some computed gravitational waveforms for events of the type described above are displayed in Fig. 24.3. There we see that different classes of events have characteristic waveforms and within classes the waveforms are often sensitive to details. For example,

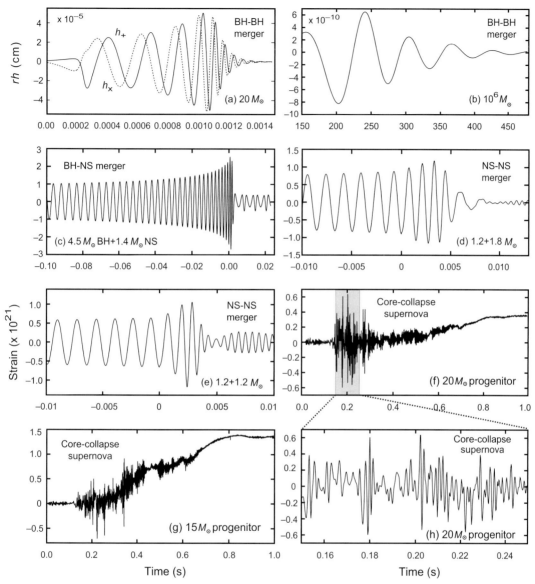

Fig. 24.3 Some computed gravitational waveforms. All are h_+ polarization except for (a), where both h_+ and $h_\times$ are shown. In the bottom six panels strain is given in dimensionless units of 10^{-21} by assuming a distance to the source. In the top two panels rh is shown, where r is the distance to the source in cm. Further details may be found in the references for each case. (a)–(b) Equal-mass black hole mergers (BH–BH) [6, 33, 35, 59]. (c) Black hole and neutron star merger (BH–NS) [6, 143] at 15 Mpc. (d)–(e) Neutron star mergers (NS–NS) [210] at 15 Mpc. Case (d) is a $1.2\,M_\odot + 1.8\,M_\odot$ merger (all masses baryonic). Case (e) is a $1.2\,M_\odot + 1.2\,M_\odot$ merger. (f)–(h) Supernova at 15 kpc for two progenitor masses; time measured from bounce [257]. Panel (h) displays the initial burst of panel (f) at higher resolution. Influence of microphysics on the form of supernova gravitational waves is discussed in [161, 256, 257]. Composited from waveforms at https://astrogravs.gsfc.nasa.gov/docs/catalog.html and Fig. 1 from K. N. Yakunin *et al. Physical Review*, **D92**, 084040 – Published 19 October 2015. Reprinted with permission from the American Physical Society.

mergers are characterized by the chirp waveform (amplitude and frequency rising rapidly near merger, followed by low-amplitude ringdown), with details depending on the type of merger, while supernova explosions are characterized by a much more complex wave pattern reflecting detailed microphysics that varies with characteristics of the progenitor star. (As a consequence, gravitational waves from mergers will likely be easier to pick out of the observational background than those from supernova events.) Hence, it may be expected that waveform templates like those displayed in Fig. 24.3 may be used to identify classes of observed gravitational wave events (see also the discussion of matched filtering in Section 24.3.2) and, with data from a sufficient number of events, the detailed waveforms might shed new light on the physics underlying each class of events.

24.2 The Gravitational Wave Event GW150914

On September 14, 2015, a century after the prediction [79, 80] that gravitational waves should exist, the two LIGO gravitational wave interferometers, one in Livingston, Louisiana, and the other in Hanford, Washington, observed a transient signal that was identified within three minutes by low-latency scans for generic gravitational wave transients as a strong gravitational wave candidate.[2] The signal had the obvious character of a compact merger event (the *chirp* pattern described below) and low-latency data pipelines scanning with matched filtering (Box 24.2) quickly ruled out the merger of two neutron stars or the merger of a black hole and neutron star, focusing attention on a gravitational wave from the coalescence of two black holes as the likely source of the event. Several months of thorough analysis confirmed with significance greater than the 5σ confidence level (corresponding to a false alarm probability $< 2 \times 10^{-7}$) that the transient labeled GW150914 was indeed a gravitational wave emitted from the merger of two black holes [10].

24.2.1 Observed Waveforms

The waveforms observed by the Hanford and Livingston detectors are shown in the upper panel of Fig. 24.4. The gravitational wave arrived first at the Livingston detector (L1) and then $6.9^{+0.5}_{-0.4}$ ms later at the Hanford detector (H1). In the top-right image the H1 wave has been superposed, shifted by 6.9 ms, and inverted to account for relative orientations of the two detectors (orientations relative to local north of L1 and H1 differ by 72°). This superposition and a 24:1 signal to noise ratio leaves little doubt that the same wavefront, traveling at lightspeed, passed first through the Livingston and then the Hanford interferometers.

---

[2] By the date of the event Advanced LIGO had been tested and the pre-run calibrations completed, but the first official observational run (denoted O1) was not scheduled to begin until September 18. Thus, the first detection of a gravitational wave on September 14 happened before Advanced LIGO was officially operational!

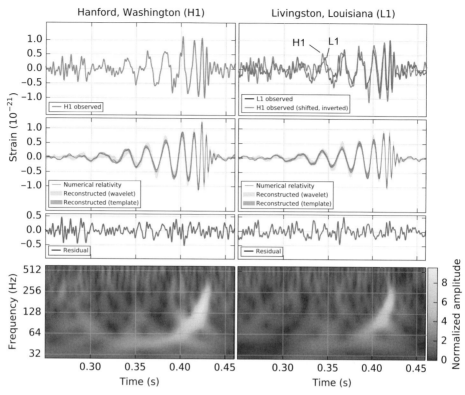

Fig. 24.4 The gravitational wave event GW150914 observed by LIGO [10]. The left panels correspond to data from the Hanford detector (H1) and the right panels to data from the Livingston detector (L1). All times are relative to 09:50:45 UTC on September 14, 2015. The top row is measured strain in units of 10^{-21}. In the top right panel the Hanford signal has been superposed on the Livingston signal as described in the text. The second row shows numerical relativity simulations [160] of the waveform assuming a binary black hole merger event projected onto each detector in the 35–350 Hz band. Shaded areas indicate 90% credible regions for two independent waveform reconstructions. The third row shows the residuals after subtracting the numerical relativity waveform in the second row from the detector waveform in the first row. The fourth row shows strain versus frequency and time, with grayscale contours indicating strain amplitude. The rapidly rising frequency pattern (chirp) is indicative of a binary merger event. For visualization purposes the data and the simulations have been filtered, as described in more detail in Ref. [10].

In the second row of Fig. 24.4, simulations of the waveform using a numerical relativity calculation [160] for merging black holes and wavelet reconstructions with and without an astrophysical black hole merger model are shown. The third row displays the result of subtracting the numerical relativity waveform for binary black hole merger in the second row from the observed waveform in the first row. The last row shows a time-frequency representation of the data, with the grayscale contours representing strain. This plot indicates that over a period of ~ 0.2 seconds the signal swept upward in frequency from about 35 to 250 Hz ("the chirp," indicative of the final rapid inspiral of a merger event), with a peak strain $\sim 1.0 \times 10^{-21}$. It may be noted that this strain changed the separation between the test masses by only about 4×10^{-16} cm, which is 0.005 times the diameter of a proton.

24.2.2 Source Localization

The light-travel time between Livingston and Hanford is 10 ms, so a valid coincidence between the L1 and H1 detectors must lie within a ± 10 ms window, depending on the orientation of the wavefront relative to the Livingston–Hanford axis. The actual difference in time for the coincidence then provides the primary directional information about the source of the wave. Amplitude and phase information in the two detectors then can be used to further refine the location (and also to infer the distance to the source and the orientation of the binary, as discussed further in Section 24.3.5) [11]. The observed difference of ~ 7 ms with the gravitational wave passing through the Livingston detector first indicates that the wave source was in the southern-hemisphere sky.

At the time of observation no other gravitational wave observatories were acquiring data (the Virgo detector was being upgraded and the GEO 600 detector – which would not have had sufficient sensitivity anyway to detect the event – was running but not in observational mode). Thus only the two LIGO detectors were available to determine source position, primarily through arrival time. This permitted the source to be localized to an area of approximately 600 deg^2, as illustrated in Fig. 24.5, which was later improved to 230 deg^2. Follow-up observations of that portion of the sky at radio, optical, near-infrared, X-ray, and gamma-ray wavelengths with ground- and space-based facilities reported no obvious counterpart to the gravitational wave event [13]. Thus GW150914 appears to have been invisible to non-gravitational observatories.

24.2.3 Comparisons with Candidate Events

The qualitative features of the data provide strong evidence that GW150914 corresponded to a binary merger event. Over a period of 200 ms the signal increased in frequency from 35 Hz to 250 Hz, and in magnitude to maximum value, over about 8 cycles. This suggests strongly the inspiral of orbiting masses m_1 and m_2 caused by gravitational wave emission. To leading order in powers of v/c, the phase information in this evolution is characterized primarily by a particular combination of the masses called the *chirp mass* $\mathcal{M}$,

$$\mathcal{M} = \mu^{3/5} M^{2/5} = \frac{(m_1 m_2)^{3/5}}{(m_1 + m_2)^{1/5}} = \frac{c^3}{G}\left(\frac{5}{96}\pi^{-8/3} f^{-11/3} \dot{f}\right)^{3/5}, \qquad (24.2)$$

where $\mu = m_1 m_2/(m_1 + m_2)$ is the reduced mass, $M = m_1 + m_2$ is the total mass, f is the observed gravitational wave frequency, $\dot{f}$ is its time derivative, and the last form follows from retaining the lowest order in a post-Newtonian expansion [17, 48, 49]. Thus the chirp mass can be deduced from the observed wave, and this in turn implies a relationship between m_1 and m_2. Estimating f and $\dot{f}$ from the GW150914 data yields $\mathcal{M} \simeq 28\, M_\odot$, implying a total mass $M = m_1 + m_2 \geq 70\, M_\odot$ in the detector rest frame (see Problem 24.2). The sum of the Schwarzschild radii for the binary components is thus

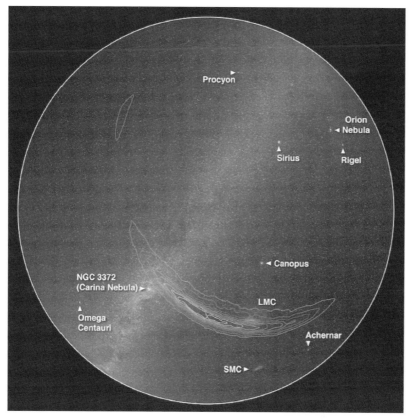

Fig. 24.5 Contours indicating approximate location of the source for GW150914 near the Large Magellanic Cloud (LMC). Inner contour is 10% confidence and outer contour is 90% confidence. The analysis localizing the source position may be found in Ref. [12].

constrained to be $2GM/c^2 \geq 210$ km. This implies that to reach an orbital frequency of 75 Hz (half the average frequency near maximum strain; recall from Section 23.2.1 that the frequency of the gravitational wave is twice the binary revolution frequency), the objects had to be very compact with small separation. For example, Newtonian point masses orbiting at 75 Hz would be only about 350 km apart.

Binaries involving white dwarfs are excluded because they are neither compact enough nor massive enough. Binary neutron stars are eliminated because they are compact enough but not massive enough. A binary consisting of a neutron star and a black hole with the observed chirp mass is eliminated because it would be required by the chirp mass to be very massive (see Problem 24.2) and so would merge at much lower frequency. Hence, the merger of two black holes with their summed masses in the vicinity of 60–70 $M_\odot$ is the only known event consistent with the data. This conclusion is reinforced by noting that the observed decay of the waveform at the end of the event was consistent with damped oscillations of a Kerr black hole relaxing to a final stationary configuration [10].

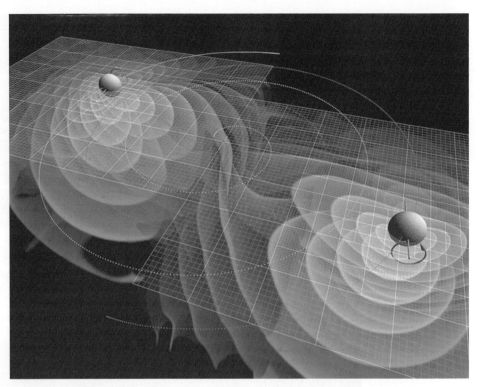

Fig. 24.6 Numerical simulation of gravitational wave emission from merger of two black holes [45]. Grayscale contours represent the amplitude of the gravitational radiation. Inspiral orbits of the black holes are indicated by dotted curves and their spins are indicated by the arrows. Credit: C. Henze/NASA Ames Research Center.

24.2.4 Binary Black Hole Mergers

The preceding inferences suggest investigating general relativistic simulations for the merger of a binary black hole system to produce a single rotating (Kerr) black hole. A visualization of such a merger simulation is shown in Fig. 24.6 for the final inspiral phase. After comparisons with many candidate event templates it was concluded that the best description of the event was the merger of two $\sim 30\ M_\odot$ black holes to form a $60\ M_\odot$ Kerr black hole having a spin $\sim 70\%$ of the maximum possible for a Kerr black hole (see Section 13.1.2), with emission of $\sim 3\ M_\odot c^2$ of gravitational wave energy. From the black hole merger model the peak gravitational wave luminosity was $\sim 3.6 \times 10^{56}\ \mathrm{erg\,s^{-1}}$, which is equal to the conversion of *$\sim$200 solar masses per second* into gravitational wave energy and is some 100, 000 times larger than the peak photon luminosity of a supernova explosion!

From the strain amplitude a luminosity distance (the distance found by comparing observed intensity with actual luminosity; see Box 19.3) of 240–570 Mpc was inferred.[3] This corresponds to a redshift $z = 0.05$–0.12, implying that the merger

---

[3] A description of how distance is determined may be found in Ref. [11]. The possibility that gravitational waves might serve as *standard sirens* analogous to standard candles for optical astronomy is discussed in Box 24.1.

24.2 The Gravitational Wave Event GW150914

Table 24.1 Properties of three LIGO black-hole merger candidates [11]

Quantity	GW150914	GW151226	LVT151012
Signal to noise ratio	23.7	13.0	9.7
False alarm rate (yr^{-1})	$< 6.0 \times 10^{-7}$	$< 6.0 \times 10^{-7}$	0.37
Significance	$> 5.3\sigma$	$> 5.3\sigma$	1.7σ
Primary black hole mass ($M_\odot$)	$36.2^{+5.2}_{-3.8}$	$14.2^{+8.3}_{-3.7}$	23^{+18}_{-6}
Secondary black hole mass ($M_\odot$)	$29.1^{+3.7}_{-4.4}$	$7.5^{+2.3}_{-2.3}$	13^{+4}_{-5}
Chirp mass ($M_\odot$)	$28.1^{+1.8}_{-1.5}$	$8.9^{+0.3}_{-0.3}$	$15.1^{+1.4}_{-1.1}$
Total mass ($M_\odot$)	$65.3^{+4.1}_{-3.4}$	$21.8^{+5.9}_{-1.7}$	37^{+13}_{-4}
Final black hole mass ($M_\odot$)	$62.3^{+3.7}_{-3.1}$	$20.8^{+6.1}_{-1.7}$	35^{+14}_{-4}
Final black hole spin	$0.68^{+0.05}_{-0.06}$	$0.74^{+0.06}_{-0.06}$	$0.66^{+0.09}_{-0.10}$
Radiated mass ($M_\odot$)	$3.0^{+0.5}_{-0.4}$	$1.0^{+0.1}_{-0.2}$	$1.5^{+0.3}_{-0.4}$
Peak luminosity (erg s^{-1})	$3.6^{+0.5}_{-0.4} \times 10^{56}$	$3.3^{+0.8}_{-1.6} \times 10^{56}$	$3.1^{+0.8}_{-1.8} \times 10^{56}$
Source redshift z	$0.09^{+0.03}_{-0.04}$	$0.09^{+0.03}_{-0.04}$	$0.20^{+0.09}_{-0.09}$
Luminosity distance (Mpc)	420^{+150}_{-180}	440^{+180}_{-190}	1000^{+500}_{-500}
Sky localization (deg^2)	230	850	1600

Masses in source frame. Multiply by $(1 + z)$ for mass in detector frame. Source redshift assumes standard cosmology. Spin given in units of spin for an extremal Kerr black hole (see Section 13.1.2). The sky localization error boxes assume 90% credible intervals (see Fig. 24.11).

event occurred 12.2–13.1 Gyr after the formation of the Universe, assuming the standard cosmology (Section 19.4.7). There is significant uncertainty in determining the luminosity distance and corresponding redshift because the orientation of the binary also affects signal strength and it is not well determined. The parameters along with their uncertainties for the black hole merger GW150914 that were determined from the waveform simulations compared with data are displayed in column two of Table 24.1, along with data for two other merger events that will be discussed further below. The analysis is in principle able to determine the spins of the initial and final black holes, with the final spin generally more tightly constrained than the initial spins. Only the spin of the final Kerr black hole is shown in Table 24.1 because in the present data analysis uncertainties for initial spins are of order 100%.

More details of the GW150914 simulation are presented in Fig. 24.7. The inset image illustrates the event horizons of the two black holes merging into that of a single final black hole. The bottom panel shows the separation of the two black holes as a function of time in units of the Schwarzschild radius $R_S = 2GM/c^2$, where M is the total mass from Table 24.1, and also their effective relative velocity $v/c = (GM\pi f/c^3)^{1/3}$, where f is the gravitational wave frequency from the numerical relativity simulation.

In Fig. 24.8 a simulation of what the black holes might have looked like from up close during the merger is shown. The dark, well-defined shapes are the shadows of the black

Box 24.1 — **Standard Sirens**

The distance to a gravitational wave source is proportional to the inverse of the observed strain amplitude and it has been proposed that if gravitational wave sources have well-understood luminosities they could act as *standard sirens*, which are the gravitational wave analogs of standard candles like those discussed in Box 19.2 to determine distances. How standardized intrinsic gravitational source strengths are has not been established fully, and the potential accuracy is limited by observational factors: pointing error, since gravitational waves are not very precisely localized by current instrumentation, dependence of the luminosity on source orientation if the source is a binary merger, and degradation of precision by gravitational lensing effects, but such standard sirens still could be useful indicators of distance. For the merger of supermassive black holes in particular, standard sirens might be rare but usable at quite large redshifts [124]. Use of such methods to determine the Hubble constant by comparing gravitational and electromagnetic waves emitted from a binary neutron star merger will be described in Section 24.7.1.

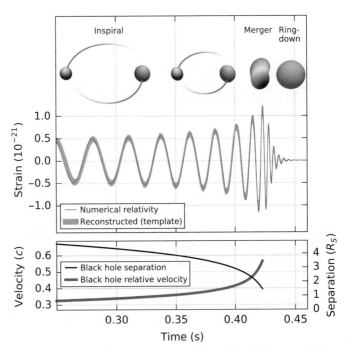

Fig. 24.7 Estimated gravitational wave strain projected onto the Hanford detector H1 for GW150914 [10]. These represent the full bandwidth of the waveforms, without the filtering of Fig. 24.4. The inset image illustrates the black hole horizons as the two black holes coalesce, as deduced from numerical relativity calculations. The bottom panel illustrates the effective black hole separation in units of the Schwarzschild radius assuming Keplerian trajectories, and the relative velocity v/c for the two black holes during the merger.

24.2 The Gravitational Wave Event GW150914

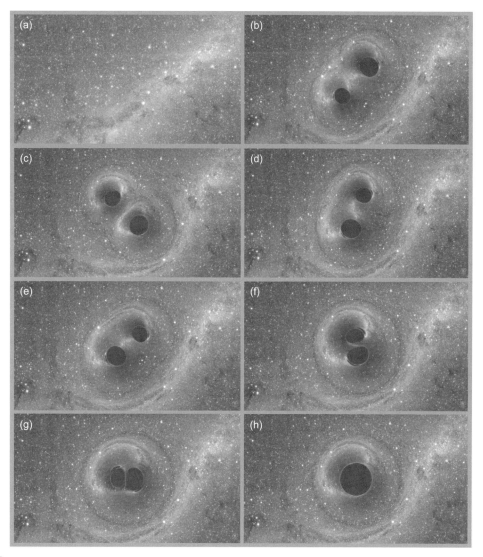

Fig. 24.8 Computer simulation of the GW150914 merger. (a) The undistorted background field of stars in the absence of the black holes. (b)–(g) Successively later times in the merger sequence. (h) The final Kerr black hole. Notice the strong gravitational lensing effects near the black holes. The background stars are fixed in position as in (a) for each panel but the gravitational lensing completely distorts their apparent positions. The ring around the black holes is an *Einstein ring* caused by strong focusing of light from stars behind the black holes. Images extracted from video in Ref. [69]. Image credit: SXS, the Simulating eXtreme Spacetimes (SXS) project (www.black-holes.orgs).

hole event horizons as they block all light from behind. The flattened dark features around them are strong gravitational lensing effects. Finally, in Fig. 24.9 the GW150914 waveform has been added to Einstein's blackboard. We suspect that Einstein would not have objected, though it should be noted that at various times Einstein expressed skepticism about the

Fig. 24.9 An iconic photograph of Einstein, modified to add GW150914 to his blackboard.

possibility of detecting gravitational waves (or even their existence), and seems to have always doubted that real black holes could form from stellar collapse [178].

24.3 Additional Gravitational Wave Events

The first Advanced LIGO observing run (which is summarized in Ref. [11]) extended from late 2015 into early 2016, and yielded three candidates for gravitational waves, all interpreted as binary black hole mergers at luminosity distances of 400 Mpc or greater. The strongest of these, GW150914 was confirmed at the 5.3σ confidence level as a gravitational wave event involving the merger of 36 $M_\odot$ and 29 $M_\odot$ black holes, as has been discussed above.

24.3.1 GW151226 and LVT151012

The second strongest event (GW151226, observed December 26th, 2015) also has been confirmed at the 5.3σ confidence level to be a gravitational wave, this time from the merger of 14 $M_\odot$ and 8 $M_\odot$ black holes at a similar distance as for GW150914. The GW151226 strain wave is shown in Fig. 24.10, the localization of the source on the sky is displayed in Fig. 24.11, and the parameters of the merger are displayed in column three of Table 24.1. Finally, Fig. 24.12 exploits artistic license to display frames from the simulation of GW150914 in Fig. 24.8 and a similar one for GW151226 in the same image.

Parameters of the third (weakest) event from the first Advanced LIGO run, LVT151012, are displayed in the fourth column of Table 24.1. The analysis suggests that this is a gravitational wave originating in the merger of 23 $M_\odot$ and 13 $M_\odot$ black holes at a (highly uncertain) distance of $\sim$ 1000 Mpc, but this was established only at the 1.7σ confidence level. Since the standard LIGO threshold for reporting a bona-fide gravitational wave is 5σ,

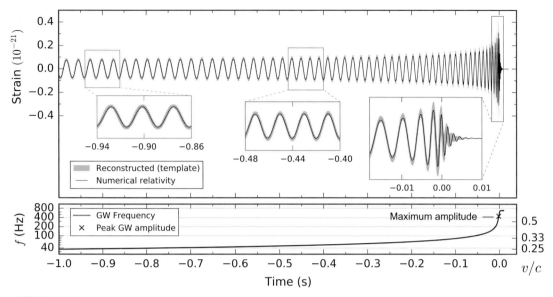

Fig. 24.10 Estimated gravitational wave strain for GW151226 projected onto the LIGO Livingston detector (full bandwidth, without filtering) [14]. *Top:* The reconstructed wave (90% credible region) is shown in gray and a numerical relativity simulation in black. *Bottom:* Gravitational wave frequency (left axis) computed from numerical relativity waveform. The cross marks the maximum amplitude, approximately coincident with merger of the black holes. The right axis gives an effective relative velocity v/c that can be related to f during the inspiral using post-Newtonian approximations (see Section 25.2.1).

this event is labeled an LVT (LIGO–Virgo Trigger) rather than a gravitational wave (GW). The analysis suggests that it was a gravitational wave of similar power as GW151226, but from a source twice as far away so the peak strain was only about 0.3×10^{-21}, compared with 0.8×10^{-21} for GW150914 and 0.35×10^{-21} for GW151226, with a signal to noise ratio more than twice as small for LVT151012 as for GW150914.

24.3.2 Matched Filtering

The strain signal and noise levels are displayed in the left panel of Fig. 24.13 for these three events. The amplitude of GW150914 is significantly larger than for the other two events and at merger it lies well above the detector noise spectrum. Hence the signal to noise ratio of ~ 24 in Table 24.1. For GW151226 and LVT151012 the amplitudes are much closer to the level of the noise spectrum at merger, accounting for the lower values of the signal to noise ratio in Table 24.1. In these events, it was necessary to rely on a signal processing method called *matched filtering* described in Box 24.2, which originated in applications such as radar and 2D image processing, to correlate a waveform model with data over the sensitive band of the detector, allowing the waveform to be extracted from the noise.

In the top panel of the figure in Box 24.2, GW151226 strain data for the two LIGO detectors filtered with a 30–600 Hz bandpass filter to suppress large fluctuations outside

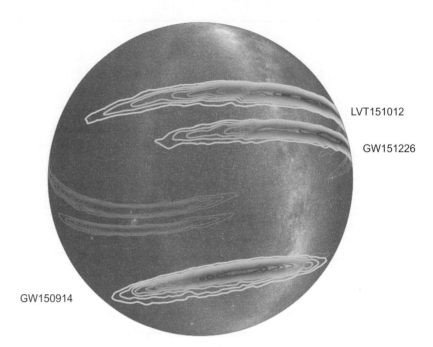

Fig. 24.11 Localization on a 3D projection of the celestial sphere for two confirmed gravitational waves (GW150914 and GW151226) and one possible gravitational wave (LVT151012) detected by LIGO in the first Advanced LIGO observational run. Outer contours are 90% confidence and inner contours are 10% confidence level. Localization is relatively poor with only two gravitational wave detectors running, and the strong event GW150914 (shown previously in Fig. 24.5) is much better localized than the other two events. With the addition in 2017 of the Virgo detector in Italy, the localization of gravitational wave events became substantially improved by LIGO–Virgo coincidences, as will be illustrated in Section 24.3.5. Courtesy Caltech/MIT/LIGO Laboratory.

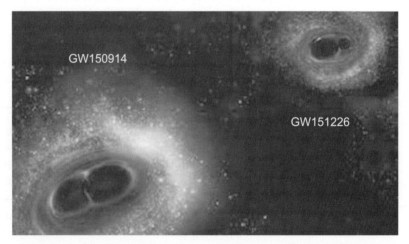

Fig. 24.12 Display of frames from simulations of the mergers leading to GW150914 (lower left) and GW151226 (upper right) in the same image. Courtesy Caltech/MIT/LIGO Laboratory.

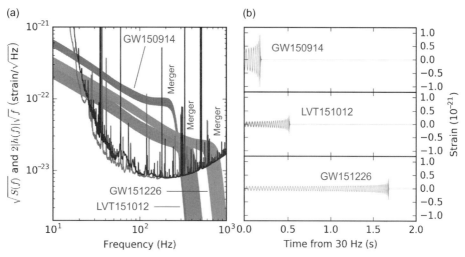

Fig. 24.13 (a) Amplitude spectral density of strain noise for the L1 (black spectrum) and H1 (gray spectrum) detectors, and recovered signals of GW150914, GW151226, and LVT151012 (bands) from the first observing run of Advanced LIGO. Units are described in Ref. [11]. (b) Time evolution of the recovered strain signals from a reconstruction. Figure adapted from Ref. [11].

this range and with band-reject filters to remove known instrumental spectral lines are plotted in gray, and a best-match template with the same filters applied is plotted in black.[4] The signal to noise ratio (SNR) is in general poor, as indicated in the next two panels, and in the frequency–time plot shown in the bottom panel almost no structure can be discerned. Unlike for the much stronger GW150914, which was identified online in low-latency generic scans as a strong gravitational wave candidate, matched filtering using merger templates was essential to identify GW151226 as a gravitational wave candidate, and to extract the results of Fig. 24.10 in offline analysis.

Matched filtering works because the waveform sought in the data has a well-defined, orderly shape. From Fig. 24.3 we see that binary merger events among some combination of black holes and neutron stars generally exhibit a very periodic waveform, but that supernova explosions may yield a gravitational wave that is much more stochastic because the supernova is an anisotropic *explosion* and not a simple inspiral. On the positive side, this may mean that a supernova gravitational wave could encode more astrophysical information than one from black hole or neutron star mergers. However, this also means that the supernova gravitational wave may be much less amenable to matched filtering, which may present a technical problem for identification and analysis of gravitational waves emitted from supernova explosions when they are eventually observed.

---

[4] Several methods are used to calculate the library of template waveforms for matched filtering of gravitational waves. In this example a combination of post-Newtonian approximations (Section 25.2.1), numerical relativity (Section 8.8.3), and *effective one-body methods* (which map the relativistic 2-body problem onto a simpler one of a single test particle moving in an effective metric [58]) were used.

Box 24.2 Matched Filter Techniques

Detection of gravitational waves relies on *matched filtering*, which correlates a known (template) signal with a measured signal to ascertain the presence of the template in the measured signal. It is considered to be the optimal linear filter for maximizing signal to noise in the presence of added random noise. In a gravitational wave application, libraries of template waveforms computed theoretically are used to analyze the detected signal for a match. Matched filtering was essential to discovery of GW151226, as illustrated in the following figure taken from Ref. [14].

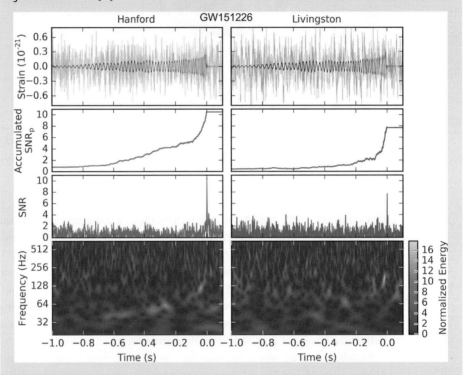

The top panels show the strain signal (gray) and a template wave (black), the middle panels indicate the signal to noise ratio (SNR), and the bottom panel shows energy versus frequency and time. As discussed further in the text, this event would have been difficult to even find in the noise without matched filtering.

24.3.3 Binary Masses and Inspiral Cycles

A comparison of Figs. 24.4, 24.10, and 24.13 indicates that many more oscillations were detected in GW151226 over a longer period than for GW150914. This is a consequence of the lower masses for the merger in GW151226 relative to GW150914, which means that the GW151226 wave spent more time in the detector sensitive band above about 30 Hz. (Notice from Fig. 22.8 that the LIGO detectors have a rather sharp loss of strain sensitivity below a frequency of about 30 Hz.) This is illustrated clearly in the right panel of Fig. 24.13, where about 55 inspiral oscillations were observed for GW151226 over a period of almost two seconds, as opposed to only about 8 cycles over a period of about 0.2 seconds for GW150914. From the rule of thumb given in Section 22.4.3, the 55 cycles observed for GW151226 extended the effective distance reach of the instrument by a factor of about $\sqrt{55}$ for that event, which was essential in identifying it as a gravitational wave. This dependence of frequency on black hole masses in mergers has some important consequences: (1) The lower-mass merger GW151226 was more difficult to detect directly because its strain was smaller than for the higher-mass GW150914, but because GW151226 was observed over many more inspiral cycles than for GW150914 matched filtering was particularly effective for it and it was more sensitive to some inspiral parameters than GW150914. (2) Even more massive black hole mergers than that for GW150914 will be increasingly difficult for LIGO to study because fewer cycles will come above the detector cutoff at $\sim$ 30 Hz.

24.3.4 Increasing Sensitivity

The second Advanced LIGO observing run commenced in late November, 2016, with the sensitivity of the Livingston detector below a frequency of about 100 Hz having been increased by 25% relative to the first run.[5] (The increased sensitivity results primarily from a reduction of scattered light in the interferometer.) This should make it more likely that a weaker event like LVT151012 could be confirmed with higher confidence as a gravitational wave. Enhanced sensitivity also will increase the distance at which events can be detected, and thus expand the volume of space for potential events (as the cube of the distance for low redshifts). For example, a 25% increase in detector sensitivity translates into a factor of about two in event rate at low redshifts, other things being equal. It is anticipated that successive observing cycles with LIGO will see incremental improvement in sensitivity until reaching full design capability in $\sim$ 2020, which will enable the gravitational wave reach indicated in Fig. 22.9. In addition, Advanced Virgo adds another detector of similar capability to the two Advanced LIGO detectors.

24.3.5 LIGO–Virgo Triple Coincidences

A network of multiple gravitational wave detectors can determine the location of a gravitational wave source by triangulation using time differences, phase differences, and amplitude ratios for arrival of the gravitational wave at difference sites. When advanced Virgo came online in 2017, further gravitational wave sensitivity resulting from simultaneous detection of a gravitational wave by the two LIGO detectors and Virgo became possible. The first such triple coincidence was realized in the detection of GW170814, which corresponded to the merger of 30.5 $M_\odot$ and 25.3 $M_\odot$ black holes to produce a final 53.2 $M_\odot$ black hole with spin $\sim$ 70% of maximal for a Kerr black hole [18]. The merger was at a luminosity distance of 540 Mpc (redshift $z \sim 0.11$) and

---

[5] By the time of this writing in December, 2017, three additional gravitational waves from black hole mergers had been reported by LIGO in the second observing run: GW170104, GW170608, and GW170814.

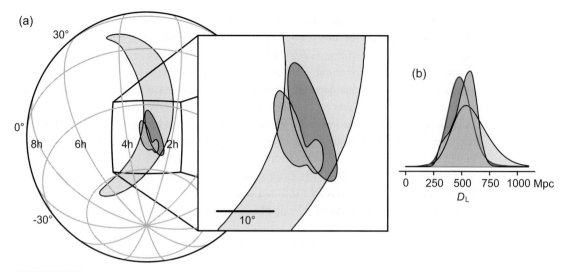

Fig. 24.14 Localization of GW170814 by a LIGO–Virgo triple coincidence [18]. Lightest gray indicates LIGO rapid localization, darkest gray represents initial LIGO–Virgo coincidence, and intermediate gray represents the triple coincidence after full analysis. All contours are 90% confidence level. (a) Position on the celestial sphere. (b) Luminosity distance.

released a total of $\sim 2.7 M_\odot c^2$ of gravitational wave energy, reaching a peak luminosity of $\sim 3.7 \times 10^{56}$ erg s^{-1}.

The increased localization both in angular position and luminosity distance afforded by the triple coincidence relative to detection by LIGO alone is illustrated in Fig. 24.14. The localization in angular position on the celestial sphere was improved from an uncertainty box of 1160 deg^2 in the LIGO rapid response alone to 60 deg^2 in the final analysis of the LIGO–Virgo triple coincidence. This also permitted the inferred luminosity distance to be sharpened from 570^{+300}_{-230} Mpc in the initial LIGO rapid response to 540^{+130}_{-210} Mpc in the final triple-coincidence analysis. The overall effect was to decrease by an order of magnitude the volume of space in which the event could have occurred relative to that inferred from LIGO alone. A total of 25 facilities observing neutrinos, gamma-rays, X-rays, optical, and IR signals found no events coincident with GW170814 in this error volume.

24.4 Testing General Relativity in Strong Gravity

General relativity has passed all tests thus far but they have all been in the regime of relatively *weak gravity*. The most interesting predictions of general relativity – and the ones that provide the most stringent tests of the theory – are in the dynamical, highly nonlinear, strong-field regime such as the merger of black holes, where gravitational fields are large, space is strongly and dynamically curved, and velocities approach the speed of light.

No observations were available to test general relativity directly in very strong gravity[6] before the detection of the gravitational wave event described in Section 24.2, since all prior evidence for black holes was indirect, typically depending on electromagnetic radiation emitted far from the event horizon and thus likely from relatively weak gravity. Because GW150914 has been interpreted as the merger of two $\sim$ 30 solar mass black holes to form a $\sim$ 60 solar mass Kerr black hole, it is expected that the gravitational wave signature near peak strain and in the ensuing ringdown bears imprints of the strong field near the event horizons of the merging and final black holes. Hence, the observation of GW150914 is perhaps the most direct evidence yet presented of the existence of black holes, and of whether their properties are consistent with the general theory of relativity.

The preliminary analysis of the first gravitational waves detected by LIGO is broadly consistent with the predictions of general relativity, with no obvious signals of a breakdown of that theory [11, 14, 15].[7] To give one example, general relativity predicts that gravitational waves travel at the speed of light, implying that the putative graviton mediating the gravitational force is massless. Hence any deviations of gravitational wave speed from c may be parameterized in terms of the value of a finite mass for the graviton. If the graviton had a finite mass it would travel at a speed less than c and there would be a dispersion of the wave corresponding to lower frequencies traveling slower than higher frequencies. The LIGO and Virgo collaboration analysis of GW170104 found no evidence for such an effect and placed an upper limit for the mass of the graviton $m_{\rm g} \leq 7.7 \times 10^{-23}$ eV/c^2, at the 90% confidence level [19]. Thus GW170104 provided compelling evidence that the speed of gravity is indeed equal to the speed of light. An even more stringent limit on the speed of gravity will be discussed in Section 24.7.1.

The five gravitational waves from merging black holes that have been detected at the time of this writing in 2018 involved merging and resultant black holes ranging in mass from 7 $M_\odot$ to greater than 60 $M_\odot$. The satisfactory description of these events as understood thus far suggests that the basics of general relativity in strong fields are correct. However, these strong-field tests are still based on limited data. If LIGO, Virgo, and future gravitational wave detectors are able to detect events on a regular basis, even more stringent tests of general relativity should be forthcoming in the strong-field, highly dynamical limit.

24.5 A New Window on the Universe

The most enduring contribution of the gravitational waves detected in Advanced LIGO's first observing run may be that it opens a new window on the Universe for

---

[6] The best strong-gravity tests before GW150914 were in high-gravity binary pulsar systems; see Section 10.4.5. While these involved considerably stronger fields than earlier solar system and binary pulsar tests, they still represent relatively weak gravity compared with that near the event horizons of black holes.

[7] GW150914 and GW151226 taken together provide stronger constraints than that of GW150914 alone. Although GW150914 had a stronger signal, only about 8 of its oscillations were observed in the sensitive band above 30 Hz for LIGO, while approximately 55 inspiral cycles occurred in the sensitive band for GW151226 because it involved the merger of less massive black holes than for GW150914; see the right panel of Fig. 24.13.

strong-gravity events. For example, the detection of GW150914, the enormous radiated power (a peak luminosity of more than 10^{56} erg s^{-1}) inferred from the data, and that no reproducible electromagnetic observations registered the event despite a concerted effort, indicate that there could be many incredibly powerful events occurring in our Universe that leave no obvious electromagnetic footprint because they don't emit photons, or because any that are emitted are absorbed in intervening matter. If so, gravitational wave detectors (and perhaps neutrino detectors with greatly improved sensitivity) will be in the vanguard exploring these phenomenal.

24.6 Multimessenger Astronomy

It has been proposed that gravitational waves from mergers involving neutron stars might determine the neutron star equation of state better than our present understanding, and that gravitational waves from core collapse supernovae might constrain the supernova mechanism more rigorously than from present data. This would be particularly true if such events were observed in more than one way, such as detection of both gravitational waves and neutrinos emitted from a core collapse supernova, or detection of gravitational waves from a neutron star merger that is also observed as a short-period gamma-ray burst (Section 15.7.8). The simultaneous detection of multiple signals from the same event is termed *multimessenger astronomy* [29]. While technically demanding, it has the potential to lead to much deeper understanding of objects like neutron stars and core collapse supernovae.

With present detection efficiencies, observation simultaneously of both neutrinos and gravitational waves from a core collapse supernova is likely only for a supernova within the galaxy, or in very nearby galaxies like the Magellanic Clouds. The expected rate for such events is about once every 50 years in a large galaxy. Neutron star mergers are orders of magnitude less common than supernova explosions in a given galaxy, so significant probability for detecting gravitational waves from such an event requires detection sensitivity in a large spatial volume (see Fig. 22.9). Gamma-ray bursts are very powerful so they can be detected at large distances. However, for the short-period burst mechanism discussed in Section 15.7.8 to give an observable signal the highly beamed burst must have an orientation allowing detection on Earth. The first observed multimessenger gravitational wave event will be discussed in Section 24.7.

24.7 Gravitational Waves from Neutron Star Mergers

On August 17, 2017, the LIGO–Virgo collaboration detected gravitational wave GW170817, which had a very different signature than that of the previously detected black hole merger events. The signal built slowly in amplitude and frequency with more than 3000 wave cycles recorded over almost 100 seconds before peak (recall that the signal for the first gravitational wave detected, GW150914, lasted only a quarter of a second). This new kind of gravitational wave would be quickly interpreted as originating in the merger of two neutron stars [20], but the show wasn't over yet! Approximately 1.7 seconds after the peak strain of the gravitational wave both the Fermi Gamma-ray Space Telescope (Fermi) and the International Gamma-Ray Astrophysics Laboratory (INTEGRAL) observed a gamma-ray burst of two seconds duration in the same part of the sky as the source of the gravitational wave, and within hours various observatories discovered a new point source in the irregular/elliptical galaxy NGC 4993 lying within the position error box for the gravitational wave. In the ensuing weeks a multitude of observatories studied the transient afterglow in NGC 4993 (named officially AT 2017gfo) intensively at various wavelengths. Thus was the discipline of multimessenger astronomy born (see Section 24.6).

The gravitational wave was identified by matched filtering against post-Newtonian waveform models and corresponded to the loudest gravitational wave signal observed to that date, with a signal to noise ratio of 32.4 [21]. The coincidence of the gravitational wave and the gamma-ray burst is illustrated in Fig. 24.15 and the sky localization of the event is illustrated in Fig. 24.16. The final combined LIGO–Virgo sky position localization corresponded to an uncertainty area of 28 deg^2. The total mass determined for the binary was between 2.73 and 3.29 $M_\odot$, and the two individual masses were in the range 0.86 $M_\odot$ to 2.26 $M_\odot$, with the ranges in masses caused by uncertainties in the parameters of the

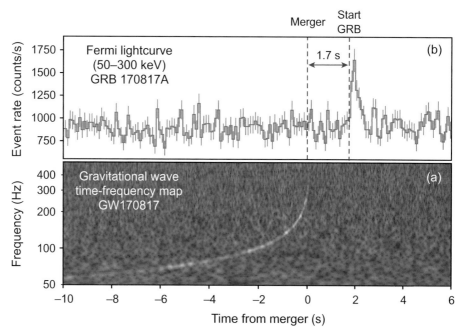

Fig. 24.15 (a) Gravitational wave GW170817 (LIGO) and (b) gamma-ray burst GRB 170817A (Fermi satellite) [21]. The source was at a luminosity distance of 40 Mpc (130 Mly) and the gravitational wave and gamma-ray burst arrived at Earth separated by only 1.7 seconds.

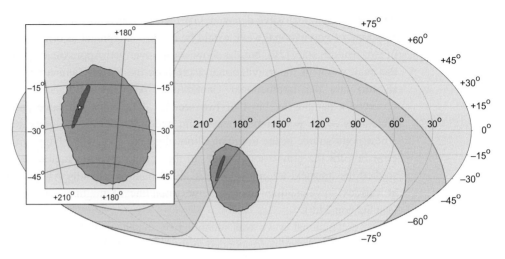

Fig. 24.16 Localization of gravitational wave GW170817 and gamma-ray burst GRB 170817A [21]. The 90% contour for LIGO–Virgo localization is shown in the darkest gray. The 90% localization for the gamma-ray burst is shown in intermediate gray. The 90% annulus from triangulation using the difference in GRB arrival time for Fermi and INTEGRAL is the lighter gray band. The zoomed inset shows the location of the transient AT 2017gfo (small white star) that was observed at various wavelengths. Axes correspond to right ascension and declination in the equatorial coordinate system.

waveform models. These masses and the waveform indicate that the two compact objects that merged were neutron stars.

24.7.1 New Discoveries Associated with GW170817

The rapid localization of the gravitational wave source and the coincidence with the gamma-ray burst gave a relatively narrow search area in which various electromagnetic observers were able to locate the candidate afterglow within hours, as a new point of light about 2 kpc from the center of the lenticular/elliptical galaxy NGC 4993 (in the southern constellation Hydra). The location of the afterglow is indicated by the small white star in the error box of Fig. 24.16. The luminosity distance was 40^{+8}_{-14} Mpc, which represents the closest gravitational wave event yet detected and is consistent with the known distance to the host galaxy NGC 4993. The resulting multimessenger nature of GW170817 proved to be a treasure trove of discoveries having fundamental importance in astrophysics, the physics of dense matter, gravitation, and cosmology [104]:

Viability of multimessenger astronomy: The event confirmed that the gravitational wave detectors could see and distinguish events that did not correspond to merger of two black holes, which had been the interpretation of all previous gravitational waves. It also demonstrated for the first time that electromagnetic signals could be detected in coincidence with a confirmed gravitational wave event, with sufficient source localization that it could be observed at many different wavelengths. All told, more than 70 facilities observed the event at optical, radio, X-ray, gamma-ray, infrared, and ultraviolet wavelengths.

Mechanism for short-period GRBs: The interpretation of the event as the merger of binary neutron stars and the coincident (short-period) gamma-ray burst provided the first conclusive evidence for the hypothesis that short-period gamma-ray bursts are produced in the merger of neutron stars (see Section 15.7.8). The gamma-ray burst was relatively weak (and likely would have attracted no special attention except for the coincident gravitational wave from the same part of the sky), which was provisionally interpreted as evidence that the gamma-ray burst beam was not pointed directly at Earth. (See the discussion of ultrarelativistic beaming for gamma-ray bursts in Section 15.7.4.) Strong confirmation of this hypothesis came two weeks after the initial event when radio waves and X-rays characteristic of a gamma-ray burst were detected. This evidence taken together represents the first definite association of a gamma-ray burst with a progenitor.

Site of the r-process: The signature of heavy-element production in the event demonstrated that neutron star mergers are one (perhaps the dominant) source of the rapid neutron capture or r-process thought to make many of the heavy elements (see Fig. 24.17). Now we have a quantitative way to investigate the relative importance of the two primary candidate sites for the r-process: core collapse supernovae, and neutron star mergers, but already it is clear

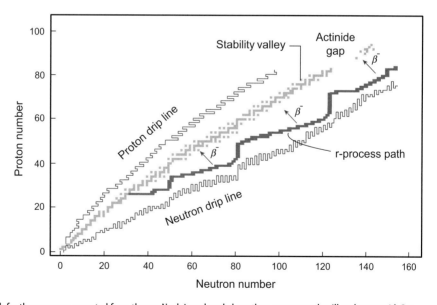

Fig. 24.17 Path for the r-process expected from theory. Nuclei produced along the r-process path will undergo rapid β^- decay back toward the stability valley, thus producing most of the neutron-rich and some of the β-stable isotopes, as well as all the actinide nuclei found in nature. (The β-stable isotopes beyond iron but below the actinide gap can be produced also in the slow neutron capture or s-process in red giant stars.) The neutron and proton drip lines denote the boundaries beyond which a nucleus becomes unstable against spontaneous emission of neutrons or protons, respectively. A more extensive discussion of the r-process in the context of stellar evolution may be found in Section 20.5 of Ref. [105].

that the dominant attitude of not very long ago that the r-process was associated mostly with core-collapse supernovae is probably not correct.[8]

Observation of a kilonova: The expanding radioactive debris was observed at UV, optical, and IR wavelengths, giving the first direct evidence for the *kilonova* (also termed a *macronova*) that had been predicted to occur following such mergers as a result of radioactive heating by newly synthesized r-process nuclei. The direct nucleosynthesis of r-process species likely ceases after a second or two, but most initially synthesized isotopes would be highly radioactive and the cloud of debris can be kept warm (10^3–10^4 K) by radioactive decay for as long as weeks. That the gamma-ray burst was emitted off-axis may have been essential in allowing the kilonova associated with the radioactivity of heavy elements produced in the merger to be observed. As illustrated in Fig. 24.18, if the GRB is seen nearly on-axis, it is likely that the GRB afterglow will mask the kilonova emission at all times, explaining why only suggestive evidence for a kilonova associated with a gamma-ray burst has been reported before.

Nuclei far from stability: The r-process runs far to the right (neutron-rich) side of the β-stability valley in the chart of the isotopes shown in Fig. 24.17, where little definitive information exists concerning the structure of nuclei because the isotopes cannot be made in traditional accelerators. The lightcurves for the kilonova are a statistical mix of contributions from many neutron-rich nuclei with no sharp lines because of the high velocities (as large as $\sim 0.3c$) for the ejecta. However, they carry information about the average decay rates and other general properties of these largely unknown r-process nuclei that could provide future constraints on theories of nuclear structure far from β stability.

The speed of gravity: Arrival of the GRB within 1.7 seconds of the gravitational wave from a distance of 40 Mpc established conclusively that the difference of the speed of gravity and the speed of light lies between -3×10^{-15} and $+7 \times 10^{-16}$ times c (that is, no larger than 3 parts in 10^{15}) [21]. Thus it took 1.7 seconds of observation to eliminate from contention theoretical alternatives to general relativity for which gravity does not propagate at c.

Neutron-star equation of state: The multimessenger nature of the event indicates that neutron star mergers will provide an opportunity to make much more precise statements about the neutron-star equation of state because – for example – the merger wave signature is sensitive to the tidal deformability of the neutron star matter near merger. This is of fundamental importance for our understanding of dense matter because prior observations have been unable to constrain candidate equations of state sufficiently to understand the maximum mass of a neutron star and the minimum mass of a black hole to better than an uncertainty of about a solar mass.

---

[8] One common theme for understanding the origin of r-process nuclei is to ask whether they were produced in a few rare events (neutron star mergers are relatively rare, occurring maybe only once every 10^5 or 10^6 years in a large galaxy), or in many much more common events (core collapse supernovae are much more common than neutron star mergers, occurring about once every 50-100 years in a large galaxy). Some evidence had been accumulating that at least some r-process nuclei were produced in rare events. The neutron star merger leading to GW170817 and GRB 170817A gives direct evidence for the production of significant amounts of r-process nuclei in a single rare event.

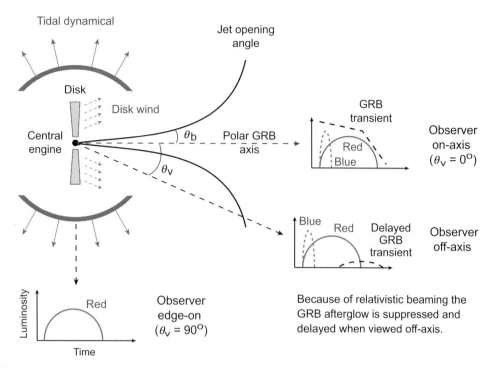

Fig. 24.18 Geometry of GW170817 afterglows [128, 187, 236]. The neutron-rich ejected matter labeled "Tidal dynamical" emits a kilonova peaking in the IR (solid arrows and solid curves labeled "Red" in the time–luminosity diagrams) associated with production of heavy r-process nuclei and high opacity (the *red kilonova*). Additional mass is emitted by winds along the polar axis (dotted arrows and dotted curves labeled "Blue") that is processed by neutrinos emitted from the hot central engine, giving matter less rich in neutrons and a kilonova peaking in the optical that is associated with production of light r-process nuclides and lower opacity (the *blue kilonova*). The usual GRB afterglow is indicated by dashed curves in the plots. It dominates all other emission when viewed on-axis but when viewed off-axis it appears as a low-luminosity component delayed by days or weeks (until $\theta_v < \theta_b$), which permits the kilonova to be seen.

Demographics of neutron-star binaries: The observation provides quantitative information about the probability that neutron star binaries form in orbits that can lead to merger in a Hubble time (a probability that has been uncertain to this point). The rate currently inferred corresponds to 0.8×10^{-5} mergers per year in a galaxy the size of the Milky Way. This has large implications for our understanding of stellar evolution, the site of the r-process, and the expected rate of gravitational wave detection from such events.

Determination of the Hubble constant: The multimessenger nature of the event provides an independent way to determine the Hubble constant H_0 by comparing the distance inferred from the gravitational wave signal with the redshift of the electromagnetic signal [219]. (See the discussion of standard sirens in Box 24.1.) Presently, different methods yield a value for H_0 in the range of about $67-73 \text{ km s}^{-1} \text{Mpc}^{-1}$, with analyses of the cosmic microwave background (CMB) tending to give values nearer the lower end and traditional "distance-ladder" methods like Cepheid variables giving values nearer the higher end. Analysis of the GW170817 multimessenger event suggests a value in the middle, $H_0 \sim 70^{+12}_{-8}\,\mathrm{km\,s^{-1}\,Mpc^{-1}}$, though still with relatively large uncertainties [22]. We may expect an accumulation of multimessenger events to yield a precise, independent determination of H_0. For example, it has been estimated that 100 independent gravitational wave detections with host galaxy identified as in GW170817 would determine H_0 with an uncertainty of 5% [22].

Off-axis gamma-ray bursts: The initial observation of the kilonova followed days later by observation of X-ray and radio emission provides strong corroborating evidence for the beamed nature of gamma-ray bursts [236] and represents the first clear detection of a weak, off-axis GRB and its slowing in the interstellar medium. Systematic studies of such events should greatly enrich our understanding of gamma-ray bursts, which previously were understood primarily in terms of those bursts beamed more directly at us.

There is insufficient space to explore all of these in detail but we will elaborate further on one topic, the kilonova powered by the production of radioactive r-process nuclei.

24.7.2 The Kilonova

General relativistic simulations of neutron star mergers identify two mechanisms for mass ejection [128]. (1) Matter may be expelled dynamically by tidal forces on millisecond timescales during the merger itself, and as surfaces come into contact shock heating at the surfaces may squeeze matter into the polar regions. (2) On a longer (~ 1 second) timescale matter in an accretion disk around the merged objects can be blown away by winds. As ejected matter decompresses, heavy elements may be synthesized by the r-process. If the matter is highly neutron-rich, repeated neutron captures form the *heavy r-process nuclei* ($58 \leq Z \leq 90$), while if the ejecta is less neutron-rich, *light r-process nuclei* ($28 \leq Z \leq 58$) are formed. The matter ejected in the tidal tails is cold and very neutron rich, and tends to form heavy r-process nuclei. The disk winds and ejecta squeezed dynamically into the polar regions may be subject to neutrino irradiation from the central region, which converts by weak interactions some neutrons to protons, making the winds less neutron-rich and favoring the light r-process.

The photon opacity of the r-process ejecta may play a central role in the observable characteristics of kilonova events. The photon opacity is generated largely by transitions between bound atomic states (*bound–bound transitions*). For light r-process nuclei the valence electrons typically fill atomic d shells. In contrast a substantial fraction of heavy r-process species produced by simulations (often 1–10% by mass) are *lanthanides* ($58 \leq Z \leq 71$), for which valence electrons fill the f shells. These have densely spaced energy levels and an order of magnitude more line transitions than for the d shells in light r-process species. As a consequence, the opacity of heavy r-process nuclei is roughly a factor of 10 larger than the corresponding opacities for light r-process species, and they have correspondingly long photon diffusion times [128].

Hence the cloud of light r-process species is considerably less opaque with shorter diffusion times, and tends to radiate in the optical and fade over a matter of days. In contrast, the cloud of heavy r-process species radiates in the IR with lightcurves that may last for weeks because of the high opacity and long diffusion times. This accounts for the observed characteristics of the transient AT 2017gfo, which differed essentially from all other astrophysical transients that have been studied: It brightened quickly in the optical and then faded but a quickly growing IR emission remained strong for many days, and only after a period of weeks did X-ray and RF signals begin to emerge.

The above considerations suggest a general picture of the geometry of GW170817 that is illustrated in Fig. 24.18. The kilonova transient AT 2017gfo that followed the gravitational wave GW170817 and associated gamma-ray burst GRB 170817A had two distinct components. The tidal dynamical ejection flung out on millisecond timescales very neutron-rich matter at high velocities $v \sim 0.3c$ that underwent extensive neutron capture to produce heavy r-process species and extremely high opacity because of the lanthanide content. In addition, winds ejected matter from the disk region on a timescale of seconds. This matter was subject to shock heating and to irradiation by neutrinos from the hot center, both of which could increase the proton to neutron ratio. Nucleosynthesis in this less neutron-rich matter was likely to produce light r-process matter of lower opacity, since there weren't enough neutrons to produce lanthanides and other heavy r-process nuclei.

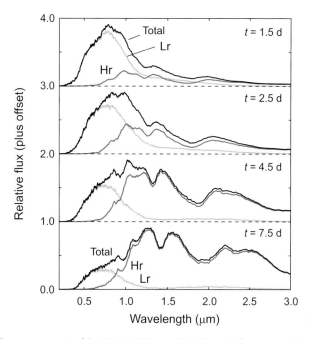

Fig. 24.19 Evolution of the different components of the GW170817 kilonova [128]. The total flux is a sum of two spatially separated components: the dominantly optical emission from light r-process isotopes (the "blue kilonova", labeled Lr) and the dominantly infrared emission from heavy r-process isotopes (the "red kilonova", labeled Hr). Reprinted with permission form Springer Nature, *Nature*, **551**(7678), Origin of the Heavy Elements in Binary Neutron-Star Mergers from a Gravitational-Wave Event, D. Kasen *et al.*, copyright 2017.

This picture is supported by model simulations that are displayed in Fig. 24.19. These simulations exhibit clearly the early emergence and rapid decay of the optical component associated with the light r-process (the *blue kilonova*), followed by the longer-lived IR component associated with the heavy-r process (the *red kilonova*), which grows within several days to dominate the lightcurve. The color evolution, spectral continuum shape, and IR spectral peaks of this composite model resemble the observed time evolution of AT 2017gfo. In the model suggested by Fig. 24.18, the red and blue components are visible only because the gamma-ray burst was seen off-axis, which suppressed the GRB afterglow because of relativistic beaming so that the underlying kilonova was visible.

24.8 Gravitational Waves and Stellar Evolution

The interpretation of GW150914 as resulting from merger of two $\sim 30 M_\odot$ black holes poses some challenging questions for theories of stellar evolution. For example, how does a binary composed of $30 M_\odot$ black holes even form? Presumably either (1) a binary formed with two stars having large masses and the binary survived successive core collapses for each star, or (2) the black holes formed independently through core collapse of massive stars in a dense cluster and then were captured by mutual gravity into a binary orbit and ejected dynamically from the cluster. Neither scenario is easy to realize without assumptions that are not well tested. Therefore, we may expect that future detection of gravitational wave events from merger of two black holes, merger of a black hole and neutron star, merger of two neutron stars, and core collapse supernovae will shed considerable light on – and pose substantial challenges to – our general understanding of stellar evolution, particularly for the neutron star and black hole endpoints for massive-stars evolution.

The implications of gravitational waves for understanding stellar evolution are discussed more extensively in Ref. [105] and part of that discussion is summarized in the following. Comprehensive simulations [43, 227] indicate that the binary black holes responsible for the gravitational waves observed thus far by LIGO could have formed in isolated binary star evolution, with a predicted merger rate consistent with the (still highly uncertain) observed merger rate, provided that the binaries formed in regions having low concentration of elements heavier than helium (regions of low *metallicity*; see Box 24.3).

24.8.1 A Possible Evolutionary Scenario for GW150914

Figure 24.20 illustrates a possible scenario for the production of GW150914 [43]. A massive binary formed about 2 billion years after the big bang (redshift $z \sim 3.2$), with initial main sequence masses of $96.2\, M_\odot$ (star A) and $60.2\, M_\odot$ (star B), a metal content of 0.03 times that of the Sun, average separation $a \sim 2500\, R_\odot$, and orbital eccentricity $e = 0.15$. Star A (being more massive) evolved quickly, expanded, and transferred more than half of its mass to the other star (B) by Roche lobe overflow as star A evolved through

> **Box 24.3** **Metallicity and Stellar Mass**
>
> A major factor limiting stellar masses is stability with respect to ejection of matter (and generation of pulsational instabilities) by radiation pressure. Radiation pressure grows rapidly with star mass since it depends on the fourth power of the temperature and more massive main sequence stars are hotter. Photon opacity has a major influence on how massive a star can become before being destabilized by radiation pressure, and opacity is greatly increased by the presence of metals (elements with atomic number greater than two) because they produce so many electrons when ionized and photons scatter strongly from free electrons. Low metal content can lead to the birth of more massive stars because metals aid in cooling. The higher temperatures favored by low metallicity then favor collapse of more massive gas clouds, and reduced stellar winds due to low opacity can then keep these more massive nascent stars from expelling their mass, allowing them to grow even larger by accretion.
>
> The big bang produced almost no metals, so metallicities in the first generation of stars (called *Population III* or *Pop III*) were very low and simulations indicate that typical stars could have had masses as large as several hundreds or even thousands of solar masses, and could have produced black holes of comparable mass when their cores collapsed at the end of stellar evolution. Subsequent generations of stars formed from material enriched in metals produced in supernova explosions of earlier stars, so they had increasingly larger metallicities until in today's Universe few stars grow more massive than 50–100 $M_\odot$.

the Hertzsprung gap to core helium burning. Star A then collapsed directly to a black hole of mass 35.1 $M_\odot$, with no ejection of a supernova remnant, but with 10% of the mass carried off by neutrinos during the collapse. By the time the first black hole had formed, star B had grown by accretion to 84.7 $M_\odot$ and it evolved quickly off the main sequence to core helium burning. The expansion of star B initiated a common envelope phase with the black hole that formed from star A, during which the average separation of the binary components was reduced from $a \sim 3800\, R_\odot$ to $a \sim 45\, R_\odot$. At the end of the common envelope phase the mass of the black hole formed from star A was 36.5 $M_\odot$ and star B was now a helium star of mass 36.8 $M_\odot$. Star B then collapsed directly to a black hole, leaving a binary black hole system with masses of 36.5 $M_\odot$ and 30.8 $M_\odot$, respectively, and orbital separation $a = 47.8\, R_\odot$. This system then spiraled together through gravitational wave emission for 10.3 billion years, merging about 1.1 billion years ago ($z \sim 0.09$) to produce GW150914.

The simulations described above paint a compelling picture but they entail large uncertainties because of assumptions such as the metallicity, and because accretion and (in particular) common envelope evolution are the least-well understood aspects of binary evolution. Tests of these assumptions and increasingly strong constraints on models of massive binary star evolution may be expected as gravitational wave astronomy matures and large data sets for black hole mergers are accumulated. One crucial feature of the mechanism outlined above is direct collapse of massive stars in a binary to black holes, without ejecting a supernova remnant and without giving a strong *natal kick* to the black hole that is formed, so that it remains in the binary. Box 24.4 gives the first direct observational evidence that such *failed supernovae* may occur in nature.

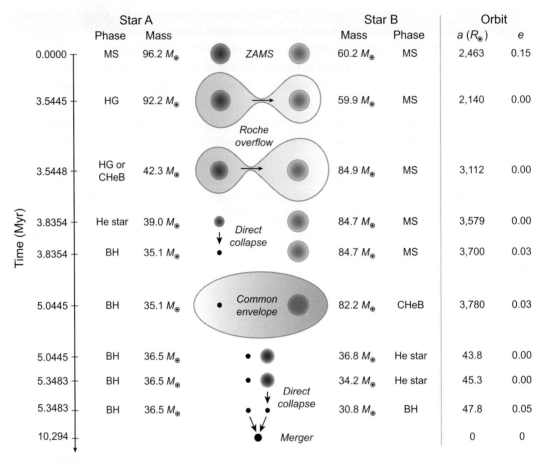

Fig. 24.20 A scenario for evolution of the massive black hole binary leading to GW150914 [43]. ZAMS means zero age main sequence (the time when the star first enters the main sequence), MS means main sequence, HG means a star evolving through the Hertzsprung gap (the evolutionary region between the main sequence and the red giant branch), CHeB means core helium burning, a He star is a star exhibiting strong He and weak H lines (indicating loss of much of its outer envelope), and BH indicates a black hole. Time is measured from formation of the binary and the scale is *highly nonlinear*. The separation of the pair is a and the eccentricity of the orbit is e. A more extensive discussion may be found in Ref. [105]. Reprinted with permission from Springer Nature: *Nature*, **534**(7608), The First Gravitational-Wave Source from the Isolated Evolution of Two Stars in the 40–100 Solar Mass Range, Belczynski K. et al., copyright 2016.

24.8.2 Measured Stellar Black Hole Masses

Black holes are one conjectured endpoint for stellar evolution. We have discussed two reliable methods of identifying stellar black holes and determining their masses: (1) the mass-function analysis of X-ray binary systems described in Section 14.4, and (2) the analysis of gravitational waves from binary black hole mergers discussed in this

Box 24.4 Direct Collapse of Massive Stars to Black Holes

A survey by the Large Binocular Telescope and followup observations with the Hubble Space Telescope (HST) identified a strong candidate for direct collapse to a black hole (a *failed supernova*) that is illustrated in the following figure [24].

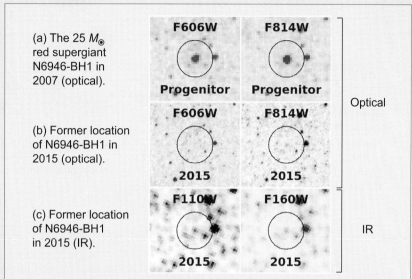

(a) The 25 $M_\odot$ red supergiant N6946-BH1 in 2007 (optical).

(b) Former location of N6946-BH1 in 2015 (optical).

(c) Former location of N6946-BH1 in 2015 (IR).

From S. M. Adams *et al.* the Search for Failed Supernova with the Large Binocular Telescope : Confirmation of a Disappearing Star. *MNRAS* (2017) **468**(4), 4968–4981. Reprinted with permission from Scott Adams.

In 2007 the 25 $M_\odot$ red supergiant N6946-BH1 in NGC 6946 appeared as a bright star in HST optical images [center of circles in panel (a)]. In 2009 this star underwent a weak optical outburst and then faded rapidly over a matter of months. Images from 2015 [panel (b)] indicate that N6946-BH1 has disappeared at optical wavelengths but a faint IR signal remains [panel (c)]. These results are consistent with the birth of a black hole in a failed supernova, with the IR emissions caused by weak accretion on the black hole [24]. As indicated in Fig. 24.20, failed supernovae may be crucial to formation of binary black holes because they may permit core collapse of the progenitor stars to black holes without disrupting the binary.

---

chapter. Figure 24.21 summarizes masses for 37 black holes in the range ∼5–65 $M_\odot$ that were determined from these two types of analysis. The data from X-ray binaries and gravitational waves from black hole coalescence (along with analysis of the orbit of the star S0-2 around the supermassive black hole at the center of the Milky Way described in Section 14.5.1, and analysis of the masers of NGC 4258 described in Section 14.5.2), may constitute the strongest evidence currently available for the existence of black holes. Clearly it is difficult to account for the properties of these objects using known or even conjectured physics through any hypothesis other than that of the black holes predicted by general relativity.

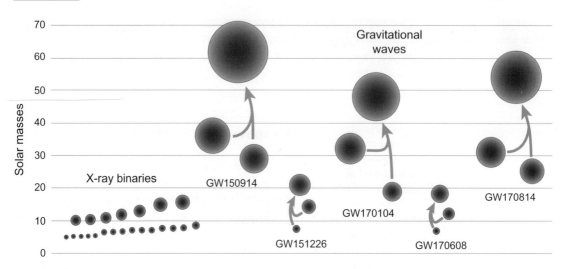

Fig. 24.21 A summary of black hole masses determined from X-ray binary and gravitational wave data as of November, 2017 (suggested by Ref. [7]). Arrows indicate black hole mergers.

24.8.3 Are Stellar and Supermassive Black Holes Related?

Also of interest is whether there is a connection between the formation of stellar-mass black holes and of supermassive black holes. Two extremes have been discussed. (1) The *light seeds model*, in which supermassive black holes formed by successive merger of black holes created by core collapse of stars. (2) The *heavy seeds model*, in which $10{,}000$–$100{,}000\,M_\odot$ seeds for supermassive black holes may have formed by direct collapse of gas clouds, or possibly by rapid merger of black holes created by very massive first-generation stars. Thus, it is possible that evolution of massive stars ending in the creation of stellar black holes also has implications for supermassive black holes.

Background and Further Reading

See the LIGO discovery paper [10], the comments on the discovery paper in [45], and the paper on astrophysical implications of the discovery [16] for an overview of GW150914. Details of the analysis of GW150914 in terms of basic physics may be found in Ref. [17]. The modeling of strong sources for gravitational waves requires large-scale computer simulations: numerical relativity for black hole mergers, and sophisticated hydrodynamics and neutrino transport for core collapse supernova explosions and neutron star mergers. Numerical relativity references include Refs. [26, 33, 35, 40, 160], the supernova mechanism and emission of gravitational waves are discussed in Refs. [127, 137, 138, 155, 161, 173, 257], and neutron star mergers are reviewed in [44, 86]. Reliable simulation in numerical relativity of binary black hole spacetimes is a relatively new development. Early attempts at such simulations were plagued with instabilities and it was only around 2005 that several groups began to achieve stable simulations from which observables could be extracted [33, 34, 35, 61, 194]. Short-period gamma-ray bursts are reviewed in Berger [44] and an overview of the r-process in neutron star mergers may be found in Thielemann, Eichler, Panov, and Wehmeyer [233].

Problems

24.1 From the chirp mass for GW150914 and the observed frequency of 150 Hz at peak amplitude for the gravitational wave, what is the lower limit on the sum of the Schwarzschild radii for the two colliding black holes and (assuming for a rough estimate Keplerian trajectories), what was the separation between the two black holes when emitting gravitational waves at peak strain amplitude?

24.2 Use the chirp mass determined from the gravitational wave frequency and its time derivative to confirm the arguments in Section 24.2.3 that GW150914 corresponded to the merger of two black holes. *Hint*: The chirp waveform says it was a merger. The power of the gravitational wave says the merger involved massive compact objects. The only realistic possibilities are: (a) merger of two neutron stars, (b) merger of an neutron star and black hole, or (c) merger of two black holes. Use the observations to rule out the first two possibilities, leaving a binary black hole merger as the only viable source. ***

24.3 How would you counter the statement that the direct detection of gravitational wave GW150914 by LIGO in 2015 is not that important because the binary pulsar data discussed in Section 23.2.3 already show that gravitational waves exist, and anyway it is just another of many experimental and observational tests that general relativity has passed?

24.4 The quadrupole formula for gravitational wave luminosity discussed in Section 23.1.6 is expected to be a good approximation if the wavelength is much larger than the characteristic source size. Estimate the wavelength of GW150914 detected at peak strain and compare with the characteristic source size, assuming the interpretation of merging black holes for the origin of the wave. Estimate the peak luminosity for GW150914 using the quadrupole formula and data from the merger, and compare with the value quoted in Table 24.1.

24.5 Estimate the chirp mass $\mathcal{M}$ for GW150914 from the gravitational wave data. *Hint*: Use the bottom panel of Fig. 24.4 to estimate the frequency f and its time derivative $\dot{f}$ at peak strain.

24.6 Gravitational wave emission from a binary system is powered by extracting the energy of orbital motion. Assuming the validity of Newtonian mechanics and Newtonian gravity, show that the total energy of orbital motion for a binary is $E = -Gm_1 m_2/2a$, where a is the average separation. ***

PART V

GENERAL RELATIVITY AND BEYOND

25 Tests of General Relativity

This final part concerns the future of gravitational theory. Since general relativity is the best current theory of gravity, it is appropriate to begin with an analysis of how well it has stood the observational test. Accordingly, this chapter gathers in one place a list of some observational and experimental tests of general relativity, and the current status of those tests. The chapter is relatively brief because the list of observables for which general relativity differs significantly from Newtonian gravity is itself brief, and because many of these tests have been discussed in preceding chapters. Our primary concern will be comparison of general relativity with observation and experiment. However, it is important also to compare the predictions of general relativity with alternative modern theories of gravity, which are described briefly in Box 25.1. It is convenient to divide these tests into four broad categories: (1) *the classical tests,* (2) *the modern tests,* (3) *the strong-field tests,* and (4) *cosmological tests.* We shall introduce also the *parameterized post-Newtonian* (PPN) formalism, which is a standard framework used to compare the predictions of general relativity with data and with predictions from other theories of gravity.

25.1 The Classical Tests

The tests proposed originally by Einstein when the theory of general relativity was introduced are sometimes termed the *classical tests* of general relativity. These are usually considered to be:

1. *Precession of the perihelion for Mercury* (Section 9.4) was confirmed already at the publication of the theory and was a source of great joy for Einstein when he found it. Much larger precession effects have since been measured for systems such as the Binary Pulsar (4.2° per year) and the Double Pulsar (17° per year), in agreement with the amount predicted by general relativity.
2. The *bending of light in the Sun's gravitational field* (Sections 6.4 and 9.8) was confirmed in 1919 during Solar eclipse observations [78], and was largely responsible for making Einstein an international celebrity. These measurements have since been repeated more precisely and extended to a variety of gravitational lensing phenomena, with results consistent with the theory of general relativity.
3. The *redshift of light in a gravitational field,* or equivalently *gravitational time dilation* (Sections 6.5 and 9.2) was suggested initially in the spectra of white dwarfs and confirmed later in direct experiments on Earth [191, 192, 193]. You also confirm it

implicitly every time you use the Global Positioning System (GPS) for location and navigation; see Problem 6.3.[1]

4. The *existence of gravitational waves* was confirmed indirectly by observations on the Binary Pulsar (Section 10.4.1) and directly – a century after the prediction that they could exist – by the LIGO detection of gravitational waves described in Chapter 24.

It is fair to conclude that the classical tests of general relativity have now all been passed with a rather high level of confidence.

25.2 The Modern Tests

In addition to the classical tests of general relativity there are various newer ones that we may term the *modern tests* of general relativity. In many cases the feasibility, accuracy, and precision of this class of tests has been aided by rapid advances in detection and measurement technology, and by the advent of routine spaceflight to place observational and experimental platforms in space. The analyses of these modern tests of general relativity often systematize results in terms of the *parameterized post-Newtonian (PPN) formalism* Accordingly, let us preface a discussion of the results of modern tests with a brief introduction to the PPN formalism.

25.2.1 The PPN Formalism

In comparing general relativity with other theories of gravity and with observation and experiment, it is useful to have a method expressing the nonlinear Einstein equations in terms of the lowest-order deviations from Newtonian gravitational theory. Such approaches may be applied to any theory of gravity that derives from a spacetime metric and embodies the strong equivalence principle, but are most useful in weak gravity since they are based on expansions.[2] Most viable alternative theories of gravity (see Box 25.1) fit this prescription [251]. Under conditions of weak gravity the differences between the predictions of general relativity and Newtonian gravity can be expressed in terms of expansions in powers of a small parameter that is the ratio of characteristic velocities in the system to the speed of light. This *post-Newtonian expansion* (so-called because it adds corrections that go beyond Newtonian gravity) reduces to Newtonian gravity in the limit that the speed of light is assumed to be infinite.

---

[1] The existence of a gravitational redshift is a necessary condition for general relativity to be valid but this is sometimes viewed as a less-stringent condition than others in this list because it follows from the equivalence principle alone (See Section 6.5 and Problem 6.2). Thus, any alternative theory of gravity that implements the equivalence principle would predict a gravitational redshift. However, it has been argued that general relativity is the only viable metric theory (one for which gravity derives from geometry) that implements the strong equivalence principle completely [251].

[2] In stronger gravity PPN expansions can be carried to higher order but it is often better to solve the gravitational equations completely by numerical means to provide a basis for comparisons in strong gravity.

> **Box 25.1** **Alternative Theories of Gravity**
>
> Numerous alternatives to general relativity as the modern theory of gravity have been proposed. There are various motivations but two have assumed increasing importance in recent years:
>
> 1. The incompatibility of general relativity with quantum mechanics, suggesting that general relativity is not complete.
> 2. The distaste of some for the mysterious nature of dark matter and dark energy in the standard cosmology, which has motivated a view that they may be fictions masking a failure of the gravitational theory.
>
> Perhaps the best-known example of an alternative to general relativity is the *scalar–tensor theory* of Brans and Dicke [56].
>
> **Alternatives and Observations**
>
> Most alternatives to general relativity are ruled out quickly by observations (for example, they may fail to predict the correct deflection of light in a gravitational field). However, some alternatives are not obviously falsified by current observations. Typically the gravitational force in these still-viable theories derives from the spacetime metric, as for general relativity, but they differ in that additional fields are added – for example, a long-range scalar (spin-0) field supplementing the spin-2 tensor field of general relativity. For such theories to be consistent with the observed properties of gravity the relationship of the additional fields to matter is one-way: matter may create the fields and thereby make contribution to the curvature of the spacetime metric, but the fields do not act back on the matter because matter is assumed to respond only to the metric [251].
>
> **Tests with Gravitational Waves**
>
> One distinction between general relativity and alternatives may be in gravitational wave emission. For example, in scalar–tensor theories scalar and dipole gravitational waves may be emitted in addition to the quadrupole ones predicted by general relativity. This could modify the rate of gravitational wave emission late in the inspiral for merging neutron star, neutron star – black hole binaries in an observable way [30, 250]. Various alternatives to general relativity require also that gravitational waves travel at a speed different from that of light. As discussed in Section 24.7.1, the finding from GW170817 that the speed of gravity can differ by no more than three parts in 10^{15} from that of light would seem to pose serious problems for such theories.

We now illustrate how this works in more detail. The most general static, spherically symmetric metric can be expressed in the form (see Appendix C)

$$ds^2 = -A(r)(cdt)^2 + B(r)dr^2 + r^2 d\theta^2 + r^2 \sin^2\theta d\varphi^2. \quad (25.1)$$

In the limit of flat spacetime this becomes the Minkowski metric (4.5) expressed in spherical coordinates,

$$ds^2 = -(cdt)^2 + dr^2 + r^2 d\theta^2 + r^2 \sin^2\theta d\varphi^2. \quad (25.2)$$

Thus $A(r)$ and $B(r)$ parameterize the deviation of the metric from flat spacetime. This suggests that for weak gravity the departure from a flat metric can be described by an expansion in an appropriate small parameter. Any viable relativistic gravitational theory should agree with the results of Newtonian gravity in the weak-field limit, as described in Section 8.1. Comparing with Eq. (25.1), agreement with Newtonian gravity in the weak-field limit requires that to lowest order (Problem 25.1)

$$A(r) = 1 - \frac{2GM}{rc^2} + \cdots \qquad B(r) = 1 + \cdots, \qquad (25.3)$$

where the dots indicate higher-order corrections. Equation (25.3) suggests that the small dimensionless parameter appropriate for an expansion of the metric in weak gravity is GM/rc^2. This is not surprising, since it was shown already in Section 6.4.2 that $\epsilon \equiv GM/rc^2$ is a dimensionless natural measure of the strength of gravity. Expanding $A(r)$ and $B(r)$ in powers of GM/rc^2 gives for the metric coefficients [111]

$$A(r) = 1 - \frac{2GM}{rc^2} + 2(\beta - \gamma)\left(\frac{GM}{rc^2}\right)^2 + \cdots \qquad B(r) = 1 + 2\gamma\left(\frac{GM}{rc^2}\right) + \cdots, \qquad (25.4)$$

where the particular form of the expansion coefficients in terms of the parameters β and γ is conventional. By virtue of the discussion in Section 8.1, the lowest-order corrections to the flat-space spherical metric given in Eq. (25.3) may be termed the *Newtonian corrections*. Hence the higher-order terms in Eq. (25.4) are called the *post-Newtonian corrections,* and their size is characterized by the *post-Newtonian parameters* β and γ in this example.

More generally, the *parameterized post-Newtonian (PPN) formalism* isolates the differences between general relativity and other theories of gravity in the values of a set of parameters associated with coefficients of terms in the post-Newtonian expansion of the metric. There are ~10 such parameters in typical applications, each tested by particular experiments or observations. For example, in the PPN formulation of Will [251] the parameter γ is associated with light deflection and time delay, and the parameter β is associated with orbital perihelion shift. A more detailed exposition may be found in Section 3.2 of Ref. [251]. General relativity predicts $\beta = \gamma = 1$ for the PPN parameters β and γ, whereas a different metric theory of gravity might predict values different from unity. Thus, modern experimental tests of gravity are often reported in terms of measured differences of PPN parameters like β and γ from their values expected in general relativity.

Example 25.1 Frequency shifts of radio signals to and from the Cassini spacecraft as it passed near the Sun on its way to Saturn in 2002 were used to constrain the PPN parameter γ to the range $\gamma = 1 + (2.1 \pm 2.3) \times 10^{-5}$ [46]. Using the precise Cassini value of γ coupled with lunar rangefinding data, the PPN parameter β has been constrained to $\beta = 1 + (1.2 \pm 1.1) \times 10^{-4}$ [254].

The extremely precise determination of the PPN parameter γ using Cassini approaches the precision required to see deviations from $\gamma = 1$ predicted for the newest versions of scalar–tensor gravity (see Box 25.1).

25.2.2 Results of Modern Tests

We now summarize the results of some modern tests of the general theory of relativity.

1. *Shapiro time delay* (Section 9.9) is a test in the same spirit as the classical tests and has been confirmed for various observations in the Solar System and beyond. For example, both the time delay and the deflection of light in a gravitational field are proportional to a PPN factor $\gamma + 1$, where $\gamma = 1$ for general relativity (see Section 25.2.1). Thus deviations of γ from unity would measure the degree to which gravity is not the purely geometrical effect predicted by general relativity. Radio ranging to the Viking spacecraft on Mars verified the light delay prediction of general relativity for light travel to and from Mars in the gravitational field of the Sun to an accuracy of 0.1% and found $\gamma = 1 \pm 0.002$ [200], while (as noted above) the frequency shifts of radio signals to and from the Cassini spacecraft found $\gamma = 1 + (2.1 \pm 2.3) \times 10^{-5}$ [46].
2. *Gravitational lensing* (Section 17.7.4), which is the generalization of the deflection of light in the Sun's gravitational field to a variety of light-deflection and lensing phenomena. A varied set of lensing effects consistent with general relativity has been found in deep-field observations.
3. *Frame dragging and geodetic effects* (Box 9.3 and Section 13.2.2), which have been confirmed by Gravity Probe B measurements for gyroscopes in polar Earth orbit.
4. *Modern equivalence principle tests,* which fall into several categories:

    (a) Experiments have been performed with very high precision using torsion balances to rule out composition-dependent or mass-dependent forces. The original Eötvös experiments established the equivalence of inertial and gravitational mass with a sensitivity of one part in $\sim 10^9$ and modern variations have extended that to one part in $\sim 10^{13}$ [217].

    (b) The strong equivalence principle requires that self-gravitating bound objects should follow the same paths in gravitational fields. This *Nordtvedt effect* [163, 164] has been tested precisely by lunar-ranging experiments that have continuously monitored the distance between stations on Earth and reflectors on the Moon to centimeter accuracy since 1969 [165]. These measurements confirm that the Earth and Moon fall toward the Sun at the same rate to one part in $\sim 10^{13}$ [254].

    (c) Another consequence of the strong equivalence principle is that the gravitational constant G should not vary with time or place. The lunar ranging measurements place an upper limit on variation of the gravitational constant, $\dot{G}/G = (4 \pm 9) \times 10^{-13}$ yr^{-1} [254]. Furthermore, the precision with which $\dot{G}/G$ can be measured is growing quadratically with the elapsed time for lunar-ranging measurements [165].

We may conclude from these examples that the modern tests of general relativity have provided strong confirmation of the theory within the limits of their precision, which is often quite high.

25.3 Strong-Field Tests

The tests of general relativity described above provide abundant evidence for the correctness of the theory, but they are all carried out in relatively weak gravitational fields where the nonlinearities are small. Until fairly recently, it was not easy to test general relativity in the strong-gravity, highly nonlinear regime because of the relative weakness of all gravitational fields in the Solar System. This has changed, with the availability now of three classes of tests in fields stronger than those discussed above: (1) tests at intermediate field strength using pulsars in binary and triple star systems, (2) tests at intermediate field strength by tracking stars near the central black hole of the Milky Way, and (3) tests at higher field strengths through direct detection of gravitational waves.

1. *Pulsars in binary and triplet systems* having compact objects in close orbits generally sample stronger gravitational fields. Tests discussed in Section 10.4 based on the Binary Pulsar, the Double Pulsar, and PSR J0348+0432 support the validity of general relativity in stronger fields, and have confirmed indirectly the emission of gravitational waves from such systems at the expected rates.
2. *Orbits of stars around the supermassive black hole at Sgr A** described in Section 14.5.1 provide tests of general relativity for gravitational source masses $\sim 10^6$ times larger and for fields ~ 100 times stronger than for tests of general relativity by binary pulsars.
3. *Direct observation of gravitational waves* has now been demonstrated, and this promises a whole range of general relativity tests in the strong gravity (binary neutron star merger and core collapse supernova), and very strong gravity (binary black hole merger) regimes. In particular, analysis of GW150914 (see Section 24.2) suggests that black holes exist – which confirms a fundamental strong-gravity prediction of general relativity – and that the extreme gravity near their event horizons (about a million times stronger than for that sampled in binary pulsars or by gravitational deflection of light by the Sun) is correctly described by general relativity.

The characteristic field strength versus source mass for some experimental and observational tests of general relativity are summarized in Fig. 25.1, with the field strength characterized by the quantity $\epsilon = GM/Rc^2$ introduced in Section 6.4.2.

Example 25.2 A typical modified form of gravity assumes that it is described by a metric tensor as in the Einstein theory, but that there is an additional field that alters the gravitational interaction over some distance range. The modified gravitational potential is often parameterized phenomenologically in the *Yukawa form*,

$$U = \frac{GM}{r}\left(1 + \alpha e^{-r/\lambda}\right),$$

where the parameter α characterizes the strength and λ the range of the hypothesized additional contribution to the Newtonian gravitational potential GM/r. Analysis of orbital data from the stars S0-2 (period $P = 15.92\,\text{yr}$ and eccentricity $e = 0.89$) and S0-38 ($P = 19.2\,\text{yr}$ and $e = 0.81$) over complete orbits around the supermassive black hole at

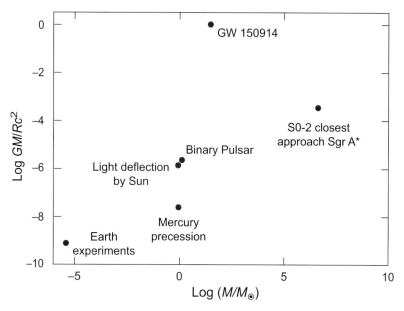

Fig. 25.1 Field strength versus mass for some tests of general relativity. Gravitational waves from merging stellar-size black holes and the orbits of test stars around the $\sim 4 \times 10^6 \, M_\odot$ black hole at the center of the Milky Way test gravity in fields that are 10^2 to 10^6 times stronger than that for all previous tests. Adapted from Ref. [116].

Sgr A* (see Section 14.5.1) finds no deviation from the predictions of general relativity and places an upper limit $|\alpha| < 0.016$ for $\lambda = 150 \, \text{AU}$, at the 95% confidence level [116]. As may be seen from Fig. 25.1, these observations probe the correctness of general relativity in a previously untested regime.

It may be expected that increasingly stringent tests of general relativity in stronger gravity will accumulate through further monitoring of binary pulsars and stars orbiting the supermassive black hole at the center of the Galaxy, and through the detection of gravitational waves from various binary merger and supernova events within the volumes illustrated in Figs. 22.9 and 22.10.

25.4 Cosmological Tests

In principle cosmological observations provide tests of general relativity. Indeed, some would argue that the observed expansion of the Universe is a confirmation of general relativity. However, although cosmology has grown into a precise observational science in recent years, the lack of understanding for the source of dark matter and dark energy undermines attempts to confirm general relativity from cosmological observations. To take one extreme example, while the mainstream has accepted that dark matter and dark energy are necessary components of a Universe correctly described by general relativity, a small minority argues that the observations of effects attributed to the presence of "dark matter" and "dark energy" are not evidence of unseen matter and energy at all, but rather a failure of current gravitational theory to describe the interactions of normal matter and energy (see Box 25.1).

Recall that in the history of astronomy an apparent failure of gravitational predictions typically has been ascribed to one of two alternatives: either (1) gravitational theory is correct but there is unknown mass affecting the system, or (2) gravitational theory is failing to describe the interactions of known masses and must be replaced. Historically, each alternative has been right at different times: the first in the discovery of Neptune through its perturbations on the orbit of Uranus using Newtonian gravity with an assumed unknown mass orbiting outside the orbit of Uranus, and the second by Einstein's demonstration that the precession of Mercury's perihelion was not caused by an unknown mass in Newtonian gravity but instead indicated that a better theory of gravity was required.

The evidence favors strongly the mainstream view that general relativity is the correct classical theory of gravity, and thus that dark matter and dark energy are real. However, until there is a deeper understanding of the source of dark matter and dark energy the judicious point of view must be that cosmological observations are consistent with general relativity but are not yet a rigorous and precise test of it. Stated slightly differently, dark matter and dark energy are key components of the new cosmology and at present the only evidence for their existence is gravitational. Hence it is clear that the new cosmology cannot be used as a rigorous *test* of general relativity until more is understood about the intrinsic nature of dark matter and dark energy.

Background and Further Reading

An overview of testing general relativity systematically and the parameterized post-Newtonian (PPN) formalism is given in Will [251]. A concise overview of general relativity with astrophysical applications that includes in the formalism the possibility of an additional scalar field may be found in the lecture notes by Wagoner [241].

Problems

25.1 Show that consistency of general relativity with Newtonian gravity in the weak-field limit implies Eq. (25.3). *Hint*: See Section 8.1. ***

25.2 Derive the values for data points displayed in Fig. 25.1 by making reasonable assumptions about the gravitational field being probed in each case.

26 Beyond Standard Models

The preceding chapters have described gravitational physics within the context of "standard models" of particular disciplines: for example, classical general relativity as the standard model of gravity and the Glashow–Salam–Weinberg unified gauge theory of the electroweak and strong interactions as the Standard Model of elementary particle physics. Although these theories have been extremely useful and remarkably correct in application to a broad range of phenomena, there is evidence that they are incomplete and require modification to describe the full range of physical phenomena. For example, the fundamental observation that the principles upon which general relativity and quantum mechanics rest are logically incompatible, and the quantum information paradox associated with black holes and entropy, point toward the need for a theory of quantum gravity but do not suggest very clearly what that theory should look like. As a second example, the abundant evidence that neutrinos undergo flavor oscillations and therefore have finite mass indicates that the Standard Model of elementary particle physics is incomplete, but again does not indicate unequivocally how to extend it. In this chapter some of the still-developing ideas for extending our current understanding of gravity and its interactions with elementary particles will we introduced. The discussion of these ideas will necessarily be rather qualitative, but will serve to orient the reader to some of these issues.

26.1 Supersymmetry

Observations suggest that most of the matter in the Universe is not seen by any standard probe except gravity. This unseen mass is termed *dark matter,* implying that it does not couple to visible light, though it is to be emphasized that so far *no non-gravitational probe* (electromagnetism, the weak interactions, or the strong interactions) has "seen" dark matter at any energy or wavelength. The evidence is that a dilute gas of dark matter obeys an equation of state similar to that of normal matter, but its exact nature is presently unknown. Likewise, there is empirical evidence that the Universe contains a mysterious "dark energy" that permeates all of space and is causing the expansion of the Universe to accelerate. The dark matter appears at this point to be distinct from the dark energy. For example, dark energy seems to require an equation of state that differs fundamentally from that of any known particle or of dark matter (negative pressure and antigravity effects). Potential explanations of these surprising yet apparently dominant components of our Universe often invoke a property of fundamental interactions that is conjectured theoretically but for which there is not yet any evidence called *supersymmetry*.

Accordingly, we now give a brief introduction to the basic ideas of supersymmetry as a prelude to further discussion concerning the nature of dark matter and dark energy.

26.1.1 Fermions and Bosons

All elementary particles can be divided into two classes, *bosons* and *fermions*, with the two classes exhibiting fundamentally different statistical properties. Fermions carry half-integer spins and obey *Fermi–Dirac statistics*, implying that fermion wavefunctions are completely antisymmetric with respect to exchange of two identical fermions. The most notable implication of fermionic statistics is the *Pauli Exclusion Principle*: no two identical fermions can occupy the same quantum state. Bosons carry integer spins and obey *Bose–Einstein statistics*, implying that boson wavefunctions are symmetric with respect to exchange of two identical bosons. The most notable implication of bosonic statistics is *boson condensation*: many identical bosons can (indeed, may prefer to) occupy the same quantum state. There is no fundamental explanation of why fermions and bosons have these properties but it is a well-documented fact that they do, and that all elementary particles discovered so far can be classified as either bosons or fermions.

26.1.2 Normal Symmetries

Normal symmetries in quantum field theory relate bosons to bosons or fermions to fermions, but not bosons to fermions. For example, the symmetry called *isotopic spin* is important in nuclear physics and particle physics.[1] One implication of isotopic spin symmetry is that there is a relationship between protons and neutrons implied by the symmetry such that – in a certain sense – the neutron and the proton are really just different manifestations of the same fundamental particle. Particles that are related in this way by an isotopic spin symmetry are termed *isotopic spin multiplets*.

This is a very powerful idea but in this example isotopic spin symmetry, just as for any ordinary symmetry, relates particles that are of the same quantum statistical type: the neutron and the proton both behave as spin-$\frac{1}{2}$ fermions at low energy. Another example of isotopic spin symmetry is that there are three kinds of π-mesons (often called *pions*), and each is related approximately to the other two by isotopic spin symmetry. Thus, in a sense the three pions are different manifestations of the same particle because of the isotopic spin symmetry. But again the isotopic spin symmetry relates particles of the same quantum statistical type: in this case, all three members of the pion isotopic spin multiplet behave as bosons at low energy. Other sets of strongly interacting particles form isotopic spin multiplets, but in every case the particles in a multiplet related by isotopic spin are either all fermions or all bosons. There is no known instance of a set of particles that appear to be related by an isotopic spin symmetry in which some of the particles are bosons and some are fermions.

---

[1] Experts will note that isotopic spin symmetry is not exact and not fundamental, and that the example particles given here (nucleons and mesons) are composite and not elementary particles. Nevertheless, it is useful at our level of discussion to introduce the ideas of normal symmetries in a simple way using isotopic spin symmetries with nucleons and mesons considered as elementary particles (which effectively they are, at low energy).

26.1.3 Symmetries Relating Fermions and Bosons

There are many attractive theoretical reasons to believe that the Universe will eventually be found to exhibit a fundamental symmetry that goes beyond normal symmetries and relates fermions to bosons. This conjectured "super" symmetry is termed, appropriately, *supersymmetry*. In the supersymmetry picture every fundamental fermion in the Universe has a partner boson of the same mass, and every fundamental boson in the Universe has a corresponding fermion partner of the same mass. For example, the Universe contains electrons that are spin-$\frac{1}{2}$ fermions with a mass–energy of 511 keV. According to the supersymmetry idea, there is also a bosonic partner of the electron (called a *selectron*) having the same mass but with integer spin (a spin of 0, to be precise), and obeying bosonic statistics. No one has ever seen a selectron, or the conjectured supersymmetric partner of any other known elementary fermion or boson. However, some theoretical arguments suggest that supersymmetry may be a broken symmetry such that the masses of the supersymmetric partners of known elementary particles are pushed up to a value high enough that they could not have been produced in any accelerator experiments carried out so far. Thus, there is strong theoretical prejudice that (broken) supersymmetry should exist in the Universe.

The most-discussed explanation for dark matter is that it consists of elementary particles that are broken supersymmetric partners of known particles. Then the very weak coupling to ordinary matter and fields, and thus the "darkness" of the dark matter, could follow naturally as a quantum-mechanical selection rule effect: the supersymmetric particles would be expected to carry quantum numbers reflecting their supersymmetric nature that are different from those of ordinary matter, so the quantum-mechanical probability for interaction of these supersymmetric particles with ordinary matter could be strongly suppressed by conservation laws and selection rules associated with these quantum numbers.[2] As will be discussed further below, supersymmetry also may be important in relating dark energy to quantum fluctuations of the vacuum (vacuum energy), though it will be seen that our present understanding of this issue is very far from satisfactory.

Experiments as of late 2017 at the Large Hadron Collider that were predicted by various educated theoretical guesses to have high enough energy to produce supersymmetric particles have failed to find them. This does not rule out broken supersymmetry as the origin of dark matter but it complicates the issue by suggesting that the simplest ideas about supersymmetry and dark matter may have to be discarded. Thus, there is abundant gravitational evidence that dark matter exists, but at present there isn't a scintilla of evidence in support of what has been the leading candidate to explain it.

26.2 Vacuum Energy from Quantum Fluctuations

Strong evidence suggest that the Universe contains a "dark energy" that permeates all of space and is causing the expansion of the Universe to accelerate. What is the source of this remarkable behavior, which is equivalent to the presence of antigravity in the Universe?

---

[2] For those lacking a background in quantum mechanics, this can be translated roughly as "supersymmetric matter and ordinary matter are so different that they don't even know how to interact with each other".

Given the successes of relativistic quantum field theory in elementary particle physics (the Standard Model of gauge-field interactions), it is natural to seek the source of the dark energy in terms of relativistic quantum fields.[3]

No dilute gas of known particles can exhibit an equation of state similar to that inferred for dark energy so, if the source of dark energy lies in quantum fields, those fields must be associated with as-yet undiscovered elementary particles. But the empirical evidence also suggests that dark energy is a property of space itself, and would still exist even in *empty space* (no particles and no fields). Quantum physicists term the ground state (lowest energy state) of a system the *vacuum state*. In a normal classical picture we might expect that the vacuum state of the Universe would consist of a state with no matter or fields, and no energy. But quantum field theory complicates this issue in two ways:

1. Because of what is known in the jargon as *spontaneous symmetry breaking* or *hidden symmetry,* the vacuum state of a system (that is, the lowest-energy state) need not be a state devoid of fields. Various examples are known where the state with no fields present is actually higher in energy than a state with a particular combination of fields, so that the lowest-energy state has fields present.
2. Classically, if there are no fields present the system has zero energy. But because of the quantum uncertainty principle, even a state that has no fields present on average will have non-zero energy ΔE because of fields that fluctuate into and out of existence over a period Δt such that the uncertainty principle relation $\Delta E \cdot \Delta t \geq \hbar$ is satisfied. These are termed *vacuum fluctuations* and the energy associated with them is termed the *zero-point energy of the vacuum* (see also Section 12.2). The reality of the zero-point vacuum energy is demonstrated indirectly by the many successes of the Standard Model of elementary particle physics, and directly through phenomena measurable in the laboratory such as the *Casimir effect*.

In light of point 2, the vacuum state of the Universe ("empty space") will have a non-trivial content because of vacuum fluctuations. Let us attempt a quantitative estimate of the energy density of empty space associated with vacuum fluctuations to see if these effects plausibly could account for dark energy and the accelerated Universe.

26.2.1 Vacuum Energy for Bosonic Fields

Because bosonic fields are easier to work with mathematically than fermionic fields, we first get the lay of the land by assuming the zero-point energy of the vacuum to be associated with bosonic fields. It is a standard result from quantum field theory that bosonic fields may be expanded in terms of an infinite number of quantum-mechanical harmonic oscillators. Assuming all fields in the Universe to be bosonic and expanding the fields as a collection of harmonic oscillators, the total energy in the fields is

---

[3] We will follow the usual practice of referring loosely to either particles or the fields associated with them using similar terminology. In quantum field theory, interactions are described fundamentally in terms of fields, but those fields have associated with them particles (sometimes termed the quanta of the fields). Since particles imply fields and fields imply particles, it is usual to refer either to a particle or to its associated field interchangeably, with the context determining whether one really means the particle or the field.

Box 26.1	**Zero-point Energy of Quantum Fields**

For readers lacking a systematic background in quantum physics, note that in Eq. (26.1) even if all the n_i are zero (no bosons in the Universe) there still is a finite energy associated with the sum over the $\frac{1}{2}$ term inside the parentheses. This peculiar feature is characteristic of quantum systems and is called *zero-point energy*. It may be viewed as a consequence of the Heisenberg uncertainty principle. A collection of classical oscillators can in principle have zero energy but if a collection of quantum oscillators had precisely zero energy the uncertainty principle relation $\Delta E \cdot \Delta t \geq \hbar$ could not be satisfied. Hence the lowest possible state of a collection of oscillators is required by the principles of quantum mechanics to have non-zero energy and the corresponding quantum fields are said to have *zero-point fluctuations*, even in their ground state.

$$E = \sum_i (n_i + \tfrac{1}{2})\hbar\omega_i, \qquad (26.1)$$

where ω_i is the frequency and $\hbar\omega_i$ is the energy of oscillator i, and the bosonic occupation number n_i can be zero or any positive integer. Equation 26.1 shows explicitly that even if the Universe is "empty" (no field quanta), there is a vacuum energy associated with the zero-point energies of the fields (a qualitative discussion of zero-point energy is given in Box 26.1)

$$E = E_\Lambda = \tfrac{1}{2}\sum_i \hbar\omega_i. \qquad (26.2)$$

Since this sum is over harmonic oscillators defined at each point of space, the vacuum energy density is expected to be constant over all space. By standard methods of quantum field theory, converting the sum in Eq. (26.2) to any integral over momentum using

$$\hbar\omega = \sqrt{p^2c^2 + m^2c^4} \qquad (26.3)$$

and the appropriate momentum-space volume element, and neglecting the masses, gives for the constant energy density ϵ_Λ^b associated with the vacuum fluctuations for bosonic fields,

$$\epsilon_\Lambda^b \propto \int_0^\infty p^3 dp = \infty. \qquad (26.4)$$

Thus, our first simplistic attempt to estimate the energy density of the vacuum leads to nonsense.

However, it is likely that general relativity breaks down on the Planck scale (see Section 26.3), so it is plausible that there is new physics on that scale that introduces a momentum cutoff in the preceding integral. Setting the upper limit of the integral to the Planck momentum $P_{\rm pl} \simeq 10^{19}\,{\rm GeV}\,c^{-1}$ gives

$$\epsilon_\Lambda^b \simeq \frac{(cP_{\rm pl})^4}{16\pi^2\hbar^3 c^3} \simeq 0.824 \times 10^{118}\,{\rm MeV\,cm^{-3}}, \qquad (26.5)$$

which is very large but offers a glimmer of hope because at least it is finite. Alas, that hope is fleeting because it already has been seen that observations (high-redshift Type Ia supernovae and fluctuations in the cosmic microwave background) indicate that the vacuum energy density causing acceleration of the Universe is the same order of magnitude as the critical (closure) density, which is only about 10^{-2} MeV cm^{-3} (see Section 17.2). Therefore, this simple estimate of the vacuum energy density gives a value that is about *120 orders of magnitude too large* to account for the observed acceleration of the Universe! This estimate of the vacuum energy density has been termed the most spectacular failure in all of theoretical physics. It is not yet clear whether this means that vacuum fluctuations are not responsible for the acceleration of the Universe, or whether it means that modern physics suffers from a (monumental) lack of understanding of how to properly calculate the vacuum energy density.

Some improvement in our estimate is afforded by assuming the Universe to be composed of fermionic fields in addition to bosonic ones, and to assume that there is a supersymmetry (see Section 26.1) that relates the boson and fermion fields. A serious treatment of fermionic quantum fields and supersymmetry would be too advanced for this discussion but in the next sections some simple estimates will be made of the effect on the vacuum energy density that may be expected if supersymmetry between fermions and bosons exists.

26.2.2 Vacuum Energy for Fermionic Fields

Because fermionic wavefunctions are antisymmetric and bosonic wavefunctions are symmetric under exchange, it may be shown that fermionic fields have a spectrum that is like a sum over harmonic oscillators in Eq. (26.1) for bosons, except that (1) the sign of the $\frac{1}{2}\hbar\omega_i$ term is negative and (2) the occupation numbers n_i are restricted to values of 0 or 1 (the Pauli principle; for fermions a state may be unoccupied or singly occupied, but not multiply occupied):

$$E = \sum_i (n_i - \tfrac{1}{2})\hbar\omega_i, \qquad (n_i = 0, 1). \tag{26.6}$$

Thus, the zero-point energy for fermionic fields corresponding to $n_i = 0$ is *negative*. This looks even less promising as an explanation of the accelerating universe than bosonic vacuum fluctuations.[4] However, let us now consider what happens if there is a supersymmetry between bosonic and fermionic fields.

26.2.3 Supersymmetry and Dark Energy

The potential relevance of supersymmetry to the vacuum energy discussion is clear from the preceding observation that the zero-point energy of fermion fields is opposite in sign relative to that of boson fields. If by supersymmetry every elementary fermion field has a partner supersymmetric elementary boson field, their contributions to the vacuum energy density can cancel each other. Indeed, a detailed theoretical treatment of supersymmetry suggests that if supersymmetry were an exact symmetry the zero-point energy of all boson

---

[4] The possibility of vacuum fluctuations with negative energy density is intriguing for three staples of science fiction: warp drives, wormholes, and time machines. It can be shown that warp drives and stable wormholes are at least hypothetically possible if exotic material having a negative energy density were at our engineering disposal; unfortunately, no such material is known. In principle, if stabilized wormholes existed they also might be used to build a time machine. See Thorne [234] for a general discussion.

and fermion fields would exactly cancel each other, leaving a vacuum energy density equal to zero. But it has already been argued above that supersymmetry (if it exists) must be at least partially broken to be in accord with observations. Therefore, this raises the possibility that, if the Universe has a broken supersymmetry, the contributions of fermion and boson fields to zero-point energy could almost, but not quite, cancel, leaving a net vacuum energy density more in accord with observation than our previous naive estimates based on boson and fermion fields alone.

A supersymmetric quantum field theory calculation using perturbation theory to first order in the boson and fermion contributions gives for the vacuum energy density associated with zero-point motion of the fields [65]

$$\epsilon_\Lambda = \left(\frac{m_f^2 - m_b^2}{P_{pl}^2}\right) \epsilon_\Lambda^b, \tag{26.7}$$

where ϵ_Λ^b is the bosonic estimate from Eq. (26.5) and $\Delta m^2 \equiv m_f^2 - m_b^2$ is the difference in the squares of the mass scales between bosonic (b) and fermionic (f) supersymmetric partners, respectively. (If the supersymmetry were exact, this difference would be zero; but the supersymmetry is broken by hypothesis, so Δm^2 is expected to be finite.) Hence, our absurdly high estimate in Eq. (26.5) will be reduced by the initial factor in (26.7) involving the mass square difference Δm^2. But the failure to observe supersymmetric particles in any experiment to date places a lower limit on the possible values of Δm^2 such that $\Delta m^2/P_{pl}^2 \geq 10^{-34}$, and the new estimate of ϵ_Λ provided by Eq. (26.7) is still more than 80 orders of magnitude too large to be in accord with observation. Even by cosmological standards that is a little beyond the pale, suggesting rather emphatically either that dark energy is not associated with quantum vacuum fluctuations, or that vacuum energy is not very well understood at all. The general issue of the quantum vacuum energy predicted by quantum field theory being many orders of magnitude larger than the observational constraints from astronomy on a cosmological constant has been reviewed by Weinberg [247].

26.3 Quantum Gravity

As the history of the Universe is extrapolated backward in time the big bang theory tells us that the Universe becomes more dense and much hotter, and the relevant distance scales for particle interactions become shorter. But if the distance scales become small enough the principles of quantum mechanics come into play. Therefore, in any extrapolation back in time to the beginning of the Universe we eventually must reach a state where a fully quantum-mechanical theory of gravity is required. This is called the *Planck era,* and the corresponding scales of distance, energy, and time are called the *Planck scale.* Quantities that are characteristic of the Planck scale were discussed in Section 12.6 and are listed again for reference in Table 26.1.

It is instructive to compare the Planck scale with the scale on which actual data for the properties of elementary particles exist at present. As of 2017, in the largest accelerators

Table 26.1 The Planck scale	
Quantity	Value
Planck mass	1.2×10^{19} GeV/c^2
Planck length	1.6×10^{-33} cm
Planck time	5.4×10^{-44} s
Planck temperature	1.4×10^{32} K

particles having a rest mass of several hundred GeV can be produced. The temperature of a gas having particles of this average energy is approximately 10^{15} K and the time after the big bang when the temperature of the Universe would have dropped to this value is about 10^{-10} s. Therefore, all speculation about the Universe at times earlier than this is based on particle properties that can be inferred only theoretically, because there are no direct experimental data for higher energies. Clearly the Planck scale lies far beyond our present or foreseeable ability to probe it directly, but presumably it was relevant in the very beginning of the big bang.

At the Planck scale the principles of quantum mechanics must be applied to the gravitational force but the best theory of gravity is general relativity and it does not respect the principles of quantum mechanics, nor does quantum mechanics respect the principles of general relativity. For example, free particles follow well-defined classical trajectories (geodesics) in general relativity but no such classical paths are permitted in quantum mechanics because of the uncertainty principle, and the standard equations in the present formulations of quantum field theory can be made Lorentz covariant but not generally covariant. What is required then is a theory of gravitation that also is consistent with quantum mechanics – a theory of *quantum gravitation*. Unfortunately, no one has yet understood how to accomplish this very difficult task so there is no internally consistent theory of quantum gravity. The most promising candidate is superstring theory but it is not yet clear whether it can provide a correct picture of quantum gravity.

26.3.1 Superstrings and Branes

A description of the microscopic world using quantum mechanics or quantum field theory assumes that elementary particles have no internal structure and exist at a point in space. A point has zero dimension, since it has neither breadth, width, nor height. This feature of quantum mechanics leads to very serious technical problems. In the modern description of the strong, weak, and electromagnetic interactions a complex mathematical prescription has been worked out that avoids these problems. The technical term for this prescription is *renormalization*. It systematically removes infinite quantities that would otherwise crop up in the theory, leaving a logically consistent description of these forces.

But gravity is different. Because of its properties, the renormalization procedure that works for the other three fundamental interactions fails for gravity.[5] This means that

---

[5] The reason is highly technical, having to do with the electromagnetic, weak, and strong interactions being mediated by spin-1 fields (photons, the three intermediate vector bosons, and gluons, respectively), but gravity being mediated by a spin-2 field (gravitons).

26.3 Quantum Gravity

attempting to apply ordinary quantum field theory to gravity leads inevitably to observable physical quantities that become infinite. Since no one has figured out how to deal with these infinite quantities mathematically, quantum gravity based on point-particle quantum mechanics leads to a theory that is not logically consistent (*not renormalizable*, in the jargon) and cannot be used reliably to calculate observable quantities. This has two related implications. First, no way presently exists to describe gravity on the Planck scale. Second, there is no way to join gravity in its present form with the other three forces into a unified description of all forces based on point-particle quantum mechanics.

The starting point for superstring theory is to change the assumption that the elementary building blocks of the Universe are point particles. Instead, superstring theory assumes that the fundamental building blocks of the Universe are tiny (Planck-length size) objects that are not points but instead have a length. That is, they are like strings. These strings are also assumed to possess a supersymmetry (see Section 26.1.3), so these elementary string-like building blocks are called *superstrings*. Because they have a length, they are 1-dimensional, as opposed to the 0-dimensional particles of ordinary quantum mechanics. The top portion of Fig. 26.1 illustrates. Although this may not seem to be a very large change, it is. In particular, it can be shown mathematically that the assumption that the basic building blocks are not 0-dimensional points permits many of the troublesome infinities to be avoided.

After the basic idea of superstrings was introduced in the early 1980s, five different versions of string theory were developed based on five different fundamental symmetries. In the 1990s it was realized that these five versions of string theories were closely related to each other. They appeared to be different because the discussion to that point had been based on approximate solutions of the five theories (the mathematics of superstrings is very difficult, making exact solutions of the equations hard to obtain). When more exact solutions were obtained it was realized that the five existing versions of superstrings could be unified in a single more general theory. This more general theory is called

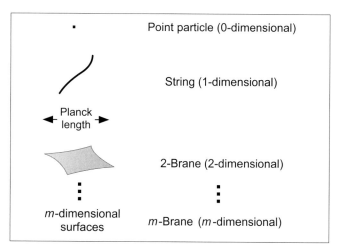

Fig. 26.1   Schematic illustration of points, strings, and branes.

m-brane theory (often shortened to just *M-theory*) because it implies a generalization of the superstring idea of fundamental particles having one dimension to their having m dimensions (on the Planck scale). The resulting geometrical surfaces are called *m-branes*, with the integer m signifying the number of dimensions. The lower portion of Fig. 26.1 illustrates a 2-brane; that is, a 2-dimensional object with dimensions comparable with the Planck length (superstrings may be called 1-branes).

Superstring theory and its generalization M-theory imply that what are currently called "elementary particles" (electrons, photons, quarks, ...) are not really elementary. They have an internal structure visible only on the Planck scale. This internal structure consists of a set of elementary building blocks that are not point particles but are instead 1-branes (strings), 2-branes, 3-branes, and so on (the theory suggests that branes up to 9-branes can exist). Since there is no hope in the foreseeable future to probe the Planck scale directly, the challenge is whether these ideas can make any testable scientific predictions at energies below the Planck scale. As discussed in Box 26.2, there might be hope for such predictions, but a large amount of mathematical development is required first. One promising potential application is the *AdS/CFT correspondence* discussed in Box 26.3.

26.3.2 How Many Dimensions?

Brane theories indicate that even our ideas about the number of dimensions in spacetime may require revision. They suggest that spacetime has more dimensions than the four evident in daily life. However, the extra dimensions are conjectured to be invisible except at distances close to the Planck length. This is not a completely crazy idea, since it is easy to be fooled about the number of dimensions for a space if it cannot be resolved on a sufficiently microscopic scale. A simple analogy is a cylindrical pipe, which is a 2-dimensional surface but if the pipe is viewed from a distance it looks like a line, which is a 1-dimensional surface. Only up close can the "hidden" extra dimension be seen.

26.3.3 Spacetime Foam, Wormholes, and Such

The domain of quantum gravity is presumably bizarre. Since there isn't an adequate theory it is difficult to make precise statements but qualitatively there is reason to believe that on this scale even space and time may become something other than our usual conceptions. For example, spacetime may develop "wormholes," such as the one illustrated in Fig. 26.2 for the more easily visualized 2-dimensional case. A wormhole could connect two regions of the Universe without going through the space in between the two points.

Even more disconcerting for our ordinary sensibilities about space and time is the possibility that the two are no longer even continuous below the Planck scale. Gravity differs fundamentally from the other basic forces, which act *in* spacetime. For the other three basic forces, spacetime often serves to a good approximation as a passive "stage" for physical events. But gravity distorts spacetime itself and is in turn generated by distortion of spacetime. Therefore, quantum fluctuations of the gravitational field don't occur on a passive spacetime stage. Rather, it is the spacetime stage itself (that is, the metric) that

> **Box 26.2** **Testing Branes and Superstrings**
>
> The present discussion implies that a logically consistent quantum theory of gravity and a unification of all four fundamental forces may be possible based on M-theory. It is difficult to be certain because its mathematical methods are challenging and still being developed. As a consequence, it has not yet been possible to use the theory to make a quantitative prediction that can be tested by currently feasible experiments. Recall that a hallmark of modern science is experimental verification of hypotheses: a theory must pass the experimental test to be acceptable as a description of nature. It is hoped that M-theory will be capable of a quantitatively testable prediction within a decade or so, but it is not yet clear that this is feasible.
>
> One idea leading to a qualitative testable prediction concerns the strength of the gravitational force. As noted below, according to brane theory the true number of dimensions for spacetime is more than the four in evidence. It has been proposed that the well-known weakness of gravity is an artifact of our observations being confined to four spacetime dimensions. In this proposal, gravity really is very strong but most of its strength has "leaked" into other dimensions that are not visible in our low-energy world. Thus, as particle accelerators probe higher energies and therefore shorter length scales, eventually they might begin to see evidence for additional dimensions well before the Planck scale. As a consequence, the effective strength of gravity might suddenly begin to grow as the energy of the probe is increased.
>
> With any foreseeable particle-accelerator technology there is no hope of compressing matter enough to produce a tiny black hole if the intrinsic strength of gravity varies in the same manner measured at larger scales in the laboratory. But if the strength of gravity were to grow more rapidly than expected at very short distances it would be much easier to make a small black hole. Hence the appearance of mini black holes in high-energy particle collisions (which could be detected through their rapid decay in a burst of Hawking radiation; see Section 12.2) would be indirect confirmation of the brane-theory hypothesis. The highest-energy collisions presently available are for the Large Hadron Collider (LHC) near Geneva, and for the highest-energy cosmic rays. There is no evidence thus far for the production of black holes in such collisions.

fluctuates at the quantum level of the gravitational field. (Notice that the detection of gravitational waves discussed in Chapter 24 indicates that time-dependent macroscopic fluctuations of the spacetime metric are real and observable.) This fluctuation of spacetime itself has been described rather poetically as the dissolution of the spacetime continuum into a frothing and bubbling "spacetime foam." Relativity implies that space and time are not what they seem, but with relativity we could at least retain the idea of spacetime as continuous. With quantum gravity, even that may not be possible.

26.3.4 The Ultimate Free Lunch

In quantum mechanics even "empty space" is fluctuating and particle–antiparticle pairs can materialize as excitations of the vacuum. The strangest of all the strange ideas associated with quantum gravity is that perhaps the Universe itself is a fluctuation in the "spacetime vacuum" (which corresponds classically to no space or time). That is, perhaps at creation an expanding spacetime appeared out of "nothing" as a quantum fluctuation, giving birth

> **Box 26.3** **The AdS/CFT Correspondence**
>
> One development that has received considerable attention is called *AdS/CFT correspondence* or more generally *gauge/gravity duality*, which is a conjectured relationship between a theory of quantum gravity in a particular spacetime and a quantum field theory formulated on the *boundary* of that spacetime (hence on a manifold with one less dimension than the spacetime in which the gravity is defined) that was proposed by Argentinian physicist Juan Maldacena. The AdS/CFT correspondence is a specific example of the holographic principle discussed in Section 12.7.1.
>
> AdS refers to a 5-dimensional *anti-de Sitter space*, which corresponds to a metric formulated in five dimensions that is like the de Sitter metric discussed in Section 19.3.1, but with the *opposite sign* for the cosmological constant. It has constant negative curvature, which implies an attraction rather than repulsion in vacuum. The boundary of the 5-dimensional AdS is 4-dimensional, and it is proposed that a *conformal field theory (CFT)*, which is a quantum field theory in four dimensions that is invariant under conformal transformations (see Section 18.8.1), lives there that is *dual* to the theory of gravity in AdS. In this picture the anti-de Sitter space is called the *bulk space* and the quantum field theory lives in the *boundary space*. Five-dimensional AdS is particularly useful because it possesses symmetries analogous to those of conformal transformations in four dimensions.
>
> The AdS does not resemble our Universe, since the observed accelerated expansion is more suggestive of de Sitter than anti-de Sitter space. The importance of AdS lies instead in the *duality* that maps a string theory living in AdS to a quantum field theory living on its 4-dimensional boundary. This implies that solutions of the 4D quantum field theory can be obtained from solutions of 5D gravity in AdS through the duality relations. The particularly interesting aspect is that if the theory on one side of the duality is *strongly coupled* (hence difficult to solve) then the theory on the other side is *weakly coupled* (hence easier to solve).
>
> Thus, it has been proposed that the AdS/CFT duality might allow quantum field theories for say quantum chromodynamics (QCD) or correlated electron systems (which both are strongly coupled and very difficult to deal with in normal field theory) to be solved using the weakly coupled dual in anti-de Sitter space. In current implementations of the correspondence the gravity side does not look like real gravity and the CFT side does not look like real quantum field theories (for example, the CFT side has a high degree of supersymmetry not exhibited by real QCD), but *some important features* of the CFT resemble those of actual 4D quantum field theories.

to our Universe. This idea has been dubbed the *ultimate free lunch,* since it corresponds to creation of a universe from literally nothing, not even space or time [107].

26.3.5 Does the Planck Scale Matter?

Since a consistent wedding of general relativity with quantum mechanics is not yet at hand, the presently understood laws of physics may be expected to break down on the Planck scale. Therefore, our picture of inflation followed by the standard big bang says nothing about the Universe at those very early times preceding inflation. In this respect then, it is relatively certain that the current laws of physics are not complete. However, the Planck

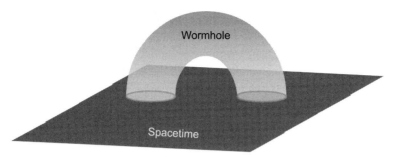

Fig. 26.2 A wormhole that connects two regions of the same approximately flat space. The wormhole illustrated in Box 9.1 connects two different asymptotically flat spacetimes.

scale is so incredibly small that this may have been significant only in the first moments corresponding to the creation of the Universe, and raises the question of whether the Planck scale even matters in the current Universe.

One viewpoint is that no method is available to probe the Planck scale, so it is of no significance for a general understanding of science in the here and now. However, if the Universe passed through the Planck scale early in its history, it is possible that the Universe is the way that it is because of events that happened at the Planck scale. For example, perhaps the "ground rules" governing physical law in our Universe were laid down at the Planck scale and those rules could have been different if different things had happened at the Planck time. If that were true, the Planck scale would certainly be scientifically relevant to a present understanding of the Universe, even if it cannot be studied directly today, because of delimiting the possibilities through setting initial conditions.

As an example mentioned earlier in this chapter, M-theory implies that the actual number of spacetime dimensions is greater than the four that are perceived in our low-energy world, but that the "extra" dimensions are not visible to low-energy probes. (In quantum mechanics there is an inverse relationship between the energy of the probe and the distance scale that it can resolve – the Heisenberg uncertainty principle again!) One paradigm for rendering the extra dimensions invisible is *compactification*, where the extra dimensions are "rolled up" on the Planck scale to such small dimensions that they can be seen only by probes having Planck-scale energies. (The effect of compactification on the particle spectrum is addressed in Problems 26.1 and 26.2.) In this regard, recall the analogy of whether a pipe appears to be 1-dimensional or 2-dimensional depending on spatial resolution in Section 26.3.2.

In condensed matter physics it is often possible to find or construct materials that effectively have fewer than three spatial dimensions because their interactions in the directions defined by other dimensions are negligible. There are many examples from effective 1-dimensional and 2-dimensional materials that the number of spatial dimensions for a many-body system can have a profound effect on the types of collective (*emergent,* in the jargon) states that exist at low energy. Since the properties of the everyday world are strongly influenced by such emergent states, the nature of reality at the scale of daily existence (primarily governed by nonrelativistic quantum mechanics) may depend fundamentally on the number of spacetime dimensions that got compactified at the Planck scale.

Finally, there might be undiscovered objects in the present Universe that carry information about the Planck scale. For example, if Hawking mini black holes were created early in the big bang, the potentially observable endpoint for evaporation of those black holes through Hawking radiation could probe the Planck scale, even in the present Universe (see Section 12.6). If that were true, then even science in the here and now might be impacted directly by the Planck scale.

Background and Further Reading

Greene [102] gives a readable popular-level introduction to many of the ideas touched upon in this chapter. Zwiebach [259] provides an introduction to string theory at a mathematical level comparable to that of this book, except that a stronger background in quantum mechanics is required than has been assumed here.

Problems

26.1 This problem and Problem 26.2 illustrate the effect of a compactified extra dimension on the particle spectrum of a theory. Consider the standard 1-dimensional square-well problem in nonrelativistic quantum mechanics illustrated in the following figure.

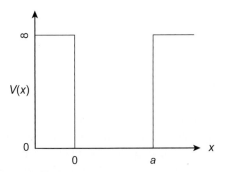

As shown in any quantum mechanics textbook, if the potential well is taken to be infinitely deep a particle of mass m is confined to the interval 0 to a, the Schrödinger equation is

$$-\frac{\hbar^2}{2m}\frac{d^2\psi}{dx^2} = E\psi,$$

and its solutions and energies are

$$\psi_k(x) = \sqrt{\frac{2}{a}} \sin\left(\frac{k\pi x}{a}\right) \qquad E_k = \frac{\hbar^2}{2m}\left(\frac{k\pi}{a}\right)^2 \qquad (k = 1, 2, 3, \ldots, \infty),$$

where the boundary conditions exclude $k = 0$. Now add a second dimension y, which is assumed compactified on an interval $2\pi R$, where R is a parameter with dimension length. That is, $(x, y) \sim (x, y + 2\pi R)$, where the notation means that y and $y + 2\pi R$ are to be identified (the y dimension is rolled up in a circle of radius R). Solve this 2-dimensional square well problem using separation of variables, $\psi(x, y) = \psi(x)\varphi(y)$, and show that solutions and energies are given by

$$\psi_k(x) = c_k \sin\left(\frac{k\pi x}{a}\right) \qquad \varphi_\ell(y) = a_\ell \sin\left(\frac{\ell y}{R}\right) + b_\ell \cos\left(\frac{\ell y}{R}\right)$$

$$E_{k,\ell} = \frac{\hbar^2}{2m}\left[\left(\frac{k\pi}{a}\right)^2 + \left(\frac{\ell}{R}\right)^2\right] \qquad (k = 1, 2, 3, \ldots, \infty;\ \ell = 0, 1, 2, \ldots, \infty).$$

Hint: The boundary conditions in x are the same as in the 1-dimensional problem but the boundary conditions in y must incorporate the $2\pi R$ periodicity. Thus the solution acquires new states because there are now two quantum numbers, k and ℓ. Physical implications of this spectrum will be explored further in Problem 26.2. ***

26.2 To understand how states associated with compactified extra dimensions could escape detection thus far, consider the solution obtained in Problem 26.1 for one regular and one compactified spatial dimension. Show that if $\ell = 0$ the original spectrum of the 1-dimensional square well is recovered. Show that the lowest-energy new state introduced because of the compactified dimension is at energy

$$E_{\min}^{\text{new}} = \frac{\hbar^2}{2m}\left[\left(\frac{\pi}{a}\right)^2 + \left(\frac{1}{R}\right)^2\right].$$

Demonstrate that if $R \ll a$, this state corresponds in energy with a state having $k = a/\pi R \gg 1$ in the 1-dimensional problem. Therefore, such states would lie at very high energy and would have little direct influence on the low-energy world. ***

Appendix A Constants

Fundamental constants

Gravitational constant: $G = 6.67408 \times 10^{-8}$ dyn cm^2 g^{-2}
$= 6.67408 \times 10^{-8}$ g^{-1} cm^3 s^{-2}
$= 6.67408 \times 10^{-8}$ erg cm g^{-2}
$= 2.960 \times 10^{-4} M_\odot^{-1}$ AU^3 days^{-2}
$= 1.327 \times 10^{11} M_\odot^{-1}$ km^3 s^{-2}

Speed of light in a vacuum: $c = 2.99792458 \times 10^{10}$ cm s^{-1}

Planck's constant: $h = 2\pi\hbar = 6.6261 \times 10^{-27}$ erg s
$= 4.136 \times 10^{-21}$ MeV s
$\hbar = 1.0546 \times 10^{-27}$ erg s $= 6.5827 \times 10^{-22}$ MeV s
$\hbar c = 197.3$ MeV fm $= 197.3 \times 10^{-13}$ MeV cm

Electrical charge unit: $e = 4.8032068 \times 10^{-10}$ esu
$= 4.8032068$ erg$^{1/2}$cm$^{1/2}$
$= 4.8032068$ g$^{1/2}$ cm$^{3/2}$s^{-1}

Fine structure constant: $\alpha = (137.036)^{-1} = 0.0073$

Weak (Fermi) constant: $G_F = 8.958 \times 10^{-44}$ MeV cm^3
$= 1.16637 \times 10^{-5}$ GeV^{-2} $[G_F/(\hbar c)^3; \hbar = c = 1]$

Mass of electron: $m_e = 9.1093898 \times 10^{-28}$ g
$= 5.4858 \times 10^{-4}$ amu
$= 0.5109991$ MeV/c^2

Mass of proton: $m_p = 1.6726231 \times 10^{-24}$ g
$= 1.00727647$ amu
$= 938.27231$ MeV/c^2

Mass of neutron: $m_n = 1.6749286 \times 10^{-24}$ g
$= 1.0086649$ amu
$= 939.56563$ MeV/c^2

Atomic mass unit (amu) $= 1.6605390 \times 10^{-24}$ g

Avogadro's constant: $N_A = 6.0221409 \times 10^{23}$ mol^{-1}

Boltzmann's constant: $k = 1.38065 \times 10^{-16}$ erg K^{-1}
$= 8.617389 \times 10^{-5}$ eV K^{-1}

Ideal gas constant: $R_{gas} \equiv N_A k = 8.314511 \times 10^7$ erg K^{-1} mole^{-1}

Stefan–Boltzmann constant: $\sigma = 5.67051 \times 10^{-5}$ erg cm^{-2} K^{-4} s^{-1}

Radiation density constant: $a \equiv 4\sigma/c = 7.56591 \times 10^{-15}$ erg cm^{-3} K^{-4}
$= 4.7222 \times 10^{-9}$ MeV cm^{-3}K^{-4}

Planck mass: $M_P = 1.2 \times 10^{19}$ GeV/c^2

Planck length: $\ell_P = 1.6 \times 10^{-33}$ cm

Planck time: $t_P = 5.4 \times 10^{-44}$ s

Planck temperature: $T_P = 1.4 \times 10^{32}$ K

Solar quantities

Solar (photon) luminosity: $L_\odot = 3.828 \times 10^{33}$ erg/s

Solar absolute magnitude $M_v = 4.83$

Solar bolometric magnitude $M_{bol}^\odot = 4.74$

Solar mass: $M_\odot = 1.989 \times 10^{33}$ g

Effective surface temperature: $T_\odot^{eff} = 5780$ K

Solar radius: $R_\odot = 6.96 \times 10^{10}$ cm

Central density: $\rho_\odot^{core} \simeq 160$ g/cm^3

Central pressure: $P_\odot^{core} \simeq 2.7 \times 10^{17}$ dyn cm^{-2}

Central temperature: $T_\odot^{core} \simeq 1.6 \times 10^7$ K

Color indices: $B - V = 0.63 \qquad U - B = 0.13$

Solar constant: 1.36×10^6 erg cm^{-2} s^{-1}

General quantities

1 tropical year (yr) = 3.1556925×10^7 s = 365.24219 d

1 parsec (pc) = 3.0857×10^{18} cm = 206,265 AU = 3.2616 ly

1 lightyear (ly) = 9.4605×10^{17} cm

1 astronomical unit (AU) = 1.49598×10^{13} cm

Energy per gram from H $\to$ He fusion = 6.3×10^{18} erg/g

Thomson scattering cross section: $\sigma_T = 6.652 \times 10^{-25}$ cm^2

Mass of Earth $M_\oplus = 5.98 \times 10^{27}$ g

Radius of Earth $R_\oplus = 6.371 \times 10^8$ cm

Useful conversion factors

1 eV = $1.60217733 \times 10^{-12}$ ergs = $1.60217733 \times 10^{-19}$ J

1 J = 10^7 ergs = 6.242×10^{18} eV

1 amu = $1.6605390 \times 10^{-24}$ g

1 fm = 10^{-13} cm

0 K = $-273.16°$Celsius

1 atomic unit (a_0) = 0.52918×10^{-8} cm

1 atmosphere (atm) = 1.013250×10^6 dyn cm^{-2}

1 pascal (Pa) = 1 N m^{-2} = 10 dyn cm^{-2}
1 arcsec = 1″ = 4.848 × 10^{-6} rad = 1/3600 deg
1 Å = 10^{-8} cm
1 barn (b) = 10^{-24} cm^2
1 newton (N) = 10^5 dyn
1 watt (W) = 1 J s^{-1} = 10^7 erg s^{-1}
1 gauss (G) = 10^{-4} Tesla (T)
1 g cm^{-3} = 1000 kg m^{-3}
Opacity units: 1 m^2 kg^{-1} = 10 cm^2 g^{-1}

Conversion between normal and natural units: see Appendix B.

Appendix B Natural Units

In astrophysics it is common to use the CGS (centimeter–gram–second) system of units. However, in many applications it is more convenient to define new sets of units where fundamental constants such as the speed of light or the gravitational constant may be given unit value. Such units are sometimes termed *natural units* because they are suggested by the physics of the phenomena being investigated. For example, the velocity of light c is clearly of fundamental importance in problems where special relativity is applicable. In that context, it is far more natural to use c to set the scale for velocities than to use an arbitrary standard (such as the length of some king's foot divided by a time unit that derives historically from the apparent revolution of the heavens!) that has arisen historically in nonrelativistic physics and engineering. Defining a set of units where c takes unit value is equivalent to making velocity a dimensionless quantity that is measured in units of c, as illustrated below, thus setting a "natural" scale for velocity.

The introduction of a natural set of units has the advantage of more compact notation, since the constants rescaled to unit value need not be included explicitly in the equations, and the standard "engineering" units like CGS may be restored by dimensional analysis if they are required to obtain numerical results. This appendix outlines the use of such natural units for problems encountered in this book.

B.1 Geometrized Units

In gravitational physics it is useful to employ a natural set of units called *geometrized units* or $c = G = 1$ units that give both the speed of light and the gravitational constant unit value. Setting

$$1 = c = 2.9979 \times 10^{10} \text{ cm s}^{-1} \qquad 1 = G = 6.6741 \times 10^{-8} \text{ cm}^3 \text{ g}^{-1} \text{ s}^{-2}, \qquad (B.1)$$

one may solve for standard units like seconds in terms of these new units. For example, from the first equation

$$1 \text{ s} = 2.9979 \times 10^{10} \text{ cm}, \qquad (B.2)$$

and from the second

$$\begin{aligned} 1 \text{ g} &= 6.6741 \times 10^{-8} \text{ cm}^3 \text{ s}^{-2} \\ &= 6.6741 \times 10^{-8} \text{ cm}^3 \left(\frac{1}{2.9979 \times 10^{10} \text{ cm}} \right)^2 \\ &= 7.4261 \times 10^{-29} \text{ cm}. \end{aligned} \qquad (B.3)$$

Table B.1 Conversions between geometrized units and standard units [111]

Quantity	Symbol	Geometrized unit	Standard unit	Conversion
Mass	M	$\mathscr{L}$	$\mathscr{M}$	GM/c^2
Length	L	$\mathscr{L}$	$\mathscr{L}$	L
Time	t	$\mathscr{L}$	$\mathscr{T}$	ct
Spacetime distance	s	$\mathscr{L}$	$\mathscr{L}$	s
Proper time	τ	$\mathscr{L}$	$\mathscr{T}$	$c\tau$
Energy	E	$\mathscr{L}$	$\mathscr{M}(\mathscr{L}/\mathscr{T})^2$	GE/c^4
Momentum	p	$\mathscr{L}$	$\mathscr{M}(\mathscr{L}/\mathscr{T})$	Gp/c^3
Angular momentum	J	$\mathscr{L}^2$	$\mathscr{M}(\mathscr{L}^2/\mathscr{T})$	GJ/c^3
Luminosity (power)	L	dimensionless	$\mathscr{M}(\mathscr{L}^2/\mathscr{T}^3)$	GL/c^5
Energy density	ϵ	$\mathscr{L}^{-2}$	$\mathscr{M}/(\mathscr{L}\mathscr{T}^2)$	$G\epsilon/c^4$
Momentum density	π_i	$\mathscr{L}^{-2}$	$\mathscr{M}/(\mathscr{L}^2\mathscr{T})$	$G\pi_i/c^3$
Pressure	P	$\mathscr{L}^{-2}$	$\mathscr{M}/(\mathscr{L}\mathscr{T}^2)$	GP/c^4
Energy / unit mass	ϵ	dimensionless	$(\mathscr{L}/\mathscr{T})^2$	ϵ/c^2
Ang. mom. / unit mass	ℓ	$\mathscr{L}$	$\mathscr{L}^2/\mathscr{T}$	ℓ/c
Planck constant	$\hbar$	$\mathscr{L}^2$	$\mathscr{M}(\mathscr{L}^2/\mathscr{T})$	$G\hbar/c^3$

The standard unit of length is $\mathscr{L}$, the standard unit of mass is $\mathscr{M}$, and the standard unit of time is $\mathscr{T}$. To convert equations to standard units from geometrized units, replace quantities in column 2 with quantities in the last column. To convert from standard to geometrized units, multiply by the factor of G and c appearing in the last column.

So both time and mass have the dimension of length in geometrized units. Likewise, from the preceding relations

$$1 \text{ erg} = 1 \text{ g cm}^2 \text{ s}^{-2} = 8.2627 \times 10^{-50} \text{ cm} \tag{B.4}$$

$$1 \text{ g cm}^{-3} = 7.4261 \times 10^{-29} \text{ cm}^{-2} \tag{B.5}$$

$$1 M_\odot = 1.477 \text{ km}, \tag{B.6}$$

and so on. Velocity is dimensionless in these units (that is, v is measured in units of c). In geometrized units, all explicit instances of G and c are then dropped in the equations. When quantities need to be calculated in standard units, appropriate combinations of c and G must be reinserted to give the right standard units for each term. Example 5.1 shows how this works. While conversion between geometrized and standard units can be handled through dimensional analysis on a case by case basis, it is convenient to tabulate the conversion factors for common quantities. These are given in Table B.1, which is adapted from a similar table in Hartle [111].

To convert equations from geometrized units to standard units, replace all quantities in column 2 of the table with those in the last column. For example, replace all masses M with GM/c^2 and all times t with ct. To convert from standard to geometrized units, multiply by the factor of G and c indicated in the last column. For example, to convert a mass in standard units of grams to geometrized units of centimeters, multiply it by GM/c^2:

$$M(\text{cm}) = \frac{G}{c^2} M(\text{g}) = 7.4 \times 10^{-29} \text{ cm g}^{-1} M(\text{g}),$$

where $M(g)$ is the mass in grams and $M(cm)$ is the mass in centimeters. Thus, the mass of the Sun is

$$M = 1\,M_\odot = 7.4 \times 10^{-29}\,\text{cm}\,\text{g}^{-1} \times 1.989 \times 10^{33}\,\text{g} = 1.47\,\text{km},$$

when expressed in geometrized units, and the Schwarzschild radius $r_s = 2M$ can be read off by inspection as 2.95 km for the Sun.

B.2 Natural Units in Particle Physics

In relativistic quantum field theory the explicit role of gravity in the interactions can be ignored (except on the Planck scale), but the equations expressed in standard units are populated by a multitude of the fundamental constants c (expressing the importance of special relativity) and $\hbar$ (expressing the importance of quantum mechanics). It is convenient in this context to define natural units where $\hbar = c = 1$. Using the notation $[a]$ to denote the dimension of a and using $[L]$, $[T]$ and $[M]$ to denote the dimensions of length, time, and mass, respectively, for the speed of light c,

$$[c] = [L][T]^{-1}. \tag{B.7}$$

Setting $c = 1$ then implies that $[L] = [T]$, and since $E^2 = p^2 c^2 + M^2 c^4$,

$$[E] = [M] = [p] = [k], \tag{B.8}$$

where $p = \hbar k$. Furthermore, because

$$[\hbar] = [M][L]^2[T]^{-1} \tag{B.9}$$

one has

$$[M] = [L]^{-1} = [T]^{-1} \tag{B.10}$$

if $\hbar = 1$. These results then imply that $[M]$ may be chosen as the single independent dimension of our set of $\hbar = c = 1$ natural units. This dimension is commonly measured in MeV (10^6 eV) or GeV (10^9 eV). Useful conversions are

$$\hbar c = 197.3\,\text{MeV\,fm} \qquad 1\,\text{fm} = \frac{1}{197.3}\,\text{MeV}^{-1} = 5.068\,\text{GeV}^{-1}$$

$$1\,\text{fm}^{-1} = 197.3\,\text{MeV} \qquad 1\,\text{GeV} = 5.068\,\text{fm}^{-1}, \tag{B.11}$$

where $1\,\text{fm} = 10^{-13}$ cm (one fermi or one femtometer). For example, the Compton wavelength of the pion is

$$\lambda_\pi = \frac{1}{M_\pi} \simeq (140\,\text{MeV})^{-1}$$

in $\hbar = c = 1$ units. Utilizing Eq. (B.11), this may be converted to

$$\lambda_\pi = \left(\frac{1}{140}\,\text{MeV}^{-1}\right) \times 197.3\,\text{MeV\,fm} = 1.41\,\text{fm}$$

in standard units.

B.3 Natural Units in Cosmology

In cosmology we often employ a set of $\hbar = c = k_B = 1$ natural units, where k_B is the Boltzmann constant. Then from $E = k_B T$ and $k_B = 8.617 \times 10^{-14}\,\text{GeV}\,\text{K}^{-1} = 1$,

$$1\,\text{GeV} = 1.2 \times 10^{13}\,\text{K}, \tag{B.12}$$

where K denotes kelvins. From Section 12.6 we then have for the Planck mass, Planck energy, Planck temperature, Planck length, and Planck time in these natural units,

$$M_P = E_P = T_P = \ell_P^{-1} = t_P^{-1}.$$

To convert to standard units, note that from Eq. (12.16) the gravitational constant may be expressed as

$$G = \frac{1}{M_P^2}, \tag{B.13}$$

where the Planck mass is

$$M_P = 1.2 \times 10^{19}\,\text{GeV}. \tag{B.14}$$

From Eqs. (12.17), (B.14), and (B.11), the corresponding Planck length is

$$\ell_P = \frac{1}{M_P} = 1.6 \times 10^{-33}\,\text{cm}, \tag{B.15}$$

multiplying by c^{-1} gives the corresponding Planck time,

$$t_P = 5.4 \times 10^{-44}\,\text{s}, \tag{B.16}$$

and from Eqs. (B.12) and (B.14),

$$T_P = 1.4 \times 10^{32}\,\text{K} \tag{B.17}$$

is the Planck temperature.

Appendix C Einstein Tensor for a General Spherical Metric

This appendix summarizes for reference the construction of the Einstein tensor for a general spherical metric (see Problem 8.2). Take a line element parameterized in the form

$$ds^2 = -e^\sigma dt^2 + e^\lambda dr^2 + r^2(d\theta^2 + \sin^2\theta\, d\varphi^2),$$

where $\sigma = \sigma(r,t)$ and $\lambda = \lambda(r,t)$. The corresponding metric is

$$g_{\mu\nu} = \text{diag}(-e^\sigma, e^\lambda, r^2, r^2\sin^2\theta)$$

$$g^{\mu\nu} = g_{\mu\nu}^{-1} = \text{diag}\left(-e^{-\sigma}, e^{-\lambda}, \frac{1}{r^2}, \frac{1}{r^2\sin^2\theta}\right).$$

From Eq. (7.30) the non-vanishing Christoffel symbols are

$$\Gamma^0_{00} = \tfrac{1}{2}\dot{\sigma} \quad \Gamma^0_{01} = \tfrac{1}{2}\sigma' \quad \Gamma^0_{11} = \tfrac{1}{2}e^{\lambda-\sigma}\dot{\lambda} \quad \Gamma^1_{00} = \tfrac{1}{2}e^{\sigma-\lambda}\sigma' \quad \Gamma^1_{01} = \tfrac{1}{2}\dot{\lambda}$$

$$\Gamma^1_{11} = \tfrac{1}{2}\lambda' \quad \Gamma^1_{22} = -re^{-\lambda} \quad \Gamma^1_{33} = -re^{-\lambda}\sin^2\theta \quad \Gamma^2_{12} = 1/r$$

$$\Gamma^2_{33} = -\sin\theta\cos\theta \quad \Gamma^3_{13} = 1/r \quad \Gamma^3_{23} = \cot\theta$$

where dots indicate time derivatives and primes indicate derivatives with respect to r. The Riemann curvature tensor may then be calculated from Eq. (8.14); the non-zero components are

$$R_{0101} = \tfrac{1}{2}e^\sigma \sigma'' - \tfrac{1}{4}e^\lambda \dot{\lambda}^2 + \tfrac{1}{4}e^\lambda \dot{\sigma}\dot{\lambda} - \tfrac{1}{2}e^\lambda \ddot{\lambda} + \tfrac{1}{4}e^\sigma (\sigma')^2 - \tfrac{1}{4}e^\sigma \sigma'\lambda'$$

$$R_{0202} = \tfrac{1}{2}e^{\sigma-\lambda}\sigma' \quad R_{0212} = \tfrac{1}{2}r\dot{\lambda} \quad R_{0303} = \tfrac{1}{2}re^{\sigma-\lambda}\sigma'\sin^2\theta$$

$$R_{0313} = \tfrac{1}{2}r\dot{\lambda}\sin^2\theta \quad R_{1212} = \tfrac{1}{2}r\lambda'$$

$$R_{1313} = \tfrac{1}{2}r\lambda'\sin^2\theta \quad R_{2323} = r^2\sin^2\theta(1 - e^{-\lambda})$$

The Ricci tensor $R_{\mu\nu}$ then follows from Eq. (8.16), with non-vanishing components

$$R_{00} = -\tfrac{1}{2}e^{\sigma-\lambda}\sigma'' - \tfrac{1}{4}e^{\sigma-\lambda}(\sigma')^2 + \tfrac{1}{2}\ddot{\lambda} + \tfrac{1}{4}\dot{\lambda}^2 - \tfrac{1}{4}\dot{\sigma}\dot{\lambda} + \tfrac{1}{4}e^{\sigma-\lambda}\sigma'\lambda' - e^{\sigma-\lambda}\lambda'/r$$

$$R_{11} = \tfrac{1}{2}\sigma'' + \tfrac{1}{4}(\sigma')^2 - \tfrac{1}{2}e^{\lambda-\sigma}\ddot{\lambda} - \tfrac{1}{4}e^{\lambda-\sigma}\dot{\lambda}^2 + \tfrac{1}{4}e^{\lambda-\sigma}\dot{\sigma}\dot{\lambda} - \tfrac{1}{4}\sigma'\lambda' - \lambda'/r$$

$$R_{01} = -\dot{\lambda}/r \quad R_{22} = \tfrac{1}{2}re^{-\lambda}\sigma' - \tfrac{1}{2}re^{-\lambda}\lambda' + e^{-\lambda} - 1 \quad R_{33} = R_{22}\sin^2\theta$$

The Ricci scalar R may then be constructed from Eq. (8.17),

$$R = 2\frac{e^{-\lambda}}{r^2} + 2\frac{e^{-\lambda}}{r}\sigma' - 2\frac{e^{-\lambda}}{r}\lambda' + e^{-\lambda}\sigma'' + \tfrac{1}{2}e^{-\lambda}(\sigma')^2$$

$$- e^{-\sigma}\ddot{\lambda} - \tfrac{1}{2}e^{-\sigma}\dot{\lambda}^2 - \frac{2}{r^2} + \tfrac{1}{2}e^{-\sigma}\dot{\sigma}\dot{\lambda} - \tfrac{1}{2}e^{-\lambda}\sigma'\lambda',$$

and the Einstein tensor then follows from Eq. (8.20):

$$G_{00} = \frac{e^{\sigma}}{r^2} - \frac{e^{\sigma-\lambda}}{r^2} + \frac{e^{\sigma-\lambda}}{r}\lambda' \qquad G_{01} = -\frac{\dot{\lambda}}{r} \qquad G_{11} = -\frac{e^{\lambda}}{r^2} + \frac{1}{r^2} + \frac{\sigma'}{r}$$

$$G_{22} = \tfrac{1}{2}r\lambda' e^{-\lambda} - \tfrac{1}{2}r\sigma' e^{-\lambda} - \tfrac{1}{2}r^2 e^{-\lambda}\sigma'' - \tfrac{1}{4}r^2 e^{-\lambda}(\sigma')^2 + \tfrac{1}{2}r^2 e^{-\sigma}\ddot{\lambda}$$
$$+ \tfrac{1}{4}r^2 e^{-\sigma}(\dot{\lambda})^2 - \tfrac{1}{4}r^2 e^{-\sigma}\dot{\sigma}\dot{\lambda} + \tfrac{1}{4}r^2 e^{-\lambda}\sigma'\lambda'$$

$$G_{33} = G_{22}\sin^2\theta.$$

Appendix D Using arXiv and ADS

Some journal articles referenced in this book are published in journals with limited public availability. They will likely be available from university libraries but readers without immediate access to a university library may still be able to access many of these papers free of charge by using the *arXiv preprint server* or the *ADS Astronomy Abstract Service*. Where possible, references to journal articles in the bibliography include sufficient information to allow arXiv and/or ADS to be used to retrieve copies of the articles according to the following instructions.

arXiv access: An arXiv reference will be of the general form *arXiv:xxxx*. Typing the string *xxxx* into the Search or Article-id field at http://arxiv.org and clicking the search icon will return an abstract with links to the article in PDF form, and a more general search on arXiv can be implemented by clicking the Form Interface button.

ADS access: The ADS interface may be found at http://adsabs.harvard.edu/bib_abs.html.

 (i) If a DOI number of the form *DOI: yyyy* is given for a reference, the article often can be accessed through the ADS interface by putting the string *yyyy* into the Bibliographic Code Query box and clicking Send Query.
 (ii) If a BibCode reference *BibCode: zzzz* is given, the article can be accessed through the ADS interface by typing the BibCode string *zzzz* into the Bibliographic Code Query box and clicking Send Query. Alternatively, the BibCode string can be used directly in a Web browser. For example, *BibCode: 1971Natur.232..246B* can be accessed as http://adsabs.harvard.edu/abs/1971Natur.232..246B.
 (iii) A search for a general article may be implemented with the ADS interface by giving the Journal Name/Code (there is a link on the page to the list of codes for standard journals; for example *The Astrophysical Journal* is ApJ), Year, Volume, and beginning Page of the article, and clicking Send Query.

Articles in ADS are scanned so the quality is not high, but they are generally quite readable.

Example D.1 Reference [33] corresponds to an article on numerical relativity published by Baker *et al.* in the journal *Physical Review* D. If you do not have access to that journal,

1. Reference [33] also gives an arXiv reference *gr-qc/0602026*. Putting gr-qc/0602026 into the Search or Article-id field at http://arxiv.org and clicking the search icon should return an abstract with links to a PDF version of the journal article.

2. Likewise, Ref. [33] also lists the DOI number *DOI: 10.1103/PhysRevD.73.104002*; putting *10.1103/PhysRevD.73.104002* into the Bibliographic Code Query box at http://adsabs.harvard.edu/bib_abs.html and clicking Send Query should return links to the journal article.

Other references that list a DOI, BibCode, or arXiv number may be retrieved in a similar way as in the above examples.

References

[1] **Journal Access** Some journal articles referenced in this bibliography are published in journals with limited public access. Readers with university affiliations often will have access through the university library but to ensure broad availability for the general reader, references to the *arXiv preprint server* or to the *ADS Astronomy Abstract Service* are included where available for journal articles. Instructions for using these arXiv and ADS references to retrieve journal articles are given in Appendix D.

[2] www.spacetelescope.org/images/cygx1_illust_orig/.

[3] http://chandra.harvard.edu/press/17_releases/press_010517cdfs.html.

[4] The SIMBAD Astronomical Database (http://simbad.u-strasbg.fr/simbad/).

[5] www.astro.ucla.edu/~wright/CosmoCalc.html.

[6] http://astrogravs.gsfc.nasa.gov/docs/catalog.html.

[7] www.ligo.caltech.edu/image/ligo20160615e.

[8] *2015—General Relativity's Centenial.* http://journals.aps.org/general-relativity-centennial.

[9] Special issue on Gravity Probe B. 2015. *Class. Quant. Gravity*, **32** (22).

[10] Abbott, B. P., *et al.* 2016a. *Phys. Rev. Lett.*, **116**, 161102 (DOI: 10.1103/PhysRevLett.116.061102).

[11] Abbott, B. P., *et al.* 2016b. *Phys. Rev.*, **X6**, 041015 (DOI: 10.1103/PhysRevX.6.041015).

[12] Abbott, B. P., *et al.* 2016c. *Phys. Rev.*, **D93**, 122004 (arXiv: 1602.03843).

[13] Abbott, B. P., *et al.* 2016d. *Astrophys. J. Lett.*, **L13**, 826 (arXiv: 1602.08492).

[14] Abbott, B. P., *et al.* 2016e. *Phys. Rev. Lett.*, **116**, 241103 (DOI: 10.1103/PhysRevLett.116.241103).

[15] Abbott, B. P., *et al.* 2016f. *Phys. Rev. Lett.*, **116**, 221101 (arXiv: 1602.03841).

[16] Abbott, B. P., *et al.* 2016g. *Astrophys. J. Lett.*, **818**, L22 (DOI: 10.3847/2041-8205/818/2/L22).

[17] Abbott, B. P., *et al.* 2017a. *Annalen der Physik*, **529**, 1600209 (arXiv: 1608.01940; DOI: 10.1002/andp.201600209).

[18] Abbott, B. P., *et al.* 2017b. *Phys. Rev. Lett.*, **119**, 141101 (arXiv: 1709.09660).

[19] Abbott, B. P., *et al.* 2017c. *Phys. Rev. Lett.*, **118**, 221101 (arXiv: 1706.01812).

[20] Abbott, B. P., *et al.* 2017d. *Phys. Rev. Lett.*, **119**, 161101 (arXiv: 1710.05832).

[21] Abbott, B. P., *et al.* 2017e. *Astrophys. J. Lett.*, **848**, L13 (DOI: 10.3847/2041-8213/aa920c).

[22] Abbott, B. P., *et al.* 2017f. *Nature*, **551**, 85 (DOI: 10.1038/nature24471).

[23] Abe, P. A. R., *et al.* 2016. *Astron. and Astrophys.*, **594**, A13 (arXiv: 1502.01589).

[24] Adams, S. M., Kochanek, C. S., Gerke, J. R., Stanek, K. Z., and Dai, X. 2017. *Mon. Not. R. Astr. Soc.*, **468**, 4968 (DOI: 10.1093/mnras/stx816).

[25] Albrecht, A. and Steinhardt, P. J. 1982. *Phys. Rev. Lett.*, **48**, 1220 (DOI: 10.1103/PhysRevLett.48.1220).

[26] Alcubierre, M. 2008. *Introduction to 3 + 1 Numerical Relativity*. Oxford University Press.

[27] Amanullah, R., *et al.* 2010. *Astrophys. J.*, **716**, 712 (DOI: 10.1088/0004–637X/716/1/712).

[28] Anderson, L., *et al.* 2014. *Mon. Not. R. Astr. Soc.*, **441**, 24 (DOI: 10.1093/mnras/stu523).

[29] Ando, S., *et al.* 2013. *Rev. Mod. Phys.*, **85**, 1401 (DOI: 10.1103/RevModPhys.85.1401).

[30] Antoniadis, J., *et al.* 2013. *Science*, **340**, 446 (arXiv: 1304.6875).

[31] Ashby, N. 2003. *Living Rev. Rel.*, **6**, 1.

[32] Bahcall, J. N, *et al.* 1995. *Astrophys. J. Lett.*, **452**, L91 (DOI: 10.1086/309717).

[33] Baker, J. G., Centrella, J., Choi, D., Koppitz, M., and van Meter, J. 2006a. *Phys. Rev.*, **D73**, 104002 (arXiv: gr–qc/0602026; DOI: 10.1103/PhysRevD.73.104002).

[34] Baker, J. G., Centrella, J., Choi, D., Koppitz, M., and van Meter, J. 2006b. *Phys. Rev. Lett.*, **96**, 111102 (DOI: 10.1103/PhysRevLett.96.111102).

[35] Baker, J. G., Campanelli, M., Pretorius, F., and Zlochower, Y. 2007. *Class. Quant. Grav.*, **24**, S25 (arXiv: gr–qc/0701016).

[36] Balbus, S. A. and Hawley, J. F. 1991. *Astrophys. J.*, **376**, 214 (DOI: 10.1086/170270).

[37] Balbus, S. A. and Hawley, J. F. 1998. *Rev. Mod. Phys.*, **70**, 1 (DOI: 90/10.1103/RevModPhys.70.1).

[38] Bardeen, J. M., Steinhardt, P. J., and Turner, M. S. 1983. *Phys. Rev.*, **D28**, 679 (DOI: 10.1103/PhysRevD.28.679).

[39] Barstow, M. A., Bond, H. E., Holberg, J. B., Burleigh, M. R., Hubeny, I., and Koester, D. 2005. *Mon. Not. R. Astr. Soc.*, **362**, 1134 (DOI: 10.1111/j.1365–2966.2005.09359.x).

[40] Baumgarte, T. W. and Shapiro, S. L. 2010. *Numerical Relativity: Solving Einstein's Equations on the Computer*. Cambridge University Press.

[41] Bekenstein, J. D. 1973. *Phys. Rev.*, **D7**, 2333 (DOI: 10.1103/PhysRevD.7.2333).

[42] Bekenstein, J. D. 1980. *Physics Today*, **33**, 24 (DOI: 10.1063/1.2913906).

[43] Belczynski, K., Holz, D. E., Bulik, T., and O'Shaughnessy, R. 2016. *Nature*, **534**, 512 (DOI: 10.1038/nature18322; arXiv: 1602.04531).

[44] Berger, E. 2014. *Annu. Rev. Astron. Astrophys.*, **52**, 43 (DOI: 10.1146/annurev–astro–081913–035926).

[45] Berti, E. 2016. *Physics*, **9**, 17.

[46] Bertotti, B., Iess, L., and Tortora, P. 2003. *Nature*, **425**, 374 (DOI: 10.1038/nature01997).

[47] Birrell, N. D. and Davies, P. C. 1984. *Quantum Fields in Curved Spacetime*. Cambridge University Press.

[48] Blanchet, L., Damour, T., Iyer, B. R., Will, C. M., and Wiseman, A. G. 1995. *Phys. Rev. Lett.*, **74**, 3515 DOI: 10.1103/PhysRevLett.74.3515.

[49] Blanchet, L., Iyer, B. R., Will, C. M., and Wiseman, A. G. 1996. *Class. Quant. Grav.*, **13**, 575 (arXiv:gr–qc/9602024).

[50] Blandford, R. D. and Znajek, R. L. 1977. *Mon. Not. R. Astr. Soc.*, **179**, 433 (DOI: 10.1093/mnras/179.3.433).

[51] Blandford, R. D. and Gehrels, N. 1999. *Physics Today*, **52**, 40 (DOI: 10.1063/1.882697).

[52] Boller, Th., Brandt, W. N., Fabian, A. C., and Fink, H. H. 1997. *Mon. Not. R. Astr. Soc.*, **289**, 393 (DOI: 10.1093/mnras/289.2.393).

[53] Bolton, C. T. 1972. *Nature*, **235**, 271 (DOI: 10.1038/235271b0).

[54] Bowyer, S., Byram, E. T., Chubb, T. A., and Friedman, H. 1965. *Science*, **147**, 394 (DOI: 10.1126/science.147.3656.394).

[55] Braes, L. L. E, and Miley, G. K. 1971. *Nature*, **232**, 246 (BibCode: 1971Natur.232..246B).

[56] Brans, C. and Dicke, R. H. 1961. *Phys. Rev.*, **124**, 925 (DOI: 10.1103/PhysRev.124.925).

[57] Broderick, A. E. and Narayan, R. 2006. *Astrophys. J. Lett.*, **638**, 572 (DOI: 10.1086/500930).

[58] Buonanno, A. and Damour, T. 1998. *Phys. Rev.*, **D59**, 084006 (DOI: 10.1103/PhysRevD.59.084006).

[59] Buonanno, A., Cook, G. B., and Pretorius, F. 2007. *Phys. Rev.*, **D75**, 124018 (arXiv: gr–qc/0610122).

[60] Burgay, M., *et al.* 2003. *Nature*, **426**, 531 (DOI: 10.1038/nature02124).

[61] Campanelli, M., Lousto, C. O., Marronetti, P., and Zlochower, Y. 2006. *Phys. Rev. Lett.*, **96**, 111101 (DOI: 10.1103/PhysRevLett.96.111101).

[62] Carrol, B. W. and Ostlie, D. A. 2007. *An Introduction to Modern Astrophysics*. San Francisco: Pearson/Addison–Wesley.

[63] Carroll, S. M. 2004. *An Introduction to General Relativity, Spacetime, and Geometry*. San Francisco: Addison–Wesley.

[64] Cenko, S. B., *et al.* 2015. *Astrophys. J. Lett.*, **803**, L24 (DOI: 10.1088/2041–8205/803/2/L24).

[65] Cheng, T.-P. 2005. *Relativity, Gravitation, and Cosmology*. Oxford University Press.

[66] Chow, T. L. 2008. *Gravity, Black Holes, and the Very Early Universe*. New York: Springer Science + Business Media.

[67] Clowe, D., *et al.* 2006. *Astrophys. J. Lett.*, **648**, L109 (arXiv: astro–ph/0608407).

[68] Coles, P. and Lucchin, F. 1995. *Cosmology: The Origin and Evolution of Cosmic Structure*. New York: John Wiley and Sons.

[69] Collaboration, SXS. 2016. *What the first LIGO detection would look like up close*. www.ligo.caltech.edu/video/ligo20160211v3.

[70] Collins, P. D. B., Martin, A. D., and Squires, E. J. 1989. *Particle Physics and Cosmology*. New York: Wiley Interscience.

[71] Corbelli, E. and Salucci, P. 2000. *Mon. Not. R. Astr. Soc.*, **311**, 441 (DOI: 10.1046/j.1365–8711.2000.03075.x).

[72] Cyburt, R. H., Fields, B. D., Olive, K. A., and Yeh, T.-H. 2016. *Rev. Mod. Phys.*, **88**, 015004 (DOI: 10.1103/RevModPhys.88.015004).

[73] Davis, T. M. and Lineweaver, C. H. 2003. Expanding Confusion: Common Misconceptions of Cosmological Horizons and the Superluminal Expansion of the Universe. arXiv: astro-ph/0310808.

[74] de Sitter, W. 1916. *Mon. Not. R. Astr. Soc.*, **77**, 155 (DOI: 10.1093/mnras/77.2.155).

[75] Dewi, J. D. M. and van den Heuvel, E. P. J. 2004. *Mon. Not. R. Astr. Soc.*, **349**, 169 (DOI: 10.1111/j.1365–2966.2004.07477.x).

[76] d'Inverno, R. 1992. *Introducing Einstein's Relativity*. Oxford University Press.

[77] Doeleman, S. S., *et al.* 2008. *Nature*, **455**, 78 (arXiv: 0809.2442).

[78] Dyson, F. W., Eddington, A. S., and Davidson, C. 1920. *Phil. Trans. R. Soc. London, Series A*, **220**, 291 (DOI: 10.1098/rsta.1920.0009).

[79] Einstein, A. 1916. *Sitzungsber. K. Preuss. Akad. Wiss.*, **1**, 688.

[80] Einstein, A. 1918. *Sitzungsber. K. Preuss. Akad. Wiss.*, **1**, 154.

[81] Einstein, A. and Rosen, N. 1935. *Phys. Rev.*, **48**, 73 (DOI: 10.1103/PhysRev.48.73).

[82] Eisenstein, D. J. and Bennett, C. L. 2008. *Physics Today*, **61**, 44 (DOI: 10.1063/1.2911177).

[83] Eisenstein, D. J., *et al.* 2005. *Astrophys. J.*, **633**, 560 (DOI: 10.1086/466512).

[84] Elitzur, M., and Shlosman, I. 2006. *Astrophys. J.*, **648**, L101 (DOI: 10.1086/508158).

[85] Everitt, C. W. F., *et al.* 2011. *Phys. Rev. Lett.*, **106**, 221101 (arXiv: 1105.3456).

[86] Faber, J. A. and Rasio, F. A. 2012. *Living Rev. Rel.*, **15**, 8 (arXiv: 1204.3858).

[87] Flanagan, E. E. and Hughes, S. A. 2005. *New J. Phys.*, **7**, 204 (arXiv: gr-qc/0501041).

[88] Ford, L. H. 1997 (July). Quantum Field Theory in Curved Spacetime (arXiv gr-qc/9707062).

[89] Foster, J. and Nightingale, J. D. 2006. *A Short Course in General Relativity*. New York: Springer.

[90] Frank, J., King, A., and Raine, D. 2002. *Accretion Power in Astrophysics*. Cambridge University Press.

[91] Freedman, W., *et al.* 2001. *Astrophys. J.*, **553**, 47 (DOI: 10.1086/320638).

[92] Freedman, W. L. and Turner, M. S. 2003. *Rev. Mod. Phys.*, **75**, 1433 (DOI: 10.1103/RevModPhys.75.1433).

[93] Frieman, J. A., Turner, M. S., and Huterer, D. 2008. *Annu. Rev. Astron. Astrophys.*, **46**, 385 (DOI: 10.1146/annurev.astro.46.060407.145243).

[94] Frolov, V. P. and Zelnikov, A. 2011. *Introduction to Black Hole Physics*. Oxford University Press.

[95] Fuller, R. W. and Wheeler, J. A. 1962. *Phys. Rev.*, **128**, 919 (DOI: 10.1103/PhysRev.128.919).

[96] Gehrels, N., *et al.* 2005. *Nature*, **437**, 851 (DOI: 10.1038/nature04142).

[97] Gillessen, S., Eisenhauer, F., Fritz, T. K., Bartko, H., Dodds-Eden, K., Pfuhl, O., Ott, T., and Genzel, R. 2009. *Astrophys. J. Lett.*, **707**, L114 (DOI: 10.1088/0004-637X/707/2/L114).

[98] Glendenning, Norman K. 1997. *Compact Stars*. Springer.

[99] Glendenning, Norman K. 2007. *Special and General Relativity*. Springer.

[100] Goobar, A., *et al.* 2017. *Science*, **356**, 291 (DOI: 10.1126/science.aal2729; arXiv:1611.00014).

[101] Gou, L., *et al.* 2011. *Astrophys. J.*, **742**, 85 (arXiv: 1106.3690; DOI: 10.1088/0004-637X/742/2/85).

[102] Greene, B. 2010. *The Elegant Universe*. New York: W. W. Norton.

[103] Guidry, M. W. 1991. *Gauge Field Theories: An Introduction with Applications*. New York: Wiley Interscience.

[104] Guidry, M. W. 2018. *Sci. Bull.*, **63**, 2–4, (DOI: 10.1016/j.scib.2017.11.021).

[105] Guidry, M. W. 2019. *Stars and Stellar Processes*. Cambridge University Press.

[106] Guth, A. H. 1981. *Phys. Rev.*, **D23**, 347 (DOI: 10.1103/PhysRevD.23.347).

[107] Guth, A. H. 1997. *The Inflationary Universe*. Reading: Perseus Books.

[108] Guth, A. H. and Kalser, D. I. 2005. *Science*, **307**, 884 (DOI: 10.1126/science.1107483).

[109] Guth, A. H. and Pi, S.-Y. 1982. *Phys. Rev. Lett.*, **49**, 1110 (DOI: 10.1103/PhysRevLett.49.1110).

[110] Hamuy, M., *et al.* 1996. *Astronomical J.*, **112**, 2408 (DOI: 10.1086/118192).

[111] Hartle, J. B. 2003. *Gravity, An Introduction to Einstein's General Relativity*. Addison–Wesley.

[112] Haswell, C. A., Robinson, E. L., Horne, K., Steinig, R. F., and Abbott, T. M. C. 1993. *Astrophys. J.*, **411**, 802 (DOI: 10.1086/172884).

[113] Hawking, S. 1974. *Nature*, **248**, 30 (DOI: 10.1038/248030a0).

[114] Hawking, S. 1976. *Phys. Rev.*, **D13**, 191 (DOI: 10.1103/PhysRevD.13.191).

[115] Hawking, S. Jan. 1977. *Scientific American*, **236**, 34 (DOI: 10.1038/scientificamerican0177-34).

[116] Hees, A., *et al.* 2017. *Phys. Rev. Lett.*, **118**, 211101 (arXiv: 1705.07902).

[117] Henry, R. C. 2000. *Astrophys. J.*, **535**, 350 (DOI: 10.1086/308819).

[118] Herbert, P. 2005 (December). *On the history of the so-called Lense-Thirring effect;* http://philsci-archive.pitt.edu/2681/.

[119] Hjelling, R. M. and Wade, C. M. 1971. *Astrophys. J. Lett.*, **168**, L21 (DOI: 10.1086/180777).

[120] Hjorth, J., *et al.* 2003. *Nature*, **423**, 847 (DOI: 10.1038/nature01750).

[121] Hobbs, G. 2010. Pulsars as Gravitational Wave Detectors. arXiv: 1006.3969.

[122] Hobson, M. P., Efstathiou, G., and Lasenby, A. N. 2006. *General Relativity: An Introduction for Physicists*. Cambridge University Press.

[123] Holberg, J. B. 2010. *J. History Astronomy*, **41**, 41 (BibCode: 2010JHA....41...41H).

[124] Holz, D. E. and Hughes, S. A. 2005. *Astrophys. J.*, **629**, 15 (DOI: 10.1086/431341).

[125] Islam, J. N. 1992. *An Introduction to Mathematical Cosmology*. Cambridge University Press.

[126] Israel, W. 1967. *Phys. Rev.*, **164**, 1776 (DOI: 10.1103/PhysRev.164.1776).

[127] Janka, H.-T. 2012. *Annu. Rev. Nucl. Part. Sci.*, **62**, 407 (DOI: 10.1146/annurev–nucl–102711–094901).

[128] Kasen, D., Metzger, B., Barnes, J., Quataert, E., and Ramirez-Ruiz, E. 2017. *Nature*, **551**, 80 (DOI: 10.1038/nature24453).

[129] Kelly, P. L., *et al.* 2015. *Science*, **347**, 1123 (DOI: 10.1126/science.aaa3350).

[130] Kennefick, D. 1997. Controversies in the History of the Radiation Reaction Problem in General Relativity (arXiv gr–qc/9704002).

[131] Kerr, R. P. 1963. *Phys. Rev. Lett.*, **11**, 237 (DOI: 10.1103/PhysRevLett.11.237).

[132] Kim, A. 2004. LBNL Rep. No. 56164.

[133] Kiziltan, B., Baumgardt, H., and Loeb, A. 2017. *Nature*, **542**, 203 (DOI: doi:10.1038/nature21361; arXiv:1702.02149).

[134] Kogut, J. B. 2001. *Introduction to Relativity*. San Diego: Academic Press.

[135] Kolb, E. W. and Turner, M. S. 1983. *Annu. Rev. Nucl. Part. Sci.*, **33**, 645 (DOI: 10.1146/annurev.ns.33.120183.003241).

[136] Komatsu, E., *et al.* 2011. *Astrophys. J. Suppl.*, **192**, 18 (arXiv: 1001.4538; DOI: 10.1088/0067–0049/192/2/18).

[137] Kotake, K., Ohnishi, N., and Yamada, S. 2007. *Astrophys. J.*, **655**, 406 (DOI: 10.1086/509320).

[138] Kotake, K., Ohnishi, N., and Yamada, S. 2013. *Comptes Rendus Physique*, **14**, 318 (arXiv: 1110.5107).

[139] Kramer, M. and Stairs, I. H. 2008. *Annu. Rev. Astron. Astrophys.*, **46**, 541 (DOI: 10.1146/annurev.astro.46.060407.145247).

[140] Kramer, M, *et al.* 2006. *Science*, **314**, 97 (arXiv: astro–ph/0609417).

[141] Kruskal, M. D. 1960. *Phys. Rev.*, **119**, 1743 (DOI: 10.1103/PhysRev.119.1743).

[142] Lambourne, R. J. A. 2010. *Relativity, Gravitation, and Cosmology*. Cambridge University Press.

[143] Lee, W. H. 2001. *Mon. Not. R. Astr. Soc.*, **328**, 583 (DOI: 10.1046/j.1365–8711.2001.04898.x).

[144] Lense, J. and Thirring, H. 1918. *Z. Phys.*, **19**, 156.

[145] Liddle, A. 2003. *An Introduction to Modern Cosmology*. Chichester: Wiley.

[146] Linde, A. D. 1982. *Phys. Lett.*, **108B**, 389 (DOI: 10.1016/0370–2693(82)91219–9).

[147] Lindley, D. 2005. *Phys. Rev. Focus*, **15**, 11 (The article may be accessed at the website http://physics.aps.org/story/v15/st11).

[148] Lyne, A. G., *et al.* 2004. *Science*, **303**, 1153 (DOI: 10.1126/science.1094645).

[149] MacFadyen, A. 2004. *Science*, **303**, 45 (DOI: 10.1126/science.1091764).

[150] Marsh, T. R., Robinson, E. L., and Ward, J. H. 1994. *Mon. Not. R. Astr. Soc.*, **266**, 137 (DOI: 10.1093/mnras/266.1.137).

[151] McClintock, J. and Remillard, R. 1986. *Astrophys. J.*, **308**, 110 (DOI: 10.1086/164482).

[152] McClintock, J. E, Petro, L. D., Remillard, R. A., and Ricker, G. R. 1983. *Astrophys. J.*, **266**, L27 (DOI: 10.1086/183972).

[153] Merloni, A., *et al.* 2014. *Mon. Not. R. Astr. Soc.*, **437**, 3550 (DOI: 10.1093/mnras/stt2149).

[154] Meyer, L., *et al.* 2012. *Science*, **338**, 84 (arXiv: 1210.1294).

[155] Mezzacappa, A. 2005. *Annu. Rev. Nucl. Part. Sci.*, **55**, 467 (DOI: 10.1146/annurev.nucl.55.090704.151608).

[156] Misner, C. W., Thorne, K. S., and Wheeler, J. A. 1973. *Gravitation*. W. H. Freeman.

[157] Modugno, G. 2014. *Nature Phys.*, **10**, 793 (DOI: 10.1038/nphys3141).

[158] Moore, C. J, Cole, R. H., and Berry, C. P. L. 2015. *Class. Quantum Grav.*, **32**, 015014 (arXiv: 1408.0740; DOI: 10.1088/0264–9381/32/1/015014).

[159] Morris, M. S., Thorne, K. S., and Yurtsever, U. 1988. *Phys. Rev. Lett.*, **61**, 1446 (DOI: 10.1103/PhysRevLett.61.1446).

[160] Mroué, A., *et al.* 2013. *Phys. Rev. Lett.*, **111**, 241104 (DOI: 10.1103/PhysRevLett.111.241104).

[161] Müller, B., Janka, H.-T., and Marek, A. 2013. *Astrophys. J.*, **766**, 43 (DOI: 10.1088/0004–637X/766/1/43).

[162] Netzer, H. 2015. *Annu. Rev. Astron. Astrophys.*, **53**, 365 (DOI: 10.1146/annurev–astro–082214–122302).

[163] Nordtvedt, K. 1968a. *Phys. Rev.*, **169**, 1017 (DOI: 90/10.1103/PhysRev.169.1017).

[164] Nordtvedt, K. 1968b. *Phys. Rev.*, **170**, 1186 (DOI: 90/10.1103/PhysRev.170.1186).

[165] Nordtvedt, K. 2003. *Lunar Laser Ranging – a comprehensive probe of post-Newtonian gravity*. Proceedings of Villa Mondragone International School of Gravitation and Cosmology, September 2002 (arXiv: gr-qc/0301024).

[166] Oda, M., Gorenstein, P., Gursky, H., Kellogg, E., Schreier, E., Tananbaum, H., and Giaconni, R. 1971. *Astrophys. J.*, **166**, L1 (DOI: 10.1086/180726).

[167] Oesch, P. A., *et al.* 2016. *Astrophys. J.*, **819**, 129 (DOI: 10.3847/0004–637X/819/2/129).

[168] Oke, J. B. 1977. *Astrophys. J.*, **217**, 181 (DOI: 10.1086/155568).

[169] O'Neill, B. 1995. *The Geometry of Kerr Black Holes*. Wellesley: A. K. Peters.

[170] Oppenheimer, J. R. and Snyder, H. 1939. *Phys. Rev.*, **56**, 455 (DOI: 10.1103/PhysRev.56.455).

[171] Oppenheimer, J. R., and Volkoff, G. M. 1939. *Phys. Rev.*, **55**, 374 (DOI: 10.1103/PhysRev.55.374).

[172] Orosz, J. A., McClintock, J. E., Aufdenberg, J. P., Remillard, R. A., Reid, M. J., Narayan, R., and Gou, L. 2011. *Astrophys. J.*, **742**, 84 (DOI: 10.1088/0004–637X/742/2/84).

[173] Ott, C. D. 2009. *Class. Quantum Grav.*, **26**, 063001 (arXiv: 0809.0695).

[174] Padmanabhan, T. 2002. *Theoretical Astrophysics, Vol. III: Galaxies and Cosmology*. Cambridge University Press.

[175] Padmanabhan, T. 2006. *An Invitation to Astrophysics*. Singapore: World Scientific.

[176] Padmanabhan, T. 2010. *Gravitation: Foundations and Frontiers*. Cambridge University Press.

[177] Page, D. N. 2005. *New J. Phys.*, **7**, 203 (DOI: 10.1088/1367–2630/7/1/203).
[178] Pais, A. 2005. *Subtle is the Lord: The Life and Science of Albert Einstein*. Oxford University Press.
[179] Peacock, J. 1999. *Cosmological Physics*. Cambridge University Press.
[180] Peebles, P. J. E. 1993. *Principles of Physical Cosmology*. Princeton University Press.
[181] Penrose, R. 1963. *Phys. Rev. Lett.*, **10**, 66 (DOI: 10.1103/PhysRevLett.10.66).
[182] Penrose, R. 1965. *Phys. Rev. Lett.*, **14**, 57 (DOI: 10.1103/PhysRevLett.14.57).
[183] Perkins, D. 2003. *Particle Astrophysics*. Oxford University Press.
[184] Perlmutter, S. 2003. *Physics Today*, **56**(4), 53 (DOI: 10.1063/1.1580050).
[185] Peterson, B. 1997. *An Introduction to Active Galactic Nuclei*. Cambridge University Press.
[186] Philbin, T. G., *et al.* 2008. *Science*, **319**, 1367 (DOI: 10.1126/science.1153625).
[187] Pian, E., *et al.* 2017. *Nature*, **551**, 67 (DOI: 10.1038/nature24298).
[188] Piran, T. 1999. *Phys. Rep.*, **314**, 575 (DOI: 10.1016/S0370–1573(98)00127–6).
[189] Piran, T. 2004. *Rev. Mod. Phys.*, **76**, 1143 (DOI: 10.1103/RevModPhys.76.1143).
[190] Piran, T. and Shaviv, N. J. 2005. *Phys. Rev. Lett.*, **94**, 051102 (DOI: 10.1103/PhysRevLett.94.051102).
[191] Pound, R. V., and Rebka, G. A., Jr. 1959. *Phys. Rev. Lett.*, **3**, 439 (DOI: 10.1103/PhysRevLett.3.439).
[192] Pound, R. V., and Rebka, G. A., Jr. 1960. *Phys. Rev. Lett.*, **4**, 337 (DOI: 10.1103/PhysRevLett.4.337).
[193] Pound, R. V. and Snider, J. L. 1964. *Phys. Rev. Lett.*, **13**, 539 (DOI: 10.1103/PhysRevLett.13.539).
[194] Pretorius, F. 2005. *Phys. Rev. Lett.*, **95**, 121101 (DOI: 10.1103/PhysRevLett.95.121101).
[195] Price, D. J. and Rosswog, S. 2006. *Science*, **312**, 719 (DOI: 10.1126/science.1125201).
[196] Qin, Y., *et al.* 2000. *Publ. Astron. Soc. Jpn.*, **52**, 759 (DOI: 10.1093/pasj/52.5.759).
[197] Raine, D. and Thomas, E. 2005. *Black Holes: An Introduction*. London: Imperial College Press.
[198] Ransom, S. M., *et al.* 2014. *Nature*, **505**, 520 (DOI: 10.1038/nature12917).
[199] Raychaudhuri, A. K., Banerji, S., and Banerjee, A. 1992. *General Relativity, Astrophysics, and Cosmology*. Berlin: Springer–Verlag.
[200] Reasenberg, R. D., *et al.* 1979. *Astrophys. J.*, **234**, L219 (DOI: 10.1086/183144).
[201] Rezzolla, L, *et al.* 2012. *Astrophys. J. Lett.*, **732**, L6 (arXiv: 1101.4298; DOI: 10.1088/2041–8205/732/1/L6).
[202] Ricci, C., *et al.* 2017. *Nature*, **549**, 488 (DOI:10.1038/nature23906).
[203] Riles, K. 2013. *Prog. Part. Nuc. Phys.*, **68**, 1 (arXiv: 1209.0667).
[204] Rindler, W. 1977. *Essential Relativity: Special, General, and Cosmological*. Berlin: Springer–Verlag.
[205] Rindler, W. 2001. *Relativity: Special, General, and Cosmological*. Oxford University Press.
[206] Robson, I. 1996. *Active Galactic Nuclei*. Chichester: Wiley.

[207] Roos, M. 2005. *Introduction to Cosmology*. Chichester: Wiley.

[208] Rosswog, S. 2004. *Science*, **303**, 46 (DOI: 10.1126/science.1091767).

[209] Rubin, V. 2006. *Physics Today*, **59**, 8 (DOI: 10.1063/1.2435662).

[210] Ruffert, M., Ruffert, H. Th. and Janka, H. Th. 2006. *Astron. Astrophys*, **380**, 544 (arXiv: astro–ph/0106229).

[211] Ryden, Barbara. 2017. *Introduction to Cosmology, 2nd edn*. New York: Cambridge University Press.

[212] Ryder, L. 2009. *Introduction to General Relativity*. Cambridge University Press.

[213] Sakharov, A. D. 1967. *JETP Lett.*, **5**, 24 (BibCode: 1967JETPL...5...24S).

[214] Sanad, M. R. and Bobrowsky, M. 2014. *New Astronomy*, **29**, 47 (DOI:10.1016/j.newast.2013.12.004).

[215] Saulson, P. R. 2011. *Gen. Relativ. Gravit.*, **43**, 3289 (DOI: 10.1007/s10714–011–1237–z).

[216] Schiff, L. I. 1960. *Phys. Rev. Lett.*, **4**, 215 (DOI: 10.1103/PhysRevLett.4.215).

[217] Schlamminger, S., Choi, K.-Y., Wagner, T. A., Gundlach, J. H., and Adelberger, E. G. 2008. *Phys. Rev. Lett.*, **100**, 041101 (arXiv: 0712.0607).

[218] Schödel, R., *et al.* 2002. *Nature*, **419**, 694 (DOI: 10.1038/nature01121).

[219] Schutz, B. F. 1986. *Nature*, **323**, 310 (DOI: 10.1038/323310a0).

[220] Schutz, B. 1980. *Geometrical Methods of Mathematical Physics*. Cambridge University Press.

[221] Schutz, B. 1985. *A First Course in General Relativity*. Cambridge University Press.

[222] Semenov, V., Dyadechkin, S., and Punsly, B. 2004. *Science*, **305**, 978 (DOI: 10.1126/science.1100638).

[223] Shapiro, S. L. and Teukolsky, S. A. 1983. *Black Holes, White Dwarfs, and Neutron Stars: The Physics of Compact Objects*. New York: Wiley Interscience.

[224] Siegel, D. M., Ciolfi, R., Harte, A., and Rezzolla, L. 2013. *Phys. Rev.*, **D87**, 121302 (arXiv: 1302.4368; DOI: 10.1103/PhysRevD.87.121302).

[225] Stairs, I. H. 2003. *Living Rev. Rel.*, **6**, 5 (arXiv: astro–ph/0307536).

[226] Steinhaur, J. 2014. *Nature Phys.*, **10**, 864 (DOI: 10.1038/nphys3104).

[227] Stevenson, S., *et al.* 2017. *Nature Comm.*, **8**, 14906 (DOI: 10.1038/ncomms14906; arXiv:1704.01352).

[228] Straumann, N. 2004. *General Relativity with Applications to Astrophysics*. Berlin: Springer–Verlag.

[229] Susskind, L. and Lindesay, James. 2005. *An Introduction to Black Holes, Information, and the String Theory Revolution: The Holographic Universe*. Singapore: World Scientific.

[230] Tauris, T. M. and van den Heuvel, E. P. J. 2014. *Astrophys. J. Lett.*, **781**, L13 (DOI: 10.1088/2041–8205/781/1/L13).

[231] Taylor, E. F. and Wheeler, J. A. 2000. *Exploring Black Holes: Introduction to General Relativity*. San Francisco: Addison Wesley Longman.

[232] Taylor, J. H. and Weisberg, J. M. 1982. *Astrophys. J.*, **253**, 908 (DOI: 10.1086/159690).

[233] Thielemann, F.-K., Eichler, M., Panov, I.V., and Wehmeyer, B. 2017. *Annu. Rev. Nucl. Part. Sci.*, **67**, 253 (arXiv: 1710.02142).

[234] Thorne, K. S. 1994. *Black Holes and Time Warps*. New York: W. W. Norton.

[235] Tonry, J. L., *et al.* 2003. *Astrophys. J.*, **594**, 1 (DOI: 10.1086/376865).

[236] Troja, E., *et al.* 2017. *Nature*, **551**, 71 (DOI: 10.1038/nature24290).

[237] van Dokkum, P., *et al.* 2016. *Astrophys. J. Lett.*, **828**, L6 (DOI:10.3847/2041-8205/828/1/L6).

[238] Vessot, R. F. C., *et al.* 1980. *Phys. Rev. Lett.*, **45**, 2081 (DOI: 10.1103/PhysRevLett.45.2081).

[239] Visser, M. 2007. The Kerr Spacetime: A Brief Introduction (arXiv: 0706.0622).

[240] Vito, F., *et al.* 2016. *Mon. Not. R. Astr. Soc.*, **63**, 348 (arXiv: 1608.02614).

[241] Wagoner, R. V. 2000. *Relativistic Gravity and Some Astrophysical Applications*. Notes for lectures given at the Summer School on Astroparticle Physics and Cosmology, Trieste, 12–30 June (2000) (www.slac.stanford.edu/gen/meeting/ssi/1998/media/wagoner.pdf).

[242] Wald, R. M. 1984. *General Relativity*. University of Chicago Press.

[243] Wald, R. M. 1994. *Quantum Field Theory in Curved Spacetime and Black Hole Thermodynamics*. University of Chicago Press.

[244] Walecka, J. D. 2007. *Introduction to General Relativity*. Singapore: World Scientific.

[245] Webster, B. L. and Murdin, P. 1972. *Nature*, **235**, 37 (DOI: 10.1038/235037a0).

[246] Weinberg, S. 1972. *Gravitation and Cosmology*. New York: John Wiley and Sons.

[247] Weinberg, S. 1989. *Rev. Mod. Phys.*, **61**, 1 (DOI: 10.1103/RevModPhys.61.1).

[248] Weisberg, J. M., Nice, D. J., and Taylor, J. H. 2010. *Astrophys. J.*, **722**, 1030 (arXiv: 1011.0718).

[249] Weisberg, J. M., Taylor, J. H., and Fowler, L. A. Oct. 1981. *Scientific American*, **245**, 74 (DOI: 10.1038/scientificamerican1081-74).

[250] Will, C. M. 1994. *Phys. Rev.*, **D50**, 6058 (DOI: 10.1103/PhysRevD.50.6058).

[251] Will, C. M. 2006. *Living Rev. Rel.*, **9**, 3 (DOI: 10.12942/lrr–2006–3).

[252] Will, C. M. 2008. *Astrophys. J. Lett.*, **674**, L25 (arXiv: 0711.1677; DOI: 10.1086/528847).

[253] Will, C. M. 2015. *Class. Quant. Gravity*, **32**, 1 (DOI: 10.1088/0264–9381/32/22/220301).

[254] Williams, J. G., Turyshev, S. G., and Boggs, D. H. 2004. *Phys. Rev. Lett.*, **93**, 261101 (arXiv: gr–qc/0411113; DOI: 10.1103/PhysRevLett.93.261101).

[255] Woosley, S. E. and Bloom, J. S. 2006. *Annu. Rev. Astro. Astrophysics*, **44**, 507 (DOI: 10.1146/annurev.astro.43.072103.150558).

[256] Yakunin, K. N., Marronetti, P., Mezzacappa, A., Bruenn, S. W., Lee, C.-T., Chertkow, M. A., Hix, W. R., Blondin, J. M., Lentz, E. J., Messer, O. E. B., and Yoshida, S. 2010. *Class. Quant. Grav.*, **27**, 194005 (arXiv: 1005.0779).

[257] Yakunin, K. N., Mezzacappa, A., Marronetti, P., Yoshida, S., Bruenn, S. W., Hix, W. R., Lentz, E. J., Messer, O. E. B., Harris, J. A., Endeve, E., Blondin, J. M., and Lingerfelt, E. J. 2015. *Phys. Rev.*, **D92**, 084040 (arXiv: 1505.05824).

[258] Zee, A. 2013. *Einstein Gravity in a Nutshell*. Princeton University Press.

[259] Zwiebach, B. 2004. *A First Course in String Theory*. Cambridge University Press.

Index

ΛCDM models, 444

absolute derivatives, 58, 61
accretion
    and binary spinup, 205
    as energy storage reservoir, 281
    disk, 264, 276, 281, 283, 285, 293, 297–299, 302, 313, 315, 317
    on black holes, 276, 277, 280, 281, 283, 295
    on compact companion, 262
    on neutron stars, 205
    Roche lobe overflow, 532
acoustic scale, *see* cosmic microwave background (CMB)
action, 98
active galactic nuclei (AGN), 292
addition theorem (spherical harmonics), 429
ADS astronomy abstract service, 571
AdS/CFT correspondence, 240, 556, 558
aether, 6
affine parameters, 98
afterglows, *see* gamma-ray bursts, afterglows
AGN, *see* active galactic nuclei
angular size distance, 433, 434
anholonomic basis, *see* non-coordinate basis
anti-de Sitter space, 558
area theorem, *see* Hawking area theorem
arXiv preprint server, 571
asymptotically flat solutions
    and local energy conservation, 132
    Kerr spacetime, 244
    Schwarzschild spacetime, 159
AT 2017gfo, 522
atlas, 33

BAO, *see* baryon acoustic oscillations (BAO)
baryogenesis
    and absence of antimatter in Universe, 453
    and excess of photons over baryons, 453
    baryon non-conservation, 457
    conditions for baryon asymmetry, 457
    leptogenesis, 459
    origin of baryons, 457
    problem in big bang, 453
baryon acoustic oscillations (BAO), 408, 438, 439
baryon number, 239, 457, 458
baryon to photon ratio η, 426

baryonic mass, 199
baryonic matter, 355
    and baryon number, 457, 458
    as "ordinary matter", 329
    as dark matter, 356
    density constrained by big bang, 425
basis
    and directional derivatives, 37–39
    anholonomic, 37–39
    coordinate, 27, 37–39, 67, 126
    dual, 18
    for a vector space, 42
    holonomic, 37–39
    Lie bracket of basis vectors, 67
    non-coordinate, 37–39, 67
    orthonormal, 18, 36
    tangent, 16
    transformation of basis vectors, 68
Bianchi identity, 150
big bang, 412
    and inflation, 444, 453
    anisotropies in CMB, 429
    baryogenesis problem, 453
    baryon to photon ratio η, 413, 426
    beyond standard big bang, 448
    cast of characters, 419
    conditions for baryon asymmetry, 457
    cosmic microwave background (CMB), 426
    cosmic neutrino background, 427
    dark ages, 422, 423
    decoupling from equilibrium, 417
    decoupling of photons, 422, 423, 427, 429
    decoupling of weak interactions, 418
    deuterium bottleneck, 421
    equilibrium in expanding Universe, 416
    evolution of early Universe, 414
    evolution of scale factor, 412
    flatness problem and fine tuning, 451
    history of Universe, 419
    horizon problem and causality, 449
    hot, dense initial state, 329
    incomplete explanations, 449
    mass-5 and mass-8 bottleneck, 421
    matter and radiation density, 413
    matter-dominated evolution, 412
    modification by inflation, 457

big bang (cont.)
  modifying the first second, 453
  monopole problem, 451
  neutron to proton ratio, 424
  no carbon production by triple-α, 421
  nucleosynthesis, 420, 421
  nucleosynthesis constraints on baryon density, 425
  origin of the baryons, 457
  production of ^{4}He, 424
  production of the light elements, 420
  radiation-dominated evolution, 412
  recombination transition, 422
  reionization transition, 422, 423
  singularity, 258
  structure and smoothness dichotomy, 452
  successes, 448
  thermodynamics, 414
  unexplained initial conditions, 449
  vacuum energy problem, 452
Binary Pulsar, 200
  emission of gravitational waves, 202, 497
  merger timescale, 204
  orbital properties, 201
  origin and fate, 203
  precession of periastron, 172, 202
  precision test of general relativity, 202, 544
  time dilation, 202
binary star systems
  circularization of orbits by tidal interactions, 262
  contact binary, 491
  high-mass X-ray binaries (HMXB), 262
  mass function, 262, 263, 265–267
  mass of, 262, 263, 265–267
  radial velocity curve, 262
  spectroscopic, 262, 263, 265–267
  X-ray binaries, 257, 262
Birkhoff's theorem, 226
BL Lac objects, 295
  blazars, 295
  emission of high-energy photons, 302
  inverse Compton scattering, 303
  Markarian 421, 302
  variable brightness, 295
BL Lacertae objects, *see* BL Lac objects
black hole central engines, 280
  accretion, 280, 281, 283
  accretion efficiencies, 283
  accretion on supermassive black holes, 295
  active galactic nuclei, 292
  BL Lac objects, 295
  Blandford–Znajek mechanism, 280
  causality and source size, 291
  Centaurus A, 301
  gamma-ray bursts, 304
  limits on accretion rates (Eddington limit), 282
  magnetic fields, 280

quasars, 285, 290
  radio galaxies, 292
  Seyfert galaxies, 293
  unified model, 295
black hole laser, 235
black holes
  accretion, 280, 281
  analog, 235
  and asymmetric gravitational collapse, 259, 260
  and charge, 225, 246
  and gravitational waves, 532
  and information, 239
  and Penrose processes, 253
  and stationary observers, 251
  and the holographic principle, 240
  apparent horizons, 260
  area of event horizon, 247
  as central engines, 280, 281
  as endpoint of stellar evolution, 532
  at center of Milky Way (Sgr A*), 268, 532, 544
  Birkhoff's theorem, 226
  candidates in X-ray binaries, 266
  cosmic censorship, 224
  Cygnus X-1, 267
  energy sources, 280
  entropy, 236, 237
  evaporation of quantum black holes, 231, 559
  event horizons, 213, 215, 260
  evidence for stellar-mass black holes, 262
  extraction of rotational energy, 253
  extremal Kerr, 245
  feeding, 301
  four laws of black hole dynamics, 237
  frame dragging for rotating, 249
  from direct stellar collapse, 531, 533
  generalized second law of thermodynamics, 236
  gravitational waves from binary merger, 510
  Hawking, 229, 230, 234, 559
  imaging event horizons, 276
  in the early Universe, 274
  in X-ray binaries, 262
  intermediate-mass, 272
  Kerr, 243
  Kerr–Newman, 225, 245, 246
  known masses of, 532
  miniature, 234
  natal kick, 531
  no-hair theorem, 225
  observational evidence for, 257, 261, 275
  origin of name, 214
  prediction of, 214
  quantum, 229
  Reissner–Nordström, 225, 246
  rotating, 243
  Schwarzschild, 213
  singularity theorems, 224, 258–260

skepticism by Einstein and Eddington, 257
sonic, 235
spherical, 213
static limit, 251
supermassive in the cores of galaxies, 266, 275
theorems and conjectures, 224
thermodynamics, 236, 258
water masers of NGC 4258, 270, 271, 532
Blandford–Znajek mechanism, 280
blazars, 295, 302
boost transformations, 75
Bose–Einstein statistics, 548
boson condensation, 548
branes, *see* superstrings and branes
Brans–Dicke (scalar–tensor) theory of gravity, 541
Bullet Cluster, 354

cartesian product, 36, 45
Casimir effect, 550
causal structure of spacetime, 79, 80, 120
causality and source size, 291, 311
CDM (cold dark matter), *see* dark matter
central engines, *see* black hole central engines
Chandra deep field, 274
Chandrasekhar limiting mass, 115, 193
charge conjugation symmetry C, 457, 458
Christoffel symbols, 58
 are not tensors, 58, 59, 65
 equivalence with connection coefficients, 135, 136
 of the first kind, 59
 of the second kind, 59
 transformation law, 58
Christoffel, Erwin, 136
closed timelike loops, 81
closure, 42
CMB, *see* cosmic microwave background (CMB)
collapsar model of gamma-ray bursts, 315
 and long-period bursts, 315
 and metallicity, 313
 and Wolf–Rayet stars, 315
 jets, 315
common envelope evolution, 531, 532
commutator, 39, 67
comoving coordinates, 371
comoving frame, *see* comoving observer
comoving observer, 367, 371, 378
compactification, *see* superstrings and branes
concordance model, 406
confinement of quarks and gluons, 419
conformal factor, 376
conformal time, 376, 377
conformal transformation, 376, 385
conformally flat spaces, 377
congruence, 62, 63
connection coefficients
 and differential geometry, 140

and preservation of scalar product under parallel transport, 140
and tangent spaces, 135
compatibility demands, 140
determination from metric tensor, 140
equivalence with Christoffel symbols, 135, 136
for Friedmann cosmologies, 379
for Friedmann cosmologies (table), 380
interpretation of indices, 135
static solutions, 381
the affine connection, 135
uniqueness of the affine connection, 140
contracted Bianchi identity, 150
contravariant vectors, *see* vectors
coordinate basis, 37–39
coordinate curve, 37–39
coordinate patches, 33
coordinate systems, 14
 basis vectors, 15
 Boyer–Lindquist, 245
 dual basis, 18
 Eddington–Finkelstein, 218
 euclidean, 14
 Kruskal–Szekeres, 221
 non-orthogonal, 18
 orthogonal, 18
 parameterizing, 14
 spacelike components, 32
 tangent basis, 16
 timelike components, 32
correlation function, 429
cosmic censorship, 224
 and extremal Kerr black holes, 246
 and naked singularities, 224, 246, 259
cosmic inflation, *see* inflation
cosmic microwave background (CMB), 426
 acoustic oscillations, 435, 436
 acoustic scale, 436
 adiabatic fluctuations, 437
 and baryon acoustic oscillations, 438
 and cosmological parameters, 405
 anisotropies, 330, 373, 429
 blackbody spectrum, 427
 cause of temperature fluctuations, 433
 COBE, 429
 correlation function, 429
 cosmological parameters (table), 441
 decoupling time for photons, 427, 429
 dipole component, 373
 discovery, 427
 fluctuations, 330, 405, 448
 interpretation, 427
 last scattering surface, 427, 429, 433
 Planck satellite, 429
 power spectrum, 432
 precise measurement of cosmology parameters, 439

cosmic microwave background (CMB) (cont.)
   preferred angular scale, 437
   preferred distance scale, 437
   Sachs–Wolfe effect, 433, 435
   seeds for structure formation, 441
   sound horizon, 436
   spectrum, 427
   spherical harmonic expansion, 429
   temperature, 330, 373, 427, 448
   WMAP, 429
cosmic time, 366, 367, 370
cosmological constant, *see* dark energy; Friedmann cosmologies
Cosmological Principle, 327, 448
cosmological proper time, *see* cosmic time
cosmology
   and elementary particle physics, 329
   closed universe, 343
   concordance model, 406
   Cosmological Principle, 327, 365
   cosmological redshift, 327
   critical density, 342
   dark energy, 329, 382
   dark matter, 329, 348
   deceleration and density parameters, 361
   deceleration parameter, 359
   density parameters, 358
   density parameters (table), 359
   dominance of dark energy and dark matter, 444
   energy densities (current), 398
   energy densities and changes in scale, 389
   evolution of scale factor with time, 344
   expanding balloon analogy, 372
   expansion governed by general relativity, 328
   expansion interpretation of redshifts, 332
   flat universe, 343
   Friedmann cosmologies, 365
   Hubble law, 327, 331
   Hubble parameter, 327, 331, 337
   Hubble radius, 376
   Hubble time, 335
   in Newtonian picture, 337, 341, 361, 383
   inflationary, 453
   lookback times, 346
   most matter is not baryonic, 329
   open universe, 343
   possible expansion histories for dust model, 344
   precision measurements of parameters, 439
   the big bang, 329
   the Universe is expanding, 327
   the Universe is observed to be flat, 442
   Type Ia supernovae and the accelerating Universe, 404
   what is expanding?, 328
cotangent bundle, *see* fiber bundle, cotangent bundle
cotangent space, 36
covariance, 32
   manifest, 87
   of Maxwell equations, 87
   principle of general covariance, 106, 125
covariant derivative, 26, 56, 58
   and Christoffel symbols, 58
   and connection coefficients, 135
   and parallel transport, 61, 132, 134, 135
   and vacuum energy, 61
   implications, 61
   is non-commuting operation, 60, 65
   Leibniz rule for derivative of product, 60, 67
   of metric tensor vanishes, 61
   rules for, 60
covariant vectors, *see* dual vectors (one-forms)
covectors, *see* dual vectors (one-forms)
CP symmetry, 457, 458
CPT symmetry, 457, 458
critical density, 342
curvature
   and dimensionality of spacetime, 149
   and general covariance, 125
   and tangent spaces, 35
   density, 359, 390
   distance intervals in curved spacetime, 129
   Gaussian, 117, 126, 128
   intrinsic and extrinsic, 117, 126, 149
   of spacetime, 11, 126
   radius of curvature, 126
   Ricci flat spaces, 151
   Ricci scalar, 150
   Ricci tensor, 150
   Riemannian curvature tensor, 148
   vectors in curved space, 35
curvature density Ω_k, 359, 390
cyclic identity, 149
Cygnus X-1, 267

d'Alembertian operator
   for electromagnetic waves, 87
   for gravitational waves, 470
dark ages, 422, 423
dark energy
   a property of space itself, 550
   and acceleration of the scale factor, 358
   and cosmological constant, 381
   and evolution of the Universe, 329
   and supersymmetry, 552
   and the cosmic scale factor, 343
   as vacuum energy, 382, 452, 549
   equation of state, 358, 388, 550
   summary, 442
   vacuum energy for bosonic fields, 550
   vacuum energy for fermionic fields, 552
dark matter
   and baryonic matter, 355

and formation of structure, 442
as remnant of big bang, 426
baryonic matter, 356
bulk of matter in Universe, 329
can't be primarily baryonic matter, 356
cold dark matter, 357
cold dark matter and bottom-up structure, 442
Coma Cluster, 350
evidence for, 348
galaxy rotation curves, 348
gravitational lensing, 350
hot dark matter, 357
hot dark matter and top-down structure, 442
hot gas in clusters of galaxies, 350
in the Bullet Cluster, 354
masses of large galaxy clusters, 349
nonbaryonic candidates, 356
summary, 442
supersymmetry, 547
WIMPS, 357
de Sitter curvature precession,
  see geodetic precession
de Sitter space, 392
deconfinement of quarks and gluons, 419
deflection of light in gravitational field
  and equivalence, 110
  elevator experiment, 110
  gravitational curvature radius, 111
density parameters Ω, 358
  baryonic Ω_b, 355, 356, 441
  cold dark matter Ω_{CDM}, 441
  curvature Ω_k, 358, 359, 390
  massive neutrino Ω_ν, 441
  matter Ω_m, 358, 359, 390
  radiation Ω_r, 358, 359, 390
  vacuum energy Ω_Λ, 358, 359, 390, 441
derivative, see differentiation
differentiation
  absolute, 58, 61
  covariant, 58
  in spaces with position-dependent metrics, 26
  Lie, 62
  of tensors, 49, 56
  partial, 49, 57
Dirac notation (quantum mechanics), 46
direct product, see tensors, tensor product
directional derivatives, 37–39
Double Pulsar
  as test of general relativity, 204
  merger timescale, 204
  orbit of, 204
  origin of, 205
  precession of periastron, 204
  properties of, 204
dragging of inertial frames, see frame dragging
Dragonfly 44, 354

dual vectors (one-forms), 49
  and Dirac bra, 46
  and row vectors, 43
  as maps to real numbers, 21, 42
  co-varying quantities, 50
  defining in curved space, 21, 35
  duality with vectors, 21, 42, 43, 51

Eddington, Arthur, 115, 257, 381, 469
Eddington–Finkelstein coordinates, 218
Einstein equations
  alternative form, 151
  definition, 150, 151
  finding solutions, 154
  for general spherical metric, 569
  in vacuum, 151
  numerical solutions, 156
  sign conventions, 154
  solutions with high symmetry, 155
  weak-field limit, 145, 155
Einstein ring, see gravitational lensing
Einstein summation convention, 20, 34
Einstein tensor, 150
Einstein–de Sitter universe, 348
Einstein–Rosen bridge, 222, 223
embedding
  diagram, Schwarzschild metric, 164
  not required to compute curvature, 130
energy conditions in general relativity, 131
energy–momentum tensor, see stress–energy tensor
equation of state for universe, 387, 441
equivalence principle, 10, 106
  alternative statements, 109
  and acceleration of Earth–Moon system by Sun, 107, 543
  and Einstein's remarkable intuition, 11, 109
  and event horizons, 230
  and gravitational constant, 543
  and lunar rangefinding, 107, 543
  and Nordtvedt effect, 543
  and Riemannian geometry, 117
  and the path to general relativity, 110, 121
  and weak-field solutions, 155
  deflection of light in gravitational field, 110
  Eötvös experiment, 107
  Eötvös parameter, 107
  elevator experiments, 108
  gravitational redshift, 112
  inertial and gravitational mass, 107
  local inertial frames, 118
  locality and tidal forces, 119
  principle of strong equivalence, 108, 109, 207
  strong equivalence and weak equivalence, 108
ergosphere, 251
  and Penrose processes, 253
  motion of photons, 251

ergosphere (cont.)
    no stationary observers, 251
    relation to horizon, 251
    stationarity condition, 251
Euler–Lagrange equation, 97, 98
event horizons
    and equivalence principle, 230
    and lightcones, 216
    area of, 247
    cosmic censorship conjecture, 224
    cosmological, 375, 376
    for black holes, 213, 218, 245
    in Eddington–Finkelstein coordinates, 220
    in Kruskal–Szekeres coordinates, 223
    Kerr spacetime, 245, 247
    Schwarzschild spacetime, 213

failed supernovae, 531, 533
Fermi–Dirac statistics, 548
Feynman diagrams, 418
fiber bundle
    and local gauge invariance, 35
    as local product space, 36
    base space, 36
    cotangent bundle, 35, 36, 42
    example of non-metric space, 26
    fiber space, 36
    for S^1, 36, 37
    tangent bundle, 35–37, 42
    trivial and nontrivial, 36
fluence, 306
foliation of spacetime, 367
formation of structure
    cold dark matter and bottom-up formation, 442
    hot dark matter and top-down formation, 442
    inflationary explanation, 456
    role of dark matter, 442
    scale-invariant density fluctuation spectrum, 456
    seeds from CMB fluctuations, 441
    summary, 442
    timescales, 442
frame dragging
    and impossibility of stationary observers, 251
    and Lense–Thirring effect, 184, 186, 249
    angular velocity of zero angular momentum particle, 250
    in Kerr spacetime, 249
    in Schwarzschild spacetime, 183, 186
    measured by Gravity Probe B, 184
Friedmann cosmologies, 365
    a standard model, 390
    calculation of observables, 392
    calculations with benchmark parameters, 409
    comoving observers, 378
    concordance model, 406
    conformal time and horizons, 376, 377
    connection coefficients, 379
    connection coefficients (table), 380
    constant negative curvature, 369
    constant positive curvature, 368
    cosmological constant, 381
    cosmological principle, 365
    dark energy, 381, 382
    de Sitter solution, 392
    density parameters, 358, 359, 390, 441
    Einstein equations for the RW metric, 378, 380
    energy densities (current), 398
    energy densities and scale factor, 389
    equation of state, 387, 388
    evolution of density components, 389
    evolution of early Universe, 414
    evolution of scale factor, 412
    evolution of Universe, 386, 410
    flat universes with radiation or matter, 396
    flat, single-component universes, 391
    fluid equation, 387
    Friedmann (Einstein) equations, 380, 386, 389, 397
    general solutions, 394
    history of, 381
    homogeneous, isotropic spaces, 366, 368
    horizons, 396, 397
    Hubble constant, 398, 402
    matter energy density, 404, 413
    matter-dominated equation of state, 412
    numerical solution, 397
    parameters of minimal model, 398, 402, 408
    particle and event horizons, 374
    proper distance, 373
    pure matter universes, 399
    pure radiation universes, 400
    pure vacuum energy universes, 400
    pure vacuum energy with curvature, 392
    radiation energy density, 404, 413
    radiation-dominated equation of state, 412
    resolving Newtonian difficulties, 383
    Ricci scalar, 380
    Ricci tensor, 380
    Robertson–Walker metric, 369, 370
    single component with curvature, 399
    single-component solution observables (table), 394
    solution dominated by vacuum energy, 392
    solution of evolution equations, 391
    solution of minimal standard model, 397
    solutions with multiple density components, 401
    stress–energy tensor, 378
    the universe is observed to be flat, 442
    thermodynamics, 414
    vacuum energy, 382
    vacuum energy density, 404
    variation of energy densities with scale, 389
    visible Universe, 396
    zero curvature, 369

Friedmann, Alexander, 370, 381
Friedmann–Lemaître–Robertson–Walker (FLRW)
     metric, *see* Robertson–Walker metric
Friedmann–Robertson–Walker (FRW) metric, *see*
     Robertson–Walker metric
functional, 98
fundamental observer, *see* comoving observer

gamma-ray bursts, 304
    afterglows, 308, 314, 522, 528
    and energy conservation, 311
    and gravitational waves, 204, 522
    and Kerr black holes, 281
    and multimessenger astronomy, 522
    and star-forming regions, 312
    association with core collapse supernovae, 312, 315
    association with galaxies, 312
    association with neutron star mergers, 315, 317, 522
    beamed emission, 311, 522, 528
    breaks in lightcurve, 311
    causality and source size, 311
    central engine, 315
    characteristics, 314
    collapsar model, 313
    cosmological origin, 314
    discovery, 304
    duration, 314
    fireball model, 308, 314
    fluence, 306
    isotropic distribution on sky, 314
    localization, 307, 522
    long-period, 281, 306, 314, 315
    long-period bursts and Wolf–Rayet stars, 312
    Lorentz γ-factor, 315
    nomenclature, 314
    nonthermal emission, 314
    off-axis emission, 528
    optical depth, 308
    power-law spectrum, 309
    short-period, 306, 314, 315, 522
    the gamma-ray sky, 304
gauge transformations, 86
    choice of gauge in general relativity, 471
    comparison of electromagnetism and gravity, 472
    gauge-fixing constraint, 86
    in electromagnetism, 86
    in general relativity, 471
gauge/gravity duality, *see* AdS/CFT correspondence
Gauss, Karl Friedrich, 117
Gaussian curvature, 128
    and circumference of circles, 128
    and the *Theorema Egregium*, 117
    generalization to Riemann curvature, 148
    intrinsic determination, 130
    intrinsic property of the space, 130
gaussian fluctuations, 429, 456

general relativity
    alternative theories of gravity, 541
    and Binary Pulsar, 200, 202
    and energy conservation, 132, 133
    and Newtonian gravity, 133, 145
    and PSR J0337+1715, 207
    and PSR J0348+0432, 205
    and spacetime symmetries, 132
    and the Double Pulsar, 204
    Bianchi identity, 150
    classical tests, 539
    deflection of light in gravitational field, 178, 539
    Einstein equations, 150, 151, 569
    Einstein equations for general spherical metric, 569
    Einstein tensor, 150
    gauge transformations, 471
    gravitational waves, 202, 465
    incompatibility with quantum mechanics, 554
    limiting cases of the Einstein tensor, 154
    linearized Einstein equations, 468
    modern tests, 540
    neutron stars, 193
    nonlinearity of gravity, 152, 466
    numerical relativity, 156
    parameterized post-Newtonian (PPN) method, 540, 542
    precession of orbits, 172, 174
    precession of perihelion for Mercury, 174, 539
    pulsar tests of, 200, 202, 204
    recipe for constructing, 147
    redshift of light in gravitational field, 113, 115, 539
    Ricci flat spaces, 151
    Ricci scalar, 150
    Ricci tensor, 150
    Riemannian curvature tensor, 148
    role of pressure, 133
    Schwarzschild solution, 159
    Shapiro time delay, 179, 352, 353
    sign conventions, 154
    solving the Einstein equations, 154
    stress–energy tensor, 129, 194
    summary of tests, 539
    systematically stronger than Newtonian gravity, 198
    testing in the strong-gravity limit, 466, 520, 544
    weak-field limit, 145, 155
    what couples to gravity?, 133
    wormholes, 161, 162
geodesic equation, 138
geodesic incompleteness, 258
geodesics, 95
    and Killing vectors, 103
    and quantum uncertainty, 229
    euclidean space, 95
    Euler–Lagrange equation, 97
    geodesic equation, 138, 180

geodesics (cont.)
   incompleteness, 258
   Minkowski space, 95
   photons and massless particles, 98
   principle of extremal proper time, 96
   quantities conserved along, 103
geodetic precession
   and de Sitter curvature precession, 180–182, 184
   measured by Gravity Probe B, 182–184
   of Earth–Moon "gyroscope", 181
   of gyroscopes in orbit, 180
   tested by lunar rangefinding", 181
geometrical object, 15, 44
geometrized units, 92, 565, 566
geometry
   and metric tensor, 26
   euclidean, 14
   non-euclidean, 27
global positioning system, *see* GPS and relativity
global topological techniques, 258, 259
GP-B, *see* Gravity Probe B
GPS and relativity, 122
Grand Unified Theories (GUTS)
   and baryogenesis, 459
   baryon nonconserving reactions, 459
gravitational charge, 185
gravitational curvature radius
   and deflection of light, 111
   and strength of gravity, 111, 122
gravitational lensing, 350
   and dark matter, 350, 354
   by black holes, 513
   Einstein cross, 351, 352
   Einstein ring, 277, 513
   in the Bullet Cluster, 354
   microlensing, 351
gravitational mass, 199
gravitational redshift
   and energy conservation, 122
   as a time dilation, 115
   weak-field limit, 112
gravitational time dilation
   and GPS, 122
   weak-field limit, 115
gravitational waves, 465
   a new window on the Universe, 521
   amplitudes, 491
   amplitudes and event rates, 492
   and asymmetric gravitational collapse, 259
   and gamma-ray bursts, 204
   and late stellar evolution, 531
   and mass of graviton, 521
   approximate energy densities, 489
   as a test of general relativity, 466, 544
   as probe of dark events, 467
   as standard sirens, 512, 527
   as test of strong gravity, 205, 520
   coherence, 477
   comparison with electromagnetic polarization, 475
   comparison with electromagnetic waves, 476
   detection technology, 468
   emission from pulsar–white dwarf binary PSR J0348+0432, 205
   frequency, 484
   from Binary Pulsar, 202, 497
   from binary system, 494
   from contact binary, 491
   from core collapse supernovae, 503
   from merger of black hole binaries, 502, 506, 510, 532
   from merger of black-hole, neutron star binary, 503
   from merger of supermassive black holes, 504
   from neutron star mergers, 318, 500, 522
   from weak sources, 489
   GW150914, 465, 506, 521, 532, 544
   GW150914 source location, 508
   GW150914 waveform, 506
   GW151226, 514, 521
   GW170104, 519, 521
   GW170814, 519–521
   helicity components, 474
   history of idea, 469
   influence on binary orbit, 496
   laser interferometer detectors, 480
   LIGO, 480
   linearized Einstein equations, 468
   linearized Einstein equations with sources, 490
   luminosity from binary systems, 494
   matched filter technique, 477, 506, 515, 518
   multimessenger astronomy, 483, 522
   multipolarity of, 490
   no monopole or dipole, 490
   polarization tensor, 473
   power, 493
   pulsar timing arrays, 484
   quadrupole formula for power, 493
   ranges of detectable strains and frequencies, 483
   ratio of wavelength to source size, 476
   reduced quadrupole tensor, 493
   response of test masses, 477, 479
   significance, 465
   spectrum, 465, 466
   states of polarization, 473, 480
   strain, 477, 479, 484, 493, 505–507
   strong-field sources, 500
   template waveforms, 504
   testing nonlinearity of general relativity, 466
   the deepest probe, 467
   trace-reversed amplitude, 472
   transverse–traceless (TT) gauge, 473
   weak-field, 473
   weak-field solution in TT gauge, 474
gravitomagnetic effects, 185
graviton, 152, 383, 475, 477, 521, 554

Gravity Probe B
  and frame dragging, 183, 184
  and geodetic precession, 182–184
GRB, *see* gamma-ray bursts
Grossmann, Marcel, 118
GUTS, *see* Grand Unified Theories (GUTS)
GW150914, 506
  properties of (table), 511
  source location, 508
  test of strong-gravity general relativity, 520
  waveform, 506
GW170817, 522
gyroscopes
  and gravitomagnetic effects, 180
  and rotating spacetime, 183
  gyroscopic equation, 180

Harrison–Zel'dovich spectrum, 456
Hawking area theorem, 224, 247, 258
Hawking black holes, 229, 235
  and information, 239
  and quantum fluctuations, 231
  blackbody temperature, 232
  endpoint of evaporation, 239
  entropy, 236
  generalization of area theorem, 236
  generalized second law of thermodynamics, 236
  mass emission rates, 232
  surface gravity, 232
  temperature, 236
  thermodynamics, 236
Hawking radiation, 230, 235
Heisenberg uncertainty principle, 456, 554, 559
Hertzsprung gap, 532
Hilbert space, 46
Hilbert, David, 11, 109
HMXB, *see* binary star systems - high-mass X-ray binaries (HMXB)
holographic principle
  and black holes, 240
  and information, 240
  and the Universe, 240
holonomic basis, *see* coordinate basis
horizons
  analog, 235
  apparent, 260
  event horizons, 375, 376
  horizon problem in big bang, 449
  particle horizons, 375
  relationship of particle and event horizons in cosmology, 378
Hubble expansion, 327
Hubble law, 331
  and expansion interpretation of redshifts, 332, 448
  and Friedmann cosmology, 409
  and scale factor, 331
  deviations from, 409

Hubble parameter, 331, 337, 402, 528
Hubble radius, 376
Hubble time, 335
Hubble, Edwin, 381
Hulse–Taylor binary, *see* Binary Pulsar
hypersurface, 33

indefinite metric, 69
inflation, 453
  and density fluctuations, 442, 456
  and elementary particle physics, 454
  and flatness of the Universe, 444
  and flatness problem of big bang, 451
  and formation of structure, 452, 456
  and gaussian fluctuations, 456
  and horizon problem of big bang, 449
  and monopole problem of big bang, 451
  consequences of, 454
  expansion in de Sitter phase, 453
  exponential growth of scale factor, 453
  gaussian fluctuations, 429
  modifies but does not replace big bang, 457
  scale-invariant density fluctuation spectrum, 456
  solution of the flatness problem, 455
  solution of the horizon problem, 455
  solution of the monopole problem, 456
  versions of, 454
inflationary cosmology, *see* inflation
information
  and black holes, 239
  and entropy, 239
  and the holographic principle, 240
inhomogeneous Lorentz transformations, *see* Poincaré transformations
innermost stable circular orbit (ISCO)
  binding energy, 283
  Kerr solution, 283
  Schwarzschild solution, 172
integration
  area of 2-sphere by invariant integration, 57
  covariant volume element, 56
  invariant, 26, 56
  of tensors, 56
intrinsic derivative, *see* absolute derivatives
inverse Compton scattering, 303
ISCO, *see* innermost stable circular orbit (ISCO)
isometries
  Kerr metric, 244
  Killing vectors, 64, 100, 166
  quantities conserved along geodesic, 103
  Schwarzschild metric, 166

Jacobian determinant, 56
Jacobian matrix, 51
jets
  and rotating magnetic fields, 285
  apparent superluminal motion, 285

jets (cont.)
    collapsar model, 315
    counterjets, 285
    from AGN, 292
    from quasars, 290
    Lorentz γ-factor, 315
    radio jets from Centaurus A, 302
    radio jets from NGC 4258, 270
    relativistic beaming, 285

Kerr black holes
    accretion, 283
    and gamma-ray bursts, 304
    and Penrose processes, 253
    area of event horizon, 247
    as central engines, 280
    extremal, 245
    extremal and cosmic censorship, 246
    observational evidence for extremal, 246
    source of quasar power, 290
Kerr spacetime, 243
    asymptotically flat solutions, 244
    Boyer–Lindquist coordinates, 245
    ergosphere, 251
    event horizons, 247
    frame dragging, 249
    interpretation of parameters, 244
    metric, 243
    motion of light in, 248
    motion of particles in, 248
    reduction to Schwarzschild solution, 244
    singularity and horizon structure, 245
    solutions of, 243
    symmetries and Killing vectors, 244
    vacuum solutions, 244
Killing vectors, 101
    and isometries, 100
    and Lie derivatives, 64, 101
    and spacetime symmetries, 244
    for the Kerr metric, 244, 248
    for the Schwarzschild metric, 166, 230
    spacelike or timelike character, 230, 254
kilonova, 526, 528, 529
Kretschmann scalar, 227
Kronecker delta, 22, 23, 41, 43, 51
Kronecker product *see* tensors, tensor product
Kruskal diagrams, 221
Kruskal–Szekeres coordinates, 221

Lagrangian, 98
lanthanides and opacity, 528
last scattering surface, *see* cosmic microwave background (CMB)
Lemaître, Georges, 370, 381
Lense–Thirring effect, *see* frame dragging
leptogenesis, 459

Levi-Civita connection, *see* connection coefficients
Levi-Civita symbol, *see* completely antisymmetric 4th-rank tensor
Levi-Civita, Tullio, 136
Lie bracket, 39, 67
Lie derivative, 62
    and covariant differentiation, 58
    and Killing vectors, 64, 101
    contrasted with covariant derivative, 58, 63
    explicit expressions, 63
    isometries, 64
    Lie transport, 64
    more primitive than covariant derivative, 64
    switching partial and covariant derivatives, 64, 68, 101
Lie dragging, 63
Lie transport, 64
LIF, *see* local inertial frame (LIF)
lightcone, 77
    and causality, 79, 80, 120, 216
    and simultaneity, 78
    and the constant speed of light, 79
    for black hole spacetimes, 216
    global organization of lightcones, 120
    in Eddington–Finkelstein coordinates, 219
    in general relativity, 120
    in Kruskal–Szekeres coordinates, 223
    invariance of, 216
    lightlike intervals, 84
    null intervals, 84
    spacelike intervals, 84
lightlike intervals, *see* lightcone, null intervals
LIGO, 480, 482
line element, 24
    euclidean, 25
    for plane polar coordinates, 25
    in curved space, 129
local inertial frame (LIF), 109, 118–121, 138
lookback times, 346
Lorentz covariance
    and special relativity, 69
    of Maxwell equations, 84
Lorentz factor
    for gamma-ray bursts, 311
    in special relativity, 77
Lorentz transformations, 7, 73
    and spacetime diagrams, 80
    as rotations in Minkowski space, 74
    boosts between inertial systems, 74, 75, 77, 82
    Lorentz group, 125, 127
    spatial rotations, 74
Lorentzian manifold, 71
LS (last scattering), *see* cosmic microwave background (CMB), last scattering surface
LSS (last scattering surface), *see* cosmic microwave background (CMB), last scattering surface

luminosity distance, 406, 433, 434
lunar rangefinding, 107, 543

macronova, *see* kilonova
magnetic fields
    and jets, 285
    and rotating black holes, 285
    Blandford–Znajek mechanism, 280
magnetic monopoles
    and inflation, 451
    and magnetic charge for black holes, 225
manifold, 33
    atlas, 33
    charts, 33
    coordinate patches, 33
    differential, 33
    geodesically complete, 222, 259
    maximal, 222
    Riemannian, 33, 116
    spacetime, 32
mapping, 21
mass
    as gravitational charge, 185
    equivalence of active and passive, 124
    equivalence of gravitational and inertial, 107, 123
mass function, 262, 263, 265–267
matched filtering, 506, *see* gravitational waves
Maxwell equations, 84
    and aether, 6
    and Galilean invariance, 6
    covariance, 87
    gauge transformations, 86
    in Heaviside–Lorentz units, 84
    Lorentz covariance, 84
    scalar and vector potentials, 85
Mercator projection, 83, 304, 376
metallicity
    and formation of massive binaries, 530, 531
    and gamma-ray bursts, 312
    and photon opacity, 312, 531
metric
    indefinite, 71, 76
    Kerr, 243
    pseudo-Riemannian, 116
    Riemannian, 116
    Robertson–Walker, 369, 370
    Schwarzschild, 159
    signature, 9, 119
    slowly-rotating, 183
    static, 160, 244
    stationary, 160, 244
    symmetries of (isometries), 64, 100
metric space, 26
metric tensor, 53, 54
    and geometry of space, 26
    and line element, 24, 53
    and scalar products, 70
    as gravitational potential, 139
    as source of gravitational field, 139
    connection coefficients computed from, 140
    contravariant components, 23
    covariant components, 23
    covariant derivative vanishes, 61
    for Schwarzschild spacetime, 159
    in euclidean space, 23
    in Minkowski space, 70
    indefinite metric, 69
    Kerr, 243
    properties, 23
    quantum fluctuations of, 556
    signature, 71
    singularities, 160
    used to raise and lower indices, 53, 54, 71
Michelson–Morley experiment, 7
Minkowski space, 69
    4-velocity and 4-momentum, 94
    and causality, 79
    and spacetime, 8
    event, 71
    geodesics, 95
    indefinite metric, 69
    invariance of spacetime interval, 71
    lightcone structure, 77
    lightlike intervals, 84
    line element, 8, 70
    Lorentz transformations, 73
    Lorentz-invariant dynamics, 95
    metric signature, 9, 71
    metric tensor, 8, 70
    null (lightlike) intervals, 78, 84
    observers, 99
    rotations, 76
    scalar product, 70
    spacelike intervals, 78, 84
    tensors, 72
    timelike intervals, 78
    worldline, 71, 94
Minkowski, Hermann, 8, 118
Mollweide projection, 304
multimessenger astronomy, 483, 522

naked singularities, *see* cosmic censorship
natural units, *see* geometrized units
neutrinos
    cosmic neutrino background, 427
    decoupling scale, 418
    leptogenesis, 459
neutron stars, 193
    equation of state from gravitational waves, 522
    formation of binary neutron stars, 203
    gravitational mass and baryonic mass, 199

neutron stars (cont.)
  interpretation of mass parameter, 198, 199
  mass and density, 193
  Oppenheimer–Volkov equations, 196
  pulsars, 200
  simple estimates, 193
  size, 193
  solution of Einstein equations, 194–196
  stress–energy tensor, 194
NGC 4258, 270, 532
no-hair theorem, 225
Nobel Prizes for relativity, 203
Noether's theorem, 102
non-coordinate basis, 37–39
nonthermal emission, 287
  from AGN, 292
  from gamma-ray bursts, 308, 314
  from quasars, 288
  implications of, 287
  polarization, 287
  requires optically-thin medium, 309
  synchrotron radiation, 287
Nordtvedt effect, 543
null energy condition, 131
numerical relativity, 12, 156
  3+1 formalism, 156
  and apparent horizons, 260
  and initial data, 156
  and strong gravity, 156
  matched filter waveforms, 517
  plagued initially by instabilities, 534
  simulation of GW150914, 506, 507, 512
  simulation of GW151226, 515

observers, 99
  comoving, 367, 371
  inertial observers, 118
  no stationary in ergosphere, 251
one-forms, see dual vectors (one-forms)
Oppenheimer–Volkov equations, 196
  differences from Newtonian hydrostatics, 197
  interpretation, 197
optical depth
  altered by relativity, 311
  and opacity, 310
  definition, 310
  in gamma-ray burst, 308, 309

parallel transport, 134
  and absolute derivative, 136
  and connection coefficients, 135
  and covariant derivative, 135
  dependence on path, 134
parameterization
  of curves, 15
  of surfaces, 15

parameterized post-Newtonian (PPN) formalism, 515, 540, 542
parity symmetry P, 457, 458
Pauli exclusion principle, 548
Penrose processes, 237, 253
perfect cosmological principle, 327
periapsis, 268
Planck scale, 238
  and initial singularity, 258
  breakdown of current physical laws?, 338, 558
  Planck era, 338, 553
  Planck scale parameters (table), 554
  quantum gravity, 238, 338, 553
Poincaré transformations, 125, 127, 139
Poisson equation, 145
Pop III stars, 423, 531, 534
Pound and Rebka experiment, see redshift, gravitational for Earth
PPN, see parameterized post-Newtonian (PPN) formalism
preferred frames of reference, 373
principle of extremal proper time, 96
principle of general covariance, 106
principle of relativity, 5, 10, 32
principle of strong equivalence, 109
proper time, 70
pseudo-euclidean manifold, see Lorentzian manifold
pulsars, 200
  as test of general relativity, 200, 202
  Binary Pulsar, 200, 202
  discovery, 257
  nomenclature, 200
  pulsar–WD–WD triplet PSR J0337+1715, 207
  pulsar–white dwarf binary PSR J0348+0432, 205
  the Double Pulsar, 204

QSOs, see quasars
quantum gravity, 238, 553
  and renormalization, 554
  and superstrings, 554
  and the Planck scale, 553
  inconsistency of quantum mechanics and general relativity, 554
  quantum fluctuations of the metric, 556
  superstrings and branes, 556
  the Universe as a quantum fluctuation of the spacetime metric, 557
quantum mechanics
  and black holes, 229, 258
  and dark energy, 549
  and geodesics, 229
  and gravity, 238
  and Hawking radiation, 230
  fermions and bosons, 548
  Feynman diagrams, 418
  incompatibility with general relativity, 554

quantum gravity, 553
quantum vacuum fluctuations in inflation, 456
renormalization of quantum field theory, 554
vacuum fluctuations, 231, 549, 557
zero-point energy, 550, 551
quasars, 285
   black hole central engines, 257, 290
   characteristics, 288
   discovery and interpretation, 288
   energy source, 295
   evolution of density with time, 289, 301
   jets, 290, 315
   Lorentz factors of, 315
quasistellar objects, *see* quasars
quasistellar radio sources, *see* quasars
quotient theorem, 66, 67

r-process
   heavy r-process nuclei, 528, 529
   in neutron star mergers, 528, 529
   kilonova, 528, 529
   light r-process nuclei, 528, 529
radio galaxies, 292
recombination transition, 422
redshift
   and deceleration of Universe, 359
   as a time dilation, 115
   cosmological, 327, 331
   expansion interpretation, 332
   for Schwarzschild solution, 165
   gravitational, 112, 113, 115
   gravitational for Earth, 114, 123
   gravitational for Sirius B, 114, 115
   measuring for white dwarf, 115
   near a black hole, 218
   weak-field limit, 167
reionization transition, 422, 423
relativity principle, *see* principle of relativity
repeated indices, *see* Einstein summation convention
Ricci flat space, 151
Ricci scalar, 150
Ricci tensor, 150
Riemann zeta function, 415
Riemann, Bernhard, 117
Riemannian curvature tensor, 148
   and Gaussian curvature, 149
   as commutator of covariant derivative, 157
   dependence on dimensionality, 149
   symmetries, 148
Riemannian manifold, 116
   and the equivalence principle, 117
   locally euclidean, 116
Robertson, Howard, 370
Robertson–Walker metric
   a consequence only of geometry, 371
   and cosmological principle, 370
   and event horizons, 376
   and Hubble law, 371, 374
   and line element for homogeneous, isotropic spacetime, 369, 370
   and particle horizons, 375
   comoving coordinates, 371
RW metric, *see* Robertson–Walker metric

S0-102, *see* Sgr A*
S0-2, *see* Sgr A*
Sachs–Wolfe effect, *see* cosmic microwave background (CMB)
Sakharov conditions for baryon asymmetry, 457
scale factor of Universe, 331–333, 335, 370–372, 412
scale-invariant density fluctuations, 456
Schwarzschild black hole, 213
   area of event horizon, 247
   event horizon, 213, 215
   lightcone diagrams, 216
Schwarzschild metric, *see* Schwarzschild spacetime
Schwarzschild spacetime, 159
   asymptotically flat solutions, 159
   classes of orbits, 169
   conserved quantities, 167
   coordinate distance, 163
   coordinate time, 164
   deflection of light by gravity, 178
   Eddington–Finkelstein coordinates, 218
   embedding diagrams, 164
   escape velocity, 174
   form of metric, 159
   gravitational redshift, 165
   innermost stable circular orbit, 172
   interpretation of radial coordinate, 160, 161
   isometries, 166
   Killing vectors, 167
   Kruskal diagrams, 221
   Kruskal–Szekeres coordinates, 221
   lightcones, 216
   orbits for light rays, 177
   particle orbits, 167–169
   physical distance and time, 162
   precession of orbits, 172, 174
   proper distance, 163
   proper time, 164
   radial fall of test particle, 175
   Shapiro time delay, 179
   singularities of the metric, 160, 176
   solution, 159
   vacuum solutions, 159
seeds for structure formation in early Universe, 441
Seyfert galaxies, 293
   brightness variability, 294
   Seyfert 1 galaxies, 294
   Seyfert 2 galaxies, 294

Sgr A*
    and the star S0-102, 269
    and the star S0-2, 268, 532, 544
    evidence for a supermassive black hole, 268
    testing general relativity, 544
    testing the no-hair theorem, 269
Shapiro time delay, 179, 352, 353
singularities
    and cosmic censorship, 246
    and Eddington–Finkelstein coordinates, 218
    and global topological techniques, 258, 259
    and Kruskal–Szekeres coordinates, 221
    coordinate, 162
    initial cosmological, 329
    naked, 246
    of Kerr spacetime, 245
    of Schwarzschild metric, 160, 176, 218, 221
    physical, 162
    theorems, 224, 258, 259, 261
singularity theorems, 224, 258, 259, 261
Slipher, Vesto, 381
Sloan Digital Sky Survey (SDSS), 408, 438
sonic black holes, 235
sound horizon, *see* cosmic microwave background (CMB)
space
    anti-de Sitter, 558
    de Sitter, 392, 453
    dimensionality and curvature, 149
    metric, 26
    Minkowski, 8, 69
    non-metric, 26
    Ricci flat, 151
    Riemannian, 116
    spacetime foam, 556
    with wormholes, 556
spacelike surface, 78, 258
spacetime foam, 556
special relativity
    and Lorentz covariance, 69
    event, 71
    Lorentz γ-factor, 315
    Lorentz transformations, 73
    proper time, 70
    relativity of simultaneity, 78, 82
    space contraction, 82
    time dilation, 72, 82
    twin paradox, 82
    worldline, 71
spectral index, 308
standard and standardizable candles
    Cepheid variables, 404
    Type Ia supernovae, 404
standard candle, 434
standard cosmological model, 327
Standard Model of elementary particle physics
    GUT extension, 459
    magnetic monopoles, 451
    particles important for big bang, 419
    particles of (table), 417
    role in cosmology, 329
standard ruler, 434
standard sirens, 512, 527
stress–energy tensor, 129
    energy conditions, 131
    for comoving Robertson–Walker observer, 378
    perfect fluid, 143, 194
strong energy condition, 131
structure formation, *see* formation of structure
summation convention, *see* Einstein summation convention
superluminal motion (apparent), 90, 285
supermassive black holes
    energy source for quasars and AGN, 290
    evidence for, 266, 268–270, 272, 273, 275
    heavy seeds model of formation, 274, 534
    light seeds model of formation, 274, 534
    relationship with stellar black holes, 534
supernovae
    as standard candles, 405
    association with gamma-ray bursts, 312
    lightcurves, 317
    Type Ia, 405
    Type Ib, 312
    Type Ic, 312
superstrings and branes, 554
    brane theory, 555
    compactification, 559
    experimental tests of, 557
    number of spacetime dimensions, 556, 559
    quantum gravity, 556
    superstrings, 555
supersymmetric particles, 549
supersymmetry
    and dark matter, 547
    and new particles, 549
    broken, 549
    no experimental evidence for, 549
    symmetry uniting fermions and bosons, 549
surface gravity, 232
symmetries
    group theory, 127
    isometries, 100
    isotopic spin, 548
    Killing vectors, 100, 101, 166
    Lorentz group, 125, 127
    Noether's theorem, 102
    non-supersymmetric, 548
    of the metric, 100, 166
    Poincaré group, 125, 127
    solution of the Einstein equations for high symmetry, 155

spontaneously broken or hidden symmetry, 550
supersymmetry, 547

tangent bundle, *see* fiber bundle, tangent bundle
tangent space, 35–37
    and parallel transport, 35, 134
    and vectors in curved space, 35
tensor density, 52
tensors
    and covariance, 32
    and form invariance of equations, 64
    antisymmetric (skew symmetric), 54
    antisymmetrizing operation, 54
    as linear maps, 41, 42, 45, 46
    as operators, 41
    calculus, 56
    completely antisymmetric 4th-rank tensor, 52
    contravariant, 41
    covariant, 41
    defined by their transformation law, 41, 48, 52
    differentiation, 56
    dual vectors (one-forms), 49
    Einstein summation convention, 20, 34
    Einstein tensor, 150
    higher-rank, 52
    horizontal placement of indices, 53
    in linear algebra, 43
    in Minkowski space, 72
    in quantum mechanics, 43, 46
    index-free formalism, 41, 42, 45, 46
    integration, 56
    Kronecker delta, 41
    metric tensor, 53, 54
    mixed, 41
    rank, 40
    rank-2, 52
    Riemann curvature, 52
    scalars, 48
    symmetric, 54
    symmetrizing operation, 54
    tensor density, 52
    tensor product, 45
    torsion, 140
    transformation laws (table), 52
    type, 40, 45, 67
    vectors, 48, 49
    vertical placement of indices, 18, 20, 32, 34
Theorema Egregium, 117
tidal forces, 119, 120
time machines, *see* time travel
time travel, 80, 81, 162, 346, 552
Tolman–Oppenheimer–Volkov equations, *see* Oppenheimer–Volkov equations
topological defects and magnetic monopoles, 452
torsion, 140
TOV equations, *see* Oppenheimer–Volkov equations

trace-reversed amplitude, 472
transformations, 14
    active, 63
    between coordinate systems, 28
    between inertial systems, 5
    boosts, 75
    Galilean, 5, 29
    gauge in electromagnetism, 86
    Lie dragging, 63
    Lorentz, 7, 30
    of derivatives, 40
    of fields, 40
    of integrals, 40
    of scalars, 48
    of vectors, 49
    passive, 34
    Poincaré, 125, 127
    rotations, 29
    rotations in euclidean space, 73
    rotations in Minkowski space, 74
    spacetime, 34
    vectors, 49
transverse–traceless gauge, 473
trapped surfaces, 258–260
TT gauge, *see* transverse–traceless gauge
twin paradox, 82
Type Ia supernovae
    and the accelerating Universe, 404
    as standardizable candles, 404
    lightcurves, 405
    measuring large distances with, 404

UDG, *see* ultra-diffuse galaxies (UDG)
ultimate free lunch, 557
ultra-diffuse galaxies (UDG), 354
unified model of AGN and quasars, 295
    accretion disk and dusty torus, 299
    anisotropic ionization cones, 298
    black hole central engine, 297, 299
    effect of orientation, 300
    high-velocity clouds and broad emission lines, 299
    low-velocity clouds and narrow emission lines, 299
    relativistic jet outflow, 299

vacuum energy, *see* dark energy
vacuum solutions
    gravitational waves, 468
    Kerr spacetime, 244
    Schwarzschild spacetime, 159
vector space, 21, 42
vectors, 49
    and column vectors, 43
    and Dirac ket, 46
    and tangent spaces, 35
    angle between two vectors, 385
    as geometrical objects, 15

vectors (cont.)
    as maps to real numbers, 21, 42
    contra-varying quantities, 50
    defining in curved space, 21, 35
    dual vector spaces, 21, 42, 43
    dual vectors, 49
    duality with dual vectors, 21, 42, 43, 51
    expansion in basis, 20
    scalar product, 20, 51
    vector space, 21, 42
vertical position of indices, *see* Einstein summation convention
virial theorem
    and weak gravitational waves, 492
    applied to galaxies, 272
    evidence for supermassive black holes, 273
    for a set of stars, 270
    mass of large galaxy clusters, 349
    virial masses, 274
visible Universe, 396
VLBA, 276

Walker, Arthur, 370
warp drives, 552
weak energy condition, 131
white holes, 222, 235
Wolf–Rayet star, 312–315, 317
wormholes, 161, 162, 222, 223, 552, 556

Yukawa potential, 544

zero age main sequence (ZAMS), 532